Reading Darwin in Arabic, 1860–1950

Reading Darwin in Arabic, 1860–1950

MARWA ELSHAKRY

The University of Chicago Press
Chicago and London

Publication of this book has been aided by a grant from the Bevington Fund.

The University of Chicago Press, Chicago 60637
The University of Chicago Press, Ltd., London

Paperback edition 2016
Printed in the United States of America

22 21 20 19 18 17 16   2 3 4 5 6

ISBN-13: 978-0-226-00130-2 (cloth)
ISBN-13: 978-0-226-37873-2 (paper)
ISBN-13: 978-0-226-00144-9 (e-book)
DOI: 10.7208/chicago/9780226001449.001.0001

Library of Congress Cataloging-in-Publication Data

Elshakry, Marwa, 1973–
Reading Darwin in Arabic, 1860-1950 / Marwa ElShakry.
pages. cm.
Includes bibliographical references and index.
ISBN 978-0-226-00130-2 (cloth : alk. paper) — ISBN 978-0-226-00144-9 (e-book) 1. Evolution—Religious aspects—Islam. 2. Darwin, Charles, 1809–1882. 3. Islam and science. I. Title.
BP190.5.E86E47 2013
297.2′65—dc23
2013009170

∞ This paper meets the requirements of ANSI/NISO Z39.48-1992 (Permanence of Paper).

CONTENTS

Acknowledgments / vii

INTRODUCTION / 1

ONE / The Gospel of Science / 25

TWO / Evolution and the Eastern Question / 73

THREE / Materialism and Its Critics / 99

FOUR / Theologies of Nature / 131

FIVE / Darwin and the Mufti / 161

SIX / Evolutionary Socialism / 219

SEVEN / Darwin in Translation / 261

Afterword / 307
Notes / 319
Bibliography / 391
Index / 419

ACKNOWLEDGMENTS

Much of the time that went into this book was spent reading texts through layers of translation: Darwin in Arabic, or Haeckel, originally in German but rendered into Arabic from a French translation. One American missionary translating the Bible into Arabic in the mid-nineteenth century complained of having piled words "six or seven deep and high, above and below nearly every important word in the line," which he then carefully vetted with a native speaker. All translations are multiple, and so I, too, like some of the characters in this book, often had to weigh literal fidelity against meaning when translating. If, as I quickly learned, writing is a collective effort, reading in and writing about translation is doubly so. Over many years, everyone from language tutors to my mother was conscripted to gloss over translations, read texts with me, and check on embedded meanings. It was they who often helped make sense of particular idioms, expressions, verses, or references that would have otherwise escaped me. Of course, all oversights and errors remain mine.

I have also relied on the ideas and scholarship of many people. I would like to thank in particular Peter Buck, Graham Burnett, Angela Creager, Omnia El Shakry, Khaled Fahmy, Carol Gluck, Thomas Gluck, Matthew Jones, Mark Mazower, James Moore, Steven Shapin, Pamela Smith, Daniel Stoltz, Bob Tignor, and many other conference participants, students, and colleagues who read drafts of chapters. Conversations with Tamara Griggs, Samira Haj, Shruti Kapila, Rashid Khalidi, Suzanne Marchand, Elizabeth Merchant, Gyan Prakash, Samah Selim, Pamela Smith, and numerous others also helped me. Echoes of the late Gerald Geison can be found throughout the text, and I shall always regret that I was not able to show it to him in its final iteration.

A special thanks goes to my favorite book dealer, Muhammad ʿAli, for tracking down many rare and scarce items in Cairo. My thanks, too, to librarians and archivists at the Egyptian National Library (particularly the periodicals room), the American University of Beirut, and the Zahiriya Mosque at Damascus, as well as at the archives and special collections at Amherst, Yale, Harvard, and Princeton. I would also like to gratefully acknowledge Ahmed Ragab and Ayşe Ozil for their help with researching and collecting materials for this project. Ayşe located files and manuscripts in the Ottoman Archives for chapter 4, while Ahmed referenced a number of newspaper and journal articles toward the end of this project's revisions. Ahmed also helped me read through stacks of handwritten Arabic materials from the khedival collection at Durham University for chapter 5.

For their generous funding and support I thank the Carnegie Corporation, the Radcliffe Institute for Advanced Studies, the British Academy, Harvard University, and Columbia University. The Social Science Research Council, the American Research Center in Egypt, and Princeton University were critical in the early phase of my research. The editors at the University of Chicago Press and especially my long-suffering copy editor, Pamela Bruton, have been wonderfully supportive. My thanks too to Marc Saint-Upery for reviewing the transliterations, and to June Sawyers for the index.

Finally, I was fortunate to have family on both sides of the Atlantic who always managed to strike the right balance between forbearing curiosity and discreet encouragement. My parents' and sisters' patience and good humor were critical; in many ways, they have lived with this book as long as I have. Omnia in particular offered inspiration from the start. Lastly, I owe my greatest debt to Mark. He had to take on a dizzying number of roles during much of the research and writing of the book—interlocutor, critic, confidante, and much, much more. Meanwhile, at least two of our creative collaborations have also provided the best distraction to it. My debt to him, to Selma, and to Jed is beyond words.

INTRODUCTION

"This journey in the East, especially in Egypt and Greece, marked a new epoch in my thinking," wrote the American diplomat and educator Andrew Dickson White after his travels there in 1888–89. "I became more and more impressed with the continuity of historical causes, and realized more and more how easily and naturally have grown the myths and legends which have delayed the unbiased observation of human events and the scientific investigation of natural laws."[1] White had recently stepped down from serving as the first president of Cornell University and was a figure of some renown: his serialized study of the "warfare of science with theology in Christendom" had (along with fellow American John Draper's 1874 *History of the Conflict between Religion and Science*) sealed the view for many that Darwin's theory of evolution constituted the "final victory of science" in its clash with "myths and legends."[2] Amid excursions to the usual sights (fig. I.1) and while visiting with English, Ottoman, and Egyptian officials, White made the acquaintance of "an especially interesting man of a different sort"[3]—one who turned out to know quite a lot about White already.[4]

This "interesting man" was a recent émigré from Ottoman Syria called Faris Nimr. White was "amazed to find in his library a large collection of English and French books, scientific and literary—among them the 'New York Scientific Monthly' containing my own articles, which he had done me the honor to read" (fig. I.2).[5] Indeed Nimr's extensive subscription list included the (New York–based) *Popular Science Monthly*, as well as many other English and American science magazines and numerous book titles in English and French.

Nimr and White turned out to share a surprising amount. Both men were avid bibliophiles, for instance. The fifty-six-year-old widower was

Fig. I.1. The Pyramids, White in Egypt, 1888.

traveling with Willard Fiske, the librarian of Cornell University, aboard a "modest ship . . . with an excellent library," and not the least of the purposes behind the trip was to collect books and manuscripts for their new institution (fig. I.3).[6] Both Nimr and White combined faith in science with a particular reverence for religious faith. Despite being labeled an "infidel and atheist" and attacked by "careless and unkind" readers, White insisted that he was "not actuated by any hostility to religion." Indeed, much like Nimr, he hoped to raise people to "the loftier conviction" that "there is a God in this universe wise enough to make all truth-seeking safe, and good enough to make all truth-telling useful."[7] Both men saw understanding the

Fig. I.2. Faris Nimr in his office.

Fig. I.3. Fiske's "modest ship," with library, anchored in Egypt.

laws of nature as compatible with religious belief and necessary for the formation of an educated society whose citizens could question authority when needed and yet still accept the rationale and rule of law. (Both men were also interested in Freemasonry.)[8]

Above all, White found that Nimr was a fellow enthusiast of Darwin and of what White called the "Darwinian hypothesis": "I found that he had been, at an earlier period, a professor at the college established by the American Protestant missionaries at Beyrout; but that he and several others who had come to adopt the Darwinian hypothesis were on that account turned out of their situations, and that he had taken refuge in Cairo where he was publishing, in Arabic, a daily newspaper, a weekly literary magazine, and a monthly scientific journal."[9] White had in fact stumbled on the editorial powerhouse of contemporary science publishing in Arabic: this press was responsible more than any other for popularizing throughout the region the ideas of both Darwin and White himself. Nimr was one of the proprietors of a Cairo-based firm that published—as White notes—the "daily newspaper" *Al-Muqattam*, the "literary" (and Masonic) magazine *Al-Lata'if*, and the popular-science monthly *Al-Muqtataf* (The digest). Of these, *Al-Muqtataf* was the first to be published, and by the time of White's visit it was already by far the most popular Arabic science periodical—a status it retained for the near century it was in print—and one of the most important sources of information about Darwin. Indeed, thanks to the controversy that White mentioned, which took place at the Syrian Protestant College (renamed the American University of Beirut in 1920), where the journal was first published, it was critically involved in the first public debate over the so-called "Darwinian hypothesis" in the Arabic press.[10]

White's encounter with Nimr encapsulates in miniature much of the process through which Darwin's ideas traveled in Arabic. White's *History of the Warfare of Science with Theology in Christendom* was not translated into Arabic until 1929; yet long before that, the editors of *Al-Muqtataf* were familiar with his "conflict" thesis from their reading of the *Popular Science Monthly* and offered partial and summary translations in their own magazine.[11] Like Darwin, White was also read in Arabic alongside an eclectic range of sources and inspirations: singing the book's praises when the final two volumes appeared in 1896, *Al-Muqtataf*'s article on White borrowed characteristically from a biographical sketch of the author that had been published a few months earlier in the American *Monthly*.[12] Unlike White, Darwin himself of course never went to Egypt or indeed to any of the major cities of the eastern Mediterranean. Yet his ideas—and others attached to his name—did and, thanks to actors like Nimr and *Al-Muqtataf* and to the global network of science popularizers of which they formed a part, they did so with a rapidity and ease that many might still find surprising.[13] In Darwin's case far more than White's, this matrix of readership, translation, and cross-border interpretation meant that Darwin was not only translated and read alongside other writers but also read through other readers. The Arabic science journal *Al-Muqtataf* was itself not the final link in this chain; on the contrary, as we dig through the metatextual layers and try to analyze the compound discourses that emerged, we find the journal (and others like it) being used in their turn as sources of ideas in yet other texts and interpretations.

Inevitably then, this book also pays attention to conjunctions and contrasts—to whom or what else was read alongside Darwin: not only White and Draper but also Herbert Spencer and Ernst Haeckel, Aristotle and al-Jahiz, as well as a vast range of popular works (both European and Arabic) on natural history and social thought. Other figures—from Thomist and Anglican natural theologians to the esotericist order the Ikhwan al-Safa' (Brethren of Purity) and the rationalist theologian Ibn Rushd (Averroës)—make their appearance. Of course, these references must also be understood against the institutional and geopolitical background that brought them into circulation. The book looks, in other words, at the layers of discourse—texts, contexts, and contingencies alike—that formed part of the interpretive frameworks within which reading Darwin in Arabic took place.

Reading, of course, involves a variety of related actions: interpretation, explanation, appropriation, and omission, among others. It is at once a private and a highly public act. What gets read is movable and transmissible. And yet it also involves and often helps to define specific communities, with distinct epistemic, didactic, and literary styles, often with their own linguistic and conceptual vocabularies. Rather than mining readings of Darwin in order to measure them against the original texts, this book explores the array of interpretations, inspirations, and intellectual metatexts that were generated and that encompassed social thought, political commentary, pedagogy, theology, and law.[14] Borrowing from translation and reading studies, and bringing together the history of science and intellectual history, it emphasizes the creative tensions involved in the negotiation of meaning across languages and locales.

This is the view of a "polyglot" Darwin, a perspective that can help us to rethink the scope and range of meanings that were attached to his name by attending to his readers through a focus on translation.[15] "Translation," writes de Man, "shows in the original a mobility, an instability, which at first one did not notice."[16] A lie, a betrayal, a seduction—the many associations of translation point to the fundamental ambiguities involved in the problem of making meanings stable across languages and idioms as much as through conceptual perspectives and metatextual frameworks. This is the same problem with "reading" too, of course, for no two readers approach a text in the same fashion.[17] In short, this approach to *reading* Darwin is one that views him not as the template against which to measure subsequent readings but rather as part of an extended system of meanings, references, and significations. Tracing the web of meaning produced by these readings is central to what this book sets out to do.

The Global Darwin

Darwin's renown was truly global, and he was as much shaped by new global realities of travel, trade, and empire as he was an influence upon the lands they reached.[18] The nineteenth century was, as contemporaries were well aware, an era in which ideas as much as peoples and goods were moving and intersecting across lands and oceans with unprecedented speed: Darwin's own voyage around the world had brought him into contact with tradesmen, colonial officials, amateur naturalists, and local informants and exposed him to a vast array of material objects and social vistas that would profoundly impact his later work. Yet his ideas traveled farther and longer.

In his autobiography, published in 1887, Darwin noted with some surprise that *On the Origin of Species* (1859) was being discussed as far afield as Japan; he had even seen an essay on it in Hebrew showing that "the theory is contained in the Old Testament!" In three decades, editions of *Origin of Species* came out in French, German, Italian, Russian, Danish, Hungarian, Polish, Spanish, Dutch, Norwegian, Japanese, and Chinese; many more would be published in the following years. Summaries, popularizations, and digests of Darwin's writings appeared even more quickly and were no less important: indeed, this was how many readers across the globe most readily gained access to his ideas.

New technologies lay behind this rapid movement. For Victorian commentators it became a commonplace that modern science had in fact "annihilated time and space."[19] The first telegraph cables were laid across the Atlantic seafloor the year before *Origin of Species* was published, and the next two decades saw Europe connected in the same way to India, China, and Australasia. Darwin's fame coincided precisely with the emergence of a series of new print technologies—such as the introduction of cylinder and powered platen machines and the mechanization of paper manufacture—as well as with rising international literacy rates and the subsequent expansion of a global print culture. Together with the telegraph, railways, and steamships, this led to a new kind of information economy that helped to move ideas at unparalleled speeds over vast distances.

Yet technology alone cannot explain the tremendous global appeal of Darwin's ideas. Another reason for their popularity was the ease with which they could be assimilated into local traditions of thought, whether theological, ethical, or cosmological, and they often served as the framework through which ideas of evolution—as much cosmic as human, natural as moral—were understood. While Darwin himself may have found such a reconciliation surprising—as his reaction to attempts to locate evolutionary theory in Hebrew scriptures suggests—this was certainly not as unusual as he imagined.

Whether in the case of religious beliefs, cosmological worldviews, or political loyalties, Darwin was brilliantly ambiguous: to materialists, he could be a revolutionary or a scourge of faiths; more commonly, and for the political mainstream, whether Christian or Muslim, he was a revivifier of traditions, straddling worlds between the moderns and the ancients, giving a new lease of life to ancient philosophers, ethical debates, and even dynastic loyalties. This was how Darwin at once captured and captivated the world—not by ridding it of the forces of enchantment, of faith, or even of God, but by revitalizing traditions of belief and reenchanting them.

Darwin himself was indefinite and at times inconsistent on the question of religion in his writings, both published and personal. He famously left the ultimate origin of species ambiguous in the last line of *Origin of Species*—deploying there a key Christian metaphor for creation—and he often cast himself as an agnostic in his letters. Hence, like so many of Darwin's readers in Asia and the Levant, plenty of Protestants and Catholics believed they could reconcile their beliefs with his theory. Indeed, this confidence spurred many to revisit and revise their own interpretation of scripture.

We might therefore as readily cast Darwin as a force for religious resurgence and theological revivification as for skepticism. Chinese intellectuals in the late 1890s melded together Confucian beliefs in the perfectibility of the cosmic order with modern evolutionary principles;[20] and Darwin's ideas were used by late nineteenth-century Bengali intelligentsia to support long-standing Hindu cosmological beliefs, with a number of people arguing that both positivism and evolution had echoes in Hindu theories of creation.[21]

The first Muslim readers did the same, embracing much of Darwin's message as their own. Supporters and critics alike pointed out that Muslim philosophers had long referred to the idea that species or "kinds" (as the Arabic term *anwa*ʿ suggests) could change over time. The notion of transmutation was also recalled in these discussions, and early Muslim philosophical and cosmological texts were cited whenever Darwin was discussed in Arabic, Farsi, or Urdu. Analogies were drawn with earlier notions of a hierarchy of beings, from matter and minerals to flora and fauna, and finally to humanity itself. That some medieval works also argued that apes were lower forms of humans provided more evidence for nineteenth-century Muslims that Darwin's theory was "nothing new."

Mystical and rational theological discussions of creation, whether ex nihilo or as prime causes, were similarly seen by many Muslim contemporaries as recalling older debates. Questions of man and creation were folded into long-standing theological and philosophical debates over such issues as the nature of the soul and the problem of causality and creation, and scholars made frequent references to the atomistic views of the *mu*ʿ*tazila* or the epistemic claims of the *mutakallimun*. Ibn Rushd's view of causality and al-Ghazali's occasionalism were thus contrasted with the tradition of a natural philosophical view of cause and effect extending to Hume and Darwin. Darwin's Arabic translator, Ismaʿil Mazhar, for example, paired the ideas of the Ikhwan al-Safaʾ with those of Ernst Haeckel and the ideas of Ibn Khaldun with those of Lamarck and Spencer. This was rather like what scholars in Calcutta, Tokyo, and Beijing were doing—constructing lineages for the theory of evolution by tracing some aspect of the theory to older and more familiar schools of thought while claiming ownership of what they saw as the precursors to these ideas. Darwin's Arabic readers were fascinated by such connections, and *Al-Muqtataf*, for instance, received regular correspondence on the topic.[22]

Indeed, the theological controversy over evolution in Europe—publicized by White and others—could actually be used against European claims of superiority. With the well-publicized opposition to Darwin often regarded from outside Europe as symptomatic of Christianity's peculiar difficulty with science, many felt motivated to show just how easily Darwin's ideas could be embraced within their own tradition. We will hear of how the Sufi shaykh Husayn al-Jisr praised the unique rationality of Islam along these lines; while Muhammad ʿAbduh, the grand mufti of Egypt, extended the argument and even helped to inaugurate a new theological genre to demonstrate it. Just as missionaries, Orientalists, and diplomats were claiming that it was Islam that hindered social and intellectual development, ʿAbduh made the reverse case: it was Christianity rather than Islam that was the real obstacle. In his 1904 articles on "science and civilization in Islam and Christianity" he argued that the history of Muslim civilization showed it to be free of the violent conflicts that plagued Christendom—drawing for support on popularizations of works like White's *History* and, like White, focusing on Galileo and Darwin in particular.[23]

Despite the specificities of the Muslim and Christian discursive traditions in Arabic, it would be too simple to assume that we can separate their readings, however. Postwar historians of Arabic intellectual thought have often treated Christians and Muslims as separate theological, as much as epistemic and political, communities.[24] Yet following Darwin's Arabic readers paints a different picture. Tied together by networks of print, through associations like Freemasonry and societies for the "advancement of science," and, later, language academies, many of the figures in this book formed a single, if loosely articulated community of readers, often borrowing from each other's interpretations, commentaries, and references, forming an extra metatextual layer to their readings.[25] This is not surprising: they read, discussed, and debated each other's work. Muhammad ʿAbduh's own 1904 treatise, for instance, is best understood against the background of his debates over the legacies of Ibn Khaldun with the socialist (Christian) writer Farah Antun—debates that saw Azhari shaykhs side with Antun's view over that of their grand mufti.

Muslim and Christian Arab thinkers alike turned to evolution as the solution to the problem of progress—perhaps the key concern for colonial intellectuals at this time. Evolution was a common discourse. Yet their theological works drew upon one another too; and the response of their theologians to evolution often ran along similar lines. When Husayn al-Jisr revived a Muslim tradition of scholastic theology around the problem of creation, he did so by referring to Catholic, Protestant, and Anglican writings on natural theology. In short, neither theology nor politics separated Muslim and Christian readers of Darwin for much of this period. In fact, the real divide would prove to be one of social standing—between this cross-confessional elite and a later, both more popular and populist, readership. This divide would grow over time, with decisive consequences for the way in which Darwin was read in Arabic.[26]

Empire and Civilization, Renaissance and Reformation

The connections between science and empire go back a long way.[27] As scholars of the colonial world have shown, European trade and imperial networks were critical from the sixteenth century onward in the constitution of a modern "world order of knowledge."[28] Yet this was transformed by the sudden expansion of empire in the late nineteenth century and the rapid exchange of information that accompanied it. It was not really until the nineteenth century, after all, that this "world order of knowledge" actually went global. Perhaps the chief spur to the global interest in science at the end of the nineteenth century was the idea that it was the secret behind Europe's rise and its industrial and imperial success. How else to explain that by the end of the nineteenth century, Europeans traded with and held dominion over more than three-quarters of the globe? It is no coincidence that it was around this time, too, that the very notion of "Western science" was itself invented. It was thanks to the new empires of the late nineteenth century that "science" acquired a new sense of its place in the world: at once Western, modern, and universal.[29]

Reading Darwin in Arabic coincided with the intensification of imperial rivalries, the erosion of Ottoman power, and the onset of the "Scramble for Africa." Indeed, for many popularizers, evolution was understood as *the* preeminent doctrine of empire (which also helps to explain the later eclipse of Darwin—particularly by Marx—in the age of decolonization). After all, the most intense phase of the debate over Darwin in Arabic coincided precisely with the intensification of the Eastern Question after 1876 and the British occupation of Egypt in 1882. It was at this time that the laws of evolution—from natural selection to the struggle for survival—made their way into discussions of contemporary affairs. As one Egyptian commentator put it in 1899—at the height of the Scramble for Africa—what was unfolding was "a process of natural selection": "Western civilization, speeded by steam and electricity, is advancing and has expanded from its origins to all parts of the earth. Hardly any country . . . has been left untouched. . . . The weak members [of the species], however, unable to compete successfully in the struggle for survival, are eliminated. . . . This analysis leads us to conclude that preparing for this battle is the only way by which a country can avoid elimination and destruction."[30]

Nineteenth-century evolutionary discourse never strayed far from the anxieties of empire or from the conjoined fascination with the fate of "civilizations." This was the heyday of the civilizational discourse in imperial metropoles, of course, and many Arabic readers associated empire with civilization and thence with a certain conception of historical time. Wherever the rise of European empires was discussed outside Europe, it was accompanied by a parallel discourse of fall. From the mid-nineteenth century onward, Arabic discussions of the rise and fall of an earlier, ecumenical Arabic or Islamic civilization gained ground. Naturally, these evoked a range of thinkers, from Ibn Khaldun to Guizot and Mill.[31] Thanks to the imperial context, the notion of "civilization" became a key concern for the figures in this story, and their understanding of social evolution was almost always bound up with the concept.

It is no coincidence that many of them feature prominently in both contemporary and subsequent discussions of the "Renaissance" of the Arab lands, for questions of rise and fall were always likely to lead to new narratives of revival. These writers were among the first to call for an "Awakening" of the East, and the idea that they formed part of a new Nahda (or Renaissance) of the Arabic arts and sciences was key to their self-presentation.[32] Unlike Europe's Renaissance, however, the Arab version was by no means built around a revival of classical thought. Rather, it was viewed as a *translatio studii et imperii*, a translation of both power and knowledge from the West to the East (once more) that would enable the recovery of past glory.[33]

Jurji Zaydan was a key Arab evolutionist and one of the first to reconstruct the history of an Arab "civilization" against a universal timeline.[34] The basic schema is a triptych that charts the progression of humanity from barbarism and tribalism. [35] In this, the story of Islamic civilization occupies the place of a universal middle ages.[36] If Zaydan was "thinking with history," for him, as for many others in this story, the Arabic Golden Age was inseparable from the question of translation, for it owed its emergence above all to a vast translation movement, that movement of ideas from West to East that was the very source of the first Arab efflorescence.[37] This was the first Arab Nahda. The second—whose influence, according to Zaydan, extended to his own day—began both in Egypt and in Syria: in the former with the French occupation of 1798 and the destruction of Mamluk despotism (another favorite theme in Zaydan's popular historical writings) and in the latter once European and American missionary activity laid the foundations for a new intellectual "awakening." The idea that this Nahda would eradicate the centuries of "torpor" and overcome the senescence of its own classical tradition through translation was key. For Zaydan, as for other Nahdawi thinkers of the time, history thus proved that translation was the key to cultural revival and universal progress.

In short, for many colonial intellectuals of the late nineteenth century and after, uncovering, reclaiming, and even redefining one's own classical or golden age was crucial for claiming a prominent place in the hierarchy of civilizations. And in this context it was thanks to new notions of science that the very understanding of "civilization" was transformed. Such a connection was forged at the very moment that post-Napoleonic Europe was itself giving a new prominence to what one French commentator called "that new divinity—civilization." Take the ideas of Rifaʿa Rafiʿ al-Tahtawi, who was a religious scholar who spent five years in France as part of an early Egyptian educational "mission" in the late 1820s (where he read Rousseau, Montesquieu, Voltaire, and many works on "universal history"): "the more you go back in time, the more you can see the backwardness of people in terms of human skills and civil sciences. Conversely, as you descend and observe the slope of time (to the present), you generally see the elevation and progress of these."[38]

A later generation built upon these formulations. They inclined to Mill's distinction in his own work on "civilization" between moral perfection, on the one hand, and wealth and worldly power, on the other.[39] Yet translation also disarmed the racialized readings of Le Bon's discussion of "the evolution of races" and transformed it into a debate about the perfection of peoples (*ʿumam*).[40] They read Guizot on the importance of religion in cementing social solidarity and recalled Ibn Khaldun, whose much earlier discussion of similar themes helped naturalize the entire debate.[41]

This emphasis on civilization as social progression explains why so many of Darwin's Arabic readers seized upon his ideas as evidence for the evolution of peoples. Reading arguments about Europe's civilizational superiority not only led many of them to refer back to their own once civilizationally glorious past in new and more hopeful terms but also explains why they understood evolution itself to be a doctrine of progress. After all, Darwin himself clearly saw moral progress as one of the outcomes of human evolution. This vision, spelled out in his *Descent of Man*, arguably the more influential of his works for readers throughout the world, chimed with Arabic readers in particular. For many of them, evolution was about much more than biology and more even than history: the new universal history for which it provided the blueprint was one with the power to recast the future.

Natural Theologies

While Darwin was typically used by Arab evolutionists to give support to their particular ethical or political views, his ideas also raised specific problems for them. It was the issue of man's special creation that emerged as the most contentious for Arabic readers, as for others elsewhere.[42] In time, this led both Christian and Muslim theologians to revisit former theological and hermeneutical methods, often transforming them in the process. The Ottoman shaykh Husayn al-Jisr offers a good example.

Al-Jisr moved between the rules of literal interpretation, or *tafsir*, and metaphoric or "embedded" ones, or *ta'wil*. Deploying this distinction meant that certain debates, on the age of the earth, for instance, or on the variation, adaptation, and transformation of species over time, could be found to be backed by scripture. The metaphoric or extended reading of the word for "days" in "days of creation" by a number of canonical medieval authors and ideas of a "progression toward perfection" also extended this trend.

In essence, the engagement with evolution allowed theologians to resuscitate classical arguments about the ontology of the cosmos or of matter while adopting the empirical and even rational claims of contemporaries. Al-Jisr's book is a brilliant exposition of this, as its thick text is interlaced with long summaries of current paleontological, archeological, zoological, and botanical findings (mostly gleaned from *Al-Muqtataf* and other popular Arabic science reports) alongside fairly traditional commentaries on Muslim dictates, principles, and even practices.[43] Through al-Jisr, one is reminded that reading Darwin led to the revival of traditions of natural theology in Arabic.[44]

Part apology, part exposé, al-Jisr had little difficulty accepting the evolution of species among lower orders, but also he left room for an exceptional spiritual human evolution. In this way, he and others managed to come to terms with the theological problem of the origin of morality raised by the new evolutionary sciences. This meant eschewing a materialist interpretation of evolution. Indeed, al-Jisr counted modern evolutionary materialism as among those schools that the prophet Muhammad had confronted when delivering God's message, and he compared contemporary evolutionary materialists to Meccan infidels or to atomists who denied a belief in a First Cause.

In essence, he treated current evolutionary debates themselves as part of a longer tradition of theological deliberations on the order of nature and the idea of creation *ex nihilo* or the creation of the soul. Many other Arabic readers did the same: the *Al-Muqtataf* group, whom al-Jisr borrowed from, also blended the theistic traditions of their own: in this case, bringing together a neo-Baconian Anglo-American vision of science together with Anglican natural theology and Protestant philosophical theism. Indeed, their reading of Darwin in this light arguably influenced all the other figures covered in this book. Weaving together Spencerian positivism with revived, and often intertwined, theological traditions, they debated old problems like the nature of creation and the problem of materialism in new terms. As the example of Arabic variations on the idea of materialism (a term that encompassed a broad range of meanings in many other languages of this time) shows, these variations were flexible and ambiguous, attached to parallel philosophical, cosmological or ontological varieties and bringing together equally complex genealogical cross-referents.

Knowledge and Belief

The challenge of assessing the theological implications of evolution brought up a familiar problem: the distinction between knowledge and belief. When Farah Antun, borrowing from Ernest Renan, attempted to praise Ibn Rushd for his emphasis on secondary causes in nature, Grand Mufti ʿAbduh merely swatted away the distinction between secondary causes and an occasionalist view of a chain of effects with an appeal to al-Ghazali (that carried echoes of the new empiricism of his time). The primary focus for ʿAbduh and others was the nature of knowing itself, and debates on Darwin helped to reanimate older arguments about the nature of knowledge and belief. As the frequent references to Ibn Rushd and al-Ghazali suggest, at least two genres of older philosophical and theological writings were invoked in these discussions: *kalam* and *falsafa*. The one was an essentially theological genre, and the other derived from ancient Greek philosophy, but both were concerned with describing what it meant to know or to believe as states of consciousness.[45] Mostly they presented this as a matter of trust: knowledge came from trustworthy persons such as from prophets or competent men and their books.[46] But there was also a long-standing rationalism that extended to questions of belief as well as knowledge. The appeal of the rationalizing early medieval *mutakallimun* for late nineteenth-century Muslim intellectuals, for instance, was precisely that they brought reason and belief in line with modernist distinctions without discounting the foundations of religious faith. The epistemological foundation of rationality in the *kalam* tradition was also gained through doubt or skepticism.[47] Al-Jisr, for instance, made the argument that scientific knowledge-claims themselves—including the "laws of evolution"—needed first to be subjected to the tests of certitude that *kalam* and other forms of Arabic logic demanded with regard to all forms of belief.

Over time, however, many began to raise the counterargument that belief was certain but inexplicable or, in other words, that reason itself had its limits. The idea that reason was the absolute, only, or final arbitrator of consciousness was thus increasingly subjected to dispute. Stressing the limits of reason highlighted an alternative definition of faith that distinguished sharply between knowledge as mere factual belief—organized around the "laws of phenomena"—and an intuitive and nonrational kind of truth.[48] Once again, early Arabic (and here especially Sufi) writings proved highly relevant. As al-Ghazali, the most frequently cited in this regard, argued, "there are many degrees of knowledge." He continued: "owing to the different degrees of perception in people, disputes must arise in tracing effects to causes. Those whose eyes never see beyond the world of phenomena are like those who mistake servants of the lowest rank for the king. The laws of phenomena must be constant, or there could be no such thing as science."[49]

It is hard to avoid discussions of these ideas when reading about Darwin in Arabic. Debates over the phenomenal, noumenal, and mystical nature of knowledge were repeatedly recalled in these conversations. This was one reason why so many of Darwin's Arabic readers drew on the views of al-Ghazali, Ibn Rushd, and others. Darwin's translator, Ismaʿil Mazhar, offered a striking critique of the idea that all knowledge-claims, of which he classed belief as a subset, had a purely rational basis, for example. Heavily influenced both by the works of British idealism and by early Arabic philosophy, Mazhar argued that much scientific knowledge was itself based upon postulates that could not be substantiated. Hence non-demonstrable beliefs were themselves crucial to the claims of modern science.

With the claim that even scientific knowledge could be said to be ultimately indeterminate or "unknowable," Mazhar left room for the notion that religious belief rested on similarly transcendent grounds. Here Spencer was also recalled: at least as important as Darwin in bringing together issues of evolution with epistemology, his somewhat abstruse notion of the Unknown chimed very well with transcendental positivists like Mazhar and others. Indeed, as elsewhere, in the Arab East the turn to positivism and evolutionism was typically accompanied by transcendentalism.[50]

The introduction of positivism, like evolutionary thought itself, had initially been bound up with American Baconianism and Scottish Common Sense philosophical theism. Arabic discussions of positivism drew on Protestant thinkers from both sides of the Atlantic, and their responses to the challenges of positivism, both in the Comtian and in the Spencerian vein, were closely followed in Arabic writings.

In many such writings, the notion of "religion"—as much as "belief"—was quietly transformed: increasingly, religious faith became associated with nonpropositional statements about the nature of the world, with "belief" in other words; while propositional or positive knowledge was said to be specific to knowledge of the phenomenal world, or to knowledge acquired through science. Meanwhile, discussions of religion were also increasingly attached to notions of the evolution of the "idea of God" and assigned a "social function" that could be considered and compared across a universal historical timeline. The coming of Darwin in Arabic was thus accompanied by reconceptualizations of the meaning of "religion" itself.[51]

Science in Translation

"Science" acquired a similarly novel meaning. *ʿIlm*—the broadest word in Arabic for "knowledge"—was the vehicle of this shift. In the classical lexicon, it had typically referred to the knowledge of "definite things."[52] Rifaʿa Rafiʿ al-Tahtawi was one of the first to hint at an emergent conceptual divide between Arabic and European categories of knowledge. In his 1834 account of his stay in Paris, he wrote: "Those who are called *ʿulama* [scientists] in France are those who are well versed in the temporal sciences. Thus, when they say *ʿalim* in France, they do not mean that he is well versed in religion but that he is well versed in another subject."[53] Of course, it was not long before what it meant to be an *ʿalim* changed in an Arabic context too. Over time, the rise of new civil and technical schools and the spread of the press transformed the position of the *ʿulama*, long counted as the traditional holders of knowledge, who were slowly displaced by a new class of civil servants, technocrats, educators, and even popularizers who laid claim to the methods, vocabulary, and authority of what they now defined as "science."

So while science may have emerged in the nineteenth century as a global category, it was nevertheless discussed and debated in very local terms. The mechanics of actual translations often provides the most fascinating examples.[54] The very contour of debates over the precise meaning and range of his ideas in translation shows how they were extended, transformed, and appropriated in novel ways. Translation has not always constituted a major theme in the history of science: yet, thinking through translation, with its primary search for equivalences—for concepts as much as objects—raises interesting questions, whether about the nature of meaning or about categories of things.[55]

Take the problem of species for instance. Initially, there was no specific Arabic term for species as distinct from varieties or kinds. (Other orders in nature proved equally ambiguous, such as phyla and order.) Hence, translators had to somehow reconcile the problem of writing about the evolution of species without being able to define "species" exactly. Darwin himself, of course, claims he was in the same predicament: writing in *Origin of*

Species, he admitted that a precise concept of "species" was not something one could invariably fix upon.[56]

"Natural selection" forms another instructive example: rendered as "nature's election" or "nature's elect" as much as "selection," it too carried different shades of meaning across different translations. Similarly, the notion of "struggle," as Mazhar noted in his introduction to *Origin of Species*, could be understood to refer to a "fight unto death" between individual members of a species or kind, or it could simply imply a collective competition for resources, in which case the struggle involved not intraspecies competition but, rather, a more generalized competition with one's environment.

As these brief examples also demonstrate, evolutionary concepts rested upon larger assumptions about the state of nature. The very language Darwin used (struggle, competition, selection, species, and races) and the examples he gave (pigeon fancying and dog breeding) drew upon a host of previous ideas and references with which most Victorian readers would have been familiar. Arabic readers had their own references and conceptual predecessors, too, from horse breeding to classical Arabic arguments about ontological materialism or atomism and theories of transmutation and alchemical transformations.

Discussing Darwin in Arabic occurred through a process of literally grafting new terms onto older ones. In 1924, the little-known Hasan Husayn published a translation of Haeckel's *Kampf um den Entwickelungs-Gedanken* (which was in fact a translation from the English freethinker Joseph McCabe's translation of Haeckel).[57] Husayn was not particularly fond of Haeckel, or of Darwin for that matter: worried about the atheistic implications of Darwin's ideas, as he dubbed them, he nevertheless justified the translation of the work because "we live in an age overshadowed by evolution [*al-tatawwur*]." Husayn also spent a considerable amount of time thinking through the trials and tasks of a translator, and he used the word for evolution—*tatawwur*—very self-consciously: "Now some would criticize us for using this term [*al-tatawwur*] and would say that the language does not know of '*al-tatawwur*' but only of *al-taur* and *al-atwar*." But he justified his use of the neologism by arguing that the Qur'anic verse "*wa qad khalaqakum atwaran*" (He created you in stages; sura 71:14), and others like it, made the derivation here appropriate, given that a more classical cognate already anticipated the supposedly novel meaning rendered by the neologism.[58] In fact, this was merely one of many prefigurings of "evolution" that Arabic translators and readers made recourse to: medieval Arabic and Persian naturalistic and alchemical writings provided plenty more.

For some, transliteration was a means of emphasizing science's modernity; others would go down the "archaizing" path for the opposite reason. Translation was a zone of contestation, and many translators were similarly conscious of the different strategies available to them and the stakes they implied.

If we understand translation to be a creative act, or a complex process of textual arbitrage, then we can no longer assume that the original text is the only source of meaning. If we want to understand how knowledge production is "socially embedded," then we need only view the same text read in different contexts. One way to view translations, then, would be as forms of cartel knowledge invoking intertwined sources of authority and arbitration. Take the case of Bishara Zilzal's writings on evolution. His textbook of natural history, *Tanwir al-adhhan,* written sometime in the 1870s, emphasizes continuities of thought with the Arabic and Greek traditions: indeed, it begins with a discussion of Aristotle's books of *History of Animals.* Then it turns to Arabic works of natural history and zoology, ending finally with those "foreign scientists who excelled in the science of natural history," the "reasons for the advancement of this science," and the shortcomings of "Easterners" in comparison to Europeans in matters of modern science.[59] The book was originally composed for use as a handbook of natural science at the Syrian Protestant College, yet it also found eager audiences elsewhere. There was an 1879 edition of the work published by a school press in Alexandria, Egypt, while a later reedition was prefaced with a eulogy to Lord Cromer and the English for their efforts at accelerating the evolution of the state of Egypt's affairs.[60]

Progress and Pedagogy

When the Syrian Shibli Shumayyil introduced his 1884 translation of Ludwig Büchner's lectures on Darwin, he ended with a futuristic vision of the transformation of the Ottoman Arab lands. One of the main mechanisms for this was to be a sweeping change in school curricula. Among his recommendations were rigorous training in all the disciplines of the modern sciences, from the elementary level up, and the elimination of theology, law, poetry, and other "irrelevant" humanities subjects. To many, Shumayyil's vision seemed excessive, even radically antitraditionalist. Yet his enthusiasm for the introduction of training in the new sciences and his focus on education as the key to social (and hence political) reform capture a utopian zeal shared by plenty of his class and generation.

Many others shared his view that education in the sciences could be a powerful agent of reform. For missionary educators, for example, education in the sciences encouraged right reasoning, proper habits, and good morals. New disciplines like hygiene and domestic science were discussed in terms of "self-preservation," while courses in logic and in natural philosophy were presented as the best means of "training minds"; indeed, training selves and disciplining bodies stood behind much of the early missionary enthusiasm for science education.

Others drew on familiar ideas of education as moral training, as many modern pedagogical theories also reworked classical traditions of *paideia*. For instance, *adab* (proper manners, morals, and taste) came to imply new norms of civility and a new kind of moral science. The idea of *tarbiya* (education), which included such themes as child rearing and moral cultivation, came also to mean school learning. Older notions of virtue and of self-knowledge were extended because of the growing concern with the "social function" of education and its role in the creation of a harmonious "social" (and national) order. For instance, in his 1881 *Discourse on Eight Words*, Shaykh al-Azhar Husayn al-Marsafi (former teacher of Husayn al-Jisr) described *tarbiya* as "perfection by degrees toward completion"—an innate force for improvement that humans inculcated for the "benefit and aid of a society."[61] As each individual was merely a "limb," or part, of the *umma* (community or state), so he must be trained by nature and the mind to cooperate and aid others as part of a group (*ta'ifa*). For Marsafi, education referred above all to the "cultivation of morals" (or *tahdhib al-akhlaq*). And like doctors, educators "serve their country" (or *watan*, another new "keyword" in Marsafi's octet).

This emphasis on moral education was one reason why Herbert Spencer's own views on education found so many eager Arabic audiences. For Spencer, training students in "the management of the family" was the crux of a true moral education—and one of the three pillars of universal education that he felt had been sorely neglected in his time. Muhammad ʿAbduh was a fan of Spencer's and he translated the latter's *On Education* and even went to visit him in Brighton in 1904. ʿAbduh's own writings on education, including his reflections on those of Spencer, similarly emphasize morality, the family, and the state and assign education a key role in the cultivation of proper moral subjects and faithful citizens. His critique of the failings of the educational system of Al-Azhar, and even his role in the reformation of shariʿa law codes, typically rested on this argument.[62]

After the British occupation, funding schools offered both a political (and fiscal) dilemma for many and an opportunity for some. Advocating largely laissez-faire policies with respect to "native reform," most British administrators in Egypt argued that reform was a precondition for a self-determined national development, and hence the only real argument for true independence for what would later come to be known as its "veiled protectorate." Yet the British were wary of introducing mass education there; they warded off calls to found a university as long as possible; and they encouraged the introduction of school fees to restrict education to the landowning or upper middle classes, who were already moving into the higher echelons of government service. This sometimes suited the latter who often adopted a similarly paternalistic stance when discussing the state of Egypt's "masses." This vision of "liberalism without democracy" was shared by many colonial intellectuals of the time.[63] It was only after the end of the British occupation that a universal primary education act was instituted by the Egyptian state, with drastic implications for both educational curricula and for the class of readers that emerged previously.[64]

The Politics of Evolution

In the period covered by this book, science and the state emerged as bedfellows. From the beginning of the nineteenth century, the creation of technocratic expertise and the official promotion of "useful science" were found in Egyptian military and agricultural institutions, in the construction of public works, in engineering and (some) industries, and in newly established teachers' training schools. Schools, founded either by Muhammad ʿAli in the 1830s or by Khedive Ismaʿil, turned out an educated elite that formed the core of the reading public for the new Arabic press. These graduates were civil servants, writers, and professionals, who for much of the late nineteenth and early twentieth centuries shared a common vision: that of a modernized and technocratically run state. *Al-Muqtataf*'s support for the "diffusion of useful knowledge" and the creation of new bodies of scientific expertise spoke to these concerns. But this vision of a scientifically managed state was even more popular with the next generation, men like science writer Salama Musa and Darwin's translator Ismaʿil Mazhar.

Yet what precisely was spelled out by the politics of evolution was subject to endless debate. Some of Darwin's Arabic readers saw evolution as a means to support the political status quo; others, to overthrow it. The editors of *Al-Muqtataf* themselves hoped to liberate Arabs from the grip of a despotic and theocratic state (chapters 1 and 2). For others of the same class and background, however, religions were both scientifically unnecessary and socially atavistic, and hence, the empire's pretension to rule rested on retrogressive impulses (chapter 3). For Ottoman loyalists, by contrast, science provided a powerful justification for imperial reform (chapter 4). The rise of new forms of partial sovereignty, competing imperial and dynastic claims, and new ideologies of nationalism blurred the boundaries between Ottomanism and "Pan-Islamism" as much as Arab nationalism did. Then there were also the British, who occupied Egypt in 1882. In line with *Al-Muqtataf*'s anti-Ottomanism and proliberalism, the journal gave the British occupiers its support. Others looked to the British for support of another kind: Grand Mufti Muhammad ʿAbduh of Egypt had a complicated relationship with the British government, as he often used their support against the khedive and palace and sometimes even against his Azhari colleagues (chapter 5). Others found that British social science, and particularly a Fabian-inspired evolutionary socialism, offered the best political solution to current problems (chapter 6).

What they almost all shared was a fundamental gradualism, an acceptance of a pragmatic compromise with power. Whether collaborationist or anticolonialist, liberal or socialist, Darwin's Arab advocates shared the conviction that evolution implied a slow change over time. In fact, the tendency was a global one: most of Darwin's readers around the world subscribed, not to revolution, but to political reform of a very gradual sort. Hence, it was not only Arab evolutionary socialists who turned to Darwin before Marx. Change was to be managed through the rational mediation of the new technocratic classes and reflected a shared commitment to the sensibilities of what was once termed the "politics of notables."[65]

The period covered by this book was the heyday of confidence in the notion of an ongoing Nahda. By the interwar era, however, the romance of an Arabic renaissance of ideas quickly lost ground as anticolonial nationalists, among others, questioned the movement of ideas from West to East. A new generation of educated middle-class men and women now saw these Renaissance or Nahdawi intellectuals as having failed on political grounds: since the modernization of the state along technocratic lines, together with the nationalist liberation and independence it promised, had not been realized, they would seek liberation elsewhere. As the luster of the Nahda—as much as the optimism in scientific progress—faded, so too did the dominance of the evolutionary paradigm.

Publics and Counterpublics

Studies on the popular discussion of science in Europe and the United States tend to take for granted the conditions of industrialized societies: mass literacy, popular print culture, and the emergence of professional scientific communities. Conditions were very different in the Arab East. For one thing, professional experimental science in the modern sense had barely begun. For another, it should be remembered that while adult literacy rates in England were as high as 95 percent by 1900, the rate in Egypt was a mere 5.8 percent by 1897. (Of course, urban areas had much higher literacy rates than rural ones: 21.6 percent and 19 percent of all men in Cairo and Alexandria, for instance.)[66] This was one obvious constraint both on the popularity of Darwin's ideas in the Arabic-reading world and on the speed with which they circulated there. Arabic readers of the nineteenth century were not "ordinary readers" in the sense that has been used to describe the "intellectual life of the British working classes."[67] By and large, nineteenth century Arabic readers reflected the rise of a newly educated middle-class audience of notables, civil servants and professionals. These were people who thought it "possible to recognize society, and the organizations and relations that sustained it," and "to bring these into public discussion and for action by the state." In this way, a certain educated elite "came to think of itself as constituting the public."[68] As the historian Natalie Zemon Davis once wrote, "a printed book is not merely a source for ideas and images, but a carrier of relationships."[69] The rise of an Arabic press in the late nineteenth century similarly created new relations. These new reading networks were both cross-confessional and transregional. They brought the *Muqtataf* group together with Jamal al-Din al-Afghani and Muhammad 'Abduh. American and British missionaries had viewed the press as a "potent tool of civilization" and a powerful means of conversion, while Muslim scholars saw it as an effective agent of transregional Muslim reform. Shattering former conventions of reading, transcribing, and collecting texts (above all, in a society dedicated until then to chirography and the palimpsestic codex[70]), the print press forged novel literary and material networks, circulating ideas much further afield than just the Arab world—now linking Beirut, Cairo, Tripoli, and Istanbul with London, Paris, New York, and Buenos Aires.

Yet the newspapers and journals that carried these debates had a circulation that rarely surpassed several thousand at most (as opposed to western Europe, where mass circulation dailies had in some cases more than a million readers). Readers were drawn neither from the traditional scholarly classes nor from the masses but chiefly from the civil service and the political elite, both of whom preserved close links (often financial) with journalists and newspaper proprietors.[71] Of course the network of relations and ideas created by the press also extended beyond the realms of literacy itself. Hence, collective readings and discussions of newspapers and journals in village squares, cafés, and homes meant that one might encounter Darwin in any of a number of ways.[72]

Yet we should not lose sight of the fact that we need to understand the emergence of reading publics as both a historical and a conceptual problem. As a number of recent critics have pointed out, notions of the "public" tend to presume a detachment from the very "institutional and disciplinary conditions that enable it." In its classical formulation as "public sphere," it is also seen as a largely "self-organizing discourse" by assuming "a structural blindness to the material conditions of the discourses it produces and circulates, as well as to the pragmatics of its speech forms."[73]

By the 1930s, dramatic increases in literacy rates and the expansion of state education had led to the emergence of quite distinct reading communities marked by their own social and moral print economies, and a broad range of "parallel discursive arenas." This led, too, to new discussions of science, often in opposition to the one that had been dominated by *Al-Muqtataf* and others at the end of the nineteenth century. New readers grew increasingly dissatisfied with the failure of the political and technocratic promises of the past, and they sometimes deployed both the banner of science and the machinery of new media toward radically different ends.[74] The emergence of "parallel discursive arenas where members of subordinate groups invent and circulate counter-discourses to formulate oppositional interpretations of their identities, interests and needs" thus came after those written in the name of another "public."[75]

Darwin's early Arabic readers were often conscious of their public audience: they were public men of science, supporting public measures, safeguarding the public good, and so forth. Yet they were always part and parcel of a generation of literary elite who spoke in the name of their masses while still considering themselves a class apart. Thinking in these terms, the rise of new publics and then counterpublics helps us to understand both the appeal of Darwin and then the decline of that appeal over the near-century covered by this book.

ONE

The Gospel of Science

In 1876 a new Arabic science monthly published in Ottoman Syria rated the invention of the telegraph above the Seven Wonders of the World (*awabid al-dunya*). A series of articles featured a brief history of the invention alongside a quick lesson on electric currents (fig. 1.1).[1] The generation of a steady flow of electric charge was detailed, and two types of voltaic batteries, a pile and a "crown of cups," were also illustrated for the reader. Written at a time when electricity had recently entered parts of Cairo and Alexandria and was only just heard of in Beirut, the journal delighted in announcing the latest discoveries in science and technology to new audiences.[2] After all, the telegraph itself had first made its way to the Ottoman Empire only in 1855—after the Crimean War—and its entry into the Syrian provinces had caused quite a stir.[3] For the journal, however, the science of telegraphic signals offered the opportunity to mark the beginning of a new pedagogic devotion to the popularization of science and its wonders.

This new science monthly first appeared in Beirut under the title of *Al-Muqtataf* (The digest) and was distributed in several Syrian and Egyptian towns beginning in May 1876.[4] Read across the Ottoman Arab provinces (and later beyond), the journal enjoyed a long and prolific career. It was based initially in the Syrian Protestant College in Beirut (where its editors were instructors) before moving to Cairo, where it was published from 1885 until the revolutionary days of 1952. Written in a comprehensible language, full of illustrations, and replete with instructive lessons on the latest scientific advances and technical inventions, the journal was one of the first to popularize the modern sciences in the Arab provinces of the Ottoman Empire. "An enterprising and ably conducted scientific magazine," ran one American commentator's assessment, "its mission is a stimulating and timely one among the educated classes in Syria and Egypt."[5]

١٨٣ آلة تلغرافية لنقل الخط

حديد مركّز على طبقة من التراب المعروف باسم تريبولي ويوضع الجميع في فرن صغير محمّى بالحطب او بالفحم وعندما يبتدئ الذوبان يخرج اللوح من الفرن فتكون القطع المستديرة قد صارت كروية الشكل وغير المستديرة قد انحنت سطوحها وصارت سهلة التقطيع . ثم تلصق براس القضيب المعد لمسكها في آلة التقطيع وتُقرّب من دولاب من الرصاص عليهِ سنباذج فتقطع حسب المطلوب ثم تصقل على دولاب من قصدير عليهِ تريبولي جيد وماء والمواد التي ترش على الدواليب هي سنباذج وتريبولي وحجر خفان واكسيد القصدير. ويختلف استعمالها حسب صلابة الحجر

اختراع آلة تلغرافية لنقل الخط كما يحررهُ كاتبهُ

بقلم حبيب افندي فارس (بحروفها)

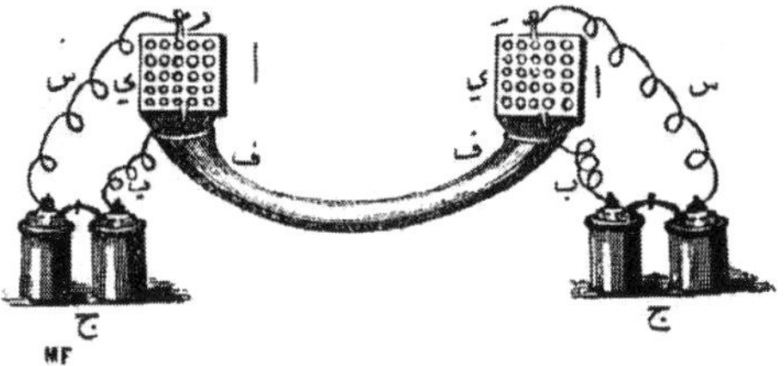

انهُ من المعلوم بان الكهربائية تقسم الى قسمين زجاجية وراتينجية وانهُ عند امتزاجهما اذا دخلتا على قطعة معدنية تجعلانها للحال مغنطيسية ما دامتا عليها وترتفع عنها القوة المغنطيسية متى ارتفعتا او ارتفعت احدهما عنها . فاذ ذاك نقول ان الآلة التلغرافية الموضوعة صورتها اعلاهُ مركبة

اولاً من حق كهربائي مزدوج متوجهة الكهربائية الزجاجية منهُ بالخيط المعدني ب

ثانيًا من خمسة وعشرين خيطًا معدنيًّا ي مصنوعة نظير فرشاية مربعة وملاصقة بعضها وكلٌّ منها ملتف بالحرير منعًا لامتداد كهربائيتهِ الى غيرهِ وجميعها ملفوفة باللستيك حتى تصير كخيط واحد ف

ثالثًا من قطعة زجاج موقعة على هذه الفرشاية ا

رابعًا من قلم ر مركب من قطعة زجاج باولهِ ومن قطعة معدن تدخل عليها الكهربائية الراتينجية بالخيط س ومعلق بشريطٍ ملتف يسمح لهُ بالنزول على قطعة الزجاج عند ما تدخل عليها الكهربائية الممتزجة وترفعهُ عنها عند انقطاع الكهربائية

والآلة الثانية التي تقبل الخط هي نظير هذه تمامًا وكيفية العمل هي ان يمسك الكاتب القلم ر عند

Fig. 1.1. "Al-talighraf" [The telegraph], *Al-Muqtataf* 1 (1877): 4.

This journal was to play a critically important role in the popularization of Darwin's ideas. Although Darwin's 1859 work *On the Origin of Species by Means of Natural Selection; or, the Preservation of Favoured Races in the Struggle for Life* was not fully translated into Arabic until well into the twentieth century, his ideas were in circulation from the 1870s onward.[6] From the early

1860s, students had heard about Darwin in American Protestant Mission schools in Beirut, and within a decade, mission tutors published science primers in Arabic that made mention of his name, among others. Similarly, ʿUthman Ghalib, an instructor at the main state medical school in Cairo (Qasr al-ʿAini), who had studied in Paris, gave zoology lectures on his return to Cairo that were full of references to Darwin, Huxley, Haeckel, and others.[7] However, Ghalib was not a popularizer, and his lectures were not published until 1886; and while missionary primers circulated throughout the Ottoman provincial schools from Tripoli to Alexandria, they were intended for young schoolboys and schoolgirls and not the general reader. It was thanks to *Al-Muqtataf* that Darwin's name moved from classrooms to readers throughout the region.

Yet, as this chapter suggests, reading Darwin in Arabic was bound up with both the classroom and the press and, in its early years, with missionary pedagogy and proselytization. For several decades, foreign missionaries in Beirut, especially American Protestants, had been teaching the "new sciences," from natural philosophy and zoology to chemistry, as part of their proselytizing efforts. The popularization of science was an outgrowth of this program, and it was one that would have decidedly unexpected consequences.

The founding aim of *Al-Muqtataf,* as the editors put it, was to announce the virtues of science to an Arabic-reading public in an attempt to encourage scientific and technical progress in the Arab East. "We hope that this journal will meet with the approval of the public," they wrote in their introduction to the first issue, "and will encourage the reader to acquire scientific knowledge and to strengthen industry."[8] Science was the key to industry, and industry the key to the success of nations. They therefore published on a wide range of scientific and industrial topics. In line with their ever-growing interests, in 1888 they changed the subtitle from "A Journal of Science and Industry" to "A Journal of Science, Industry, and Agriculture," adding "and Medicine" in 1893.[9]

Its founders were Yaʿqub Sarruf and Faris Nimr, two enterprising young Syrian Protestant College instructors who dedicated themselves to campaigning for scientific advancement.[10] "We used to regret," they wrote, recounting how the idea for a science journal occurred to them while working as tutors in Beirut, "that our Arabic language lacked a magazine that could simplify the arts and sciences." Shortly thereafter, the two young science enthusiasts embarked on their ambitious project, offering their Arabic readers summaries of "new discoveries and useful research" month after month.[11]

They were committed to popularization: articles ranged from accounts of new inventions such as the phonograph and radiometer to practical tips on how to whiten one's teeth, remove unwanted hair, and get rid of bedbugs. In one issue, one could read the editors' rebuttal of magic, encounter three varieties of zebra, ascertain the latest technologies in blood transfusion, and learn how to grow long, luxurious hair.[12] "It is well known," ran the notice they circulated throughout Greater Syria to introduce the new journal, "that science journals are the best means for diffusing useful knowledge among both the learned and the lay." Their journal was to be based on the summaries and digests of the latest scientific literature of the day, based on their "access to numerous current and reliable sources in several languages." This enterprise, in their minds, was novel not merely in terms of the works they were translating but also in its aspiration to pursue a purely scientific path. "The journal," they emphasized, "will be exclusive of religious and political matter."[13] But this was not so easy to achieve, and as we will see, their writings on science were very much bound up with both religion and politics. Their vision of science was interpreted through a determinedly popularizing (but not populist) lens, especially in the early years.

At first issues were slim compilations of the latest scientific and technical advances in Europe and North America, each month initially totaling no more than twenty-four pages. Apart from one or two original, typically unsigned, articles, the rest were usually summary translations from British and American science journals and periodicals. On *Al-Muqtataf*'s original cover was an emblem—the pen and hammer—which was the same as that found in the *American Artisan* (fig. 1.2).[14] Although the editors rarely acknowledged their sources, we know that these included *Scientific American, American Journal of Science, Popular Science Monthly*, and *Knowledge*, alongside general-interest publications like the *Economist, Spectator, Nineteenth Century, Fortnightly Review*, and the *Times*.[15] Articles derived from all these appeared, often a matter of only weeks after their original publication. Summaries of the proceedings of professional societies—such as the Royal Society of London, the Royal Asiatic Society, the International Congress of Orientalists, the Royal Geographic Society of Egypt, the Eastern Scientific Society, and the Cairo Scientific Society—were later also included.[16] This range of sources testifies not only to the editors' wide reading but to the literary resources they were able to draw on in Beirut. European and American science thinkers of the day—such as Darwin, Huxley, Pasteur, Tyndall, Bastian, Lubbock, Wallace, Spencer, Haeckel, E. B. Tylor, Henry Sidgwick, Agassiz, Faraday, Max Müller, Kropotkin, and others—were thus regularly featured, and versions of their ideas disseminated.

المقتطف

جريدة

علمية صناعية

تصدر اول كل شهر

لمنشئيها يعقوب صروف وفارس نمر

قيمة الاشتراك في بيروت ولبنان سبعة فرنكات وفي الجهات ثمانية

السنة الاولى

طبعة ثانية

Fig. 1.2. Cover of *Al-Muqtataf*.

The format—following those popular-science monthlies from England and America on which the journal drew—was standardized relatively early. A few feature articles, such as "Glassmaking," "The Microscope," "The Moon," or "Arab Astronomers," usually started off an issue; these would typically include a biography of a famous scientist, such as Galileo or Harvey. A section on useful inventions and household hints followed and then, finally, a question-and-answer column. The price was initially relatively modest—an indication of the editors' efforts to make science accessible to a general audience. The first few issues cost only seven francs; for comparison, *Al-Jinan*, another early Arabic science journal, in 1879 cost over three times the price of *Al-Muqtataf*, or twenty-three francs.[17] In 1881, however, when the editors nearly tripled the size of their journal, the price of an annual subscription rose to one Ottoman pound, which was a very expensive proposition and which points to the journal's essentially core elite readership despite its efforts to reach "the learned and the lay" alike.[18]

As a purely scientific journal, dedicated to the "diffusion of useful knowledge"—a Victorian aspiration the journal shared with Arab intellectuals of the time—*Al-Muqtataf* marked the rise of a new Arabic-reading public's interest in popular-science writings.[19] Before it, there had been only one journal of this kind in Syria: the Beirut monthly *Al-Jinan*, which was started in 1870 by another onetime Protestant missionary assistant, the Arabic lexicographer and educator Butrus al-Bustani.[20] Bustani's journal, however, was considerably more costly; relying on financial support from Khedive Isma'il in Egypt, it had a more limited circulation and lasted only fifteen years. It was also much more eclectic in its outlook and contents: *Al-Jinan* set out to "spread universal knowledge" (*al-ma'arifa al-umumi*) of all kinds—"practical, literary, historical, industrial, commercial, material and the like."[21]

Indeed, there were few comparable Arabic journals in any of the other publishing centers of the Ottoman Empire. In Cairo and Alexandria, for instance, even after the British occupation of Egypt in 1882, when there were hundreds of journals, periodicals, and newspapers in Arabic, English, French, Greek, and Italian, there were only three scientific magazines in circulation in total: *Al-Muqtataf*, which moved its operation to Cairo in 1885, and two medical journals, *Al-Shifa'* and *Al-Sihha*.[22] But this changed completely with the expansion of the press in Egypt. By the turn of the century the journal's success spawned competitors, and periodicals such as Jurji Zaydan's semimonthly journal "of science, history and literature," *Al-Hilal*, founded in 1892, and Farah Antun's *Al-Jami'a*, founded in 1899, began to cover similar ground.

Just two years after its founding, the journal expanded from twenty-four to twenty-eight pages, and that same year it included seven lithographic illustrations (the following year, this rose to nine); within another three years, it would treble in length. If this was an indication of the venture's success, it was nevertheless a relative one. With a population of about 2.7 million and a literacy rate estimated at little or no more than 5 percent, Greater Syria hardly offered a large potential readership.[23] But although the journal's subscription list remained modest (at an estimated five hundred) during its early years, it was obviously read much more widely. (In the late 1870s, for instance, we hear of a group of enthusiasts who banded together in Baghdad to buy a single subscription.)[24] Soon letters were reaching it from readers throughout the empire and beyond—from Egypt and from Iran and from further afield, including from the growing migrant diaspora in the Americas.[25] Within a few years, it had also established a network of agents who distributed it in Greater Syria (in Tripoli, Damascus, Jerusalem, Jaffa, Homs, Aleppo, Sidon, and Hama) and throughout Upper and Lower Egypt (in Cairo, Alexandria, Damanhur, Tanta, Al-Fayyum, Al-Minya, Asyut, and Tahta), as well as abroad. Syrian émigrés purchased it in Brazil, Argentina, Mexico, Cuba, Canada, and the United States, and by 1892 its circulation (at an estimated three thousand) was one of the largest of any Arabic journal.

The journal was novel and expensive and attracted contributions from renowned literary and public figures such as Ali Mubarak, Mahmud al-Falaki, Riaz Pasha, Shibli Shumayyil, Salama Musa, and Jurji Zaydan—Egyptian and Syrian intellectuals, technocrats, and politicians. It quickly became a prominent forum for what the historical novelist and popular intellectual Zaydan and others termed *al-nahda al-ʿilmiya* (the renaissance of knowledge or science). "*Al-Muqtataf*," wrote Salama Musa, later one of the journal's most prolific contributors, "opened my eyes to new and beautiful vistas of life."[26]

Many thought similarly. *Al-Muqtataf* enjoyed a wide and relatively popular readership and received considerable accolades both throughout the empire and abroad. "I have found it easily preferable to other philosophical journals which are currently published in the different centers of civilization," wrote "Prince Hechmat el-Saltaneh" from Tehran in 1883, who complimented the editors on their "interesting articles in modern philosophy and the new sciences."[27] In fact, its reception was probably more favorable outside than within its native Beirut. When the editors visited Egypt in 1880, only four years after they had launched their journal as young college tutors, they were happily astonished to realize that they had already acquired a considerable following. During their travels through Cairo and Alexandria, they were met by such dignitaries as Muhammad ʿAbduh and Khedive Tawfiq himself.[28] Their journal was also applauded by fellow journalists: *Al-Waqaʾi al-Misriya, Al-Watan, Mirʾat al-Sharq, Al-Mahrusa,* and even the revolutionary Islamist paper *Al-Urwa al-Wuthqa* sang its praises.[29] It was one of the first Arabic journals to which the newly founded Egyptian University subscribed in 1908.[30]

The journal's appeal owed something to its deliberately easy and comprehensible prose. "We shall endeavor to write it in simple language, accessible to everyone," the editors wrote in 1876.[31] Their aim was to introduce to an Arabic-reading audience the wider (both practical and theoretical) charms of science, avoiding whenever possible a narrow scholasticism. By and large they seem to have succeeded: Jurji Zaydan, another alumnus of the Syrian Protestant College, recalled in his autobiography that when *Al-Muqtataf* first came out, "some of the school teachers, who often came to us, pointed out to me an article about the eclipse in one of its issues. I read it, and when I understood it, I felt a great joy."[32]

Yet this effort to introduce modern science to a new public required conceptual as well as stylistic inventiveness. Translation involved bringing a great deal of modern scientific terminology into Arabic, either by inventing or transliterating new words. "New discoveries in science and technology require new terms," the editors noted when they first introduced their section on technical vocabulary translated from English, French, and Latin—the glossaries, or *muʿaribat,* which became a staple item in the journal.[33] Many of these terms shaped the subsequent lexicon of science writings in Arabic. The journal was the first to popularize the translation of "evolution" as *tatawwur,* for instance. The "struggle for life" became *tanazuʿ al-baqaʾ*, and Darwinism, *darwiniya*. The journal was thus one of the first to deploy a new prose style in Arabic that eventually overthrew its more traditional classical rival. As its editors later wrote, a newspaper should be written in a language that could be understood by laymen as much as the learned, adding: "We shall therefore seek to select correct and familiar words, simple phrases and uncomplicated expressions."[34]

The need to use familiar concepts where possible meant that when *Al-Muqtataf* introduced new scientific terms and ideas into Arabic, it stressed their connection to older forms, native to the region and its history. The concept of descent and transmutation, for instance, was said to be an old idea; ancient Greek philosophers had also linked the varieties of life together in such a way. Medieval Arab philosophers, they emphasized, such as Abu Bakr Ibn Tufayl and al-Khazini, also came to similar conclusions—a form of reference that many later Arab thinkers would popularize and develop.[35]

This kind of appropriation struck a chord with the journal's readers. In 1885 Amin Shumayyil, the brother of the well-known Shibli Shumayyil, wrote a letter to the editors that echoed their own earlier sentiments, declaring Darwin's theory of evolution to be nothing but a reformulation of medieval Arabic ideas: "Darwin's school of evolution is not new. Arabs, Hindus, and many others held to similar ideas," he wrote. He then mentioned Ibn Khaldun and even Arabic compendiums on animal husbandry.[36] The editors received scores of letters such as this. Eventually, however, they grew weary. In response to a question in 1887, "Was Darwin's theory known among the Arabs and Persians?," the editors replied curtly: "Darwin's theory, meaning the transformation of species by natural selection, was not known among the Arabs or the Persians or anyone else—otherwise, scientists today would not have ascribed this theory to him."[37]

Darwin's Descent

Darwin's name was first mentioned in *Al-Muqtataf* in 1876, its first year of publication, and prompted a long-running debate. Five years after the appearance of *The Descent of Man*, his ideas proved to be a provocative, and popular, topic of discussion. They were featured frequently in the journal—indeed, in nearly every issue there was some reference to them—and they invited lively responses from the journal's readers.

From the start, it was the issue of human origins that proved most controversial. In a series of short articles entitled "The Origin of Humanity" (*asl al-insan*), Rizq'allah al-Birbari, a "native tutor" at the Syrian Protestant College who was one of Ya'qub Sarruf's teachers, noted the interest that Darwin's views had provoked. "It is well known," al-Birbari began, "that people these days are constantly talking about the origin of life, the descent of man, the unity of the species, the antiquity of humanity, the essence of the soul, and other such problems."[38]

This first treatment of Darwin's ideas was—unlike many to follow—decidedly negative in tone, its inspiration drawn from Protestant refutations of evolutionary theory that were already accompanying the debate within the college. Initially, noted al-Birbari, many writers simply laughed away the suggestion that man was descended from "beasts" (*baha'im*). Surveying a range of theories on the origins of humanity from the time of the ancients until Darwin, and relying on the orthodox view of man's special creation by God (*khalq khusus*), al-Birbari based his discussion on an 1874 essay, *What Is Darwinism?*, by Charles Hodge, a noted American Protestant theologian who wrote frequently on the subject. Like Hodge, al-Birbari ridiculed the notion of man's descent from a lower form. He rejected evolution "from natural causes without the interference of a conscious power"; such a theory, he wrote definitively, "went against sound intelligence."[39]

In the following year, Dr. Bishara Zilzal, another early graduate of the Syrian Protestant College's medical department, also dismissed the theory of descent in no uncertain terms.[40] Humanity was God's highest creation. Endowed with reason, sensation, speech, thought, and memory, mankind was wholly distinct from brutes. And what of the natural scientists who would have us believe otherwise? Linnaeus, for instance, "related man to monkey and placed them in one category which he called 'primate' [*al-brimat*]."[41] Zilzal objected to this primarily because it denied humans their immortal souls. Praising Blumenbach and Buffon's classification of animals because they kept humanity in a "special class," Zilzal concluded that a true understanding of the nature of human origins, or "how [the] soul entered the body," could only be attained through revelation. The scientific study of humanity should not attempt to deal with this question.[42]

Like many natural theologians of the time, Zilzal was deeply interested in the new geology. In fact, he devoted two of his series of four articles entitled "Man" to the latest findings. Zilzal felt that science should reinforce religious truth, and he claimed recent geological evidence to be entirely in agreement with revelation. And here too he was repeating antievolutionary views expressed elsewhere, this time in the English press, drawing primarily from an article by Thomas Bendyshe in *Memoirs of the Anthropological Society* that had argued that man's appearance in the world could be dated to the fourth geological period and to an original couple in Central Asia, thus confirming the Mosaic account.[43]

Early articles on evolution in *Al-Muqtataf* clearly followed the debates and discoveries in Europe and the Americas closely. The tension between the implications of Darwin's theory and literalist Christian defenses of scripture was a frequent theme. Yet, in their own early writings, the editors of *Al-Muqtataf* deliberately sought not to fuel what they felt were essentially theological debates. They were careful in their selections of Darwin's writings and careful, too, with the arguments in defense of evolutionary theories that they chose to publish. Two years after Zilzal's defense of Genesis, the editors returned to the evidence of geology in an essay entitled "The Origin and Vestiges of Man." They addressed the concerns of critics at the outset. Did extant skeletal remains of man, they asked, predate Genesis?

Can the claims of the natural historians—"that a kind of monkey one day stood erect, developed a face, widened his mind, and nourished his powers of reasoning and his manners until he became a man"—be verified?[44]

The nineteenth-century rage for fossil discoveries—from the discovery of the "Engis skull" in 1830 to the hunt for "Neanderthals"—had spurred great interest in the possible existence of "prehistoric men." This question of ancient human remains was also attracting considerable publicity in the Arabic press, as elsewhere outside Europe. Although Huxley, in his *Man's Place in Nature* (1863), had used the skulls as evidence of man's evolutionary connection with the great apes, he also admitted that these skulls did not provide the missing link between man and ape. Despite the growing enthusiasm for "prehistoric" human fossils, skepticism still largely defined the field. Even geologists like Charles Lyell had resolved that there were no truly ancient *human* remains. Until the last decade of the nineteenth century, therefore, the fossil record offered little concrete evidence for human evolution.[45]

Surveying the latest paleontological findings in Europe, America, Asia, and Africa and detailing Lyell's, Lubbock's, and Huxley's research, the editors of *Al-Muqtataf* concluded carefully in 1879 that there was as yet not enough evidence to support the claims of evolutionists. After discussing the various skulls, bones, tools, and other indications of what appeared to be humanity's early ancestors, the authors surmised that the fossil record did not offer conclusive proof in the defense of the hypothesis of human evolution. But they were keen to keep the door open: geology remained a topic that the journal often returned to, and the editors made an effort to dispel the notion that its study was synonymous with impiety (*kufr*). As late as 1898 they were still tackling the relation between geology and Genesis and attempting to reconcile Lyell's *Antiquity of Man* (1863) with divine creation.[46]

In the absence of a convincing fossil record, most evolutionists turned instead to arguments of "social" progression for evidence of the biological development of man: more specifically, this was understood to be the progress of human societies' ability to manipulate their natural environment through material or technical advancements. Discoveries in the 1840s and 1850s of stone tools and weapons alongside extinct animal fossils generated interest in the archeological evidence of human progress. From the earliest crude stone tools to the discovery of bronze and iron, the notion of a progressive technical advancement was used to confirm the evolution of the human race itself. Lyell's *Antiquity of Man* (1863), John Lubbock's *Prehistoric Times, as Illustrated by Ancient Remains and the Manners and Customs of Modern Savages* (1869), his *Origin of Civilisation and the Primitive Conditions of Man, Mental and Social Conditions of Savages* (1870), and Sir John Evans's *Ancient Stone Implements* (1872) provided the framework for the modern view of prehistory. Throughout the late nineteenth century, many natural scientists and anthropologists elaborated an overarching sequence of civilizational epochs, linking biological notions of evolution to evidence of social progression.[47]

Sarruf and Nimr were themselves heavily influenced by such Victorian evolutionary thought, and their early interest in geology set the stage for their later ideas on progress more generally. Their 1879 essay on the archeological remains of "prehistoric man," for instance, was wary of the archeological evidence used by Huxley, Lubbock, and Lyell in support of a theory of human descent but came down strongly in favor of the evidence of humanity's technical and social development (fig. 1.3). Indeed, this would remain a leitmotif for much of *Al-Muqtataf*'s later writings on evolution, and the editors often linked their commitment to evolution to the new ethos of progress they hoped would reshape the Arab world. The study of humanity's past gave them, as well as many other evolutionists, the confidence to map out what was needed for its future.[48]

Intelligence and Instinct

Questions about the origin of man led naturally to questions about his essential nature. Here early Arabic discussions of evolution focused almost entirely upon *Descent of Man*. In the absence of a convincing fossil record, Darwin had seized upon intelligence as a central feature of adaptive change. In 1838, when Darwin began his M and N notebooks, upon which much of *Descent* is based, man's intellect seemed to him a promising testing ground for exploring the hypothesis of heritable variation. Darwin also felt that evidence of intelligence in animals was one way to analyze the apparent gulf between man and animals. "He who understands a baboon," wrote Darwin, "will do more toward metaphysics than Locke."[49]

Darwin sought to convince his readers that the difference between human and animal mental capabilities was a difference in degree but not in kind. To be sure, he was not alone in this view. Drawing not only on his own research but on previous deliberations on the subject, Darwin set out to show that each of our faculties is present, to some degree, in all the higher animal forms. He argued that animals possess mental powers beyond mere instinctual behavior; they experience the whole sweep of

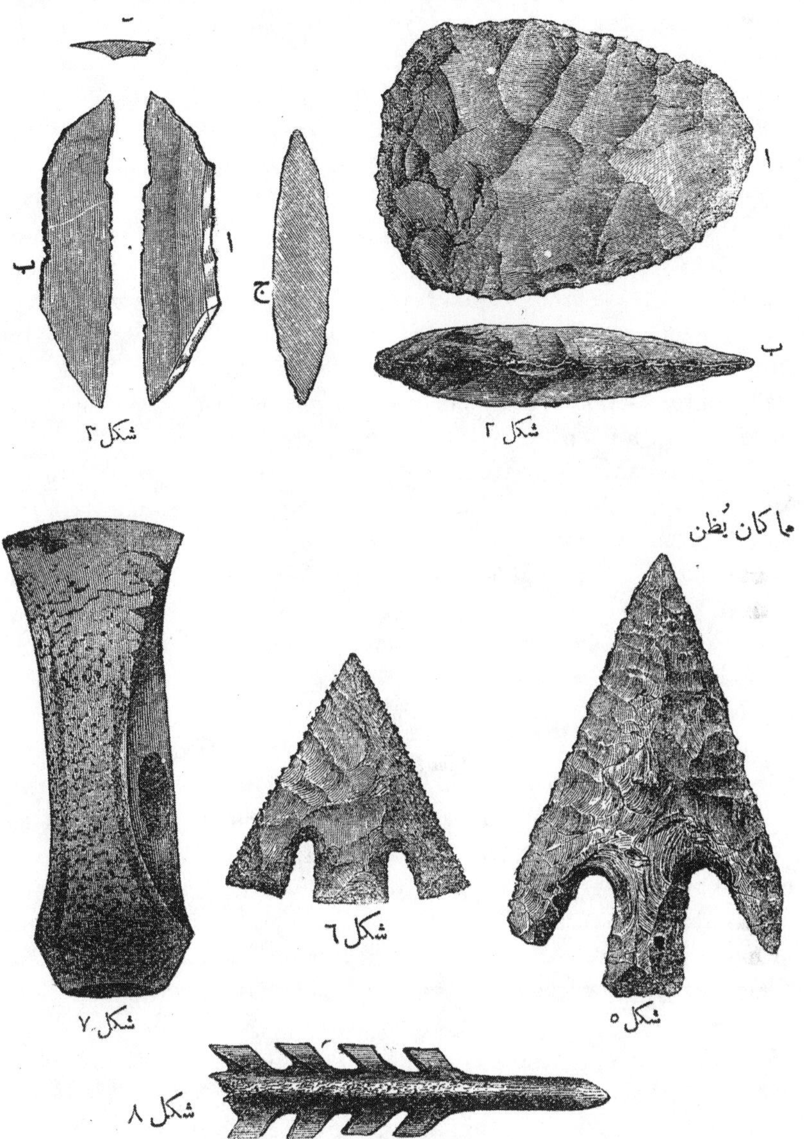

Fig. 1.3a–b. Stone tools. From "Asl al-insan wa-atharuhu" [The origins and vestiges of man], *Al-Muqtataf* 4 (1879): 90, 92.

emotions experienced by man, and they possess language skills and have a sense of beauty and moral instincts.[50]

In this debate, *Al-Muqtataf*, like many contemporary writings from America and Europe, initially focused on the question of animal intelligence. Many of its contributors greeted the very idea with considerable skepticism. If animals are endowed with intelligence, Zilzal quipped, then it was of the lowest form. For Zilzal, as for al-Birbari before him, it seemed obvious that man was unique and that that which distinguished him from beasts was his God-given reason. For Zilzal, speech was proof enough that man was endowed with a divine intelligence: man after all is "an instrument of speech" (*alat natiq*) and an animal is not, because man was specially created by God in body and soul (*al-nafs al-natiq*, the rational soul or literally "the speaking self").[51]

Two years after Zilzal's denunciation, however, Sarruf and Nimr offered a much less hostile account in which they described the classes of reason in man and animal. There are, they claimed, two schools of thought on the matter. The first holds that man's reason differs from that of animals in kind, and the second that it differs in degree only. After discussing Descartes's, Locke's, and Condillac's philosophies of mind and views on animal intelligence and instinct (*qiwa ghariziya*), Sarruf and Nimr then discussed the "latest scientific opinion." They told their readers that most scientists believed that there are three broad classes of reason in man—sensation and perception; the faculty of "understanding," which includes imagination, memory, and the ability to distinguish between appearance and reality; and, finally, intuition. Intuitive knowledge (*ʿilm badihiya*) was responsible for our moral faculties, including belief in God and religion. As though seeking to placate fears aroused by a purely materialist philosophy of mind, the editors repeatedly stressed that the consensus view among scientists was that, of the three classes of reason, it is intuition that is completely lacking in animals.[52]

However, once they moved on to those scientists who believed that animal intelligence was only a difference of degree, the question of whether man was divinely created became unavoidable. The adherents of this school, Sarruf and Nimr claimed, believe that even intuitive knowledge and moral sentiments may be present in animals to some degree. They mention Darwin as one of the adherents of this view but seem to temper its implications by praising his claim in *Descent of Man* that religious sentiments are the highest of man's mental abilities.[53]

This led to one of their other favorite subjects: animal instincts. Here, once again, the editors followed Darwin's lead and evoked instinct as evidence of animal intelligence. In an article published in 1885, they aimed to "list some of the results proven by scientists." Denouncing those who claimed that investigations into animal intelligence implied impiety, the editors stressed that such knowledge could "benefit anyone who wants to know about the wonders of creation and conduct research into animal nature." It is clear that in the animal world, they wrote, each species behaves differently, and while this at first may appear to be evidence of rational thought, it is not. A bird may build a nest and pad it with feathers for warmth, but it does not do this through reasoned thought or through experience or even through imitation. "She is forced to do it," they tell us, "by a natural force, and this force is instinct." After providing scores of examples of the instincts of animals, Sarruf and Nimr concluded that instincts in animals can be established either by heritable experience (*al-ikhtibar al-mawruth*), following G. J. Romanes's notion of "lapsing intelligence" or, following Darwin, by natural selection (*al-intikhab al-tabiʿi*).[54]

While discussion of animal intelligence and theories of a natural selection of instincts in animals may sound tame to modern ears, they had threatening religious and ethical implications. To many, the very idea of animal intelligence smacked of materialism and suggested that morality was a matter of instinct rather than God-given. Indeed, for Darwin, it seemed probable that "any animal whatever, endowed with well-marked social instincts, would inevitably acquire a moral sense or conscience, as soon as its intellectual powers had become as well developed or nearly as well developed, as in man." For Darwin, then, a moral sense in man had developed much along the same lines in which social instincts had developed among the lower species, "for the general good of the community." Only as "man advances in civilisation" do these social instincts then reemerge as religiously sanctioned moral laws. For Darwin, morality—like religion itself—was, to some degree, practical or utilitarian in nature.[55]

Such a utilitarian view of ethics, however, reached an Arabic-reading audience that was still by and large unwilling to define man's moral qualities in purely functionalist terms. Following a literary tradition of moral thought, or *adab*, many Arabic writers of the time thought the proper discussion of morality centered on refinement of character or the nature of the soul. The study of man, Zilzal wrote, for example, although a part of natural history, should also include the study of philosophy, history, and politics—which collectively he termed the study of ethics (*ʿilm al-ikhlaq*). Man's body was but a temporary home to his true essence, his intangible soul (*al-nafs ghayr mahsusa*), which shaped both his mental and his moral attributes.[56]

Seeking to overcome the objections this subject aroused, the editors returned to it often. Their articles featured numerous anecdotes about animals: "The Intelligence of Ants"; "Revenge in Animals"; "Animal Speech"; "Grief in Chimpanzees'; "The Conversations of Birds"; "Criminality in Animals"; and "What Animals May Be Taught." These colorful stories entertained their readers, who often clamored for more. In 1892, for instance, a reader wrote in: "Last year your journal published a study on 'Monkey Language' by a certain Professor Garner . . . and we understood from you that he went to Africa to continue his research there, and we would like to know more: What has he learned since then?"[57]

While entertaining their readers in this fashion, the editors also typically tried to ward off accusations of materialism by appealing to the argument from design. In their 1885 article "Animal Instincts," for instance, they defended an evolutionary view of animal instincts by reference to both natural selection and divine providence. "Instincts," the editors concluded, "grow and diversify and are established by heritable experience and natural selection," and this is a rule of nature and the universe, "like gravity and chemical attraction," established by a divine Creator, "who created the universe and its miraculous laws."[58]

Al-Madhhab al-Darwini

When they published his obituary in June 1882, the editors lavishly extolled Darwin's contributions as a scientist and struck a much more confidently supportive tone than they had done six years earlier.[59] Most scientists, Sarruf and Nimr now pointed out, broadly agree with Darwin's theory of evolution, although they sometimes disagree about the specifics. The article praised him as the "Newton of our times" but acknowledged that he had his opponents: "No doubt our article will displease those readers who disagree with Darwin's principle of the descent of man from monkey."[60] They downplayed such objections, however, pointing out that in fact many scientists and religious scholars had come round to accepting that his views were not opposed to fundamental religious beliefs. Citing Frederick Temple, the bishop of Exeter, and others, Sarruf and Nimr were keen to emphasize the compatibility of his ideas with Christian thought. Quoting James McCosh, a Free Church member and president of Princeton University, they reassured their readers that "evolution is a law of God quite as much as gravitation or chemical affinity or vital assimilation."[61]

It is therefore not surprising that their new confidence rested upon the support of Protestants on both sides of the Atlantic. The following month, in their account of "Darwinism" or "the Darwinian school of thought"—*al-madhhab al-Darwini*—the editors offered a concise and comprehensive account of the theory of natural selection based on the writings of the popular American science writer and philosophical theist John Fiske. They hoped this would give their readers some idea of "the most famous theory of our day and the most extraordinary discovery of our time." Their account followed contemporary debates in Europe and America and offered a thorough account in Arabic of the main arguments in support of a theory of descent through natural selection—ranging from the domestication of species to pangenesis and spontaneous generation.[62] But as their reliance on Fiske suggests, it was also highly colored by the latest Christian theological interpretations. In fact, the editors' own reconciliation of evolution with dogmatic theism, as well as their general conception of science, sounded a great deal like that of Fiske. As Fiske argued in his 1879 *Darwinism and Other Essays*, the proper "business of science" was merely to detail the relationship between or the sequence of phenomena, and in pursuing this, "its legitimate business," "science does not trench on the province of theology in any way." Hence, there is no conceivable occasion for conflict, for just as the "Darwinian theory" is "entirely consistent with theism," it is also "the only kind [of explanation] with which science can properly concern itself at all."[63]

One thing they were quite clear about was the difference between Darwin and Lamarck. "Darwin's theory does not require that all things progress as Lamarck's does," they wrote, "and it bases progression, not on the body's will, but on natural selection and the struggle for life."[64] They pointed out that many saw Lamarck's theory of transformism as "mere fancy" and felt it was imperative to distinguish between that and a Darwinian evolutionary perspective, though they are "very similar," pointing out that "many people who [think they] follow Darwin belong to the Lamarckian school."[65]

Darwin's theory, as they summed it up, states simply "that everything on earth, animal or plant, extant or extinct, evolved, one from another." (Here the editors used *tasalsala*, with its connotations of a "chain," for "evolved," whereas in later writings they more typically used *tatawwur*—the former implying *descent*, and the latter, *development*.) They outlined the idea of species differentiation and attacked the erroneous view that Darwin had proposed that "man was first an ox, then a donkey, then a horse, then a monkey," before attaining his final shape.[66] They also explained the distinction between those who believed in the separate creation of species and those who espoused an evolutionary position; although many people had long held the former view, the overwhelming scientific evidence favored the latter.[67] Evolution, they encouraged their readers to imagine, was like "a huge tree the size of the earth, with branches pointing to the heavens." The connections among living beings, they wrote, should thus be seen as connected branches of a single organism even if "between these branches is a vast expanse of sky—for they still come from the same root." By presenting the theory of evolution as a framework for unity in nature, the editors were once more hoping to sidestep the more thorny and radical aspects of Darwin's theory—particularly any imputation of materialism. Instead, they stressed its theoretical sophistication, universality, and applicability. Detailing problems of form and function, and following arguments of ontogeny and phylogeny, the editors described how Darwin's theory helped reveal natural links that were previously uncertain among organisms.[68]

Sarruf and Nimr's real admiration for Darwin's theory was based on their view that it answered more questions than any other theory, Lamarckian or creationist. It was its "puzzle-solving" ability that was, in their view, a sign of its true sophistication: when describing how more scientists were in favor of Darwin's theory than that of a special creation (*al-khalq al-mustaqil*), they justified it simply by claiming that "they were able to solve more puzzles with Darwin's theory than the other." It accounted for speciation, embryonic development, and the geographical distribution of life forms and had the potential, as they saw it, to uncover mysteries of nature hitherto unknown.[69]

They were not uncritical in their admiration. Like some other supporters of Darwin, they were aware of areas where the evidence of his theories was in short supply, and they were also reluctant to abandon their belief in the "uniqueness of man." On this issue they sided with Alfred Russel Wallace. Rejecting the assumption that humanity had descended from "a certain type or types of extinct ape," they told their readers that most scientists agreed with Wallace on this point. They pointed out that the theory of natural selection was discovered independently by both Darwin and Wallace but stressed that the latter had "said that all animals evolve one from the other except humans," adding that he thought that humans possess characteristics that could not be found in any other species. There was a "great barrier" between humanity and animality, which even natural selection could not pass. This they agreed with: man was a form of "special creation."[70]

They concluded that the question of the nature of the soul, looked at "from the point of view of science," was one of the most difficult issues to tackle. Ultimately they felt that it could not be solved by depending on scientific theories or even reason alone. Anyone who examined Darwin's theory on this point, they thought, would find it lacking in evidence as well as full of "contradictions, conflicts, risks, and poor judgments."[71] They recognized other difficulties with Darwin's account as well but did not allow this to change their fundamentally positive view, arguing that, "for the sake of fairness," "whenever one sees the truth, one must recognize it as such and accept it as ordained from the Lord."[72]

Their faith, in short, remained crucial to them. "The wise," they wrote, "will hold on to what God revealed and will accept from science only the highest truths." Yet, despite their many efforts to underline the complementarity of Darwin's theory of evolution with personal belief, the tension between scripture and science remained a critical and open-ended issue for the two science enthusiasts. Throughout the coming decades, they would consistently find any attempt to refute or even to reconcile scientific findings directly with revelation offensive. Ultimately, they retreated to a more classically positivist stance, and they never tired of repeating that issues of science and religion should be separated whenever possible.

Their own particular reconciliation, if we can call it that, between science and faith was, however, in many ways consistent. Following the resolution of many contemporary Protestant theologians, they also held that the process of evolution was itself evidence of design in nature, and they clung tenaciously to a faith in the unity of nature, natural laws, and the scientific method. If one held that nature was a unified system of being, operating under fixed laws and investigated by scientific proofs, and under the providence of divine benevolence, then there was no reason, they believed, for religion and science to come into conflict. Wherever possible, the two authors stressed the harmony of the laws of nature with God's laws. In this sense, they maintained a strictly antimaterialist but not quite positivist position and added to that the argument that evolution neither touched upon nor challenged first principles or final causes. "Those who chose to believe in the immortality of the soul," the editors often repeated, "and believe that God has revealed to man the way to worship Him, do not, by so doing, contravene the dictates of reason."[73] Theirs was a broadly Baconian and pragmatic vision of science, if a somewhat uneasily positivistic and ultimately a doxological one.[74] What *Al-Muqtataf*'s early presentation of evolution demonstrates in fact is just how closely bound up the journal was with contemporary Christian, and particularly British and American Protestant, philosophical and natural theological discourses. To a large extent they combined James McCosh's theocentric faith in "moral struggle" and "supernatural design" as the motivating factor of natural selection with John Fiske's theistic positivism and his humanistic emphasis on evolution as the transcendentalist "destiny of man."[75] This formed the basic metatextual background to their understanding of evolution, and it helps explain, among other things, their particular views on theism, progress, and the problem of morality in relation to evolution.[76]

The Gospel of Science

By the late 1870s Sarruf and Nimr formed part of a new reading community in the Arab provinces of the Ottoman Empire who seized upon science as a sphere of inquiry full of radical possibilities. This helped in part to construct the vision they would later promote of technocratic governance under the rule of science; science was the solution to the problem of progress and a means to escape what they increasingly saw as a repressive regime. The paradox of this story, however, is that key members of this elite in Greater Syria were actually graduates of a Christian missionary culture that was itself deeply ambivalent about the methods and modes of modern scientific inquiry. Their American mission patrons had indeed been motivated by the desire to create "new men" abroad or, in the words of an early missionary in Beirut, the Reverend James Dennis, "the embryonic norm of a new society and a new life."[77] However, the "new men" that they produced—figures such as the well-known materialist Shibli Shumayyil and science popularizer and novelist Jurji Zaydan and even "native tutors" like Sarruf and Nimr—were scarcely what they had had in mind.

Yet *Al-Muqtataf* was deeply indebted to men like Dennis and to others at the Syrian Protestant College. The very didactic aims of the journal were inspired and shaped by the missionary culture out of which it grew. Evangelism also critically shaped discussions of evolution in Arabic—not only the earlier articles covered here but indeed a whole gamut of writings, spanning several generations, from *Al-Muqtataf*'s own later coverage to those by Sufi shaykhs, Ottoman educational reformers, and socialist thinkers. To understand these connections we need to take a closer look at these missionary pedagogues and proselytes in the decades before *Al-Muqtataf* itself began.

American Protestants were not the first Christian missionaries to make their way to Greater Syria, of course. Catholic—Jesuit, Capuchin, and Carmelite—missions had been active in Syria from at least the seventeenth century.[78] Yet it was the Protestant missions that coincided with a new American interest in the "Holy Land"—those lands that stretched from the Mesopotamian valleys to the Nile basin and that were seen, as one missionary put it, as "rendered sacred in the eyes of every Christian by a thousand religious associations."[79] Guided by a new spirit of "disinterested benevolence" during the "Great Awakenings," their renewed evangelical zeal was also spurred by expectations of the Second Coming of Christ and the mass conversion of the world's peoples, particularly the Jews, who were then predicted to return to Jerusalem and thereby inaugurate Christ's millennial rule over the earth.[80]

In 1810 a fraternity known as the "Brethren" at the Calvinist Divinity College (later the Andover Theological Seminary) founded what would quickly become one of the largest interdenominational missionary societies in the northeastern United States: the American Board of Commissioners for Foreign Missions (hereafter ABCFM).[81] Within fifty years, the board had established numerous missionary establishments throughout the Mediterranean, Asia, Africa, and the Americas. From the start, theirs was a project with global ambitions: "Prophesy, history and the present state of the world," announced the ABCFM during their second annual appeal in 1811, "seem to unite in declaring that the great pillars of the Papal and Mahometan impostures are now tottering to their fall. . . . Now is the time for the followers of Christ to come forward boldly and engage earnestly in the great work of enlightening and reforming mankind."[82]

Unsurprisingly, the first generation of American missionaries began in Jerusalem, but they soon moved to Beirut in order to escape what they felt were the insurmountable suspicions of local ecclesiastical authorities there. Yet they continued to complain that their work was not much easier in Beirut, and their memoirs and letters highlight their difficulties with local bureaucrats and the obstinacy of those they hoped to convert. As the difficulties of effecting direct conversions became obvious, they increasingly turned to strategies of "indirect conversion" through schooling and the press. From the very start, they had stressed the need for general education to accompany their proselytizing efforts, believing that it was "education which forms the Mohommetan and Pagan, the Jew and Christian."[83] Teaching their charges to read and write, moreover, was particularly important given the value missionaries placed on the power of individual communion with God through the reading of scripture.

Yet it was not long before mission schools did far more than promote literacy and the rudiments of an elementary education, and their curricula grew to include a range of subjects from moral and natural philosophy to geography and astronomy. This approach proved remarkably successful. The number of students enrolled in Beirut and the surrounding mountain villages rose steadily, from about 600 in 1827 to over 5,000 in 1884.[84] Their strategy was, as one missionary put it, to "bait the hook with arithmetic." The Beirut High School for Boys (a boarding school), for example, opened in 1835 with the express aim of introducing the sciences in association with Christian doctrine. Alongside English, Arabic, and arithmetic, students were instructed in geography, astronomy, natural and moral philosophy, and logic; they also had regular readings in the Bible and attended mandatory church services (mostly prayer). Missionaries believed that such instruction would work to "enlighten" the minds of their students "respecting the prevailing and soul-destroying errors of this country."[85]

Education in the natural sciences in particular was promoted as a way to aid pupils on the path to God: geography, for instance, was said to demonstrate to students the providential control of God in nature—in adapting the climate to inhabitants or vice versa—and was seen to inculcate an appreciation for the goodness of his divine will and order, as well as to excite the sense of religious awe and wonder so central to the spirit of evangelical thought.[86] Eli Smith, a long-serving missionary in Syria who taught geography and astronomy at mission high schools, expressed his enthusiasm for the spiritual potential of these subjects.[87] Having just finished a short course of lectures on astronomy for the first time in 1835, Smith noted how, despite some of the students' initial reservations toward the "new" (i.e., Copernican) astronomy, "some of them, in the end, could not restrain their admiration at His power and wisdom."[88] Winning students over to an irrefutably rational, mathematically proven Copernican cosmology hinted at the eventual promise of conversion itself: over time, with perseverance, and through an appeal to the rationality and truth of their message, students might be similarly won over to the true faith.

Missionaries in Syria made much of the "pioneering" nature of this task: "It was perhaps the first time," wrote Smith, no doubt mistakenly, that "the Copernican system was taught in the country." He noted that he had "many objections to encounter" at first, but "all who heard me soon become advocates, and by their reports and arguments made considerable talk in the city." Whether or not the missionaries were the first to introduce Copernican astronomy to their students, "talk" was precisely the sort of thing they wanted, for it was clear that they were hoping to attract widespread interest and elicit a broad, public discussion. As early as 1837 missionaries at the school began to stage public experiments in natural science for exactly that reason. After experiments on pneumatics and electricity were carried out (with apparatus sent to the school from the United States) missionaries reported back happily that not only were these numerously attended "by the first people in Beyroot" to witness such wonders, but they had, as a result, also "excited much interest."[89]

The printing press was another key missionary instrument, a crucial "agency in the diffusion of religious truth." By 1849 the ABCFM had seventeen printing establishments worldwide for the use of its missions, with four type foundries and thirty-one presses, and published in over thirty-five languages. The presses published translations of the Old and New Testaments, as well as a variety of schoolbooks, primers, and grammar manuals. The need to establish an Arabic printing press in Greater Syria was felt to be particularly urgent—especially since the only available Arabic Bible was the translation that was made by the British and Foreign Bible Society in London and printed in Rome in 1671, which Protestant missionaries circulated without the Apocrypha.[90]

American Protestant missionaries were also the first to establish a press in Beirut, which they had transferred from Malta in 1834. Shortly afterward, the ABCFM acquired a new font of Arabic type, "so conformed to Arabic calligraphy as to suit the extremely fastidious taste of the Arabs" and developed under the superintendence of Eli Smith at Smyrna and Leipzig. In 1847 the Jesuits, in keeping with their Protestant rivals, also established a new Arabic printing press in Beirut. Between 1850 and 1880 Beirut print culture flourished. During this time about twenty-four journals were published in the city, compared with a mere fifteen in Cairo for the same period. By 1875 there were eleven printing presses in Beirut alone, including Al-Matbaʿa al-Suriya, founded in 1857, and Al-Matbaʿa a al-Umumiya, founded in 1861. By the late nineteenth century, they were among the largest and most up to date in the empire. The American Protestants alone had four steam and six hand presses, a lithography and electrotype outfit, a type foundry, and a book bindery and had published over seven hundred different works by 1903.[91]

The most consuming project of the Beirut mission's early years was undoubtedly the translation of the New Testament into Arabic, the first Arabic translation of its kind since the seventeenth century. This took Eli Smith, the Syrian lexicographer Butrus al-Bustani (and later Cornelius Van Dyck) twelve years to complete.[92] But American missionaries also devoted their energies to other more "worldly" publications. As early as the 1830s and 1840s, missionaries in Beirut had received scores of letters requesting Arabic books on geography, astronomy, history, and theology.[93] Taking advantage of this demand, members of the mission station began to publish a wide variety of materials, from student textbooks and primers to journals, calendars, and almanacs.[94] They also became increasingly concerned to produce civilizing and morally edifying literature in Arabic, especially after the French military occupation of Beirut in August 1860—following the conflicts in Mount Lebanon—led the missionaries to worry about the corrupting influence of dangerous and skeptical French literature. "What a host of priests, Jesuits and nuns to dispute every inch of ground!" exclaimed an exacerbated William Bird in 1869. "What armies of skeptics are springing up, a result of French novels and volumes of Voltaire & Co.!"[95]

Despite these efforts, however, the Syria mission station had still made very few converts by midcentury.[96] In fact, it was earning a reputation back home for directing too much of its energy to the publishing of extra-religious matter and to the opening and running of schools. Although enrollments were high, not one student from the Beirut High School for Boys had been won over to Christianity: "It is a painful truth that we do not see those manifestations of spiritual life attending our labors, which we desire," lamented one missionary in Beirut in 1841. "These youth are in a state of moral death. We have not satisfactory evidence that a single one of them is a child of God. . . . Why, we often enquire, why is no one from this school converted to Christ?"[97]

Prompted by budgetary concerns, questions of this imbalance in missionary priorities became a subject of considerable concern for ABCFM members. Rufus Anderson, the secretary of the ABCFM, pledged to cut costs and to reconsider issues of mission strategy.[98] ABCFM funding, maintained Anderson, could be justified only on the grounds of effective efforts at conversion. Despite its popularity, Anderson closed down the Beirut High School in 1842 and replaced it with a theological seminary deliberately located at ʿAbeih, a village just outside Beirut, as Anderson feared the corrupting influences of the increasingly "Frankish" city. "Civilization is not conversion," announced *The Divine Instrumentality for the World's Conversion*, a circular published in Boston in 1856 by the ABCFM. "Iron rails, steam engines, electric wires, power looms and power presses, however powerful, are no part of Christianity and can never turn men from the power of Satan unto God," for "the world will never be converted by the arts of civilized life . . . nor by the introduction of true science or an improved literature."[99]

Yet missionaries "on the ground" clearly disagreed and continued to successfully defend their commitment to a broader vision of pedagogy. Just one year after the seminary at ʿAbeih opened, they combined forces with local scholars in Beirut to inaugurate the short-lived Syrian Society of Arts and Sciences, or Al-Jamʿiya al-Suriya li-Iktisab al-ʿUlum wa al-Funun, a society that was open to members of all religious backgrounds and that aimed at "the awakening of a general desire for the acquisition of the sciences and arts, irrespective of disputed questions relative to religious rites and doctrines."[100] Along with a sizable library, the society owned a lecture hall where papers were delivered monthly and read in Arabic. Aiming to foster "a general desire for attainments in the sciences and acquisitions of knowledge," the society hosted such lectures as "On the Delights and Utilities of Science," "A Discourse on the Instruction of Women," and "On the Principles of the Laws of Nature."[101]

Even at the ʿAbeih seminary, ironically, the course of instruction grew well beyond its initial focus on religious studies and Bible training. Within a few years, in fact, its curriculum began to look suspiciously similar to that of the earlier Beirut High School for Boys, and its students received instruction in mathematics, geography, astronomy, history, and natural philosophy. The emphasis on science was something missionaries were clearly (even if cautiously) proud of: "It is, we suppose," announced the ʿAbeih station report of 1849, "the only institution in Syria where the true principles of science are taught."Nevertheless, they were quick to add the following qualification: "At the same time we desire never to lose sight of the fact that it was established with express reference to the training up of preachers of the everlasting gospel. . . . We feel the danger of merging too much the religious in the literary and scientific . . . and seek grace to make this seminary more and more a religious institution and to teach the sciences as entirely subservient to Christianity."[102]

Early Arab converts, like *Al-Muqtataf*'s Yaʿqub Sarruf and Butrus al-Bustani, betrayed a missionary zeal toward science that was the equal of that of their mentors. Al-Bustani, a former Maronite and one of the mission's early converts, was a key figure in the introduction of the "new sciences" (*al-ʿulum al-haditha*) to an Arabic-reading audience. He founded the science journal *Al-Jinan*, was a founding member of the 1847 Syrian Society of Arts and Sciences (which was later reestablished as the "Syrian Scientific Society" in 1868), and authored a dictionary and the first seven volumes of *Dairat al-maʿarif*, one of the first encyclopedias of its kind in Arabic.[103]

If Anderson's pedagogical conservatism failed to prevail, it was because it completely ignored the social and political obstacles to conversion, including excommunication and ostracism.[104] But, more importantly, it ignored the growing local demand for education and his missionaries' own anxieties about the competition to meet this. Beirut was a prosperous and growing city as capitalism transformed the eastern Mediterranean.[105] Its bureaucracy, infrastructure, and commercial sector rapidly expanded, particularly after the resolution of the civil conflicts in Mount Lebanon. After 1860 the number of schools increased dramatically, and American missionaries found themselves competing for custom not only with other Protestants but also with Jesuit, Syrian Catholic, Greek Orthodox, Maronite, Druze, and local Muslim and Ottoman schools.[106] "All the various groups are vying with each other in finding Boarding Schools for their children and youth," warned missionaries there, "and if Protestant Christianity is to maintain a foothold here, it must do its part in the great work of education."[107]

As local missionaries knew from their earlier experience with the Beirut High School, science was the great draw—and Anderson's reforms never took root precisely because he failed to acknowledge this. The transformation of Jesuit schools—with their fully equipped laboratories and latest scientific equipment—confirmed that students would flock to the schools that offered the latest in scientific and technical training. Missionaries boasted about their pioneering efforts in the diffusion of modern science, but they were in fact responding to local demand and widespread interest in new forms of technical and professional knowledge.

Postwar historians of the Middle East have often depicted these missionaries in the region as external forces for change and key figures in a long line of agents of Western modernization, beginning with Napoleon's *savants* in Egypt.[108] In fact, not unlike Napoleon's expeditionary force, missionaries in Syria were not so much reeling in a recalcitrant periphery for the forces of modernity as finding themselves enmeshed in a broader and ongoing process of Ottoman social and political-economic transformation.[109]

Yet missionaries also had theological justifications of their own for these rapid changes in their spiritual works: imparting scientific truth might itself testify to the superior rationality of the Protestant faith, demonstrating to the more "superstitious" Eastern creeds that only a truly rational and reformed faith could foster a genuine understanding of the natural and physical world. Far from seeing science as engaged in constant "warfare" with theology, some missionaries in Beirut saw the natural and allied sciences as integral to their broader spiritual commitments and theological vision. Healing and training in the medical arts and sciences might offer one way to fulfill the promise of conversion, whether by acting in the role of Christ in healing the body (and soul) and extracting somatic (and spiritual) evils or by striking at "degraded superstitions" found in "the prevalence of quackery and magic in the healing arts."[110] Missionaries also found in them a means to promote the path to salvation and to advance the reformation of minds and daily habits. Meanwhile, public experiments and demonstrations of electricity and pneumatics embodied the virtues of disciplined understandings and allowed the missionary to stage a public testimonial to the power of truth; similarly, training in engineering works provided practical skills and acted as a spur to the technological progress of nations, to free trade and commerce, and, ultimately, to just governance and universal brotherhood.[111] Paralleling developments in postmillennial American Protestant thought itself, such as the rise of the "Social Gospel" movement in the United States, some missionaries in Syria ultimately acquired a different sense of "conversion" itself.[112] As was the case in India and China during this period, science (and medicine) became ever more critical weapons in the missionaries' spiritual arsenal.[113]

Before long, all these factors came into play to push the missionary pedagogic crusade in a new and even more ambitious direction. Competition with Ottoman educators, Jesuits, and others, along with a growing confidence in the value of their scientific curriculum, led leading figures of the American mission station in Beirut to argue for the need for their own permanent institution of higher learning. Such an institution would "give a thorough scientific and professional education to the youth of Syria."[114]

Behind this vision also lay, for some, an appeal to natural theology: arguments about God's existence, character, and being could be made through rational proofs of design in nature after all.[115] Many missionaries in Syria argued that this rational approach to issues of belief exemplified the superiority of Protestant rationality over the superstitious and irrational character of the eastern churches and, of course, of Islam. (A number of Muslim intellectuals of the following generation would argue precisely the reverse.)

Yet, while some missionaries promoted scientific education and training as one of the paths to God, others increasingly saw it as a dangerous and ineffectual diversion, a waste of resources, and possibly worse. Indeed, the very establishment of an interdenominational (and transatlantic) college implied breaking with the ABCFM's own definition of mission school pedagogy, and when it came, the break was a real one.

Science and Salvation

By the 1860s American missionaries in Beirut had begun to expand their horizons. In 1856 two leading Presbyterian missionaries, Henry Jessup and Daniel Bliss, had arrived in Beirut and shortly thereafter began to talk of establishing a college for boys. The idea was discussed with members of the ABCFM, but it was quickly agreed that any future college should be administered separately since, as Anderson later put it, "the American Board could not undertake so large a *literary* work in any one mission" (emphasis in original).[116] Setting aside denominational distinctions, senior missionaries proceeded to collect funds from a wide array of Protestant organizations. Daniel Bliss, who had joined the Syria mission station in 1855, was released from his connection with the mission to raise funds for the new college.[117] Hoping to secure "the liberal cooperation of all friends and patrons of a sound Christian education in the East," he began in America and Great Britain in 1861.[118] After two years of campaigning for money in Civil War America and Europe, Daniel Bliss and a Boston-based Board of Trustees established a new college.

In their fund-raising campaigns at Protestant churches across the Northeast, Bliss and his sometime companion and fellow missionary (and author of the best-selling *The Land and the Book*) William Thomson stressed the "remarkable awakening in the Orient" and the rise of a "great anxiety for education, especially in scientific knowledge," taking hold among the "Arab races." It was, they claimed, left to the American people alone—as "their freedom from all political entanglements" gave them a decisive advantage and duty over all others—to assist them in rousing from their centuries' old "mental and spiritual torpor." They spoke of the dangerous inroads being made in Beirut by "Popish educators," with their "showy but deceptive" education. As Jessup stated in 1863 in a letter he circulated in papers and evangelical dailies throughout the Northeast: "*Syria will have a College of some sort.* If our religious influence is not thrown in to found and control it, irreligious men will do it, and we shall have an infidel breeding institution growing up as the bane and curse of Syria forever."[119] Within two years they raised over $100,000, and in December 1863 a New York–based Board of Trustees was established. The following year, the Syrian Protestant College at Beirut was officially chartered.[120] Circulars were distributed—in Arabic and in English—throughout Greater Syria shortly thereafter to announce the new boarding school.[121] According to the prospectus, there were to be six departments at the college: Arabic language and literature; mathematics, astronomy, and engineering; chemistry, botany, geology, and natural science; modern languages; medicine; and law and jurisprudence.[122]

On 3 December 1866, the Syrian Protestant College, known to local Syrians as Al-Madrasa al-Kulliya al-Injiliya (Evangelical College School), opened to admit eighteen students of various faiths, presided over by Bliss, the first president of the college, and inaugurated with a service of prayer and a scripture reading from the third chapter of First Corinthians. The college (now known as the American University of Beirut) would prove to be one of the missionaries' most ambitious—and successful—enterprises, and it expanded rapidly. By 1887 it had a chapel, a library with geological and antiquarian collections, study rooms, and dormitories, as well as a fully equipped medical building with a chemical laboratory and zoological and botanical collections. In addition, it housed an observatory that sent twice-daily telegrams to Istanbul with meteorological results.[123] By 1902 it had forty acres of land, over a dozen buildings, forty teachers, and over six hundred students (figs. 1.4–1.6).

The secret behind the college's success was without question its strong interest in science education and its commitment to the diffusion of scientific, medical, and industrial knowledge. For a moderate charge, students (all male, of course) received instruction in the latest scientific disciplines, from engineering and astronomy to medicine and natural science (figs. 1.7–1.8).[124] A medical school, or department, was founded in 1867, and those students who enrolled at the college but decided not to engage in medical studies were listed as pupils of the "Literary Department." A few years later a preparatory department for younger students was also added.[125] All students spent the first few years studying algebra, geometry, and natural philosophy, together with Arabic grammar, English, French, Latin, and Bible studies. Those in the Medical Department spent four additional years studying a variety of medical and clinical subjects; while those

Fig. 1.4. Assembly Hall of the Syrian Protestant College, 1900. From *Catalogue of the Syrian Protestant College* (Beirut: Syrian Protestant College Press, 1900–1901), 1.

Fig. 1.5. Observatory of the Syrian Protestant College, 1900. From *Catalogue of the Syrian Protestant College* (Beirut: Syrian Protestant College Press, 1900–1901), 37.

Fig. 1.6. Library of the Syrian Protestant College, 1900. From *Catalogue of the Syrian Protestant College* (Beirut: Syrian Protestant College Press, 1900–1901), 44.

Fig. 1.7. The Syrian Protestant College laboratory. From a private collection.

Fig. 1.8. The Syrian Protestant College laboratory. From a private collection.

in the Literary Department had lessons in a variety of literary and scientific subjects, from natural and moral philosophy to zoology and physiology.[126]

The college's overall emphasis on science represented a significant shift from Anderson's earlier educational priorities (Anderson, incidentally, retired from the ABCFM the year the college opened). Missionaries at the college remained committed to the "general cause of science" and believed that their wide-ranging curriculum would serve as an agent of social and political reform. Daniel Bliss, who served as the college's first president (until 1902), declared in 1862 that the work of the college should be to "cast in just as much good seed as possible, and thus prepare material for constructing good government and good society."[127]

Yet it would be quite wrong to think that evangelism had in any way faded in the minds of the missionaries involved. Faculty and members of the Board of Managers (formed from among resident American and British missionaries as well as local Protestant merchants of Syria and Egypt) insisted that the school be "conducted on strict evangelical principles." Established outside the ABCFM proper, the college nevertheless shared many of its aims. Its constitution made clear that the college would be "conducted on principles strictly Protestant and evangelical."[128] Alongside their literary and scientific or medical studies, students had regular lessons in scripture and daily readings in the Bible and were introduced to topics in philosophy and moral science, so as "to enforce the great fact upon the mind of the student that a pure morality and a rational religious faith are in accordance with the constitution of the human mind, and a necessity to its highest well-being."[129]

Although the college was administrated separately from the ABCFM, missionaries at the college maintained that it was nevertheless a "child of the Syria Mission."[130] When George Post took up a position as professor of surgery, materia medica, and botany, he was described as bringing "to the college the same earnestness of Christian purpose that made him so valuable as a missionary. It is as a *Christian* institution that the college is of special interest to other gentlemen connected with it."[131] Indeed, for Post, the new Protestant college, free from the constraints of ABCFM financing, promised to be an even more effective vehicle for enlightening men and saving souls than the mission station itself had been.[132]

Those in charge of the Beirut mission saw the college as a partner in their work. In 1870, when the Syria mission was transferred to the Presbyterian Church in the U.S.A. Board of Foreign Missions (as a result of the unification of the Old School and New School Presbyterian Churches), the PCBFM praised the college for serving as a "prominent agency in promoting the cause of missions in Syria."[133] Within three years of taking over the mission, the new organization closed down the ʿAbeih seminary, declaring that such theological training was already covered by the work of the college.[134]

Missionaries at the college and beyond thus clearly saw the boarding school as a vehicle for moral and spiritual improvement along what they considered to be evangelical lines. Yet the curriculum at the Syrian Protestant College was intended to attract not only those local Christian sects among whom American missionaries had thus far had some measure of success but also local Muslims and others who had so far remained aloof. Blaming stubborn Muslim feelings of "pride in the superiority of their own religion," James Dennis complained in 1872 of the difficulties American missionaries faced in converting them and suggested science as the way forward. "They rarely attend our preaching, rarely visit us for religious conversation," he wrote. "But Mohammedanism, as a system, is vulnerable through science. . . . To an educated mind there are in it puerilities, absurdities, glaring inconsistencies." For missionaries like Dennis, science—as a means of exercising the "reasoning powers" of the mind—would expose these absurdities and thus prepare the way for the acceptance of the true faith: "Contact with Moslem minds, so difficult through other means, is in a measure possible through education. Scripture truth may be inculcated in connection with science, and this when youthful minds are most susceptible to impressions."[135] When in 1877, the college opened its observatory, President Bliss wrote of his hope that it would "prove useful in the direct education of students and in attracting the attention of natives to the superiority of Western knowledge, thus helping to dispel ancient deep-rooted superstitions."[136]

Bliss, like other missionary instructors at the college, saw instruction above all as a vehicle for the "awakening of intellects" and the reformation of minds. As early as 1873, he published *Primary Lessons in Rational Philosophy*—a primer composed with the assistance of "native tutor" Ibrahim al-Hurani—for use in his course on logic and moral philosophy at the college. "Reason," began Bliss, "is what distinguishes men from brutes," adding that the science of reason was among the most important subjects of study at schools and "the best means to exercise, strengthen and improve minds."[137]

Within the first few years of the opening of the college, Arabic books—particularly science primers—were published by various members of the faculty. These included Cornelius Van Dyck's 1869 *Usul al-kimiya* (Principles of chemistry), George Post's 1869 *Kitab nizam al-halaqat fi silsilat dhawat al-faqarat* (Classification of vertebrates) and his 1871 *Mabadi ʿilm al-nabat* (Principles of botany), and Asʿad Ibrahim al-Shadudi's 1873 *Al-ʿarus al-badiʿa fi ilm al-tabiʿa* (The marvelous bride of natural science). Bishara Zilzal's *Tanwir al-adhhan* (Enlightenment of minds), a natural history textbook published in 1879, was perhaps the first book in Arabic to discuss

Darwin's theory at length. These books were needed because the college's language of instruction was Arabic until 1886, when after much debate it was changed to English. Modeled on American school textbooks, they were translated with the help of "native tutors" who assisted the regular faculty in their teaching duties and served as on-site translators in classroom discussions and lectures.[138] They also proved highly successful and were often used in other local Muslim and Christian schools at the time.[139]

By staging public experiments and scientific demonstrations, establishing scientific societies and debating clubs, and authoring and sponsoring new forms of publications—from science journals to calendars and almanacs—missionaries had deliberately opened new avenues for public discussion and debate about science and its particular knowledge-claims. The most important of these activities—and the one with the greatest consequences—was undoubtedly their use and sponsorship of the printing press. Yet by the 1880s the mission press in Beirut was only one of many in the Ottoman port city and the empire at large.[140] Indeed, mission publications on science were readily taken up by a growing public sphere of readers and authors who collectively constituted a new cross-confessional and interregional community of knowledge.

Missionaries of Science

One of the most successful avenues of publication developed by missionaries was journals. Indeed, it was through these that they helped to foster a new generation of Arabic journalists and print proselytes. In 1874 Post started a medical journal and in 1882, along with Van Dyck and Hurani, he founded *Al-Tabib* (The physician).[141] Trained as a doctor in Philadelphia in the late 1830s, Van Dyck (fig. 1.9) had received no formal theological training before he arrived in Syria, where he was later ordained; he joined the Syrian Protestant College as the chair of Internal Medicine and General Pathology in 1867. Devoted to the spread of science into the Arabic-speaking East, he contributed to or helped to found several journals, and he translated scores of science textbooks and other works for use at the college. He wrote on algebra, chemistry, geology, geography, and natural science, edited the missionary paper *al-Nashra al-Usbuʿiya* and was a regular contributor to *Al-Muqtataf* and *Al-Jinan*, the Arabic scientific and literary monthly founded by Butrus al-Bustani.[142]

Van Dyck had known both Sarruf and Nimr from the start of their careers at the Syrian Protestant College. It was Van Dyck who suggested to

Fig. 1.9. Cornelius Van Dyck. Photo from the American University of Beirut Archives and Manuscripts Collection, no date.

Sarruf that he undertake the translation of Samuel Smiles's *Self-Help* in the late 1870s and even helped secure an advance from the Turkish Missions Aid Society. (The text was later used as a reader at the college, where it enjoyed a lasting popularity.)[143] Van Dyck also founded the observatory at the college in order to teach astronomy and meteorology, and he later appointed Faris Nimr to teach elementary astronomy and to serve as his assistant.

But he shaped the future editors' path in other ways too. Van Dyck's emphasis on rationalism, proof, and the value of open discussion impressed itself upon Sarruf and Nimr: one of the American's favorite maxims was "Intolerance cannot be tolerated." Sarruf and Nimr followed suit and made it their aim to bring about popular "enlightenment," unaffected by sectarianism and religious dogmatism, through the publication of a new journal.[144] As the director of the Syrian mission press, Cornelius Van Dyck played a crucial role in helping Sarruf and Nimr to realize this project. It was he who suggested the name *Al-Muqtataf* ("the digest" or, literally, "selections") recommending that they take "selections" from the latest Western science journals and texts of the day, translating from English-language magazines and other journals housed at the college library. He also used the mission press to publish the journal and arranged to obtain a license from the Syrian director of publications. Throughout the journal's early Beirut years, Van Dyck remained a crucial ally; he continued, whenever he could, to ease the editors' path, and he contributed several articles to the early issues of the science monthly.[145]

Nor was Van Dyck its only supporter in the college. Senior faculty as well as board and missionary members were proud of it, and the magazine was passed around at board meetings at home.[146] Instructors at the college referred their students to it, and many saw it as an example of the superior enlightenment provided by Protestant culture and learning. Henry Jessup, for instance, referred to the journal in his Bible class when in February 1877 it published a full account of the exact timing and duration of an impending total lunar eclipse. The journal's precision impressed students at the college and beyond, and Jessup, in his letter to the PCBFM, commented happily on how his students' calm and reasoned appreciation of the eclipse contrasted with the chaos and bedlam outside the college gates: "Superstition, ignorance of God and ignorance of his works keep these poor people in bondage of fear and terror," wrote Jessup, noting with satisfaction that "of the millions of subjects of the Sultan of Turkey," it was only "our little Protestant communities" who could both appreciate and comprehend the phenomenon. "It is a comfort to know," he added, "that there is some light."[147]

Both Sarruf and Nimr were thus products of the instruction and culture of the early years of the American mission in Syria. Yaʿqub Sarruf graduated from the Syrian Protestant College in 1870, the year of the first graduating class, and afterward he taught American missionaries Arabic in Sidon and headed the American missions' school for girls in Tripoli. In 1873 he converted to Protestantism and shortly thereafter was appointed instructor in natural philosophy and mathematics (and, a few years later, chemistry) at the Syrian Protestant College. He taught at the college until 1884, when after years of controversy and conflict, the *Muqtataf* group departed for Egypt.[148]

Faris Nimr followed a path not unlike that of his colleague and collaborator. He graduated from the Syrian Protestant College a few years after Sarruf, in 1874, and was hired to teach astronomy, mathematics, and Latin at the college, having taught previously at the Prussian School for girls and the Greek Orthodox School. Although neither he nor Sarruf signed the articles they wrote for *Al-Muqtataf,* it is clear that his interests lay more in the astronomical and mathematical sciences than in the natural sciences—most of the many feature articles on astronomy in their journal were probably penned by him, as were the mathematical puzzles begun by the journal in the mid-1880s. In line with such interests, he translated E. Loomis's *Treatises on Meteorology* in 1876. Shortly after their departure to Cairo, his literary output took a slightly different turn, as he directed most of his efforts toward overseeing the newspapers *Al-Muqattam,* which began its publication in Cairo in 1889, and *Al-Sudan,* or *Sudan Times,* an Arabic-English paper founded in Cairo and Khartoum in 1903 and published twice a week. Both papers were taken on by the *Muqtataf* group as a means of subsidizing their less lucrative, if more enjoyable, scientific interests. Nimr, who carried the greater burden of these additional productions and who received the greater share of public odium for both papers, and *Al-Muqattam* in particular, often regretted the direction in which these commercial schemes led him.[149]

Shahin Makarius—the third of the original editors—was of a slightly different background. He had received no formal education after the age of eight and had a humble start working in the printing trade. Impressed with the quality of his work, Cornelius Van Dyck urged him to join the mission press in 1870. In his work as a printer, he developed a taste for literature, and even aspired to become a writer. He wrote and published several tracts on Freemasonry, and shortly after his arrival in Cairo, he even published his own monthly journal of literary and "comic stories," *Al-Lata'if,* which contained numerous literary articles, historical anecdotes, and humorous and amusing sketches, along with instructive items on scientific and industrial subjects. Nimr joined him in this venture, making regular contributions. *Al-Lata'if* was also Makarius's and his brother-in-law's key vehicle for the promotion of Freemasonry, and Masonic themes are found throughout the journal. The journal, however, enjoyed only a limited success: *Al-Lata'if* ceased publication some ten years later in 1896.[150]

In addition to their literary output and their dedication to the medium of print, Sarruf, Nimr, and Makarius saw the dissemination of knowledge and the cause of intellectual reform as dependent upon another new institutional phenomenon of the late nineteenth-century Ottoman world: the "society." In Beirut and later in Cairo, they were involved in, and helped to found, several professional (literary and scientific) and political clubs. Many of these had missionary support or patronage. The Syrian Society of Arts and Sciences, founded by Butrus al-Bustani, Eli Smith, and several other Protestant missionaries in 1847, was one early example. "The existence and prosperity of this society," wrote Eli Smith, who became its president in 1852, are "an indication, most interesting to the philanthropist and the scholar, that the culture of Western nations is exerting a great and happy influence upon minds in Syria, and even gives promise for the naturally fine intellect of the Arab race in the mould of modern civilization."[151] In a similar vein, Sarruf and Nimr started a short-lived scientific discussion group of their own in 1882, the Eastern Scientific Society (Al-Majmaʿ al-ʿIlmi al-Sharqi), which was modeled after the earlier Society of Arts and Sciences, although Sarruf and Nimr placed greater emphasis on the applied sciences.[152] Nimr and Makarius had earlier cofounded Shams al-Birr (Sun of Charity), a literary and religious society with Masonic connections that was affiliated with the YMCA, and in Beirut, Makarius also founded a society of craftsmen.[153]

Closely associated with such novel forms of sociability was political radicalism. Both Nimr and Makarius were active in the Beirut Masonic Lodge; Nimr was himself even master of the lodge for a while, and in Cairo, both remained active Freemasons. More radical was probably the short-lived so-called Secret Society that reportedly aimed—as Nimr admitted later in life—to overthrow the Ottoman sultanate, either by direct political agitation or by instructing the people in their "constitutional rights." It apparently succeeded in inspiring a great many Syrian notables and idealists during the governorship of the noted reformer Midhat Pasha until he was removed from the post in 1880 (it was rumored that the society had even been encouraged by the governor in his battle to force administrative modernization upon the Porte). Nimr, who had lost his father in the intercommunal violence that followed 1860, retained a lifelong aversion to what he felt was the despotic, schismatic, and autocratic traditions of the empire. It is significant in this context that from the mid-1880s, both Nimr and Sarruf, as we will see in the next chapter, championed a vaguely laicist liberal individualism and a staunchly laissez-faire politics.[154]

Catholics and Creationists

Rumors of Sarruf and Nimr's involvement in a secret society and hints of anti-Ottomanism must have created an air of suspicion around the *Muqtataf* group in the minds of the authorities, especially after Midhat

Pasha was forced out in 1880. But the journal had become embroiled in heated controversies over issues of religion and science from the very start. These stemmed not from Ottoman misgivings but from the protests of the Catholics, who looked upon the American Protestants and their protégés as rivals and who routinely attacked the *Muqtataf* group in their own journals. Catholic-Protestant rivalry in Beirut—a city with a significant Greek Orthodox, Roman Catholic, and Maronite population—was particularly acute, especially as the Jesuits felt that the Protestants had muscled into their territory.[155]

The Catholic missionaries in particular competed with many of the Protestants' mission projects. After the opening of the Syrian Protestant College in 1866, for instance, the Jesuits moved their seminary school from Ghazir to Beirut in 1874, and shortly thereafter the French government certified the school as a full-fledged university, the Université Saint-Joseph: keeping up with the Syrian Protestant College, it established a school of medicine in 1883 and a school of pharmacy in 1889 (figs. 1.10 and 1.11). The Jesuits established a school journal, *Al-Bashir* (the equivalent of the Protestant *Al-Nashra al-Usbuʿiya*), and a journal of the sciences and arts, *Al-Mashriq*, founded by Louis Cheikho in 1898 as competition for *Al-Muqtataf*. Each read the writings of its rivals closely, penning direct refutations or responses.[156]

Father Louis Cheikho, the editor of the journals *Al-Bashir* and *Al-Mashriq* and a professor of Arabic at the Université Saint-Joseph, was perhaps *Al-Muqtataf*'s fiercest critic. In 1878 *Al-Bashir* accused the Protestant journal of denying the existence of supernatural entities and branded the writers heretics. This was a response to numerous articles in *Al-Muqtataf* (in a no doubt deliberately anti-Catholic vein) that attacked the obscurantism implied by belief in magic, hypnotism, magnetism, and spiritualism. Other commentators in the city weighed in on both sides in this and other debates, with figures such as Shaykh Yusuf al-Asir and the Tripolitan educator Husayn al-Jisr stepping in to defend *Al-Muqtataf*.[157]

Then, in 1883, *Al-Bashir* attacked *Al-Muqtataf* for promoting dangerous ideas, referring to its articles on evolution. *Al-Muqtataf* replied by again accusing its rival of ignorance and superstition and charging it with atheism and disbelief. For Cheikho, however, the very basis of Darwin's idea of evolution—as he put it, "that the origin of man and beast came from a small germ that evolved with time to reach the present stage of development"—seemed inherently absurd. Any attempt to argue that intelligence was a concept common to animal and man seemed to him misguided. "Hardly an issue of this magazine appears," Cheikho later complained of *Al-Muqtataf*,

Fig. 1.10. Université Saint-Joseph, c. 1908. From Université Saint-Joseph, *Faculté de medecine, 1883–1908* (Beirut: Jesuit Press, 1908), 50.

Fig. 1.11. Anatomy laboratory at the Université Saint-Joseph, c. 1908. From Université Saint-Joseph, *Faculté de medecine, 1883–1908* (Beirut: Jesuit Press, 1908), 18.

"which does not indicate its preference for the beasts, may God forgive them. Dare we forget their article 'Man and Brutes,'" he continued, "in which they placed dogs in a higher position than man? These are the feeble opinions of *Al-Muqtataf*, unashamed to broadcast this as science among Easterners, when science remains innocent of them." Linking men to brutes, Cheikho argued, could only lead to a fatal materialism.[158]

These rivals retained a lifelong antipathy to one another, and the disputes between them lasted well into the next century. Cheikho may have even started his 1898 journal *Al-Mashriq* to serve as his own mouthpiece against *Al-Muqtataf*. When *Al-Muqtataf* questioned his scientific competence, however, the insult evidently stung: "Our journal," Cheikho later wrote, "just like theirs, is scientific, literary and artistic, despite the owner's supposed deficiency of scientific knowledge."[159]

Science, Knowledge, and Wisdom

In 1882 the journal acquired a much more formidable enemy than the Catholics. This time the opposition could not have been closer to home—in a dispute that erupted in the college itself. It was this episode—subsequently known as the "Lewis affair"—that brought to the surface all the ambiguities that had been inherent in the missionaries' role in the promotion of the modern sciences. The dispute began in the summer after Darwin's death when the administrators of the Syrian Protestant College raised a series of angry objections to a commencement speech delivered there by Edwin Lewis, a Harvard graduate and professor of geology and chemistry at the college. What had attracted the attention of his employers were his highly favorable references to Darwin.

Lewis's address, given in Arabic under the title "Al-ʿilm, al-maʿrifa wa al-hikma" (Science, knowledge, and wisdom), was essentially a reflection on the nature (and limits) of science. In that respect it could be seen as a typical missionary translation project—turning *ʿilm*, the broadest word in Arabic for "knowledge"—into "science." Knowledge (*maʿrifa*) announced Lewis, is "not science" (*ʿilm*), for while the former involved the patient and thorough accumulation of observations or facts, the latter was a superior epistemological system of inquiry since it involved both the gathering of facts and, through induction, the formulation of causal theories: "Science searches in nature for the causes of events and places them in their correct context." Yet Lewis was also bringing together a new, positivistic or neo-Baconian understanding of knowledge with an older one. For in line with a Greco-Latin distinction between *sophia* and *episteme* or between *sapientia* and *scientia*, he also distinguished between knowledge (*maʿrifa*) and science (*ʿilm*), on the one hand, and wisdom (*hikma*), on the other, by which he meant divinely ordained truth. As he put it, "through science man may know something about the existence of God, the Cause of all Causes, but he fails to apprehend who and what God is." For "no telescope will show us God; no microscope will show us the soul of man, and no chemistry will disclose the secret of life." Science, he concluded, is therefore "limited," for though "man may attain its highest peaks," he will "still find ahead of him peaks higher than those which he has attained."[160] Lewis thus categorized *ʿilm* as essentially knowledge of things human and *hikma* as knowledge of things divine.

Lewis's lecture highlighted competing traditions of knowledge—and wisdom—in an Arabic context and represented, more specifically, an intervention in the ongoing epistemological reorientation of the word *ʿilm*. In the Qurʾan, the frequency of the term meant that *ʿilm* came to be equated with both divine and true human knowledge. In the classical lexicon *ʿilm* thus referred to the knowledge of "definite things," a broad categorization that encompassed both revealed and acquired knowledge and that was variously equated with the knower and the known, with comprehending and obtaining, perception and apperception (*tasawwur wa tasdiq*), intuition (*al-badiha*) or believing (*ʿatiqad ʿala ma huwa bihi*). *Maʿrifa*, more closely aligned with gnosis but also referring to cognition, realization, and perception, had an equally complex genealogy and deployment; it was, moreover, typically classified as a subcategory of *ʿilm* itself.[161]

Lewis took no account of this. Glossing over centuries of philosophical debate and exegesis, he narrowed the meaning of *ʿilm*, defining it essentially in neo-Baconian terms as equivalent to the English "science."[162] Indeed, his use of all three terms, *ʿilm*, *maʿrifa*, and *hikma*, would have seemed unfamiliar to most contemporary nineteenth-century Arabic speakers, and his speech would in all likelihood have attracted little interest among the learned classes of the time had it not been for the unintended afterlife his speech enjoyed in the Arabic press of the late nineteenth century: for thanks to his unfortunate praise of Darwin—and the highly public controversy this initiated with the help of *Al-Muqtataf*—the speech soon made its way to the forefront of Arabic debates on science.[163]

In the year of Darwin's death, Lewis used him as an example of contemporary scientific achievement and greatness, "as an example of the transformation of knowledge into science by long and careful examination"—neatly setting the discussion of evolutionary theory within an exposition on modern science and its methods of verification and proof. Lewis, of course, was aware of the theory's controversial status: he claimed that it was "opposed by enemies and antagonists, and that expounders played about with it and interpreters tainted it," adding, somewhat provocatively, that this was "because his doctrine led to the nullification of certain ideas strongly adhered to by the people as though they were part of their religion." Nevertheless, Lewis was also careful to claim that, ultimately, science was not wisdom—specifically, divinely ordained wisdom—and hence that it paled in comparison to revelation and true religious conviction, and he described "real wisdom" as that which "descends upon us from above."[164]

Although Lewis's views were strikingly similar to those of any of a number of Presbyterian and Unitarian theologians and commentators on Darwin at that time, administrators and senior faculty at the college claimed that they were greatly troubled by the speech, which they felt was "an apology for Bible truth" and an acceptance of "yet unproven theories," namely Darwinism. They reacted strongly. Dr. Lewis, announced the Board of Trustees during a special meeting in December, "appeared so distinctly to favor the theories of Darwin, that several of his associates and of the Managers of the College were constrained to express alarm at the utterance of such views by a Professor of the Institution."[165] Lewis, wrote President Daniel Bliss later to his sons, "gave the impression that he was a Darwinian and that man descended from the lower animals. Dr. Dennis was up in arms—replied to Lewis' oration (the oration was printed in M. Faris' journal) and wrote very strongly to Mr. Dodge, so did Post and so did I."[166] The fact that Lewis's speech was published in *Al-Muqtataf* (along with numerous responses to it, mostly favorable) made matters all the worse. Writing to President Bliss on its publication, David Stuart Dodge asked, "How did it get in there? Is not this a point to be examined? Has [Lewis's] influence so perverted our young Tutors that their journal is to throw its influence in that direction?"[167]

Hoping to dampen this favorable reaction, James Dennis—the director of the college's theological school and one of the American missionaries who was offended by Lewis's speech—sent a letter to *Al-Muqtataf* in which he "regretted the many references made by Lewis to Darwin and his theories." Wrote Dennis: "I am shocked and grieved, that such a eulogy on Darwin should have come from such a professor before such an assembly." After all, Darwin, Dennis argued, separated religious truths from conclusions of science. He was an agnostic and a corrupter of Bible truths. Darwin's theories, Dennis claimed, had little influence among men of science, and he cited Agassiz, Wallace, Mivart, Pasteur, and Owen as some of those "scientists who rejected Darwin." Quoting Pouchet, he warned: "species are not theoretical concepts created by human intellects. Rather, they are created by the all-powerful hand of God in numerous stages. They cannot change into other species, but they change independently." Darwin's theory, he cautioned, should thus not be confused with a theory of "the progression [of species] by a divine power" (*al-irtiqa*ʾ *bi-quwat ilahiya*). Divine progression, Dennis explained, may be a law by which the Creator operates, "so long as a concept of the transmutation of species is not included."[168]

Lewis, in response to Dennis's article, wrote a rejoinder the following month. In it, he cautioned against making too swift a denunciation of Darwin's theory. "We should not," he wrote, "make a rigid judgment against the values of this theory since it has not been significantly tested yet." He also emphasized the ultimate harmony of religion and science. "It is clear that the scientific method, correctly applied, does not make men turn away from their religion." Revelation and natural science both worked to strengthen men's faith. "By studying nature we learn about the way God established it, but through revelation we learn who and what God is."[169] In this respect, like Asa Gray, whose lectures at Harvard he may have attended, Lewis argued that Darwin and design could be united.[170] Lewis also sent a letter to William Booth, president of the Board of Trustees for the college. In it, Lewis described his oration. "Showing how thoroughly it defined Darwinism, limited its authority and showing also how it could not explain the existence of the soul," wrote an incredulous Dodge to Bliss, "it would seem that the address was strongly orthodox and scriptural." After reading this, Booth himself had concluded: "it seems our friends in Syria have evidently only misunderstood each other." But Bliss felt Booth had fallen for a ruse, and he sent the letter to Jessup for corroboration: "when Bro. Henry read it, he was almost speechless." Jessup then studied the address in its original Arabic and later claimed that the "Arabic was so ambiguous that hardly anyone could fail to be led astray." It seems translation played a role here too, for Dodge and Jessup then shared their own version of the translation with Booth to convince him of their views on the matter.[171]

Soon, others outside the college joined in the conversation. Writing from Alexandria, Yusuf al-Haʿik, an avid reader of the journal, sent a letter to *Al-Muqtataf* that was published in the same issue. In it, al-Haʿik corroborated Lewis's view that "true religion does not contradict science." He saw, moreover, little need to bother over whether or not Darwin was a believer. "There is nothing wrong with a believer referring to the theory of a learned nonbeliever." After all, "if it is not correct, science will disprove it; if it is correct, humanity will not be lowered from its high station."[172]

The End of the Affair

Members of the college's Board of Managers remained unconvinced, and after a meeting with the Board of Trustees was convened in New York, Lewis was dismissed from the college. It may at first sight seem surprising that the board should have reacted so strongly, given the fact that Darwin's ideas had been discussed in a fairly open spirit in the college's prime publication for some years. Lewis himself was evidently taken aback. He had acquired enemies inside the college who suspected his orthodoxy—he had been criticized before for his love of music and wine—but this in itself was scarcely sufficient to have prompted the row that erupted. Much more important, surely, was the changed atmosphere in Beirut: the dismissal of Midhat Pasha, the growing attacks of Father Cheikho and others equating evolution with materialism, and the appearance in *Al-Muqtataf* itself of the first promaterialist pieces by their own graduate Shibli Shumayyil—no doubt all this made the college's leaders decide to take a firm stance.[173]

As a result, all the faculty members were required to sign a religious manifesto—a "Declaration of Principles" that was drafted by Dennis in conjunction with the Evangelical Alliance (an international Protestant coalition group)—and to pledge themselves "to the inculcation of sound and reverent views on the relation of God to the natural universe as its Creator and Supreme Ruler." They were to give instruction, moreover, "in the spirit and method best calculated to conserve the teachings of revealed truth, and to demonstrate the essential harmony between the Bible and all true science and philosophy." Only those who were deemed compatible with the religious "spirit of the college" were declared eligible to hold teaching positions.[174]

Several faculty resignations followed. These included *Al-Muqtataf*'s mentor, Cornelius Van Dyck, and his son William. Cornelius Van Dyck felt that his colleagues had misinterpreted Lewis and behaved improperly in suspending students; while William, who had always been dedicated to science above all else (to the extent that his employers had hesitated to employ him at all), felt that the college was moving, as he put it, in "a dangerous direction." (William Van Dyck was also among the first to introduce students at the college to Darwin's original texts: in 1880 he brought with him copies of the *Origin of Species* and *Descent of Man* for use in his class on zoology. He had also corresponded briefly with Darwin in 1882.)[175] Several others resigned from the college's medical department with similar complaints. In fact, the medical school was depleted.[176] Senior missionary and board members were not much bothered, however: "Let us stand clear on the records—'Resignations' springing from such grounds there be welcomed. Better run at half-speed—'slow her clear down'—than make sixteen knots with the aid of atheistic, materialist or non-religious boilers."[177]

Student protests and even one or two violent skirmishes erupted on campus.[178] "The medical students," wrote Jurji Zaydan in his autobiography, "learned that Dr. Lewis had resigned as of the first of December 1882 . . . and they decided to protest and boycotted the College." Forty-five medical students then held their first meeting in one of the halls of the Prussian Hospital; many of them had gathered together like this before, and they were used to meeting in the college itself, in the society Shams al-Birr, or in the Freemasons' society. The students immediately organized a petition, using the opportunity to voice other complaints about the conditions and stipulations at the Medical School (complaining, for instance, that the requisite medical diploma in Istanbul had to be taken in Turkish or French—languages not taught at the college). They rallied too in defense of Lewis and against his supposed blasphemy. "You have decided in favor of dismissing the pious, distinguished Dr. Lewis," read the petition, "accusing him . . . of making a statement on behalf of the blasphemous ideas of Darwin in his last speech. Nobody who understands his speech and who knows his proper conduct, his virtuous ability and strength as the head of the students' organization of the College and the head of our religious collegiate association, and as a leader in charitable activities, will agree to this."[179]

Despite the protests and the riots, the administrators and senior faculty held their position. "With regard to the dismissal of Dr. Lewis from the school," they wrote, in a memorandum issued to students and signed by Daniel Bliss, "we can understand what chagrin it causes you and we agree with you that you have the right to be notified. But we do not see how it gives you the right to absent yourselves from your classes." Faculty and college board members alike demanded that all dissenting students return to their classes on the condition that they sign a public apology, or else be dismissed from the college. In a letter they drafted to the faculty, students wrote in dismay: "It never occurred to the thoughtful people of Syria nor to the sons of the college, your students, that good people such as yourselves, who belong to the country of freedom, America, would pass such a verdict." Within the year, some fifteen medical students, including Jurji Zaydan, had been suspended.[180]

The Lewis affair highlighted the growing concern over the relation of science to religious pedagogy at the college. It may well have been this concern that eventually led the senior American faculty members of the college to rescind Sarruf's and Nimr's promotion from the rank of "native tutors" to "adjunct professors" shortly after the events of 1882.[181] Indeed, both parties agreed to "terminate their connection" altogether. Although it is not clear why faculty members voted for such an action, one factor may have been growing fears over the degree to which the religious aims of the college had been subordinated to science. Although the American missionaries had, at first, been enthusiastic about the didactic and pedagogic aims embodied by their new journal, they soon grew wary of their protégés' own project of scientific edification.[182]

Initially, all parties at the college had believed in the "mental and moral" benefits of scientific pedagogy and the diffusion of popular knowledge. "Mental and moral science," wrote Jessup in the missionaries' *Annual Report* of 1871, "is so intimately connected with man's spiritual nature that opportunities are continually occurring both within the classroom and without [such as through pedagogical literature] to enforce the great fact . . . that a pure morality and a rational religious faith are in accordance with the constitution of the human mind and a necessity to its highest well-being." After all, in Jessup's own words, it was only through "the advancement of education and popular knowledge" that the "absurdities" of the Mohammedan doctrines, like those of the Eastern Churches, would be exposed.[183]

Sarruf and Nimr, however, proved to be missionaries of a different kind. Whereas for Jessup, say, science would demonstrate the superiority of a reformed Christianity, for them the point was less the triumph of Protestantism than the "diffusion of useful knowledge" and scientific truth itself—including, of course, Darwin's own theory of evolution. After 1882, with Sarruf and Nimr's obstinate lobbying in defense of both Darwin and Lewis, missionary enthusiasm for their protégés' project of an enlightened diffusion of science diminished. As Jessup—another of Lewis's antagonists—later wrote: "When mere intellectual and scientific eminence become the only objects of the Christian missionary," when the sole aim of "consecrated missionaries and missionary societies . . . is to have the best astronomers, geologists, botanists, surgeons and physicians in the realm for the sale of scientific prestige and worldly reputation, then we do not hesitate to state that such a mission has stepped out of the Christian and missionary sphere into one purely scientific and worldly."[184]

In 1884, unhappy with the previous years' turmoil, bitter about the college's long-standing reluctance to promote them from the title of "native tutors" to the rank of "adjunct professors," Sarruf and Nimr left Beirut. That year Sarruf inveighed against those who were "fanatical against some of the sciences taught in their institutions" and those who "selected teachers based upon their religious beliefs and not their academic competence." Institutions of higher learning, he warned bitterly, are destined to fail unless "they renounce religious fanaticism and permit their teachers and pupils to embrace whatever religious beliefs they choose, expecting nothing from them but teaching and learning."[185] After the Lewis controversy, Sarruf and Nimr headed south. Upon their "retirement," Dodge wrote to Bliss: "I do not feel troubled by Nimr and Surroof as far as the College is concerned, but those young men will turn miserably unless grace prevents." "Going down to Egypt," he wrote, "has never proved wholesome."[186]

They went to Egypt, where one could "breathe freely," as they had noted earlier during their visit to Cairo. British rule appeared to them to provide a more liberal environment for their ideas than the combination of Ottoman officialdom and American Protestantism had done. There was little of that religious, and especially Christian, "fanaticism" that proved so oppressive in Beirut, and they were free from Ottoman censorship, which had become particularly burdensome for the two editors as the disputes with their Catholic and other rivals grew more and more unruly. Besieged by furious attacks, weary of sectarian strife and religious protests, and under pressure from Syrian-Ottoman officials to redirect their journal toward less controversial matter, the editors chose to move. At the close of 1884 Shahin Makarius and Faris Nimr left for Cairo to set up a new press. The others followed shortly in February 1885. After nearly a decade, it seemed that their true success rested with their readers and not with their patrons. In British-occupied Egypt their journal, as they had realized during their earlier visit there, enjoyed considerable popularity. Nevertheless, there too, as we will see in the next chapter, their journal did not stay free from controversy for long.

TWO

Evolution and the Eastern Question

Inspired by a belief in progress through intellectual enlightenment and self-improvement, Yaʿqub Sarruf and Faris Nimr remained dedicated to the "diffusion of useful knowledge" throughout their long careers. After their move to Egypt, they continued to promote the "new sciences" as part of their prescription for the "reformation of minds" and the "awakening" of the intellectual (if not spiritual) torpor of their Eastern readers, or, in the words of one of their former mission colleagues, for the formation of a "new society" and of "new men."[1] Even as they formally parted ways with their patrons, they continued to share many of the latter's aims and visions, particularly concerning ideas of science and social progression. Indeed, understanding the evangelical roots of nineteenth-century Arab intellectuals like Sarruf and Nimr—and their immersion in ideas of progressivism that shaped postmillenarian social discourses after the "Second Great Awakening"—goes a long way toward explaining the very conception of the "Arab Awakening" (or Nahda) that they helped to promote (if not actually to invent).[2]

While in Egypt, they also continued to engage in the sweeping reformulation and conceptual transformation of the notion of *ʿilm*. This was still largely a pragmatic and an instrumental, technical vision—one they continued to yoke to ideas of technological and industrial transformation and one that acquired a particular charge under Egypt's technocratically minded British colonial reformers and overseers. But they never lost sight of their former doxological and theistic framework, even as over time, they came to embrace a more positivist view of science.[3] For this, they turned not to Comte or to Mill but to Herbert Spencer, who increasingly guided their vision of science, cosmic evolution, and a utilitarian "social statics"—what Spencer defined as the "conditions essential to happiness."[4] Like many theists before them, particularly the American and Victorian Protestant theologians and science popularizers whom they familiarized their readers with, they "busily exaggerated theistic themes" found in Spencer and other late nineteenth-century neopositivists.[5] Yet Spencer appealed to them (as he did to so many Arab evolutionists through the early twentieth century) because he seemed to leave room for faith in God through his vague appeal to the "Unknown" as the ultimate, if yet inscrutable, metaphysical force behind the cosmic order of things. But this they merely hinted at, and they never developed this idea, leaving it to later Arab evolutionists to do so.[6]

For them, Spencer's real appeal rested with his vision of the unity and progression of science as "organized knowledge" and with the idea that society should itself be reorganized around this. A nineteenth-century neologism in Arabic as in other languages, ideas of "society" (and the search for "social laws") provided the background to *Al-Muqtataf*'s project of science popularization after their move to Egypt, and it provided the basic framework for their vision of reform, which, like Spencer's, was as much anti-statist as it was evolutionary.

For Sarruf and Nimr, science, rather than the state, was the key to social progression; nevertheless, they found that moving to Egypt opened up the possibility of an increasingly close engagement with political affairs. In Egypt, both men remained staunch critics of the Hamidian regime, which they regarded as despotic and dogmatic. Increasingly, they looked to European discourses on liberal political theory for ideals of the separation of state and church, individualism, and free enterprise, even as the advancement of science continued to serve as their particular hope for the Arabs' path to progress. For them, the British occupation was an opportunity, and they held to a belief in the inevitability of the occupation thanks to Egyptian (as much as Arab) backwardness. As another elite colonial intellectual later put it: "British imperialism was not our people's only enemy. A reactionary mentality characterized by a permanent clinging to traditions, a hatred against the modern spirit in politics and social affairs and beliefs—these were the elements that together made another enemy of the people, a formidable obstacle on its road to progress." Here, too, *Al-Muqtataf*'s reading of evolution provided inspiration. "The theory of evolution which I discovered about this time in the magazine *Al-Muqtataf* opened my eyes . . . I came to believe that science, by which Europe had been able to realize if not her happiness yet at least her supremacy, would raise us from our lowly status of thought."[7] For these men with their increasingly Anglophone orientation, evolution provided not only a theory of human origins but equally a set of powerful metaphors for political backwardness and a model for social analysis.

Yet not everyone agreed, either about the metaphors or the models, and the conjunction of foreign occupation and the spread of print culture in the Arab world opened the gates to a host of nationalist critics. In Egypt, growing dissatisfaction with the impact of British rule altered the terms upon which Darwin's ideas were debated, and although Sarruf and Nimr saw themselves as spokesmen for an Arab revival, their critics in Cairo wrote them off as suspect émigré interlopers, tied to British apron strings. The editors' emphasis on individualism, free-market capitalism, and even their vision of the occupation itself soon came under fire. As this chapter shows, reading Darwin in Arabic was inextricably bound up with the politics of empires, particularly in the global age of print.

A "Most Potent Engine of Civilization"

Behind the rise of *Al-Muqtataf* lay the rapid spread of an Arabic print culture in the last two decades of the nineteenth century. Maps, figures, and diagrams were easily incorporated into printed works; new forms of knowledge could be cataloged, codified, and disseminated far more easily than ever before. In Cairo and Beirut, as in Istanbul and the other major metropolitan centers of the empire, the press fostered a new community of Arabic readers. Described by one American observer in the late Ottoman Empire as a "most potent engine of civilization," the idea that the press could inspire social and political change was shared by many.[8] "Newspapers," wrote Shibli Shumayyil in 1897, "have the highest command in civilized states because they have a power over society which no other entity has."[9]

Although printing presses were first introduced into the Ottoman Empire in 1493, they were not used by local scholars themselves until Ibrahim Müteferrika obtained a mandate to print books dealing with science, language, and history in 1727. Most others relied on centuries' old traditions of copying and composing manuscripts in script until the late eighteenth century. Indeed, the first Qurʾan was not printed until 1874.[10]

As ruling governor of Egypt, Muhammad ʿAli was one of the first to systematically deploy the new medium in the service of state reform measures, and he sent a student on a "mission" to Europe as early as 1815 to study printing in Rome and Milan. These early publications primarily reflected the interests of his regime and brought in a number of works in translation on engineering, medicine and public health, and geography. This was one of the first large-scale state-sponsored translation movements in modern Arabic. Between 1821 and 1850, he founded a press at Bulaq that issued some eighty-one books on science in translation, and in the same period more than two hundred books were translated into Arabic and Turkish from European and Eastern languages. New circuits of readership emerged under his rule and under his successor Ibrahim Pasha's, as books printed in Cairo circulated in various Arab provinces of the Empire.[11]

Most of the early nineteenth-century ephemera, moreover, were literary or scientific journals, since the reporting of strictly political news was somewhat constrained both by the slow integration of telegraphic networks and by Ottoman and foreign diplomatic oversight. The new circulation of print matter on the sciences and the arts, however, opened the way for some news reportage and general political discussion in the press, particularly by the latter half of the nineteenth century. Early periodicals such as Butrus al-Bustani's *Al-Jinan* and Khalil al-Khuri's *Hadiqat al-Akhbar* (both issued from Beirut) published official news and reported on international affairs. The latter, which enjoyed the informal patronage of Fu'ad Pasha, the Ottoman foreign minister, printed regular columns containing official decrees and included a section on "telegraphic messages," featuring current events, while Bustani's *Al-Jinan* commonly used current affairs as a prelude to more straightforwardly didactic or historical discussions.[12]

The real expansion of the Arabic press came in the second half of the nineteenth century. Alongside *Al-Muqtataf*, many other Arabic periodicals and papers appeared, mainly in Beirut but later also in Cairo, Alexandria, and elsewhere—especially under Egypt's Khedive Isma'il (1863–79) and the Ottoman sultan Abdülaziz (1861–76). More than twenty journals were established during Isma'il's reign.[13] By 1875, there were over one hundred and fifty printing presses scattered throughout the empire, while émigré communities in Europe and the Americas also founded numerous Arabic papers and magazines. But the British occupation wrought a further dramatic transformation, making Egypt the unquestioned print capital of the Arab world between 1852 and 1880, 25 periodicals had been founded in Beirut, 13 in Cairo, and 10 in Alexandria. But between 1880 and 1908, while Beirut produced 42 periodicals and papers, Cairo had become home to 514 and Alexandria 113.[14] "In Egypt," wrote an astonished German orientalist in 1899, "literary work is daily on the increase. New printing-offices, new books, new periodicals, and new men follow one another with a rapidity which is surprising."[15]

Print Politics

The emergence of the press raised the issue of censorship, and both the Ottoman and the Egyptian authorities established new systems of regulation. The Ottoman government passed their first such law in 1857; it required the licensing of publishers and prepublication censorship. In 1863, the Egyptian government set up a special government department to ensure that the Ottoman regulations were enforced; later, a censorship bureau was created in the Ministry of Foreign Affairs. Contemporary observers made much of the heavy hand of Ottoman censorship, but we have little evidence for this before the 1880s. In practice, it was probably little if any more stringent than its counterparts elsewhere. Often, as with Khalil al-Khuri in Beirut and Muhammad ʿAbduh in Cairo, the censors themselves were journalists who encouraged a fairly loose censorship policy.[16]

Perhaps more effective than formal censorship from the point of view of the state was the provision or withdrawal of financial backing to specific papers and journals. The Arab press may have suffered, for example, in the 1870s when the khedival government was obliged to cut back its subsidies because of the debt crisis in Egypt. From the middle of that decade, Ismaʿil nevertheless began to encourage the press to defend him against encroaching European powers by featuring attacks on acts of British and French imperialism.[17] It was not long before the British too engaged with the business of financing sympathetic opinion. "The native press is in the hands of French Syrians and others hostile to England," complained Lord Cromer in 1887. "Our view of public questions is not properly put before the public." But it was hard to see what was to be done when, in his view, "editors of current papers are all scoundrels."[18] Given the extent to which the press served as the main opposition platform to the British occupation, Lord Cromer's low opinion of journalists and newspaper editors was hardly surprising. Yet it seems this was also a widely held view in Egypt, no doubt because of the extent to which all parties there, and particularly the khedive and the sultan, continued to fund their own partisans. The editor of *Al-Muʾayyad* was described in the course of a highly publicized court case in 1904 as a member of "a despicable spying profession."[19] And when Jurji Zaydan started his long journalistic career at the *Al-Muqtataf* publishing house in 1887, his father expressed regret that he had not gone into a more "decent" profession such as law or medicine.[20]

Regardless, by the end of the nineteenth century, the Arab press had emerged as a powerful medium for debating public affairs, mobilizing public opinion, and expressing political opposition. "Neither the power of kings nor that of armies nor the authority of laws," wrote an enthusiastic Shibli Shumayyil at the close of the nineteenth century, "equal the power of newspapers." Under the British in particular, the absence of legitimate institutions through which Egyptians could partake in the government of their country turned the press into a substitute for political parties and a means by which to incite anti-British nationalist sentiment.[21]

Compared with the Ottoman Empire, the British encountered unanticipated difficulties in enforcing a rigorous censorship policy, and not only because of the liberal inhibitions that Lord Cromer later claimed motivated his hands-off stance. They also had to contend with the legal ramifications of the highly ambiguous foundations of their own political authority in Egypt. The implications became clear as early as 1884 when the British attempted to suppress the French-language newspaper *Le Bosphore égyptien* triggered a minor diplomatic imbroglio with the French government. The outcome established the principle that the Ottoman capitulation agreements remained in force and that the khedival government—and by extension the British authorities—had no power over foreign-owned presses. Thus, a few years later, after the Ottoman sultan had complained about articles published in Cairo-based newspapers, Lord Cromer responded wearily that measures taken against a paper owned and edited by an Ottoman subject would simply lead to control being handed over "to some European, who in the absence of any effective press law, could not be touched." Egypt's complex legal situation was thus responsible for an unusual degree of press freedom, and its journalists took full advantage of this.[22]

Prominent among those journalists were immigrants from Syria. Martin Hartmann, a German Arabist and an early observer of the press in Egypt, referred in 1899 to "the invasion of Egypt by Christian Syrians which is still going on and has become of the utmost importance to the development of the country."[23] Many of the prominent Arabic journalists publishing in Egypt in the late nineteenth century, such as Sarruf and Nimr, Farah Antun, Jurji Zaydan, and the Taqla brothers, were Syrian émigrés. A number of these came out of Christian mission schools. In Alexandria, Shibli Shumayyil published the journal *Al-Shifa'* between 1886 and 1892. In Cairo, Jurji Zaydan published *Al-Hilal*, a semimonthly periodical on science, history, and literature (founded in 1892). But the Syrian émigré community there included prominent Muslim publishers as well; indeed, they formed a close-knit social network that crossed confessional lines, and many of them met in salons and cafés and collaborated on various literary endeavors. *Al-Hilal*, *Al-Muqtataf*, and Rashid Rida's *Al-Manar* (founded in 1898) became the three most influential journals in the Arab world, each of them also expanding to become publishing houses that printed and distributed books as well. In this, the editors of *Al-Muqtataf* blazed the way.

"Going Down to Egypt"

Despite Dodge's misgivings ("Going down to Egypt has never proved wholesome"),[24] Sarruf and Nimr's move to Egypt proved an immediate success. In 1884 Nimr and Makarius set up operations in a quarter in Cairo close to Al-Azhar and later, along with Sarruf, they set up a press for printing Arabic, as well as English, publications. In addition to the journal, their commercial press also printed books, stationery, bills, and advertisements. On the list of their first publications were the works of Shahin Makarius and Shibli Shumayyil and a history of Egypt by Jurji Zaydan, who was employed for a time by the press before becoming publisher of *Al-Hilal*, the main competitor to *Al-Muqtataf*. They also published travel books, collections of political speeches, and eventually even the annual reports of the British consul general himself.[25]

Thus, with the move to Cairo, Sarruf and Nimr entered a new phase of literary and journalistic activity. Their journal, which had already had a considerable following in Egypt, prospered, and by 1892 its circulation had increased from a monthly average of about five hundred copies to about three thousand copies.[26] Nevertheless, the financial situation must still have been uncertain because in 1889, in order to help finance *Al-Muqtataf*, they took the step of establishing a new daily, *Al-Muqattam*. In doing so, they entered a political minefield—one that ultimately tarnished their reputation and altered the reception of their ideas.

During the first few years of the British occupation of Egypt, there were few Arabic papers. The occupation itself had led to the imprisonment or deportation of many of the journalists who had played a leading part in mobilizing public opinion during the events that led to the deposition of Khedive Ismail in 1879 and to the military uprising known as the ʿUrabi revolt shortly after.[27] One or two papers continued their publication after the occupation, however, such as the popular *Al-Ahram*, which had been launched from Cairo by Syrian émigré brothers Salim and Bishara Taqla in 1875, and Jindi Ibrahim and Mikhaʾil al-Sayyid's *Al-Watan*, which first appeared in 1877. As these and other papers did little to present the British in anything but an unfavorable light, Lord Cromer was immediately bothered by the lack of a sympathetic Arabic voice. "The inconvenience of our not having a newspaper, and especially an Arabic newspaper, at our disposal here has too often been recognized," he wrote in 1883.[28] In 1887 he considered making a bid for the *Egyptian Gazette* and negotiated with its editor, Philip Graves, to have a section published in Arabic. The project turned out to be more ambitious and less feasible than he had anticipated. "I cannot say," he wrote in 1888, "that Mr. Philip [*sic*] is a genius or that his paper carries much weight."[29]

Thus, when Nimr applied for permission to publish a political weekly two years later, Lord Cromer jumped at the opportunity, although it seems he also made it clear that it would be worthwhile to accept the proposal only if it were a daily. And indeed, although it began as a weekly, within six weeks *Al-Muqattam* was being issued daily.[30]

Nimr's, Sarruf's, and Makarius's ties to the British were strong—particularly those of Nimr, who did the majority of the editorial work for the paper. In fact, he was married to the daughter of a British consul and was close friends with Oriental Secretary Harry Boyle.[31] Cromer described the daily as one of the "native papers" that "support English views and dwell on the benefits of English interference" and as "a very moderate newspaper which generally supports the British occupation."[32] They were also given the business of printing and distributing Lord Cromer's annual reports as a supplement to their newspaper as well as of publishing his various speeches.[33] In 1903 they were given an additional concession to start the *Sudan Times* in Khartoum.[34] The exact nature of the original agreement between Nimr and Lord Cromer is not known, but many contemporaries who had been aware of *Al-Muqattam*'s pro-British stance were nevertheless quite surprised to learn of its close connection to the British Agency.[35]

Thanks to this backing *Al-Muqattam* was impressively produced and sold in respectable numbers. By 1892, the evening paper was selling over 2,500 copies daily (the pro-khedival *Al-Ahram*'s circulation was estimated to be 6,000 in 1898). *Al-Muqattam* cost one piaster an issue, and its income was further boosted by substantial advertising revenue. At least a quarter of its content was given over to government announcements and commercial advertisements—revenue that was supplemented by the fact that civil servants in the British administration were forced to subscribe to it. It had correspondents in Istanbul, Beirut, and Damascus and a regular staff of over forty correspondents within Egypt. Many of its editorials and extracts were simply translated from English-language papers like the *Times*. Written for a mostly professional readership, the paper contained local, foreign, and general news reports as well as law reports, Reuters financial and commercial telegrams, and advertisements.[36]

The founding of *Al-Muqattam* had a substantial effect upon the content and profile of its editors' science journal. Despite its old policy of avoiding overtly political reportage, *Al-Muqtataf* sometimes published articles from the columns of its sister paper. In 1895 it even expanded to include a "News" section. The prosperity of the daily newspaper brought other tangible benefits to the monthly, too. In 1889 *Al-Muqtataf* increased from sixty-four to seventy-two pages a month and, in 1895, to eighty pages. Several new columns were also added, entitled Industry, Agriculture, Home Economics, Mathematics, Inventions, and Discoveries. Later, there was The Opinions of the Savants, which contained short excerpts on current literary, scientific, or political developments of the day such as "The Causes of Eastern Decline" or "The Coming Century." Another later column was News of the Day, which included local and foreign current items such as "Awards from the Sultan," "Ismail Pasha's Funeral," "Events at Al-Azhar," or, further afield, "The Armenian Question," "Macedonia," and "The War between China and Japan." Because of the expansion of the project, and due to Nimr's greater involvement with the daily paper, they also had to hire paid contributors in the 1890s.[37]

Thus, although *Al-Muqtataf* still remained faithful to its scientific mission, after 1890 the tenor of its writings on science changed. For many intellectuals, the move from Syria to British-occupied Egypt enabled them to engage publicly in politics in a way that had not been possible before, and this was certainly true for the editors of *Al-Muqtataf.* Close as never before to the centers of political power, they laid a new emphasis on the social aspects of science, and their message acquired an increasingly political tone.

The Social Organism

One consequence of this was a growing interest in the thought of Herbert Spencer, one of the key figures in this story and as important as (if not more important than) Charles Darwin in the debates over "Darwinism" (*madhhab Darwin*) in the Arab provinces, as elsewhere, at the century's close. Spencer's appeal at the time was global. A cosmic evolutionist and liberal anti-imperialist who emphasized social management and reform and who deliberated extensively on the proper (and evolutionary) function and development of the state, his ideas proved appealing everywhere from the United States to Japan, India, and the Ottoman Empire.[38]

In 1885, as they settled into their new offices in Cairo, the editors of *Al-Muqtataf* reported on an article, "The Social Organism," written by Spencer that appeared in *Popular Science Monthly* (an American periodical founded with the explicit purpose of popularizing Spencer's writings). The article drew on his *Principles of Sociology* to announce the "discovery" of the natural laws that governed society. *Al-Muqtataf* relayed the news with characteristic flair. Headlining its own column "A Prodigious Animal" (Hayawan haʿil), *Al-Muqtataf* declared: "An animal so massive that the image of it has not crossed the imagination of any of the Ancients has only now been discovered by the modern philosophers. . . . We are not speaking figuratively or in riddles. That such an animal exists is quite certain, if we are to believe the modern philosophers. You will ask us: 'And what is this strange animal?' The answer is 'Society' [*al-ijtmaʿ al-insani*, literally the 'human collective']." Speaking directly to their readers, they reminded them that "you are a member of this globule [*al-karriya*]."[39]

From the very start of its transfer to Egypt, *Al-Muqtataf* shifted its focus from Darwin to Spencer, from the laws of biology to the evolutionary laws of society, and its reliance on *Popular Science Monthly* grew. That year, at perhaps the height of the philosopher's renown, the editors dubbed Spencer "one of the greatest philosophers of the age."[40] In Britain, Spencer had popularized the view of society as a social organism: it was "a growth and not a manufacture, and has its laws of evolution."[41] For Spencer, the laws of evolution and nature governed the actions of individuals, and these cohered to produce larger social outcomes.[42] The editors of the journal were of a similar mind: We are, they wrote, but parts (*ʿadaʾ*) of the larger social body. Societies, like biological organisms, exhibit the properties of living phenomena, are subject to natural principles, and are susceptible in particular to laws of evolution and decline.[43]

Yet like Spencer, the editors of *Al-Muqtataf* would consistently look to individual action as the primary force for social change. Sarruf had translated Samuel Smiles's *Self-Help* many years before, and his library included not only Smiles's works but classics by Bentham, Guizot, Mill, and other liberal thinkers (alongside Tylor's *Anthropology* and Ibn Khaldun). Progress, in *Al-Muqtataf*'s view, could only come by way of an individualistic effort to develop men's character and conduct through self-improvement, thrift, and industry.[44]

Before long *Al-Muqtataf* was publishing articles on Spencer with extraordinary frequency. Its earlier interest in Darwin's theory of natural selection, as well as in animal intelligence and the descent of man, did not vanish but became secondary. Readers still asked about Darwin; as late as 1890 a reader in Alexandria wrote in to ask which language his writings were available in.[45] But for the magazine, the chief inspiration was now Spencer, and scarcely an issue passed without reference to the cosmic philosopher. In 1898 it published a long series of translations of extracts from his *Principles of Sociology*, culled once again from *Popular Science Monthly*. "In those years," wrote Salama Musa in his autobiography, "Sarruf . . . maintained that Spencer was the greatest thinker on earth."[46]

Spencer's educational views, which stressed the importance of science and the "appeal to individual reason," also chimed with the line espoused by Sarruf and Nimr. They remained committed—as they had in Beirut—to demonstrating the value of scientific progress and pedagogy for civilization. "The natural sciences, which help discover the laws of the universe, also have the greatest benefit for the ordering of human affairs, both material and spiritual," they wrote in 1886 in an article entitled "Science and the Good of the Nation." In this essay, they also praised Muhammad ʿAli, who, as they put it, moved Egypt from the dark ages to an enlightened age (*ʿasr al-nur*) through his dedication to the sciences and by opening new schools.[47] Years later, in 1903, Najib Bey Shuqri wrote in to announce his intention to translate into Arabic Spencer's *Education: Intellectual, Moral and Physical*—a "precious book, the likes of which has never been seen before."[48] (The following year, Grand Mufti Muhammad ʿAbduh called on Spencer in England; ʿAbduh had already, it is said, translated this work from French. Yet the book was not published in Arabic [translated by Muhammad al-Sabaʿi] until 1908.)[49]

Al-Muqtataf marked Spencer's death with several articles about him and his philosophy. They also concentrated on his opinions on education and included a detailed catalog of what they considered to be his most important books.[50] "We heard the news of the death of that great shaykh of the philosophers, Herbert Spencer, the summit of the ancients and the moderns. . . . we should mourn his passing like that of the greatest of Egypt."[51] The editors paid effusive tribute to the man whom they labeled "the last philosopher" of their century. "Spencer's death," they wrote, "has given us occasion to discuss his scientific philosophy—the likes of which we have never before seen in the Arab world," arguing that like many social and natural philosophers and scientists, they "have used him as a guide," consulting many books written on or by him.[52]

At this time, it was chiefly through *Al-Muqtataf* that most Arabic readers learned of Spencer, as they had learned of Darwin before him—there was simply no other outlet of comparable importance. And it was to the journal that impatient readers wrote in inquiring whether there were yet full translations of his works.[53] One reader in 1904, perhaps intrigued by the spate of articles that followed the philosopher's death, asked the editors: "What are the moral benefits [*fawaʾid*] of Spencer's philosophy for our day and age?" Their reply was categorical: "That the rational and material progress that Europe and America have attained during the last 50 years is based on numerous laws, some of which aid the discovery or creation of such things as the telegraph, the railroad, and vaccinations from diseases." But this optimism in the universal laws of science that the editors shared with Spencer was only part of the story; there was, in their view, another key practical benefit for those who engaged with Spencer: they thought his published works had "the greatest consequence for the enlightenment of the minds of the people who read them," arguing that they were invariably "mightier and more capable than others in bringing about benefits and deterring harm."[54]

They did admit that Spencer's manner of expression was often "overly complicated for the casual reader." Hence, to simplify his ideas, the journal relied not only on periodicals like *Popular Science Monthly* but also on early popularizing books such as Hector Carsewell MacPherson's 1900 *Spencer and Spencerism*. Indeed, the editors also announced that they would describe Spencer's philosophy in an "easily comprehensible fashion," as his writings were often difficult for the uninitiated. But of his originality, they left their readers in no doubt. Borrowing loosely from MacPherson, they highlighted his place within the history of European philosophy: "Spencer's philosophy has surpassed that of Aristotle, Spinoza, Kant, Hegel, and all others past and present," they enthused. Spencer's special contribution was to emphasize the interrelatedness of the elements of the natural world: "Before Spencer, people looked at the universe as though it was one huge mechanical tool, each of its parts manufactured separate from the other and for one purpose alone, that of the supreme Creator's." Spencer's philosophy separated science from the idea of God and was premised upon the all-powerful principle of unity in nature itself. "Spencer's comprehensive philosophy has demonstrated how the universe is one complete picture—like parts of a puzzle that fit together. The laws which apply to its parts, also consist of general laws which apply to the whole. This is the basis of his philosophy, which he achieved through much research and by way of many proofs." They praised his "synthetic philosophy" for its "unity [*tawhid*] of all the sciences," and his principle of universal progression—the movement, whether in societies or in organisms, from heterogeneity to homogeneity. "The aim of Spencer's philosophy," they maintained, "is actually very simple: namely, that all observable things in the natural world—from the celestial heavens to the earth and all that man is master of, and from the movements of the stars to those of men's minds—are created from matter and force."[55]

But this was to move into dangerous territory. Indeed, the editors felt obliged to defend Spencer against the charge of materialism. He did not say—they asserted—that all beings were composed from matter and natural forces alone, nor that they brought themselves into being, nor did he broach the subject of first causes. On the contrary, he acknowledged certain limits to the mind's understanding. It was chiefly in this sense, they claimed, that he sought to separate science, as the study of knowable things, from metaphysics and religion.[56]

These evaluations date from 1904, the year of his death. Yet only a few years earlier they had taken a more critical stance with respect to his evolutionary interpretation of religion. In a short "Commentary on Spencer," they began with cautious praise: "Any of our readers who has looked at Spencer's works or our brief translations of his works has no doubt been amazed at his breadth of knowledge, the detail of his research, and his command of the literature—a decisive and comprehensible philosophy, which is nevertheless not entirely free from doubt and contradiction." What worried them was what Spencer had written about the social evolution of customs and worship. They denied that in this area evolutionary behavior in the natural world provided any kind of a model, and they invoked Wallace against Spencer in reminding their readers that reputable scientists had disagreed with Spencer's opinion on revelation. In the evolution of his mental and bodily powers, man was an exception. One might reasonably speak of a power other than natural law controlling man; it was this that aided man's ultimate progress and ethical advancement and to whose study religious scholars devoted themselves. "In short, Spencer's philosophy is not an entirely proven one, nor is it free from doubt. Those who hold to a belief in Creation," they wrote, "do not stand against reason, even if these ideas contradict some of Spencer's claims."[57] Indeed, elsewhere they referred to his concept of the "Unknown" to show how even Spencer's own philosophy might help to reinforce a theistic faith in creation.[58]

Aside from this caveat (which was one they often made of many of the writers they surveyed, particularly when they strayed on to the topic of the evolution of religious tenets or practices), the editors' admiration for Spencer's universal philosophy proved unbounded. What it chiefly gave them, of course, was the argument that the laws that governed the natural world also ruled man's social evolution. Indeed, the 1885 article on Spencer's ideas about "the Social Organism" served as part of a commentary on a longer series of articles published earlier by one of their regular (and most controversial) contributors, Shibli Shumayyil: "The Natural History of Societies" (Tarikh al-ijtimaʿ al-tabiʿi).[59] Spencer's idea that societies, like species, progressed through a struggle for existence and the "survival of the fittest" offered them a powerful tool for presenting to their readers their thoughts on the rise and fall of civilizations. And it was here—in their interventions in the intensely political debates of the time about the reasons for Western superiority and Eastern backwardness—that *Al-Muqtataf*'s writings on evolution acquired unmistakable political overtones.

The stagnation, decline, and decay of nations, wrote Sarruf and Nimr, following Spencer, were themselves a condition of progress, for the "very life of the social organism depends upon the death of nations." The rise and fall of civilizations—such as those "glorious Eastern civilizations of the ancient Egyptians, Phoenicians, Babylonians, and Assyrians"—constituted a natural law of universal progress.[60] To be sure, progress for Spencer was far from inevitable: it was conditional upon being "unhindered, undisturbed, unrepressed, undistorted, undwarfed, undeformed, uninjured." For him, a "reversion towards the old type" was one possible outcome of a hindered or disturbed society. While Spencer feared a reversion in the case of England from an industrial society to what he dubbed a "militaristic" one, Sarruf and Nimr, by contrast, found the idea of a return to the past consoling in the Arab context.[61] "There is an absolute and invariable law in the universe," they wrote, "and that is that men's ancestral racial traits do not easily disappear. And if they do so partially, it is quite easy to bring them back by a resumption of the necessary conditions." It might be possible, in other words, to return Eastern civilization to its former glory. "The Arab East may once again be on the road to progress, if we attend to those conditions that allow for it."[62] Through Spencer, *Al-Muqtataf* thus brought evolutionist principles—and discussions of the rise and fall of civilizations—to bear on one of the key geopolitical issues of the day: the "Eastern Question."

Eastern Questions and the Race for Civilization

"If you ask a politician what is the most important issue occupying men's minds in this day and age," wrote the editors of *Al-Muqtataf* in 1888, "he would answer at once that it is the Eastern Question [*al-masʾala al-sharqiya*], for no political paper or periodical appears without a discussion of some of its implications or some of the issues it raises. And if you ask a man of learning in our times," they continued, "he would no doubt answer that it is the theory of evolution [*masʾala al-irtiqaʿ*] . . . for no scientific paper or periodical appears without some discussion of this latter question."[63] The connection between the two now needs explication.

By the late nineteenth century, the issue of the fate of the Ottoman Empire was of international concern: "The Eastern Question has by degrees assumed such large proportions that no one can be surprised at the space it occupies in all public discussions whether of the tongue or of the pen," Stratford de Radcliffe wrote in the *Times* on 9 September 1876, at the height of the Near Eastern crisis as Russian intervention threatened to whittle away the Ottoman state in the Balkans.[64] Seen from an Arab point of view, this threw into question the fate of the empire's Arab provinces. The British occupation of Egypt in 1882 and the subsequent "Scramble for Africa" reinforced concerns in the region over the fate of the empire with respect to the European Great Powers.

This only intensified over the years. The crushing defeat of the Mahdiya state in the Sudan by the British under Kitchener at the battle of Omdurman in 1898 and the establishment of the Anglo-Egyptian Condominium the following year triggered a particularly fierce debate over these issues partly because of the long-standing connection between the Sudan and Egypt and partly because it sharply intensified the British stranglehold over Egypt itself. *Al-Muqtataf* reported on every stage of the war, providing a heavily pro-British account drawn chiefly from *Nineteenth Century* and the *Pall Mall Gazette.* Kitchener's triumph served only to confirm their basic stance: they noted that while Europe continued to progress, particularly in the arts and sciences, the Arabs had for a long time "not only not progressed but reversed."[65] For others, however, it merely heightened their fears of Western encroachment. "Our present era differs from every other era in history," wrote the nationalist Qasim Amin, calling for the "liberation of women" in 1899. Western civilization, "powered by steam and electricity," had advanced and expanded in a frightening fashion. "Whenever Western civilization enters a country, it seizes the wealth and resources of that country"—following (as Amin claimed) "[what] Charles Darwin identified . . . as the law of natural selection." For Amin, like so many other Arab thinkers at that time, the encounter with the Western world was itself an example of the "struggle for life" between nations.[66]

Amin, however, drew a distinction between the impact of colonialism on the "primitive nations" inhabiting Africa and on "a country like ours—a country with a former civilization, with historical traditions, religions, laws and customs." Like most Egyptians of his class and generation, he assumed that the Sudan was, on the one hand, Egypt's—and hence Egypt's to colonize, not Britain's—and, on the other, racially, if not civilizationally, distinct.[67]

The issue of race had already appeared in the popular press in Egypt before Omdurman, but the conversation became much more contentious afterward. *Al-Muqtataf* was in the forefront of reporting the work of British racial scientists, though it followed Spencer in favoring environmentalist interpretations.[68] In line with the editors' views on the "special creation" of man, they usually assumed that all humans were descended from a common stock, and they attributed racial variation to climate and social habit. Skin color was a key concern: they described blackness as caused by tropical sun and reported on experiments to change pigmentation. It was a subject that often provided them with sensational material: for example, they reported the story of a "black man in his sixties," "deeply religious," who developed white spots and eventually became white! While Africans were often portrayed as cruel, socially backward, and uncivilized, the racial origins of the Egyptians formed a contrast: the ancient Egyptians were, they reported at one point, "a white race connected to the Semites"; elsewhere, they reported that before the pharaonic age Egyptians were "fair and white" like Europeans. (Although when readers asked where blond people in Egypt came from, the answer they got was that they originally came from Turkey and Circassia.)[69]

The issue of race remained of deep concern to Arab readers. Indeed, very similar views to those of *Al-Muqtataf* could be found in another monthly, *Al-Hilal,* a journal that had been founded in 1892 by the *Al-Muqtataf* editors' sometime collaborator and former Syrian Protestant College colleague Jurji Zaydan. *Al-Hilal* also offered environmentalist accounts of racial difference; readers from as far away as Canada and Japan wrote in to inquire about the subject, and Zaydan himself authored a popularizing study of the world's races in 1912.[70] Apart from the seeming popular appeal of this subject, another reason behind the interest in contemporary racial and anthropological sciences was that the British were conducting anthropometric surveys in both Egypt and Sudan, largely to demonstrate the utility of these sciences to imperial governance. Both *Al-Hilal* and *Al-Muqtataf* were aware of this and were essentially reporting the British point of view. It was not long before they took on board both the literary and administrative ideals of a unified Anglo-Egyptian condominium themselves: Nimr started his *Sudan Times* in 1903, and many Syrians, including many of the writers for both *Al-Hilal* and *Al-Muqtataf,* served as administrators under the British in the Sudan.[71]

The politics of racial difference was thus ambiguous. At the same time as the British sought to distinguish between Egyptians and Sudanese to reinforce their argument that the former had no claim on the latter, some Egyptian nationalists were seizing on evidence of racial differentiation to make a civilizational argument about why they were entitled to rule over their southern neighbors. For them, the Sudan represented Egypt's own civilizing mission, an idea that was reinforced simply by the fact that the terms often used to translate "colonization" and "civilization" were derived from the same root (*ʿamara*).[72] In other words, for the Egyptians, the argument about separate racial origins—the distancing from Africa in the racial hierarchy and the assertion of racial kinship with non-African groups, whether "Semites" or Europeans—formed part of a larger set of questions about the country's civilizational standing and development. In this way, the Sudan question helped crystallize a set of distinct orientations with respect to Egypt's own political identity and future. Even before 1898, the nationalist Mustafa Kamil had proclaimed Egypt the country of a "religious liberalism" and moderation that made it the perfect intermediary "between civilized Europe and fanatical Africa." For Kamil, it was not so much race as Egypt's specific brand of Islam—in contrast to the violence of the Mahdiya—that marked out its potential mission for the future. For others, still, the civilizational rationale would be made in very different terms, and ideas of the "unity of the Nile Valley" were sometimes later extended by Egyptian nationalists to advocate for the expansion of the Egyptian border into southern territories. They did this either by claiming a racial, like a geographic and a historic, unity of human habitation in valley lands or by arguing about the ambiguities of "race" as a defining character of peoples and substituting "civilization" in its stead.[73]

The debate came to a head much earlier, however, with the translation into Arabic in 1899 of an international best seller—Edmond Demolins's *A quoi tient la supériorité des Anglo-Saxons*. For Demolins, the explanation for England's superiority was straightforward: imperialism was a sign of vitality, and the "Anglo-Saxon world" now stood "at the head of the most active, most progressive, the most over-flowing civilization."[74] His book elicited particular interest in the Ottoman Empire, where it was widely discussed, especially in Egypt.[75] Only two years after its appearance, the book was translated as *Sirr taqaddum al-Inkiliz al-Saksuniyin* (The secret of the progress of the Anglo-Saxons) by Ahmad Fathi Zaghlul; brother of the future nationalist leader Saʿad Zaghlul, he translated Gustave Le Bon as well. What Demolins, editor of the Paris-based periodical *La Sciénce Sociale*, provided was a proudly social-scientific analysis of Anglo-Saxon supremacy that focused on the family, education, and the formation of national character. The Anglo-Saxons were self-reliant individualists (the English were contrasted with the "communistic" Irish and inferior French and German educated classes for this purpose), and it was above all their character that allowed them to prevail in the struggle for existence. Determining "what social state is most conducive to happiness" was key. Unlocking the secrets to this success was therefore a necessary condition of national regeneration.[76] For Demolins the success of the "Anglo-Saxon empire" was based on its "civilizing institutions," such as public schools, as well as on its "entrepreneurial spirit," embodied in the trader, the missionary, the explorer, and the engineer.[77]

Demolins's book was much cited in British-occupied Egypt, where the moral was quickly drawn. British commercial and industrial, political and military success could be attributed to their dedication to "individual enterprise" and their ability to withstand the "struggles of life." The Egyptians, by contrast, as Ahmad Fathi Zaghlul wrote in his introduction to the translation, were lacking in dignity and were slavish and overly dependent. He had translated the book in order to show his fellow countrymen that the only way to regain control over their own affairs was to become more self-reliant and to adopt an impartial and scientific attitude to their plight.[78]

It would not do, the editors of *Al-Muqtataf* wrote, citing Demolins's Arabic translation, "to point out [British] superiority merely to denounce it in Parliament or in the press, or to shake one's fists at the English like angry old women." One had to analyze the situation, "as a scientist . . . with exactness and most coolly, so as to become acquainted with its real properties." For Demolins, they approvingly noted, such an investigation (an examination of the secret of national success, "that prodigious power of expansion, of that extraordinary aptitude to civilize—and the means of doing it") was "a question of life or death."[79]

The idea that Demolins had provided a scientific explanation for the rise and fall of civilizations met with widespread acceptance. A writer in the daily *Al-Muqattam* praised Demolins's account for being based on scientific and, indeed, evolutionary principles. Others agreed. Ahmad Lutfi al-Sayyid, a well-known nationalist supporter and leading public intellectual in Egypt, later applauded the book for having "spread among the masses a scientific basis for development, so that people could apply its principles to their situation." The Young Turk leader Sabahaddin Bey later hailed its sociological use of a "scientific method similar to the methods of the natural sciences."[80]

Civilizational progress—like evolution—was thus seen as a law of nature in the age of Darwin and Spencer, and commentators who seemed to offer scientific insights into this process were seized upon by those seeking to promote their particular policies of social reform. Qasim Amin's argument for the education of women as an indispensable part of Egyptian national revival, for instance, was bolstered by references to Demolins: pointing to his account of British public schoolboys' regimen of swimming, ball playing, and horseback riding, he insisted that the idea of *mens sana in corpore sano* applied to girls as well as boys. This was because girls were the nation's future mothers: not only did children inherit traits from both parents, but the deep ties between mothers and sons were indispensable for producing "successful men." "This is the worthwhile goal that civilization has entrusted to the women of our era."[81]

Eastern Progress

After the Russo-Japanese War, many in Egypt started to look east and not just west for evidence of progress. But in fact, the rapid development and success of Japan were topics that attracted commentary as early as the 1880s.[82] Scores of articles on the country, which was described as "having only yesterday emerged from the dark ages," appeared well before 1905. The war further heightened this interest. The spectacular success of Japan, *Al-Muqtataf* wrote, "gives hope to all Easterners who strive to achieve greater accomplishments in their own civilization." As Salama Musa would later write: "[During] the war between Russia and Japan . . . public opinion was in favor of Japan, considering that it was an Eastern country like ours. We rejoiced every time we read about a Russian defeat. Russia in popular representation stood for Europe, to which Britain also belonged, whereas Japan stood for the awakening of the East."[83]

The leaders of Egypt's emergent nationalist movements in particular followed Japan's rise closely. Mustafa Kamil, the founder of the Nationalist (Watani) Party, published numerous articles on the subject in his journal *Al-Liwa*, as well as a book, *The Rising Sun*, in 1904 in which he praised the country as an exemplar of national pride. Ahmad Lutfi al-Sayyid, the leader of the more moderate People's (Umma) Party, praised the way they had synthesized the "spirit of Japan" with the "knowledge of the West." Jurji Zaydan similarly emphasized their union of Eastern morality and Western science. The lessons that these commentators drew ran along broadly similar lines: Japan formed a model of resistance to Western encroachment thanks to its commitment to universal education, parliamentary constitutionalism, scientific learning, and the development of a collective spirit of unity founded upon individual patriotism and initiative.[84]

What distinguished *Al-Muqtataf*'s coverage was its emphasis on the Japanese openness to scientific education and training: "The success of Japan is based on their firm grasp of the sciences, which they excel in as neither imitators nor mere translators. Rather, their scientists follow the surest path to knowledge, while unlocking the secrets of nature." Article after article reported on their annual literary and scientific accomplishments. *Al-Muqtataf* noted the astounding number of publications—3,792 titles in 1880 and 4,910 in 1881—and approvingly reported on the number of translations undertaken there (among which Samuel Smiles's *Character* received first notice). But with us Easterners, the editors complained, "we just go day after day, month after month, and year after year, reading the same books, while most of those we translate have been so tampered with that they no longer retain their original meanings. Meanwhile, most of our *ʿulama* are still saying a thousand times over what they have been saying for the last thousand years, like cattle chewing their cud; it makes the heart sick." While the Japanese send their students to Europe to excel in the arts and sciences there, "our government sends students abroad only to study the arts of dancing and preaching [*al-raqs wa al-daʿwa*]."[85]

"What are the causes of Eastern backwardness and our weakness in the modern sciences in comparison with the West?" a reader from Aden wrote in to ask *Al-Muqtataf*. The problem, the editors replied, is that "many of our *ʿulama* consider European knowledge worthless, and they find it difficult to overcome those prejudices they were raised with." The best means to succeed in these sciences was thus to encourage the development of such things as schools, science journals, and state sponsorship of Arabic translations from a variety of European-language scientific and literary books, along with sending students to Europe to study the new sciences—"like Japan does."[86]

As they saw it, Japan's progress had been dependent on following a Spencerian diet of universal education, the development of industry, and, above all, the separation of religion and state.[87] This was another theme that distinguished their writings on Japan. The secret to Japan's success was that there religion remained in its proper sphere, which is to say, a private matter and separated from issues of politics or governance. "What are the causes of the rise of Japan although it is geographically much farther from European countries and their civilization than we are?" the editors of *Al-Muqtataf* were asked. They answered that "their religion does not prevent them from adopting Western civilization." "It appears to us that the greatest reason for Japan's success is that its religion does not divide its people from the Europeans, and it does not prevent them from adopting their civilization and customs." If their religion were used to turn them against the Europeans, to prevent them from mingling with them at all costs and from studying their sciences, "then they would have been far from European civilization, even if they lived in Paris, London, or Berlin."[88]

Reporting on an international conference in Japan, where the idea of embracing one of the monotheistic faiths as the official religion of the state had supposedly been considered, the editors of *Al-Muqtataf* commended the Japanese for coming to the conclusion that "to force upon a nation a particular form of belief by the establishment of a state religion is very injurious to the natural and intellectual development of a people." According to Sarruf and Nimr, they had realized that those men who struggle to "remake a nation and fight their way into the front ranks of civilization have no time to devote to metaphysical questions. They do not believe that Christianity influences the statesmanship of foreign countries. . . . They have more faith in big guns, men of war, diplomacy, and political economy."[89]

Neither Sarruf nor Nimr opposed belief in religion per se: "true science and true religion," they had long held, "do not contradict one another; in fact, they are probably the same thing."[90] However, their move to Egypt had accentuated their anxieties about the public role played by religion in the Arab lands. During the 1890s, they observed with disquiet the growing spread of "pan-Islamic" ideas within the Ottoman Empire—ideas they invariably denounced as "opposed to the spirit of the age."[91] But many of their readers disliked this line of reasoning, and a number of them wrote in to complain about the way in which the editors portrayed Japan's success as predicated upon its ensuring that religion was no barrier to the adoption of Western ideas and practices. When one asked angrily which religion they had in mind, the editors responded that they had no one faith in mind but pointed out how in India, Siam, and China, for example, "religious zeal" had prevented the inhabitants from acquiring "Western civilizational practices." Nevertheless, they seem to have taken these objections to heart, since later they softened their remarks about religion, agreeing that perhaps the faiths and beliefs of the Japanese had had a positive effect in smoothing the path to progress.[92]

Still, they remained basically dubious about Muslim efforts to demonstrate how science and religion could be reconciled for the sake of civilizational advance. Although they had considerable respect for Grand Mufti Muhammad ʿAbduh—whose religious, pedagogical, and legal reform efforts they much admired and whom they described in one review as "the Leader of all Leaders in Islam and the Wisest of the Wise"—they generally looked upon his ideas and those of like-minded thinkers less favorably.[93]

Reviewing Muhammad Farid Wajid's 1899 book *Islam and Civilization*, for instance—which went to great lengths to demonstrate Islam's compatibility with modern science—Sarruf and Nimr argued that "we know from the experiences of Europeans themselves that these attempts to square scripture with science neither strengthen faith nor aid the development of civilization." These attempts, they prophesied, inevitably come to their natural end, for a new stage of knowledge acquisition is sure to follow—"one that, following general evolutionary laws, encourages the separation of science from religion altogether. . . . By this, of course, we do not mean to deny the role of religion in man's progress, neither for the past nor for the present and future. Rather, one should search for signs of faith in men's characters and deeds and not through scientific research and publications."[94]

Al-Muqtataf's faith in adopting Western science had remained unchanged and still struck a chord with their readers. But the divergence over their views on religion was the sign of a growing problem for them. They had come to British-occupied Egypt to escape Ottoman repression and missionary zealotry; for them, British rule had spelled opportunity. For most Egyptians, however, British occupation meant something very different, and as nationalist opposition rose, so *Al-Muqtataf*'s position became more precarious.

Al-Muqtataf and the British

Sarruf and Nimr had been warmly recognized by elite society in Egypt during their early years there. But in less than two decades, they came increasingly to be looked upon with suspicion. In an era of growing anticolonial politics, it was their association with the British that earned them the strongest criticism.

They took the view, crudely, that the occupation was not so bad: a stable administration, influenced and monitored by an advanced European power, was perhaps the best road to reform. "In the beginning [the occupation] had been very grievous to [the Egyptians]," said Nimr in 1891, speaking to an interviewer for the *Pall Mall Gazette*. "Proprietors of lands, merchants and poor fellahs, men in and out of office, all complained alike. . . . But when things got a little more settled," he continued, "reforms came to the relief of the Egyptians." That Nimr could so easily claim this at a time when anti-British nationalist sentiments were flaring up made his opinion all the more provocative.[95]

With articles such as "Restoring Order" and "Egypt in 1889 Compared with 1882," *Al-Muqattam* repeatedly acclaimed its new paymasters. Despite occasional criticisms of specific policies, the editors left their readers in little doubt of their basically positive attitude. "The British," they wrote in 1889, "are partners of our government, whether by advising it on foreign affairs or by working to perfect the country's irrigation, organize its army, and improve public discipline."[96] The editors even attempted to explain Britain's international and political right to occupy Egypt: in attempting to secure their route to India via the Suez Canal, they protected their natural right to self-interest; by governing Egypt's finances, they acted reasonably in trying to protect monies owed to them; in "restoring order to countries of chaos," they acted within their legal international rights.[97] Ironically, in this respect, they differed even from Spencer—who, in 1881, was a key figure in mobilizing opposition to British jingoism and aggression abroad, particularly in Egypt.[98] The editors greeted with surprise Spencer's advice to the Harvard-educated Japanese constitutionalist Baron Kaneko that Japan avoid international intercourse with Western powers: "the weak should learn to avoid the strong, until capable of withstanding the contest."[99]

Needless to say, their paper received a great deal of public odium as a result, especially with the emergence of other newspapers that were founded to promote the nationalist cause. A few months after the launch of *Al-Muqattam* in1889, the Azhari shaykh Ali Yusuf established a rival journal, *Al-Mu'ayyad*—which Lord Cromer later described as the "leading Anglophobe paper in Cairo"—specifically to oppose it.[100] *Al-Mu'ayyad* was soon joined by other nationalist opposition organs. The 1890s witnessed a rapid rise in the number of papers issued from Cairo, and the vast majority were dedicated to discussions of politics in one form or another: some ninety papers and periodicals were launched in Cairo alone.[101] Many of the more prominent of these were supported in one way or another by the khedival administration, which looked to the press to rally opinion against European imperial domination. ʿAbbas Hilmi II, in his effort to launch a public campaign against the British, was keen to regain the prestige and power of his office, looking first to the French and then to the Imperial Porte itself for support. And whereas earlier pro-Ottoman papers—such as the Taqla brothers' (also pro-French) *al-Ahram*—did not have an explicitly Islamic dimension, many of the new oppositional papers—such as *Al-Mu'ayyad*, Abdullah Nadim's *Al-Ustadh* and Ibrahim al-Muwaylihi's *Misbah al-Sharq*—did.[102] "The nationalist movement," wrote Salama Musa, "was mixed up on the one hand with religious propaganda and on the other with the desire for a return to Ottoman sovereignty."[103] Of course, all this was taking place at a time when Sultan Abdülhamid was himself publicizing his role as the caliph of the faithful. This alliance—between Ottoman loyalists and pan-Islamists—was cemented by the high-profile activities of British missionaries associated with the Church Missionary Society in Egypt. The pan-Islamic press, wrote Lord Cromer in 1900, "has seized avidly on the action of the missionaries." He complained that the "highly provocative nature of their own conduct"—they had sent out a tract entitled "Which of Them, Christ or Mohamed?" to "well-nigh every Sheikh and Omdeh [village head] in the country"—"stirs up the latent fires of Mohammedan fanaticism and renders Muslims generally more than ever disposed to look askance at the work of the proselytizing reformer."[104] In this context, "nationalism" was above all a label for anti-British sentiment and anti-imperial action, yet it clearly took many forms: from Ottoman or pan-Islamist to Arabist and Egyptian.[105] Sarruf and Nimr were widely seen as antinationalists, thanks to their pro-British writings, and were soon viewed together with the British as "foreign intruders" (*dukhala*), a label the journalists much resented: "Some people imagine that the presence of foreigners among them is bad and that the intruders are snatching their crumb of bread from them. Nothing could be further from the truth."[106]

Because of its pro-British sympathies, *Al-Muqattam* was even barred from several Ottoman provinces, including the editors' native Syria; while Sarruf and Nimr themselves soon became the object of vituperative attacks. "I would lie awake night after night," Nimr later wrote in his memoirs, "brooding over my fate, not knowing how to protect myself against the ever-increasing web of political intrigues, hiding from my two partners the death threats which descended upon me because of the policy of *Al-Muqattam*."[107] "I once saw," wrote Salama Musa, "some young men buying copies of *Al-Muqattam* and tearing them to pieces so that nobody could read them."[108]

"Some have doubted and accused us," Sarruf and Nimr wrote in an article in their daily paper in 1889, with respect to "*Al-Muqattam*'s politics," "charging us with selfish interests and base financial motives, claiming we sold our freedom for a few dinars." Hoping to explain the stance of their paper and defending themselves against the charge of being in the pay of the British, the editors excused themselves by claiming that criticizing the British was neither fair nor useful. Those opposition papers who blame the British for all the country's ailments "only detract from British efforts to reform . . . and may even delay British evacuation from Egypt."[109]

But the opposition was widespread and growing. Many readers were sending letters to the editors complaining of *Al-Muqattam*'s policy and threatening to cancel their subscriptions. "To him who is famous for his outrages," read one such letter from Alexandria (signed "one of a society whose numbers, by God, are 600 persons"), "look out for your slaughter, oh son of a she ass, oh enemy of the faith of Islam. Say what you like against our glorious Empire, which by God's favour, will be victorious in spite of you. Soon you will see yourself smitten and the printing office of your vile paper lying in ruins. In short, you are a dog and no two ways about it."[110]

It was not long before readers extended their antipathy to *Al-Muqtataf*. "In view of the constant attack in your paper El Mokattam against Islam, the Moslems and their Khalif, the Commander of the Faithful, I have ceased to subscribe to that paper and to the Muktataf," wrote one reader, Ahmed Nashid Bey, in 1896.[111] In a letter sent to the editors from Beirut (signed simply "a critic") in 1905, we learn "that many of *Al-Muqtataf*'s readers who readily admit the benefits and delights of this journal nevertheless criticize it for its predilection for the British." The critic pointed out that, because of the paper's "exaggerated love of all things British," its readers were only offered biographies of Englishmen, while such Eastern luminaries as the revolutionary Jamal al-Din al-Afghani and affluent Egyptian self-made businessman Ahmad Pasha al-Minshaway were left out of their journal. "We do not write biographies of Englishmen to bring ourselves close to them," the editors replied, "for most of those we write of are dead or else they do not read Arabic and thus have never heard of our journal." Choosing among the "best and most instructive" of articles and biographical stories from a variety of English-language papers, "we seek only to benefit our readers by reports of British knowledge and science—as we have been doing for the last 30 years." The only purpose of this, they reassured their readers, "is to provide scientific benefits to all those who may read our journal." Although many Arabic readers now turn away from all things British, this is wrong, "for even the French urge their people to follow in the footsteps of the British, particularly in matters of education and moral training, as Demolin's recently translated book demonstrates."[112] Thanks largely to Sarruf and Nimr, Darwin and Spencer were now familiar names among the Arab intellectual elite. But the journal that had been the vehicle for this, and that had started out proclaiming the impartiality of scientific truth, had now become identified as the standard bearer for pro-imperialist and antireligious opinion.

THREE

Materialism and Its Critics

> The materialists are an ancient faction who renounced the Creator, Governor, Knower, and Almighty. They claimed that the universe has always existed as it is without a creator, and that animals come from seed and seed from animals, and that this has always occurred and will occur forever. These are the atheists.
>
> —Abu Hamid al-Ghazali, *The Rescuer from Error*

> See him in a tree, see him in a stone, and see him in everything, that is God.
>
> —Ibn ʿArabi, twelfth century, cited in Shibli Shumayyil's *Falsafat al-nushuʾ wa-al-irtiqaʾ*

Materialism has a fairly long genealogy in Arabic: some of the earliest treatises on theology and philosophy use the term, and early Arabic commentators on Greek atomism and allied schools of thought make frequent mention of it. Yet it was not always distinguished from other, eclectic heretical beliefs: hence, the term also has many cognates. Shibli Shumayyil was one of the first modern Arabic authors to write in defense of "materialism," under a neologism that he no doubt hoped would distinguish it from these older associations. He also defined it in a very particular way.

Shumayyil constructed an idiosyncratic philosophy whose implications for both science and politics were to prove highly contentious. His view of materialism was as monistic as it was pantheistic. Citing liberally from contemporary German materialists, French and English evolutionists, and radical mystical Muslim thinkers such as Abu al-ʿAla al-Maʿarri and Ibn ʿArabi, Shumayyil developed a concept of unity (*tawhid*) in nature—albeit understood as the unity of matter with force rather than with spirit or mind—that also had strong mystical overtones. Yet this was a very different kind of mystical experience than that familiar to most of his Arabic readers. Indeed, Shumayyil ultimately used his materialism to argue against faith in spiritualism or supernaturalism of any kind, while his allied history of the evolution of religions allowed him to argue for the need to eliminate theocratic states like the Ottoman Empire and hence to take anti-Ottomanism to a level unmatched by any of his contemporary Arab evolutionists.

Unsurprisingly, Shumayyil's campaign in defense of materialism generated considerable publicity and a frenzied debate. This was triggered chiefly by his translation of the German materialist Ludwig Büchner's commentaries on Darwin in the early 1880s. Former colleagues at the Syrian Protestant College and theologians and educators in Beirut more widely were among the first to respond: his ideas on the primacy of matter and force over notions of spirit or soul were intended to deliberately provoke "creationists," "spiritualists," and "vitalists" (terms he also used idiosyncratically). Of course, his own brand of materialism (*al-maddiya*) was not free of its own vaguely spiritual and especially vitalist urges. Nevertheless, he also hoped that a materialist theory of life would help his compatriots in the Ottoman Empire overcome their unthinking obedience to religious authority. This in turn, he hoped, might lead to the emergence of a modern sense of political community, one that was more in keeping with the laws of nature and the evolution of peoples. His was thus a characteristically late Ottoman reading of the concept.

Over time, the debate widened. In 1886 a polemic by Jamal al-Din al-Afghani—the political and pan-Muslim revolutionary—appeared in Arabic under the title *A Refutation of the Materialists*. Translated from Urdu under the directorship of Muhammad ʿAbduh in Beirut, it was published in the midst of the growing controversy there over Shumayyil's new philosophy. Afghani's refutation made ample references to the spread of a new kind of evolutionary materialism in the Muslim world (it was originally written by Afghani when he was in India and where he presumably encountered Muslim supporters of both Darwin and of modern evolutionary materialism). But the *Refutation* was also a religious treatise of an older kind, its language steeped in classical Muslim philosophical arguments and categories: Afghani's discussion of Darwin even has him promoting a materialist view of the creation of life from "seed" (or germ plasm, *buzur*), as described by Abu Hamid al-Ghazali and others. In fact, few saw Shumayyil's materialism as novel: Muslims, Maronites, and Catholics alike revived classical theological debates of their own to combat what seemed to most a resurgent ancient foe. Not many Arab evolutionists shared Shumayyil's materialist interpretation of evolution either. Nevertheless, his ideas on progress and the evolution of state and society more broadly would prove extraordinarily influential, and the engagement with them would shape Arab attitudes toward evolutionary thought for generations to come.

On Germs and Generation

In February 1878 the editors of *Al-Muqtataf* posted a short notice on the latest scientific theories of the origin of life under the title "The Beginnings of Life and Death." The editors were not impressed with what they found. "Those who have looked closely at the scientists' disputes over the origins of life," began their brief report, "will be shocked to discover the inconceivable claims put forth by some." Despite the fact that the constraints of experimental technique should have led to caution, they noted, some scientists in Europe were making the extraordinary claim that life was self-generated (*tatawalid min nafsaha*, later described as *tawalid al-zati*, literally "self-born")[1] The editors then reviewed what they took to be the current scientific thinking on spontaneous generation. Like materialism (with which many Arabic readers associated it), this was an old and familiar term and concept.[2] Yet the Syrian editors' reporting on it served as the Arabic world's first introduction to a controversy that was then sweeping through Europe. It was also the first hint of a new, potentially threatening view of evolution—what the editors of the journal would later denounce as "the corrupt philosophy of the materialists" (*fasad falsafat al-maddiyin*).[3] From the very start of their journalistic enterprise, Ya'qub Sarruf and Faris Nimr made little effort to disguise their highly critical opinion of theories of spontaneous generation, and over the next four decades their stance scarcely changed. Yet their antipathy to these theories was not shared by all Darwin's new Arab proselytes. In fact, the editors' brief report initiated an acrimonious debate over the very nature of evolution and over just what such a view of life-generating matter might imply.

The Arabic discussion of spontaneous generation was tightly connected to debates in Europe—where a renewed curiosity in ideas of spontaneously generating life had reemerged shortly after Darwin published his *Origin of Species*. As the readers of *Al-Muqtataf* were well aware, prior to Darwin, spontaneous generation had gained the interest of natural philosophers and others and was the subject of repeated speculation. In the eighteenth century and even earlier, some explained the emergence of infusorians, algae, fungi, bacterial spores, and even parasitic worms by referring to their spontaneous creation from living or dead organic matter (later known as heterogenesis). In the early nineteenth century, such discussions became embroiled with new "germ" theories in medicine.[4] After Darwin, however, the question of spontaneous generation became concerned not just with heterogenesis but with abiogenesis, the idea that life arose from inorganic matter.

In Britain, the controversy over abiogenesis gained momentum with the publication of Henry Bastian's works *The Beginnings of Life* (1872) and *Evolution and the Origin of Life* (1874), which claimed proofs in favor of the generation of life from inorganic matter. By the late 1870s however, the supposed verification of abiogenesis was losing ground. Pasteur's prior experimental work on germs and John Tyndall's findings in the late 1870s seemed to settle the issue.[5] A mere matter of weeks after Tyndall published an open letter to Thomas Huxley with his results in *The Nineteenth Century*, Sarruf and Nimr happily reported on their significance. Indeed, the real purpose behind their 1878 report "The Beginnings of Life and Death"—its very title was a play on Bastian's *The Beginnings of Life*—was to announce the demise of certain "implausible claims" regarding spontaneous generation.[6]

The editors were clearly familiar with the experimental procedures and scientific debate surrounding spontaneous generation in Europe. Nevertheless, they used it as an opportunity to come to their own conclusions, drawing on the latest germ theory to argue in favor of a divinely willed creation. They explained how those who believed that "life was created by the Creator" had successfully challenged the claims of those who followed the theory of spontaneous generation. According to Sarruf and Nimr, the creationists had completely disproved Bastian's claim and demonstrated that such life-forms were simply the product of "plasm which entered it from the air" (*bizur min al-hawa*)—in other words, a germ. Although intense argument on the matter continued, they claimed it was now resolved thanks to Tyndall: as a result, the first school (i.e., that advocating heterogenesis, here lumped with the creationists by the editors) was, as they put it, "now more popular."[7]

Their use of germ theory may strike us as odd. But as we saw before, the youthful editors generally presented accounts of evolution and other scientific findings in ways that supported their own—and their American missionary patrons'—religious views. To be sure, after the controversy over Darwin in 1882 at the Syrian Protestant College, their positivism became much more pronounced. But even before the Lewis affair raised the stakes, there was at least one dissenter who had charged wholeheartedly into the fray.

The editors' brief and partial account of spontaneous generation had, in fact, provoked a lengthy—and argumentative—response from Shibli Shumayyil, a former classmate of theirs at the Syrian Protestant College and then a practicing physician. Writing from Tanta, in northern Egypt, where he was working, the young doctor's defense of spontaneous generation introduced him to the thrills of scientific debate and marked the beginning of his long career as a science polemicist; it also helped to introduce Arabic readers to a new "materialist" position on evolution.

The Making of a Materialist

Shibli Shumayyil was to become a key figure for at least two generations of Arab evolutionists. Many years later, an admirer described how this "stout, short man who physically resembled a wrestler" had been "characterized, or as others will say, infested, by a spirit militantly opposed to mysticism and obscurantism."[8] A Roman Catholic by birth, Shumayyil had been educated at an American mission primary school, and like the editors of *Al-Muqtataf,* Sarruf and Nimr, he too would be shaped by American missionary culture and its educational outlook. He attended the Lazarist Collège Saint-Joseph (in Antoura, Lebanon) and the Greek Catholic Patriarchal College of Beirut before entering the Syrian Protestant College in 1867, just one year after the new college had begun to introduce medical studies into its curriculum. The instructors there—including Cornelius Van Dyck and fellow science enthusiasts John Wortabet and George Post—made a strong impression on him: he often praised their dedication to scientific knowledge, philanthropy, and public service, which became a model that he himself attempted to follow.[9]

Like his classmates and cohort, Shumayyil was an ardent aficionado of the sciences, and the natural sciences in particular. He had shared a desk with Sarruf and later fondly recalled how they often performed various scientific experiments together. His graduation speech was entitled "Differences between Man and Animal in Reference to Environment, Nutrition, and Upbringing," and the same year (1871) he wrote a didactic series of articles on electricity for Butrus al-Bustani's new journal *Al-Jinan.* With a family network that extended throughout the Mediterranean and indeed as far as Liverpool, Shumayyil was fluent in French and also knew some English. After graduating from the Syrian Protestant College, he went to Paris and Lyons for further medical training, and it was in France that he first encountered references to Darwin during a course on anatomy.[10] While in France he also learned about popular materialism, then much in vogue: as he recalled, it made an enormous impression on him and uprooted his childhood religious beliefs. His new materialist orientation, no doubt, was accentuated when he went to Istanbul to take his qualifying examinations. In the 1870s the influence of the Young Ottoman movement was at its height, the empire was in crisis, and the Imperial Medical College was proving to be a hotbed of materialist ideas where young would-be reformers were avidly reading Spinoza, d'Holbach, Voltaire, Bernard, and, above all, Büchner.[11] Indeed, Istanbul was probably one of Shumayyil's most important sources of German materialist thought.

By 1876 Shumayyil had established a small but growing medical practice in Egypt—first in Tanta and later in Alexandria; by 1895 his practice had flourished to such an extent that he was able to run a full-page advertisement in the popular journal *Al-Hilal*.[12] He eventually settled in Cairo, where he lived until his death in 1917 and where he was well known as an original if not a radical writer. He frequented the capital's cafés and salons, was a leading figure in the city's Syrian community, and was on good terms, despite the controversial character of his views, with an enormous range of the country's leading cultural figures, from Mayy Ziadeh, the most famous salon hostess in Egypt, to Rashid Rida. He was, in other words, perfectly placed as another member of that early and influential generation of science popularizers. In addition to his medical work, he traveled abroad to attend scientific conferences and started his own journal in 1886, *Al-Shifa'* (Curatives). This was a conduit for the latest medical news from Europe and for "the promotion of general information on nutrition, hygiene and preventive medicine" for the sake of "public welfare." Coming at a time when the Porte was interested in scientific patronage, this earned him a decoration from the sultan for his "service to medicine and science."[13] Published by the *Muqtataf* press (Shumayyil worked on it with another Syrian Protestant College graduate, Jurji Zaydan), it was, in fact, the first serious medical journal in Arabic. It was partly underwritten by the Egyptian Health Department, and Shumayyil hoped that the journal would help to pressure the government to implement needed health reform measures.[14] However, when the government withdrew its support after only a few years, the journal was forced to close.[15] Shumayyil did not embark on another journalistic enterprise until 1914, when he helped found *Al-Mustaqbal* (The future), an irreverent political and literary paper that he coedited with the younger Salama Musa, himself an outspoken Darwin propagandist.[16] This did not last long either. It was shut down by the British, who feared it was spreading revolutionary politics; meanwhile, Christian preachers in Cairo were said to have denounced it in their Sunday sermons for its materialist views.[17]

Shumayyil did not mind. He courted controversy, and throughout his long and prolific career he saw himself as a voice for "truth." "My misfortune, if it may be called a misfortune, is that as soon as a truth appears to me, it attracts me to such an extent that I cannot refrain from expressing it."[18] In fact, he contributed scores of articles to all the major journals and newspapers, including *Al-Muqtataf, Al-Hilal, Al-Muqattam, Al-Mu'ayyad, Al-Ahram, Misr al-Fatat,* and *al-Watan*—articles that helped contribute to his growing popularity (and notoriety) well beyond the borders of Egypt.[19] A visit to Syria in 1909 revealed a great following, and supporters even helped fund the republication of some of his writings.[20] Shumayyil also published several books, from a collection of poems to a translation of Racine's *Iphigenie.* His most renowned work, however, was undoubtedly the summary of the commentaries on Darwin by the German materialist Ludwig Büchner that he published in 1884.[21]

Courting Controversy

The idea for the translation had occurred to Shumayyil in the course of the controversy over spontaneous generation in *Al-Muqtataf.* Following Sarruf and Nimr's 1878 report, Shumayyil sent a letter to the editors entitled simply "Objection." There he expressed his dismay over their dismissal of the theory of spontaneous generation: "I read . . . from your brief words regarding the origins of life . . . that the existing dispute between the scientists of your opinion [as he put it earlier, that life is a "special creation formed in its own special germ"] and those against it [that life is the "product of natural processes in its origins and in its varieties"] . . . is about to end." Shumayyil thought that this assessment obviously suited Tyndall, but Shumayyil himself was not convinced.[22]

Tyndall had claimed that when all mote-laden air was cut off from his experimental apparatus, no life arose within it. But Shumayyil objected that the explanation was obvious and that it had no real bearing on spontaneous generation itself. It was, he thought, because "there is no creation or maintenance of life without air." How can we imagine that matter will germinate without air? And how could anyone support Tyndall and others who based their "false conclusions" on such "ridiculous proofs"? Shumayyil reminded the editors that Bastian separated all mote-laden air—but not all air itself—from his experimental setup with a special apparatus, and that life nevertheless appeared. Bastian claimed that life can be self-generated, Shumayyil reminded the editors, and he found that these life-forms could survive on very little oxygen, which "God knows man cannot remove from the air!" As the air cleaned in this process was different from the outside air, Shumayyil concluded, Bastian's claims still stood. He then pleaded with his former classmates: "As we are used to you providing us with dependable facts, may you please give us more reliable information on the truth of this matter, because your intelligence regarding this is precious to both science and a host of other important matters."[23]

In their response, the editors made very clear exactly what was at stake. Sensing that not only their credentials as impartial reporters but also their commitment to a theistic view of evolution were being implicitly challenged, they advanced what would later become familiar themes to *Al-Muqtataf*'s readers—faith in a benign, gradual evolution under divine Providence. They started with the big picture: all scientists agreed that the earth was first created "empty of animals and plants" and that these animals and plants did not appear until "conditions were suitable for their creation." What they disagreed about was life's origins: was there a Creator or was life self-generated through natural, material elements? The editors emphasized where they stood on the issue, claiming that this was the consensus view: "Most scientists think that the Creator of the universe created life, and we believe that they are correct."[24]

Returning to the question of spontaneous generation, they then reviewed the entire course of the discussion. At one time people had thought that life could be self-generated. However, this had been disproved long ago, and the discovery of bacteria (*baktiriya*) had changed the terms of the debate. The only issue now in question was whether matter could be proven to self-generate under certain experimental conditions. Once again they recounted Bastian's and Tyndall's experiments, this time detailing them in greater detail—highlighting Tyndall's use of refraction techniques and his findings on heat resistance in bacterial spores. And once again, their conclusion was that most scientists had found Tyndall's evidence and experimental method to be more persuasive than Bastian's: "This [judgment] is what scientists have reached regarding the origin of life, and we have offered a forthright presentation without any purpose on our part to claim what is right or what is wrong."[25]

Yet it was perhaps not as straightforward as they thought. Historians of science have shown that the acceptance of experimental results is a matter of persuasion that involves appealing to professional, social, and even ethical norms. The Tyndall-Bastian controversy was a prime example of this, since the experimental procedures on both sides were widely questioned. Both had their supporters, and the reasons why Huxley and the "Young Guard" of evolutionists opted for Tyndall over Bastian do not seem to have been due to the compelling nature of the evidence itself.[26] The editors of *Al-Muqtataf* were similarly motivated by their own predilections, and they opted for Tyndall because his views dovetailed with their own larger concerns. As they confessed in the same article: "If what is said agrees with our religion, we will accept it, and if it does not, we will reject it."[27]

Perhaps it is not surprising that the young instructors at the Syrian Protestant College should have taken this line, but Shumayyil was not impressed by their declaration of faith, and he continued to defend spontaneous generation.[28] He seemed confident that "there is certain proof in defense of [it]," although what he took such proof to be remained unclear.[29] It was enough to identify spontaneous generation as an a priori postulate of evolution. In fact, Shumayyil seemed happy simply to take the same position that the Swiss botanist Carl Nägeli had earlier professed: "To deny spontaneous generation is to proclaim miracles."[30] Later, in the 1910 introduction to his translation of Büchner, Shumayyil would return to the controversy long after many felt the scientific consensus was unequivocally against Bastian's claims in favor of spontaneous generation. In effect, the controversy had been "closed." Shumayyil himself acknowledged this, yet he also made the argument that it did not detract from the possibility of future laboratory evidence proving the reverse. (As if to further endorse this, he pointed out that while the controversy had not contributed to the cause it had initially set out to prove, the experiment itself had opened the way to another tremendous scientific theory—namely, germ theory—offering equally radical social effects and potentials.)

Ultimately, Shumayyil was much more interested in contesting the editors' views that faith laid a certain claim to the debate. "Between our topic of research and religion," he declared in 1878, "is a *parasang*" (a long Persian measure of distance). He could not understand what harm spontaneous generation posed to religious belief and went on to say that whether or not it agreed with scripture was irrelevant; either way it should not affect belief in God.[31] (Indeed, theistic evolutionists elsewhere were sometimes open to the idea of spontaneously generating matter: even Alfred Russel Wallace was not averse to Bastian's findings.)[32]

Yet despite this apparent acceptance of the compatibility of religious belief and the theory of spontaneous generation, Shumayyil soon moved to a direct attack on faith itself. Just a few years later, he adhered to a notion of spontaneous generation as the basis for a future progressive world order completely devoid of a divine presence or supernatural control—a natural philosophy that he drew from German materialists and that he labeled innocuously "the philosophy of growth and progress" (*falsafat al-nushu' wa-al-irtiqa'*). It was also a doctrine of evolution that, for an Arabic-reading audience of the early 1880s, took Darwin's theory into uncharted territory.

A New Philosophy

A few years after his exchange with the Sarruf and Nimr, and after encountering new channels of materialist thought in Istanbul and Lyons, Shumayyil went on to describe the materialist tendency in Europe at some length.[33] According to him, this school believed that all life emerged from natural elements. Consequently, all phenomena obeyed the laws of matter and force (*al-madda wa-al-qiwa*), as "matter in motion" (*tahawul al-madda*). The universe was not governed by intelligence, purpose, or final causes but only by natural law. Shumayyil praised Darwin in particular for his scientific approach to the problem of life and for—as he saw it—putting aside ideas of final causes.[34] He believed that Darwin had provided a plausible causal account of speciation and had demonstrated how life was created "from a few origins." What he found strange, however, was Darwin's failure to tackle the ultimate origin of life. It was not until his followers—and here he mentioned Huxley, Büchner, Haeckel, and Spencer—"addressed themselves to this puzzle" that a solution had been found with the aid of a "materialist science and philosophy" (*al-ʿilm al-maddi wa-al-falsafa al-maddiya*).[35]

Although Shumayyil was silent on this, at that time materialism was an ambiguous concept, and it covered a number of distinct orientations, from hylozoism to monism. Some authors stressed epiphenomenalism over naturalism; others, mechanism. Shumayyil drew eclectically from a host of figures—not only those already mentioned but also Bernard, Bichat, Haller, and Glisson. Like most late nineteenth-century materialists, Shumayyil would have agreed with Haeckel that "a truly natural and consistent view of organisms can assume no supernatural act of creation for even the simplest of organic forms, but only a coming into existence by spontaneous generation."[36] Yet, as we will see, Shumayyil's reading of materialism was also very selective, as well as being at odds with long-standing Arabic understandings of the term. The real issue for him was spontaneous generation, which he took to be definitive proof of the nonexistence of a Creator. To deny it was to stand against nature itself: the idea that species were specially created required the belief in the fixity of species; but in fact, species were changeable, connected, and evolved from one another according to natural laws. It was thus only the misguided who "made life separate from matter, [a principle] which they called 'vitalism.'"[37] What could be said to have entered or left matter, he complained, was simply left undefined. Those who felt life was the creation of divine injunction or mysterious "vital forces" he therefore interchangeably denounced as "creationists" (*khalqiyin*), "spiritualists" (*ruhiyin*), or "vitalists" (*hayawiyin*).[38]

Shumayyil wanted to do away with all supernatural and incomprehensible forces: "There are no vital forces except natural, chemical ones."[39] All physiological processes, and hence all of life, he explained, depend on chemistry and nothing else. Even the soul was described as merely the result of material operations, such as the functioning of nerves in the brain, which resembles the mechanical workings of the stomach or chemical florescence in plants: "The soul is a mode of force inherent in matter, as the mind is a mode of matter connected with force." Life and death he saw as nothing but changes "in matter and force."[40] Thus, all observable entities and active powers in nature he held accountable to uniform natural laws—particularly to that "law of natural economy" which stipulates that everything in nature is "of, in, and for nature."[41] In consequence, he maintained that there was no significant distinction between man and animal: "Man is physiologically like an animal," he wrote, "and chemically like a material object; the difference between these two is in quantity not quality, or in appearance not essence."[42]

Shumayyil's word for this creed was *al-maddiya* (materialism)—an abstract noun derived from *al-madda* (matter or material). But Shumayyil's particular brand of materialism had very little to say about matters of central concern to most European materialists of the time—such as the relation of mind to matter or the nature of emotion, reason, or consciousness. "Shumayyil was a thinker rather than a man of science" was how one later Arab evolutionist described him. Yet he also wanted to highlight modern materialism's scientific credentials and to distance it in particular from that of older schools of thought. He had little patience for comparisons with ancient Greek atomism—repeating over and over again that the "scientific materialism" that he referred to was based conclusively on modern physics, chemistry, and other natural sciences. His use of the term *al-maddiya* was also intended to distinguish it from the more critical *al-dahriya* (this-worldliness, hence denial of the afterlife, but sometimes used by theologians to denote materialism, a point we will return to later). But the distinction between the scientific *al-maddiya* and the impious *al-dahriya* was not apparent to many of Shumayyil's readers, and it is not surprising that his opinions aroused a strong reaction. Nor was he unaware, or even unhappy, about this. "My writings," Shumayyil wrote retrospectively, "caused a great deal of controversy. . . . And all this fuss made me want to slap people awake from their deep slumber . . . so as to reach us Easterners . . . deep in torpor . . . housed in stasis, and on the margins of life, neither dead nor alive."[43]

He came closest to achieving this with his most controversial contribution—his 1884 translation from the French of Ludwig Büchner's commentaries on Darwin. He published it first as *Taʿrib li-sharh Bukhnir ʿala madhhab Darwin* (straightforwardly, "A translation of Büchner's commentaries on Darwin") and later (in 1910), with a new introduction, under the title *Falsafat al-nushuʾ wa-al-irtiqaʾ* (more euphemistically, what we might render "the philosophy of evolution and progress"). A translation of a translation, this mode of transmitting ideas was characteristic of the way in which evolutionary thought reached Arabic readers.[44]

Like many young Ottomans at the time, Shumayyil's philosophy was deeply influenced by the weltanschauung of nineteenth-century German materialism. His comment that he did not want his "philosophy of growth and progress" to concern itself merely with "the evolution of life itself but with all of nature and society" echoed Büchner's holistic ambitions.[45] Shumayyil's opinion of Büchner was such that he considered Darwin's theory of evolution incomplete without his insights. What Darwin had not achieved, in Shumayyil's eyes, was a genuinely universal view of evolutionary progress. He followed Büchner in viewing progress as much a matter of nature as it was of society and civilization: "Whether in silent nature, budding life, dumb animal, or rational man, the unfolding of matter and force was bound by one and the same fundamental laws of nature."[46]

This led him, as it did so many materialists, into an almost mystical vision of unity (although this was not something that he ever developed at length): "The melding of matter, selection in plants, comprehension in matter, and will in humans come together under one force, call it what you will—life, heat, electricity, light, movement, gravity, sensation, or love. They are all of one essence."[47] Shumayyil labeled this *tawhid* or Divine Unity—a teasing allusion to a key concept in Muslim philosophy with diverse implications, including the oneness of God, communion between God and man, and even the principles of religion. "When the idea of universal materialism settled in my mind," Shumayyil later wrote, "the advantages of the theory of growth and progress became clear to me, for this theory was established on the universal principle of natural unity [*al-tawhid al-tabiʿi*]." Other theories he described as being built on false foundations because they distinguished between the natural and supernatural worlds—the latter, for Shumayyil, being nothing more than myth. Unity in nature, he thought, precluded such an artificial estrangement.[48]

The transcendental possibilities of such a philosophy were, however, overshadowed by its heretical implications. Büchner's own *Kraft und Stoff* was better known for its atomism and atheism, including its complete rejection of God, creation, free will, and organized religions. This seemed to be precisely what had made Büchner so attractive to at least two generations of Ottoman dissidents: it was for these reasons, for example, that Beşir Fuat, a former military officer and science popularizer in Istanbul, was himself inspired to turn to the German materialist. (So taken was he with these new ideas that he even committed suicide in 1887 to make a "scientific experiment" out of death.)[49] Shumayyil's decision to translate the German philosopher's commentaries on Darwin was a similarly calculated choice. Like Büchner, Shumayyil felt that the mysteries of life should not be exempt from the scrutiny of science by sanction or sacrament. And like Büchner, Shumayyil himself was accused of using materialism for heretical ends.

A Natural History of Religion

The bulk of Shumayyil's 1884 introduction to his translation of Büchner was devoted to what he called a "natural history of religion," one of the first expressions of such a view in Arabic. Following Victorian popularizers like John Lubbock, E. B. Tylor, and James Frazer, Shumayyil presented religion as nothing but a primitive instinct in man that had evolved from fetishism, nature worship, and other forms of animism and polytheism to monotheism.[50] Adopting a comparative approach—so typical of the new social sciences of that time—he argued that there were certain principles underlying all religions. Fear of death was preeminent: it had appeared to early man "like the beginning of a mystery with no solution, and which no obstacles can conquer, until the mind conjured strange fantasies of escaping it, like a voice that whispered, 'You will not die, follow me.'" According to Shumayyil it was through such fear and fantasy that mankind had created and fashioned religions through the ages.[51] The moral was clear: incapable of proof, religion was tantamount to irrationalism and primitivism.

Fetishism was the main expression of this. "Fetishism [*al-fatishiya*]," Shumayyil wrote, "was the beginning of worship in primitive society."[52] Primitive man chose a tree or an animal, or any other object that seemingly possessed power over him, to worship, guard, protect, or favor him in life. After man worshipped one by one all that was on earth, Shumayyil wrote, "he became disgusted with this world and eventually raised his eyes to the sky." His eyes on higher aims, he then turned to the worship of the stars and the sun, moving to visions of celestial gods and eventually to monotheism and the worship of a single Supreme Being: an absolute God; a God whose wrath could be roused by the deeds and misdemeanors of men; a God who alone could satisfy men's desires; and a God whose divine injunctions bound men in this life and after to a moral community under a divinely ordained system of justice and law. It was then that men began to submit themselves to his divine will, through proscribed rites and rituals and through laws of permissible and forbidden things. But, he complained, it was no longer acceptable to hold on to such principles, which had simply evolved and been inherited from the distant past.[53]

Darwin provided him with all the necessary proof he needed and he claimed that the theory of evolution applied as much to religions as it did to species. It thus gave him a crucial analogy: all religions "evolve from one source, they change from one form to another, and they struggle with one other." Which would be the victor? The species of religion most suitable to its environment and age.[54] Whatever happens with species, he wrote, happens with religions, too. Even the extinction of certain religions or the emergence of new ones could come about through the accumulation of incremental changes—humanity's changing needs, the merging of peoples and ideas, and new systems of values. All these, Shumayyil wrote, affected the fate of religions. Religions must therefore adapt to the changing times or die.[55]

For Shumayyil, approaching religion in terms of natural history allowed him to treat all faiths as expressions of similar phenomena. They were all alike in their emphasis on duties, rewards, and punishments, in their reliance on superstition and divinity, and even in their use of "magical numbers"—as in concepts of duality, of a Trinity, of the seven levels of heaven and hell, and of the Ten Commandments. Yet they all claimed exclusive rights to truth and denied this to others. "If all religions were correct," Shumayyil wrote, "then the truth would not be divided." In support of this view, he cited the eleventh-century poet Abu al-ʿAla al-Maʿarri (a critical source of inspiration for many Ottoman anti-deists at this time and himself sometimes regarded as heretical for his audacious emphasis on the pretensions of religion).[56]

The real point for Shumayyil was that all religions were equally unreasonable in their dependence on revelation.[57] And yet "men of religion" tired themselves out by searching for rational proofs and philosophies. He gave the example of the theologians' struggle to reconcile the existence of

God with free will and concluded that "there is no meaning in their tiresome proofs or in their rotten evidence."[58] It was to these "men of religion" in fact that Shumayyil's comments seemed largely directed.

Yet Shumayyil did not apparently see himself as a crusader against religion *tout court*. We know little about his personal faith and he wrote almost nothing about it: Yaʿqub Sarruf, in his obituary for Shumayyil, insisted that he was a believer, and indeed on at least one occasion Shumayyil expressed impatience in print with those who accused him of atheism. Rashid Rida, the founder of the influential Muslim journal *Al-Manar* (and Muhammad ʿAbduh's key disciple), defended Shumayyil from the charge that he wanted to discredit religion, claiming that he only sought to provide a scientific basis for religious observance. As for Salama Musa (with whom Shumayyil collaborated closely), he recalled how he often found Shumayyil at home with a well-thumbed copy of the Old Testament in front of him.[59] It would seem, therefore, too simple to describe him, as some commentators have done, as antireligious: the point is that, as for other European materialists, his animus against religions was chiefly directed against the theologians and their role in politics. Indeed, following pages of scornful criticism of religious dogmatism and misrule in his Büchner introduction, Shumayyil denied that he intended to denounce religion or its personal value in any way. His purpose was simply "to indicate that governments should not impose beliefs on people" and that religious authorities should not constrain freedom of expression.[60]

In short, his natural history of religion was essentially a means of political and social critique. He attributed England's advance to the liberating influence of the ideas of the Reformation. This had led, first in England and then elsewhere in Europe, to a political, industrial, and commercial revolution—the establishment of laboratories, the emergence of technologies such as steam and the telegraph, and the accumulation of wealth. England itself might decline in the future—this could happen to any power—but for now, it offered a model for emulation, and indeed its takeover of Egypt was not all bad for this reason (in this respect, at least, he was not unlike the editors of *Al-Muqtataf*). For his native Syria and for the Ottoman Empire at large, the lessons were obvious: "The condition of the nation will not improve until the power of religion is weakened."[61]

In the years that followed the Büchner commentary, he appears—as so many others also did—to have lost faith in the willingness or the ability of Sultan Abdülhamid to make the necessary reforms, above all where religion was concerned. The sultan had at precisely this time begun to project a new image of himself as the leader of the Muslim world, renewing his claim to the caliphate in particular. But by the close of the century, Shumayyil's critical views had become unmistakable. In a treatise addressed to the sultan in 1896, shortly after he had been honored by him for his medical work, he described the evolution of the polity as depending not only on education, impartial justice, and the encouragement of science but also on their indispensable prerequisite, the development of a sense of political community that transcended outmoded communal and religious affiliations: "Some people are already calling for reform, from different religious backgrounds, but are they voicing the request with one voice. . . . This is an indication that this union will envelop the entire nation [*watan*] with all its varying religions." Like other political reformers at that time, in other words, Shumayyil was arguing for the emergence of an Ottoman civic identity, united in loyalty to the sultan and transcending confessional lines.[62]

Spiritualists, Vitalists, and Creationists

Shumayyil's primary importance was as an outlier. Throughout his career, he remained the subject of vituperative attacks and ridicule from those critics he dismissed as "spiritualists" (*ruhiyin*), "creationists" (*khalqiyin*), or "vitalists" (*hayawiyin*). Even at the start, few took Shumayyil's newly minted scientific materialism seriously. His 1881 translation of an article in *Science*, "Sensibility and Its Diverse Forms," for instance, aroused a certain amount of mockery.[63] The article set out to describe theories of sensation from Aristotle to the present, a genealogy that at first might seem hardly capable of exciting controversy. But its repeated description of matter as "capable of sensation" provoked contemptuous responses from several of *Al-Muqtataf*'s readers, especially as it also claimed that life was a mere property of matter itself (*al-haya khassa min khasa'is al-madda*). One reader mockingly asked: "When I light my cigarette, does it feel pain?"[64]

Sarruf and Nimr always offered Shumayyil a forum for his writings and he often borrowed freely from their writings (fig. 3.1). (In 1910, Shumayyil thanked the editors for tolerantly publishing his ideas in their journal and for taking on, as a result, a great deal of criticism on his behalf.)[65] Nevertheless, they remained steadfastly opposed to materialism, sharing the general suspicion of it manifested by almost all intellectuals in the Arab provinces. As early as 1882, Nimr delivered a speech entitled "The Corrupt Philosophy of the Materialists" at a meeting of a Syrian philanthropic society. Perhaps reflecting an anxiety about *Al-Muqtataf*'s own reputation, he was worried that Beirut's youth were following the new philosophy and wanted to set out for them the arguments against it. He attacked the

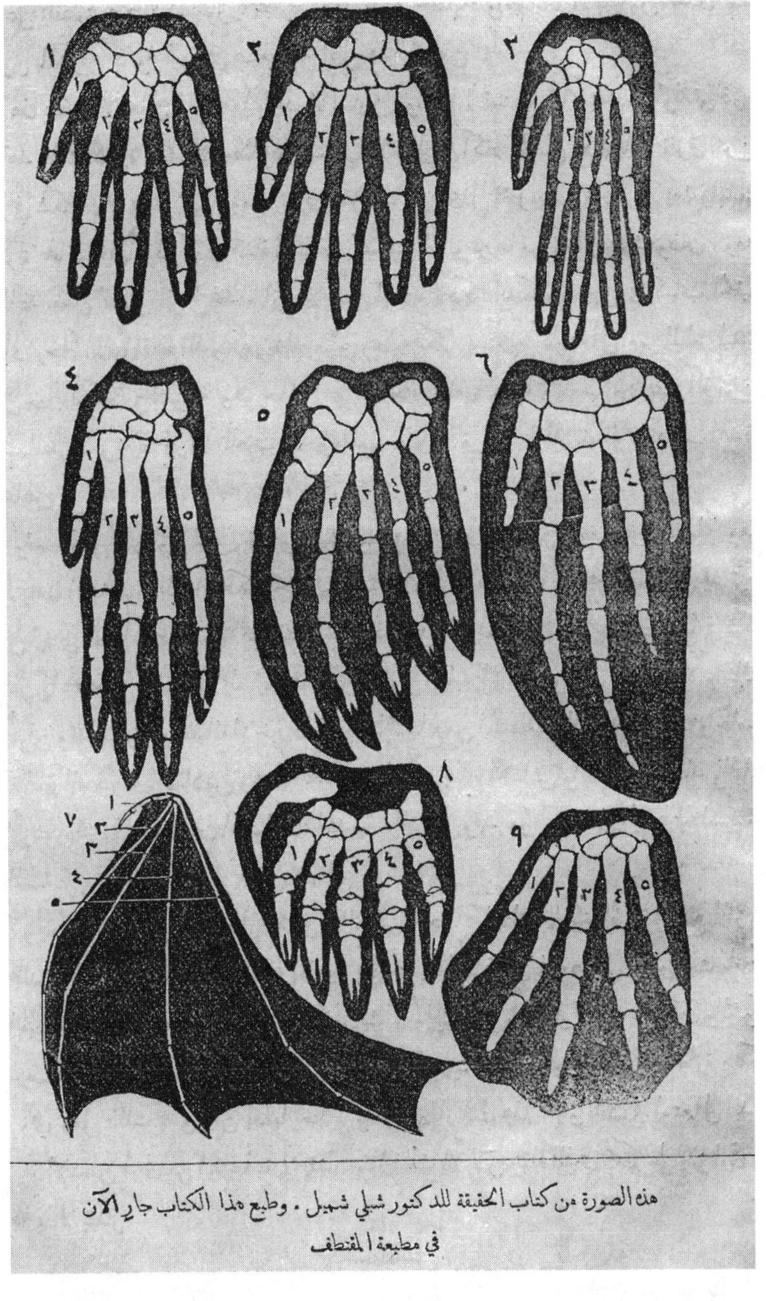

Fig. 3.1. "A Lesson on Homologous Structures in Man and Animals." First printed in "Yad al-insan wa-al-hayawan" [Forelimbs in man and animals], *Al-Muqtataf* 9 (1885): 519, and later reproduced in Shibli Shumayyil's *Falsafat al-nushu' wa-al-irtiqa'* [Philosophy of evolution and progress] (Cairo: Matbaʿat al-Muqtataf, 1910), 257.

materialists for their postulate that all of life could be reduced to physical laws, such as the conservation of energy, and cited Tyndall's opposition to the identification of mind and matter.[66]

Two years later, at about the time that Sarruf and Nimr were dismissed from the Syrian Protestant College, their review of Shumayyil's translation of Büchner appeared. They hailed the popularity of Darwin's ideas, claiming that his theory was "the most prominent scientific doctrine today" and that it had spread worldwide, but they emphasized that its hypotheses were being researched and awaited confirmation and that its implications were widely debated. As for Shumayyil's work, they praised the translator while refusing to comment on the science, but they did have one important reservation regarding Shumayyil himself. Ever the cautious empiricists, they complained that he had presented his own opinions as "scientific truths," above all in the matter of religion. They warned that "those who fear science" would take his views as confirming their suspicions and that those who "have faith in science" might take them as a definitive attack on religion.[67] As for Büchner's materialist philosophy, they characterized this as nothing but "sheer unbelief" (*kufr mahd*).[68]

They had good reason to worry, and indeed, although it cannot be definitively proven, their association with Shumayyil may have been one of the factors leading to their dismissal. For Sarruf and Nimr were certainly not the only critics of the "corruption of the materialists." With his translation of the Büchner commentaries, Shumayyil's controversial reputation spread, particularly in Beirut. In fact, only five hundred copies of his book had been printed—in Alexandria—and he later complained that "many people attacked me, even though many of them based their judgments merely on hearsay."[69] Shumayyil's early writings, however, had been avidly read for signs of blasphemy, and when his translation of Büchner came out, an array of critics were already at hand. The Jesuits, for instance, responded by publishing their own translation of an 1880 antievolutionist scholastic tract on the origins of man by a Vatican cardinal. Over the next decade and more, Catholic, Maronite, and Greek Orthodox critics all joined the fray, writing against Shumayyil's views on the origins of life and man, in particular on spontaneous generation and on the relationship between man and animal.[70]

Ironically, if there was one Christian critic to whose attacks Shumayyil owed his reputation, it was to a figure associated with and backed by his old alma mater. No doubt worried about its standing in the Arab provinces, the Syrian Protestant College was especially keen to distance itself from the heretical implications of evolutionary materialism, particularly after the Lewis affair. It therefore sponsored and widely disseminated the writings of Shumayyil's most tenacious critic, Ibrahim al-Hurani. Hurani had been an instructor of Arabic, logic, and mathematics at the college, and he was an Evangelical theologian and a senior figure in the American mission and since 1880 he had been editor of the college's *Weekly News* (*Al-Nashra al-Usbuʿiya*).[71] Hurani had actually published a defense of Lewis's speech shortly after it was given where he also praised what he claimed was Darwin's high regard for Evangelical missions around the world. But two years later, he was singing a very different tune.[72]

In September 1884, barely two months after Sarruf and Nimr's dismissal, Hurani publicly announced his regret at not having taken a tougher line against *al-madhhab al-Darwini* in the past. The reason he gave was that he had not then thought its principles threatened the Arab lands; now, however, "the germs of deception and the army of heresy are everywhere. They are prepared to attack."[73] Prompted above all in these fears by the publication of Shumayyil's Büchner commentaries, which had appeared that spring, Hurani published his own book, *A Philosophical Refutation of Evolution*.[74] When *Al-Muqtataf* reviewed this—along with Shumayyil's rejoinder—they had reservations: no doubt still bitter about their own unhappy departure from the college, they were entirely unimpressed by an argument that equated evolution with materialism and that targeted not only Shumayyil but Darwin himself.[75] Shortly after this, Shumayyil published his own response under the title *Al-haqiqa* (The truth), only for Hurani to come out with yet another book.[76]

The exchanges between the two authors were characterized by their bitter and vituperative tone. "Among my critics," Shumayyil later wrote dismissively of Hurani, "is a woodcutter, working by night and treading here and there like a person who has lost his way, uttering voluminous prattle and exceeding all limits of incitement. Be careful, O Woodcutter of a Writer; for it is often that I have listened to your speeches, thinking perhaps I would get a glimpse of light from your proofs, only to see, lo and behold, you are like a man sporting a club cut from the forests of stupidity."[77] In response, Hurani labeled Shumayyil's writings "The Treading of a Blind Camel in the Faulting of the Methodology of the Philosophers."[78]

Hurani's argument was basically that the theory of evolution lacked absolute proof. He argued, for instance, that the theory of natural selection could not provide any evidence of "intermediary forms" in nature and that there was thus a "missing link" in the supposed evolution of species, one from the other. He also objected to the period of time over which evolution was supposed to have occurred. Juxtaposing the calculations of various scientists regarding the emergence of life and the age of the earth, Hurani attempted to argue that they were claiming the impossible: namely, that the emergence of living organisms must have first occurred when the earth was a mere mass of molten matter.[79] Turning to Büchner, Hurani insisted that creation ex nihilo was thoroughly unscientific. If atoms, he wrote, were said to have come together to form all the created forms we have today, how did they do so if they themselves were uncreated?[80]

Shumayyil cast doubt on Hurani's arguments and qualifications. While Darwin, and many other scientists, had devoted themselves to careful research, he noted, Hurani's thoughts were poorly assembled and reflected no scientific knowledge.[81] In his response to Hurani's criticisms, Shumayyil corroborated the theory of the mutability of species by appealing to the evidence of vestigial organs. He defended natural selection and argued that medial forms did indeed exist, the proof of which could be found in the difficulty many naturalists faced when attempting to classify species. Huxley, he claimed, had even discovered a fossil form that connected birds with lizards.[82] On the period of time required for the evolution of life and the age of the earth, Shumayyil claimed current scientific research in this area was inconclusive, referring in particular to William Thomson and George Jackson Mivart's claims as merely tentative.[83]

Such objections failed to convince Hurani. The scientific details of Shumayyil's arguments bothered him less than their impious implications. In this he very much reflected the mood in the college in the aftermath of the Lewis affair. For Hurani, as for so many, a materialist view of evolution was synonymous with atheism. He was, however, careful to distinguish between the various camps of evolutionary thinkers, differentiating, for instance, between atheism, agnosticism (*al-la'adriya*), and deism (*al-ilahiya*). Hurani cast Büchner, Haeckel, and Fichte as atheists, while labeling Spencer, Huxley, and Tyndall agnostics. As for the final group, he distinguished those like Darwin, who insisted on man's descent from animals, from those like Wallace, who maintained belief in the special creation of man.[84] If one thing was clear for Hurani, it was that Shumayyil's own brand of evolutionary materialism meant atheism. Hurani thus concluded that the religious implications of Shumayyil's theory of evolution ran counter to Darwin himself: "Darwin claimed that the universe has a Lord. I am certain of that and consider proclaiming that fact a duty, but the writer of the imaginary *Truth* denied that and faulted those who asserted it."[85]

In this whole affair, the Syrian Protestant College had done more than merely allow Hurani the pages of their weekly newspaper: it also subsidized the publication of his articles and books. But the college's efforts completely backfired and succeeded only in making both Darwin and Shumayyil much better known than they had been before: "Even if Doctor Shumayyil had published hundreds of advertisements in local newspapers," wrote *Al-Muqtataf*, "his book would never have been promoted in the way that it has been by this opposition. Some people might think that *Al-Nashra* conspired with Doctor Shumayyil to promote his book as is done by some foreign newspapers."[86]

Refutations of Materialism

As the debate intensified, Shumayyil's defense of materialism invited response from an even wider set of critics. In fact, what perhaps proved in the long run to be the most influential attack of all came with the translation into Arabic of a recent Persian antimaterialist treatise. The Beirut connection was crucial, for the translation was published there under the title *A Treatise on the Corruption of the Materialists' School and Proof That Religion Is the Origin of Civilization and Unbelief the Source of Civilizational Decline*—later simply known as *A Refutation of the Materialists* (*Al-radd ʿala al-dahriyin*). It was translated under the supervision of the thirty-seven-year-old Muhammad ʿAbduh, who was then back in Beirut, teaching at the Madrasa al-Sultaniya, a school that had been founded precisely to confront the challenge posed by the city's new missionary establishments. ʿAbduh himself had recently arrived from Paris, where he had been working closely with the author of the *Risala*.[87] This was none other than the orator and political revolutionary, Jamal al-Din al-Afghani.

Afghani is today a much-cited figure in the literature on "Islamic modernism" and on the history of anticolonial intellectuals more broadly. Yet few of these studies have thought to connect his ideas with those of Shibli Shumayyil, or indeed with those "Christian secularists" who had long been seen as offering a parallel yet ultimately distinct modernist discourse on religion in Arabic. Bringing Shumayyil and Afghani together, however, demonstrates the many shared themes and the broader, transregional discussions they engendered. After all, the problem of materialism, with all its attendant meanings, captured the attention of Muslim readers from Calcutta and Alexandria to Beirut and Cairo. Moreover, the story of how Arabic readers in 1880s Beirut actually came to encounter Afghani's original Persian work (Afghani was in Hyderabad when he composed it) was very much connected to Shumayyil.

The title of Afghani's Persian original—*The Truth about the Naturalists and Their School* (*Haqiqat-i madhhab-i nichari va bayan hal nichariyan*)—was revealingly different from the Arabic *Refutation of the Materialists*, yet it was first translated and published in Urdu in Calcutta while Afghani was in India, and his primary target then had not been Shumayyil but the leading Indian Muslim reformer and pedagogue Sayyid Ahmad Khan. Founder of what became the Aligarh Muslim University, Khan was himself a controversial figure: he sought to reconcile natural science with Islam and taught that while the truth of the Qurʾan was undeniable, textual interpretation had always to be judged in the light of the findings of science and the laws of nature; he also deemed extra-Qurʾanic Islamic sources irrelevant. Describing Khan as a "materialist" highlights the very flexibility of the term in Arabic: indeed, Afghani's main criticism was against Khan's search for the sources of a "natural religion," and Afghani's use of the term "materialism" did not here imply Khan's subscription to hylozoism or other theories of matter and life. Nevertheless, ʿAbduh and other Muslim critics of Shumayyil clearly regarded Afghani's arguments against Khan as effective against the new brand of scientific materialism then circulating in the Ottoman lands. The Arabic version of Afghani's book would prove highly popular in particular, and it was repeatedly republished well into the following century: editions appeared in Cairo alone in 1894, 1900, 1902, 1903, and in every remaining decade of the twentieth century. Its fame overshadowed that of Shumayyil, and it easily became the principal conduit through which materialism was understood in the Arab world.[88]

Yet to understand Afghani's original text, and not just the translation, we need to consider the role of allied concepts of "nature," "naturalism," and even "materialism" in the context in which he composed it. Afghani begins by quoting a letter from a certain Muhammad Wasil, an instructor in mathematics at a madrasa in Hyderabad. "These days," Wasil wrote (the letter was dated December 1880), "the sound 'nature, nature'" (literally "*nayshir, nayshir*") reaches us from all over India—the Western and Northern states, Oudh, the Punjab, Bengal, Sind, and Hyderabad in the Deccan." Noting that "nature lovers" or "naturalists" (*nayshiriya*) seemed to be on the increase, especially among Muslims, Wasil inquired what they really stood for: were they attempting to reform civilization and were their ideas compatible with religion?[89]

The term *nayshiriya*, "naturalists," had actually been coined by Ahmad Khan's opponents to describe his followers and was a derogatory neologism. Its meaning was sufficiently vague, however, for Afghani to group such people together with atomists, evolutionists, nihilists, and materialists as "enemies of religion" and "destroyers of civilization." *Nayshir*, Afghani responded to Wasil, means "nature" (*tabiʿa*) and the "naturalist sect" (*al-tariqa al-nayshiriya*) is the same materialist sect (*al-tariqa al-dahriya*) that first appeared in ancient Greece and then spread throughout the world: "These days 'naturalism' has spread to all provinces, it is mentioned at every gathering. And all people, high and low, traffic in it."[90]

The Arabic translation, as did the Persian original, identified these naturalists preeminently as *dahriyin,* a derogatory term in classical Arabic for materialism and impiety. The Qur'an dismisses those who claim: "There is nothing save our life in this world; we die and we live, and only a period of time [*dahr,* the course of time] makes us perish." In this way, the eleventh-century Abu Hamid al-Ghazali, among others, spoke of those who "professed the *dahr*" (*al-qawl bi-al-dahr*), meaning those who believed in the eternity of the world and who denied, as a result, resurrection and an afterlife. Yet while many classical Muslim authors regarded the materialists as a school of philosophy, they did not always agree on which school was involved. For al-Ghazali, for instance, the term could be applied to Socrates, Plato, and Aristotle as well as to Ibn Sina and al-Farabi. Some referred specifically to atomists, others to Epicureanism, and in general to those who denied divine creation, reward and punishment, the existence of angels and demons, and the significance of dreams; and it could refer to those who recognized no difference between men and beasts.[91]

The modern use of *dahriya* was no less loose. Official Ottoman declarations in 1798 against the French deployed the term in reference to the French revolutionary principles of equality and liberty. "They assert," one *firman* read, "that all men are equal in humanity and alike in being men, that none has any superiority or merit over any other and that every one himself disposes of his soul and arranges his own livelihood in this life, and on the basis of these false, materialist ideas they established new principles and laws."[92] Nineteenth-century European natural science and politics similarly gave rise to new definitions. Afghani himself traced a genealogy of materialists from the ancients to Darwin—making reference along the way to the Mazdaists, Genghis Khan, the French revolutionaries, and the Ismaʿilis and Babis. Beyond this, he also included the Mormons, socialists, nihilists, and communists.[93]

What is striking is the extent to which Afghani highlighted Darwin in his attack on Khan. After a brief genealogy of materialist thought, Afghani moved straight to a consideration of "the modern materialists, especially Darwin," describing him as the leader of the current school of materialists (*madhhab al-maddiyin*). He then went into Darwin's ideas in some detail, although his interpretation of them was rather unorthodox (so much so that

Darwin's later Arabic translator would understandably wonder whether he had actually ever read him).[94]

According to Afghani, Darwin's theory depended on a kind of atomism, on the one hand, and on a vague Lamarckian evolutionism, on the other, and he used the terms "atom" and "germ" interchangeably: "They claimed that those germs transfer from one species to another and change from one form to another through the demands of time and place, according to need and moved by external forces."[95] The term *jurthum* would later become one of the main translations for "microbe," yet the concept was here used (as with its allied concept, *bizr,* as we saw at the very start of this chapter) to refer to the self-organization of matter.

Although Darwin himself was ultimately silent on the question of the origin of life in matter, Afghani folded him into a discussion of atomists or materialists. This was the extent of Afghani's engagement, however, and he showed little interest in ideas of speciation and attributed to Darwin a Büchnerian view of the origin of life, stressing "matter, force, [and] intelligence" (transliterated, as in the Arabic version of his refutation, as awkward neologisms, from the French: *matiyr, furs,* and *intallijans*!). Yet even this Afghani claimed as little more than a kind of pre-Socratic atomism, making references to Democritus and others.[96]

As all this suggests, for Afghani, that neither Darwin nor these modern materialists were terribly novel. Afghani regarded them as an extension of the ancient Greeks and hence felt that they posed the same kind of challenge to religious truth as had faced the classical Arabic philosophers who had undertaken the project of harmonizing Greek science with Muslim theology. His view of science was that it was fundamentally unchanging, and later he would make this explicit and deny that there were any "new discoveries or theories left in this world of ours." It was merely that some scientific insights had remained hidden or had been lost: "Take the libraries of Baghdad and Andalus and all that was translated in the ʿAbbasid era and all the Arabic books on philosophy, natural science, and chemistry, and then ask me if those Arab scientists of the 'school of growth and progress' [*madhhab al-nushuʾ wa-al-irtiqaʾ*] have offered us sufficient research, evidence, and proof to satisfy those subjects in the arts and sciences that they offer to us today as though they came only from the West." Afghani's discussion of materialism was thus far from the self-consciously "modern" one that Shumayyil was then proposing. For Afghani, Shumayyil's orientation toward European thought was nothing more than "imitation [*taqlid*] of the West." This was all the more profound an error in that the West itself had flourished only when its beliefs were consistent with Muslim precepts: drawing on François Guizot to stress the importance of the Reformation for the success of European civilization, Afghani claimed that Luther, for instance, was merely "following the example of the Muslims."[97]

Shumayyil's view of Afghani's scientific understanding was, not surprisingly, no more complimentary. He was familiar with Afghani from the latter's stay in Egypt, when Afghani had impressed him with his classical Arabic, his intelligent gaze, and his extraordinary eloquence: he used to hold forth in Cairo at a café near the old stock exchange. One evening the doctor and the peripatetic revolutionary had enjoyed a long evening's discussion, and even in 1912 Shumayyil remembered their meeting: "What does the Sayyid Shaykh say," he had asked him, "to the question: 'What was the first deity worshiped by man?'" Although Afghani's views were obviously in disagreement with Shumayyil's views on fetishism, the latter recalled their conversation with warmth and even praised Afghani as "one of the most famous men of his time . . . novel in spirit . . . broad in intellect . . . and one of the most engaging philosophers of the salon [*falasifat al-riwaq*]."[98]

Nevertheless, Shumayyil made it quite clear that he considered Afghani, as he put it, "a philosopher of old," quite untrained in "the new sciences" and essentially unqualified to pronounce on modern materialism: "When news reached me that Afghani was in exile from Egypt and that he wrote a treatise in Afghan [*sic*], which Muhammad ʿAbduh translated into Arabic, to deny the principles of the materialist school, I was quite surprised." Lacking the proper training, thought Shumayyil, men like Afghani could raise objections but could not provide adequate scientific proofs. For Shumayyil, Afghani was the real imitator, merely following conventional opinion.[99] In short, Afghani's conception of "science" (*ʿilm*) was at odds with the one promoted by men like Shumayyil. For Afghani, science (*ʿilm*) was a corpus of knowledge that made reference to a tradition of texts, procedures, arguments, and practices that aimed to incorporate rather than repudiate the past. For Shumayyil, by contrast, it was a system of investigation, experimentation, and discovery that implied a de facto break with, and hence repudiation of, the past.

Like "science," their views on "religion" were also at crosscurrents. For Shumayyil, it was the ultimate cause of social stagnation and retrogression; for Afghani, following Ibn Khaldun and François Guizot, it was the primary motor for civilizational progress itself.[100] Shumayyil, as we saw, had been among those early Arab evolutionists who turned to modern disciplines to highlight the evolution of religion. For him, the very category of "religion" was an abstract, innate tendency or disposition in humankind.[101]

Once again, one could find echoes of this concept in classical Arabic texts, and it is telling that one of the words he used for "instinct" in Arabic (*fitra*) reflected this. Yet Shumayyil added to that the modernist fascination with "world religions" and their place in civilizational hierarchies.[102] Afghani's own views, like those of his Arabic translator ʿAbduh, were similarly drawn from an eclectic array of sources and references, classical and modern. Both Afghani and ʿAbduh viewed "religion" or "Islam" as rooted in the value of interior experience—a key Sufi or mystical theme. Yet they also drew on new discourses of "society" and "world civilizations" (as opposed to the Khaldunian notion of "civilization").[103]

In yet other respects the distance between the two men was perhaps not as great as one might expect. For both of them, the invocation of Darwin constituted an aspect of their broader political vision. As we saw, Shumayyil's interpretation of Darwin was tied to his hopes for the reform of the Ottoman Empire, and he was part of that generation of Ottoman intellectuals who were using materialism to try to persuade citizens of the empire of the need for institutional change. Meanwhile, the translation into Arabic of Afghani's *Refutation* coincided with the author's (and translator's) efforts to appeal to Abdülhamid and others to assert their leadership over the Muslim world. In this context, Darwin's name was a way of attacking the alternative represented by Ahmad Khan, with his policy of collaborating with the British in India.

Indeed, for Afghani, there was only one key difference between materialists like Khan and his equivalents in Europe: the latter, despite losing their faith, retained a love of their country, whereas the former "disparaged their fatherland and made people consider foreign domination over them a slight thing."[104] Yet with time, Afghani's hopes in the sultan dimmed (as indeed did Shumayyil's), while his disciple ʿAbduh even returned to Egypt to pursue a policy of accommodation with the British and a theology that may not have been so different from Khan's.

The gist of Afghani's criticisms of Darwin changed subtly but tellingly over time. In 1884, drawing a different genealogy from *Refutation*, he depicted Darwin's transformationism as the latest incarnation of a theory that had first been developed by medieval Arab philosophers. A verse by ʿAbu al-ʿAla al-Maʿarri gave Afghani the opportunity to expand upon this idea: "al-Maʿarri's intentions were clear. . . . He meant evolution, taking this concept from Arab scientists before him . . . such as Abu Bakr ibn Bashrun . . . who claimed that minerals transform into plants, plants into animals, and that the last of these three transformations and the highest link in the chain is man." With respect to the theory of evolution, Afghani claimed, "such Arab *ʿulama* preceded Darwin."[105]

Where the implications of transformationism were concerned, Afghani still stood by what he had written in his *Refutation*. In the next paragraph, however, he came down, surprisingly, in defense of Darwin as a theist: "When [he] came to the crucial problem of who created the 'breath of life,' he could only stop and say that it was the Creator who gave life to the living, writing as follows: 'I see that all the living creatures that lived on this earth are from one primitive form, which the Creator gave life to.'" Afghani took this to be evidence that men like Shumayyil—the "materialist natural scientists" (*ʿulama al-tabiʿa al-maddiyin*)—pushed Darwin further than they should have done in order to "deny the existence of God and relate things solely to nature." Materialists claimed, said Afghani, that if Darwin had not done this, it was merely for fear of Christian opinion, and had had the consequence of leaving his theory incomplete.[106]

Shumayyil himself, as the most prominent "Eastern sage" to espouse materialism and use it "against Eastern customs," was very much on Afghani's mind. He returned Shumayyil's admiration, praising his "literary courage" and his "deep knowledge of philosophy," his "assertiveness" and his lack of concern for criticism from those unfamiliar with science. But Afghani accused him directly of pushing Darwin's ideas much too far: "Naturalists [*madhhab al-tabiʿiyin*] . . . trace all living things from a few groups, from which larger numbers branch off; none of this is harmful to believe in. But to claim that life appeared solely as a result of a natural power or cause yields them no benefits . . . [especially] as they have no proof for spontaneous generation. . . . And this is what I would criticize Shibli Shumayyil for, for he went against his own *imam*, Darwin."[107]

A Natural Rearing, a Natural Religion

Afghani—via ʿAbduh—was certainly not the first to criticize the new school of materialism in Arabic. One recent scholar has counted some two dozen critics of Shumayyil.[108] These included not only many local Christians already discussed, such as Ibrahim al-Hurani, Bishara Zilzal, and Louis Cheikho, but also a number of Muslim notables.[109]

But Shumayyil also had supporters. In 1910 he republished his translation of Büchner at the behest of several admirers, whom Shumayyil lists at the end (along with the amounts of their contributions).[110] And there was a new introduction, in which Shumayyil, ever the provocateur, returned to the theme of religion and the legacy of the empire. Writing two years after the Young Turk Revolution and the deposition of Abdülhamid, Shumayyil claimed that the "vague, spiritual power" once assigned to "supernatural forces" or to "God" had now been "accounted for" by "natural, material forces." Religion had robbed man of his natural instincts and inclinations and reared him on "imaginary stories and fantasies," while a "corrupt social system" had further led to a population of men "manufactured against nature." The cultivation of reason and "independent judgment" (*ijtihad*) was the only way to overcome these "acquired instincts" from this corruption.[111]

In hailing the dramatic events taking place in Istanbul, Shumayyil found himself calling for a "natural religion" to help revitalize the empire. He called this the true "religion of humanity" (*al-din al-bashariya*). Natural—not theological—sciences would show the way forward to national revival and international harmony, allowing all men to view one another as brothers in humanity. (This view would also pull him toward socialism.)[112]

For Shumayyil, citing *Al-Muqtataf*'s article on Spencer, only the cultivation of a healthy "social body" could lead to true moral principles. And only the laws established by social science, he added, could lead to the correct governance or administration of the social body. This, he predicted, would not only result in a progressive nation but also in a new international order. Natural science, as one could already see from recent "technological revolutions," will lead ultimately to an international "peaceful contagion."[113]

For Shumayyil, internationalism of this kind suited his regional concerns. Always a critic of "Ottoman tyranny" and its "theocratic laws," he could now write in anticipation of its demise and the rise of new regional orders. As the empire's breakup accelerated, the political future of the Arab provinces and of their relationship with the new regime in Istanbul was thrown into question with the Italian invasion of Libya and then the Balkan Wars. Hoping to play a part in an Arab revival, Shumayyil advocated a series of political and pedagogical reforms.

Much of Shumayyil's writings in this genre concerned the twin pillars of progress and pedagogy. Two years after the establishment of a university in Cairo, he called for the abolition of university education in the humanities, theology, and law and their replacement with natural science subjects alone. He also suggested that schools in every city and town provide a compulsory education similarly steeped in the sciences so as to improve pupils' material and moral well-being. This pedagogy would be supplemented by literature that would "instruct people how to be clean in dress, daily life, and spirit." Such a system, he prophesied, would liberate the masses from the shackles of superstition and "ensure for the future what we have not achieved in the past."[114]

But disillusioned by the Young Turks' swift repression of the press and by growing signs of their pan-Turkism, Shumayyil became a central figure in the emergence of a movement for Arab autonomy within the empire: the Ottoman Party of Administrative Decentralization (indeed, its founding meeting took place in his house). This group called for a constitutional parliament that would reflect the diversity of the empire and for Arabic to be recognized as an official language. He was particularly dismayed by the empire's inability to move beyond religious allegiances. In a speech before party members, Shumayyil underlined his familiar view: Christian Europe had advanced only once power was taken "from the hands of the propagators of religion. . . . We, however, carry the broad and lengthy flag of religion and place it squarely in thoroughfares and even crowd out the pedestrians with it."[115]

The coming of the war in 1914 and the subsequent repression of Arab nationalists by Turkish officials increased his concern, and from Egypt he organized relief efforts for his native Syria. He sought assistance from the large Syrian diaspora in the Americas and, with less success, from the US authorities. Ottoman misrule only further confirmed his view that the British occupation of Egypt had in fact been beneficial for the country, introducing press freedoms and state reforms that had been badly needed. During the war, he was accused, together with other members of the party, of supporting British plans to separate the Arab provinces from Ottoman rule. There appears to have been no evidence for this, but it was true that his evolutionary perspective led him in this direction: in his eyes, the Ottomans were by now discredited as potential leaders of an Arab revival, and he doubted that the Arabs themselves were as yet capable of creating the necessary political institutions. European tutelage—to which he also added the model of European socialism, as we will see later—thus offered itself as the means to progress and civilization. When he died in 1917, his political views had become as controversial as his scientific ones.[116]

In Shumayyil's Shadow

Shumayyil's brand of radical evolutionary materialism never caught on in the Arab lands—in contrast to Western Europe or even Ottoman Turkey. Whereas in the latter case it was one of the chief modes of expression of reformers anxious to remodel the empire and its institutions along entirely new lines, in the Arab provinces it never gained the favor of the public or the intellectuals, thanks in part to its association with impiety and perhaps also to the elite's cautious attitude to radical political change. Nevertheless, the controversial Shumayyil cast a long shadow upon later readers of Darwin in the Arab world. Ten years after his death, Ismaʿil Mazhar would recount how reading his work introduced him to evolution and plunged him into an existential crisis. It was largely thanks to Shumayyil that the concept of "materialism" itself acquired a fundamentally new evolutionary connotation in Arabic. Indeed, Shumayyil's materialism colored how many people understood evolution itself—a development that Mazhar devoted much of the rest of his career attempting to correct.

In fact, although Shumayyil's brand of materialist thought was popularly equated with unbelief, as we have seen, he remained respectful of religious faith. Seeing organized religion as a social ill, he did not deny the genius of the founders of the great faiths or the power of their texts. When Lord Cromer notoriously attributed the ills of the Arab world to Islam in his 1908 work on Egypt, Shumayyil leapt to the defense of the faith. He argued for the superior rationalism and civilizational contributions of Islam to the detriment of Christianity (enraging above all members of his own church establishment.)

The many complexities of Shumayyil's position were increasingly overlooked, however. With the coming of the First World War, the antagonism toward materialism deepened as many regarded its German credentials as suspect. And while France had given the Arab world its access to Büchner, it now also provided the main philosophical challenge to him. In April 1916 *Al-Muqtataf* published a summary of the speech Henri Bergson had delivered in the first winter of the war in which he had attributed German aggression to materialism itself. Sarruf had already made a similar argument to Shumayyil. The war was the outcome of materialism, Sarruf argued, and he advised Shumayyil to read Bergson's speech. How could Shumayyil now deny the importance of a divine force in securing cosmic order and overcoming the pursuit of individual self-interest? "What would prevent a man from killing anybody who hinders his interests," Sarruf wrote to Shumayyil, "exactly as he kills lions, wolves, and flies? Why should not a man from Paris or Berlin then kill blacks who prevent him from hunting in Africa? Is not the deterrent that keeps strong people from having a free hand with weak people a moral and not a materialist one?"[117]

Shumayyil was not persuaded. Nevertheless, he was equally dismayed by German behavior, and when Ernest Haeckel signed the 1914 manifesto of German intellectuals to defend his country against accusations of barbarism and cruelty toward civilians, Shumayyil wrote to him to protest. Unlike many others, however, he did not lose faith in materialism and certainly did not attribute the war's outbreak to it. On the contrary, he defended it in *Al-Muqtataf* shortly before his death in 1917. There he attributed the international conflict rather to German self-interest.[118] The problem was individualism not materialism. Only the abuse of power by "corrupt individuals," like the kaiser and his entourage, could have led to the war. Materialist thought, he wrote, was essentially natural philosophy, a pacific creed that taught men throughout the world to strive for the collective welfare, as in the relationship of a "living organism and its interrelated parts." "People would change their minds about materialism," Shumayyil wrote, "if they knew it was a philosophy of collective interest."[119] After his death, even some of Shumayyil's closest associates, men like Salama Musa, turned away from Shumayyil's brand of materialism toward a somewhat more positivist view of evolution. But Musa and others also embraced Shumayyil's turn to collectivism and made this the basis of the interwar reading of Darwin. In the end, Shumayyil's attempt to win support for evolutionary materialism may have been a failure, but his evolutionary socialism was not.

FOUR

Theologies of Nature

> I would have you know that a pebble proves the existence of God just as much as a mountain, and the human body is evidence as strong as the universe that contains our world.
>
> —al-Jahiz, *Book of Animals* (ninth century)

> If we saw a watch among watches that tells the time and all its parts were known to include a most wonderful construction, built upon principles of engineering, systematic measurements, and mechanical laws, exact in detail and organization, so we must conclude that it has a manufacturer that manufactured it.
>
> —al-Jisr, *Al-risala al-Hamidiya* (1887)

Historians of science have argued for some time now against the idea that the rise of evolutionary thought spelled the end of the tradition of natural theology in Western Europe.[1] The same held in the Ottoman lands. In fact, if anything, the emergence of radical materialism helped to initiate a theological resurgence and to transform earlier providentialist arguments about evidence of the existence and will of God in nature into a new genre. In the process, traditional approaches to hermeneutics and logic were brought to bear in novel ways upon contemporary scientific debates.

Of course, long before Darwin, the appeal to nature had also been used to bolster arguments in favor of religious faith; thus, Darwin's writings themselves were often read in the light of older debates: materialism, for instance, in all its varieties, emerged much earlier than the nineteenth century, and the threat posed by a purely materialist view of nature was one of the classical themes in Arabic theological treatises. As we saw, this formed a critical backdrop to contemporary readings of evolutionary materialism itself. Through accelerated cross-confessional exchanges initiated by the press, however, many Muslim thinkers also became familiar with the modern Christian tradition of natural theology, as is shown by the extent to which William Paley's watchmaker analogy made its way into Arabic, for example. Assisted by the spread of print and the rise of new readerships, natural theological references moved rapidly across confessional, conceptual, and political lines.

This created new intellectual and doctrinal interactions and cross-fertilizations. The men at the Syrian Protestant College who brought Darwin to Arabic readers did so in an idiom that combined Protestant natural theology with classical Arabic concerns. (Indeed, it was in this context that interest in classical Muslim works on natural history, such al-Jahiz's *Book of Animals*, or the legacies of Aristotelian natural philosophy and logic was rekindled.) The shock of Shumayyil's materialist reading of evolution even led some anxious college instructors to mount an antimaterialist counterattack that drew similarly on an eclectic array of inspirations, from Anglican natural theology to Catholic neo-Thomism. Outside the college, this counterattack was articulated in an influential 1888 treatise by a now relatively neglected Sufi shaykh, Husayn al-Jisr.

Al-Jisr himself was not only a close follower of these debates during his time teaching in Beirut but also a prime intellectual exponent of Ottoman theological modernism. His 1888 treatise looked forward and back at the same time: it defended Islam by confronting and refuting contemporary materialist arguments, but it also presented these views as manifestations of a much older tradition of materialist doubt stretching back to Islam's earliest detractors. It offered a staunch defense of Hamidian policies and responded directly to Orientalist and colonial criticisms of the empire and of Islam. Unsurprisingly, the work won favor with Sultan Abdülhamid; it was also widely read and translated into Turkish and Urdu. But many among the next generation of Arab thinkers would criticize al-Jisr for putting his faith in the Ottoman sultan and for failing to opt for Arab solidarity. From ʿAbduh onward, they would deploy the "new theology" in the service of Arab nationalism instead. The new theological engagement with modern science, however, endured and flourished.

An Ottomanist in Arab Lands

Husayn al-Jisr was born in 1845 to a family of scholars who were originally from Damietta, Egypt, and settled north of Beirut in Tripoli (Tarablus al-Sham) in the late eighteenth century.[2] His father, Muhammad al-Jisr, was trained at Al-Azhar in Cairo and later became the head shaykh of the Khalwatiya Sufi order in Tripoli.[3] Muhammad al-Jisr was man of local renown; he was known for his ability to perform miracles in particular: he was said, for instance, to have predicted those who would be hung after the uprising against the Egyptian occupation of Syria in 1838.[4] The younger al-Jisr wrote a biography of his father in 1888 that described his religious life and miraculous acts and included a history of the Khalwatiya movement in Syria.[5]

Orphaned at an early age, however, Husayn al-Jisr was actually raised by his pious uncle. The young al-Jisr first studied with local Muslim scholars before traveling to Egypt in 1863 for further studies at Al-Azhar (like his father before him and like so many other Ottoman *ʿulama* at that time).[6] He remained there for about four years, under the tutelage, in particular, of the well-known religious scholar and lexicographer Husayn al-Marsafi, one of the first prominent clerics at Al-Azhar to promote "Western methods" in his teaching and among the first to use the term "al-Nahda" in reference to the Arab literary revival. Like al-Jisr, al-Marsafi was dedicated to the renewal of Islam in the modern age, and he later published *Risalat al-kalim al-thaman* (Discourse on eight words), which provided a lexicon for a modern political nationalism compatible with the spirit of Islam in the wake of the ʿUrabi revolt in 1881. His eight words were *umma* (nation), *watan* (homeland), *hukuma* (government), *ʿadl* (justice), *zulm* (oppression), *siyasa* (politics), *hurriya* (freedom), and *tarbiya* (upbringing or education).[7] Al-Jisr emulated his teacher in making this last concept—*tarbiya*, which we can here loosely translate as "education"—the one he was committed to above all: "Once education is made perfect, everything else is also made perfect," al-Marsafi had argued.[8]

After his uncle became ill in 1867, al-Jisr returned to Tripoli, emerging as the leader of the Khalwatiya order there in succession to his father. This order had spread to the Ottoman province around the seventeenth century and had since developed especially strong linkages between Syria and Egypt. Intellectually, it was also distinctive, thanks to a number of incoming Persian and Kurdish scholars who brought with them an interest in grammar and semantics, logic and rhetoric. (Many also leaned toward the mystical writings of Ibn ʿArabi.) This contributed to the revival of *fiqh* (jurisprudence) and *kalam* (scholastic theology) that al-Jisr would dedicate so much of his life to and also helped revive interest in Aristotelian logic and an emphasis on a rationalist theology.[9]

Al-Jisr taught for a few years at a local Muslim school, concentrating in particular on mathematics, chemistry, and the natural sciences and drawing his material from early periodicals and books in Arabic on these subjects.[10] Aware of the growing importance of these new sciences, al-Jisr worried that traditional Muslim pedagogy was no longer able to fulfill the needs of its students. With the encouragement of the prominent Ottoman governor and administrative reformer Midhat Pasha—who was also wary of the growing popularity of the new missionary schools[11]—al-Jisr, along with a progressive landowner, al-Hajj al-Danawi, established a new school in Tripoli for foreign languages and sciences in 1879.[12] Named Madrasa al-Wataniya (National School), its pioneering curriculum included Arabic, French, and Ottoman Turkish, the traditional religious sciences (law, logic, rhetoric, and philology), as well as geometry, geography, and the natural sciences. The contemporary arts and sciences of Europe were studied in Arabic using translations that were often made from primers and textbooks produced by the American Protestant missionaries.[13] Rationalist works on logic were emphasized: under al-Jisr's directorship, for instance, al-Taftazani's works on *ʿilm al-mantiq* (logic) were taught, sometimes even in preference to more classical texts, such as Abu Hamid al-Ghazali's *Ihya ʿulum al-din*.[14] But as Muhammad Rashid Rida, a former student of al-Jisr's later described, al-Ghazali was read too, as indeed was *Al-Muqtataf*.[15]

Although the school appears to have provided Muslims with a successful alternative to the Christian missionary schools, attracting around one hundred students in its first year alone, some local *ʿulama* complained about its unusual curriculum. They were successful in convincing Ottoman officials that since the school did not offer a proper religious education, its students should not be exempt from military service, as was common for other madrasa students of the time. When military service was made mandatory for its students, many fled and it closed down after only three years.[16]

In 1882 al-Jisr received an invitation to Beirut from Shaykh ʿAbd al-Qadir al-Qabbani of the Jamiʿat al-Maqasid al-Khayriya (Charitable Aims Society), an organization founded five years earlier with the patronage of Midhat Pasha.[17] Al-Qabbani himself was a reformer of considerable significance, and he had earlier helped to establish the Jamiʿat al-Funun (Society of Arts), with the double aim of disseminating "useful knowledge" and helping the poor. Concerned about the growing influence of some of the mission journals in Beirut, this society began a weekly paper, *Thamrat al-Funun* (The fruits of knowledge), in 1875; edited by al-Qabbani himself, it lasted until 1908. Later al-Qabbani became the head of the municipality and its director of education (*mudir al-maʿarif*).[18]

Al-Qabbani was inviting al-Jisr to become head of a proposed new school.[19] Like the earlier Wataniya school, this was largely a response to the activities of Christian missionaries in Syria, and to the popularity of their establishments in Beirut in particular.[20] The founders of the new school lamented the missionaries' impact: "It is obvious to all of us in this country that for quite a long time various religious sects have formed voluntary associations to care for important matters." Among these "important matters" were listed first and foremost "opening schools for boys and girls, to teach them sciences, knowledge, and languages," but they were also responsible for establishing printing presses, publishing newspapers, and providing "free hospital services for the poor," and, in general, encouraging the spread of "knowledge of how to gain wealth, comfort, and the luxuries desired by man."[21]

The new school was to emphasize a science-oriented curriculum—the key to the "path to progress and prosperity"—in accordance with the teachings of Islam. "True," one of the founders of the Jamiʿat al-Maqasid al-Khayriya stated in his speech to the society in 1879, "we have in our country and especially in this city [Beirut] schools which teach the higher sciences—but is their teaching in accordance with the spirit of our people and government such that it will benefit them both? The answer is emphatically, No!"[22] Many Syrian *ʿulama* shared the view that the missionaries' schools went against the "spirit of the people." Yusuf al-Nabhani, a court official in Beirut, for instance, launched scathing attacks on both the Protestant and the Jesuit schools.[23] Muhammad ʿAbduh later echoed these anxieties when in Beirut—and this despite a number of prominent differences between the two *ʿulama*, as we will see in the next chapter. In a memorandum on the need for educational reform that ʿAbduh sent to the Seyhülislam (the highest religious official) in Istanbul in 1887, he wrote: "We can scarcely find a place without a school of the Americans, Jesuits, Lazarists, Frères or other European religious societies. Muslims do not shrink from sending their children to these schools. They desire that their children be instructed in the disciplines considered necessary for their livelihood and in European languages, which are seen as a prerequisite for future happiness." ʿAbduh, like so many other Muslim educators of the time, regarded these developments as dangerous. "Many of those who have attended [such] schools are no longer religious and are not acquainted with the principle of their faith."[24]

Throughout the Ottoman Empire, a number of *ʿulama* thus sought to combine Muslim pedagogy with new methods and subjects of instruction.[25] With the backing of Midhat Pasha, who arrived as governor of Syria in 1878, a renewed effort was made to reform school curricula. In Damascus, for instance, Tahir al-Jazaʾiri, a prominent member of the Maqasid charitable society and the first to found a public library in Syria, stressed the need for Muslim schools to include instruction in the modern sciences. He designed new curricula and textbooks and composed school primers in the early 1880s on grammar and religious doctrine as well as on arithmetic and astronomy.[26]

However, the sultan himself was deeply ambivalent about such initiatives, especially those developed by his bête noire Midhat Pasha and emanating from Beirut, a city that he regarded as a spearhead for European penetration of the Arab lands. Indeed, shortly after its creation, the Maqasid society was dissolved as a result of the sultan's suspicions that it was promoting anti-Hamidian sentiments: Midhat Pasha's patronage had been essential, but after he fell afoul of the sultan, this connection proved damaging, and following his enforced departure from Syria in 1880, the Maqasid society was similarly regarded as a political threat.[27] Founders of the society had also faced criticisms from local ʿ*ulama* who had opposed these reforms. People like Yusuf al-Nabhani argued in fact that these reformers were no better than the foreign missionaries, imitating their educational programs and threatening to corrupt their students with such "Christian" subjects as the new natural sciences.[28]

Despite all this, the school that the reformers had planned opened in 1883 as the Madrasa al-Sultaniya (Sultan's School). This was just at the time that the Syrian Protestant College was struggling with the aftermath of the "Lewis affair," and given that al-Jisr in particular had openly supported Sarruf and Nimr, it seems likely that he and his backers saw this as an opportunity to establish themselves. Formally, the new madrasa was an imperial lycée of the kind that had been envisioned in the 1869 Law on Public Education, but in fact it was little different in ethos from the one that the Maqasid society had wished for. A boarding school, it attracted students from all the Arab provinces; indeed, it was one of only three such schools in the entire empire in this period. Critics like al-Nabhani were scandalized by the fact that it was open to Christians as well as Muslims—the former attended Sunday church services presided over by a priest. But others may have thought the school did not go far enough: ʿAbduh, who also taught there and had attempted to institute his own innovations at the school, later criticized its excessive discipline and its continued reliance on "rote learning."[29]

Unfortunately, evidence for al-Jisr's pedagogic approach is sparse, but it is clear that when, after about a year, he left the school and returned to Tripoli, his Beirut experience had left its mark. From Beirut—a much livelier and less parochial intellectual center than Tripoli—he gained a new familiarity with contemporary scientific debates, and perhaps with the political sensitivities of pedagogy too.[30] After his departure from the highly politicized environment of Beirut, it was surely not coincidental that al-Jisr demonstrated his loyalty to the sultan in a most public fashion, entitling his work *Al-risala al-Hamidiya fi haqiqat al-diyana al-Islamiya wa haqiqat al-shariʿa al-Muhammadiya* (A Hamidian treatise on the truth of Islam and the shariʿa) after Abdülhamid and dedicating it to him. His son claims that this was necessary given the Ottoman censors' deep suspicion at this time of Arab separatism and reformism.[31] But this is probably a retrospective downplaying of the sincerity of al-Jisr's Ottomanism since he had more than a decade earlier sent the sultan an ode he had composed celebrating his ascension to the throne.[32]

Published in 1888, the *Risala* proved quickly popular and some twenty thousand copies were printed in Istanbul alone.[33] It received widespread acclaim, and Afghani, the veteran revolutionary and antimaterialist, dubbed the author the new "Ashʿari of his time" for his reconciliation of revelation and philosophy. The work was reprinted several times in Arabic and translated into such languages as Ottoman Turkish, Urdu, Tatar, and Chinese.[34]

Sultan Abdülhamid himself was among its admirers and awarded the author a prize in 1891, for which al-Jisr received an annual income of fifty liras. That the book should have received such an official accolade was unsurprising, for the *Risala* was very much in line with ʿAbdülhamid's policies at the time, in its antimaterialism and in its staunch reaffirmation of Hanafi doctrine. Rebutting current Orientalist critiques of Islam, the text explains why jihad is allowed, but why Islam is not a "religion of the sword"; why polygamy and slavery are permitted under certain circumstances, and yet why "prostitution with political foreigners" is not. And defending a highly centralized and elite conception of theological legitimation from reformist critics among the *ʿulama* (on whom, see the next chapter), the work underlined the dangers of *ijtihad* (or independent juridical and theological opinions) and argued against the view that sound Muslim theology and practice should be rooted only in the Qurʾan and not in the *sunna*.

Al-Jisr would remain a loyalist, and he continued to emphasize the compatibility of religion and science, always presenting this as in line with imperial policy. Two years after his return to Syria from Istanbul, he praised the sultan's reform project and called upon his countrymen to promote modern science, industry, and education in order to "catch up with our neighbors."[35] Western learning, he repeatedly wrote, *could* be taught alongside Islam; the only real danger was when it was taught independently of Muslim pedagogy, which was the main check against the temptations of materialism: "Whosoever observes the condition of people in the West will see the extent to which the school of materialists has been popularized and the belief of the masses in the existence of the soul and of punishment in the afterlife has diminished. They feel free to commit all and any sins."[36]

A Hamidian Treatise

Let us examine in more detail the work that made al-Jisr's reputation. Coming some three years after Shumayyil's translation of Büchner, al-Jisr's *Risala* was intended as the definitive refutation of the new evolutionary materialism that had caused such a sensation in Beirut.[37] It presents itself as an exercise in Muslim theological and juridical thought for the modern Hamidian state. The treatise begins with a traditional account of the spread of Islam and the nature of prophetic revelation and is ultimately organized around a demonstration of the wisdom of the shariʿa on both rational and practical grounds. It then enumerates eight groups—or "sects"—of people on the basis of their reactions to the shariʿa. There were, for instance, those contemporaries who had demanded proofs from Muhammad of his prophetic stature, with the result—according to al-Jisr in a claim that might seem to be at odds with the rationalism of his treatise—that the Prophet sometimes even performed miracles to convince skeptics, such as the "splitting of the moon" (in reference to Qurʾanic verses that have been variously interpreted in this light).[38] There were also those people who used their own reason and intelligence to work out the divine truth of the Qur'an.[39] Then there were those who found the new revelation in line with the spirit of the beliefs of previous "peoples of the book" (*ahl al-kitab*), finding similarities between the Qur'an, the Torah, the Gospels (*injil*) and the psalms (*mazamir*).[40] And, of course, there were those who immediately recognized the divine inspiration behind the systematic ethical code provided by what al-Jisr referred to as the "Muhammadan Law" (al-Shariʿa al-Muhamidiya), hence emphasizing both the Qurʾan and *hadith* as the mainstays of Muslim law from a Sunni perspective.[41]

Al-Jisr also describes the Muslim code of ethics in considerable detail, including the uses and practices of prayer, *khuttab* (sermons), charity, fasting, the pilgrimage, *hijab* (veiling), and laws of inheritance and punishment, within a discussion of the general purpose (or *maqasid*) of the shariʿa.[42] He explains the conditions under which polygamy was permissible and rebuts critics who claimed Muslims had historically treated slaves badly, citing passages in both the Old and New Testaments justifying slavery and arguing that by contrast Islam had improved the condition of slaves. He also answered British charges that Islam had remained fossilized by explaining that although the "gate of *ijtihad*" (reasoning against conventional interpretation) had been closed to protect the faith, in fact there were circumstances in which independent reasoning was permissible. Yet once again, in keeping with Hamidian policies, he was careful to recommend the close safeguarding of traditional interpretions unless a reinterpretation was felt to be absolutely necessary. In that case it should only be carried out by a fully competent and officially sanctioned *ʿalim*.[43] One had to be careful of unschooled "scholars," whom he labeled *ʿulama al-suq* (scholars of the souk).[44]

Al-Jisr's notion of the shariʿa was oriented toward the usual Sunni sources: the Qur'an, *hadith*, *ijmaʿ* (consensual law), and *qiyas* (analogic cases, extrapolated from the Qur'an and *hadith*). But like many Muslim reformers of his time, he added a special emphasis on the *maqasid*, or practical and ethical aims, of the shariʿa. Like other like-minded *ʿulama* in this regard—such as his own pupil Rashid Rida—he followed al-Ghazali and the Hanbalite Ibn Taymiyya and emphasized that the purpose or aim of the shariʿa was to safeguard a communal *maslaha*, or "interest"—both public and personal.[45] Hence, when describing the ritual acts of purification and prayer, for example, al-Jisr emphasizes the efficacy of these actions both in terms of religious devotion and duty and in terms of their spiritual, mental, and bodily benefits.[46] His descriptions of other shariʿa norms—including marriage, *hijab*, and the laws of punishment for adultery, drunkenness, and so forth—are couched in similar language.[47]

He ends this recapitulation by appealing to the eternal truthfulness of the shariʿa and by citing the perennial argument that it could not possibly have been the invention of an illiterate (i.e., unlearned) man (Muhammad); the text itself was evidence that Muhammad had been inspired by God.[48] Only the ignorant could argue that it "no longer suffices for our times" or suggest that "our times require that which is not in the shariʿa." Al-Jisr insisted repeatedly that the empire and indeed all Muslim nations as well as the *umma* at large can take from it all they need to meet the demands of the present.[49] Al-Jisr was certainly not the first to stress, through recourse to a long-standing tradition of flexible hermeneutics, the timelessness of the shariʿa. Emphasizing the relevance of concepts like *qiyas* (analogic reasoning) and *maslaha* (interest) allowed him to reinvigorate a faith in its efficacy and continued relevance. Al-Jisr's purpose was to demonstrate that the shariʿa contained all the necessary proofs needed to guide all people, provided that it was properly interpreted.

The work gains its novelty once al-Jisr moves on to single out what he considers the greatest challenge to Muslims now, as it was in the time of the Prophet: namely, the "atheistic naturalist-materialists" (*al-tabiʿiyin al-maddiyin al-dahriyin*). He moves casually from discussing opponents in Muhammad's time to those in the present as a means of underscoring the eternal presence of these views. The idea was to show just how easily a connection could be drawn between ancient and modern ideas of materialism, atomism, naturalism, and the "theory of evolution and progress."

Al-Jisr uses a debate between a believer and a materialist to present the core issues. His dialogue between a "most simple" Muslim *ʿalim* and an evolutionary materialist makes liberal use of dialectical and particularly syllogistic arguments typical of the *fiqh* tradition. It takes up nearly a third of what is by any standards a weighty tome, and after a series of moves and countermoves, the *ʿalim* trumps the materialist by arguing partly through reason and partly through appeal to evidence of God's design in nature—using the evolutionists' own empirical arguments against his interlocutor and showing how he too must ultimately come to admit the evidence in favor of a prime mover or first cause (i.e., God) in nature. The result was the application of generations-old techniques of Muslim exegesis, rhetoric, and reasoning to new questions of contemporary empirical inquiry and scientific debate. Indeed, for this reason it was a prime example of the *yeni ilm-i kalam*, or "new theology," that emerged across the empire at this time.[50]

Although he would certainly have known of him, al-Jisr makes no mention of Darwin in his text; nor does he mention Spencer or even Shumayyil. Nevertheless, their ideas, like modern theories of evolution more generally, filtered through the articles in *Al-Muqtataf* and the works of other contemporary Arabic writers, are the invisible presence at the heart of his text. In contrast, al-Jisr does name classical figures from within the Arab tradition, such as the logician Fakhr al-Din al-Razi, the naturalist al-Jahiz, and the theologian Abu Saʿud al-Hanafi, for the structure of his work was certainly shaped more by long-standing traditions of theological deliberation than by contemporary scientific debate. Yet what is striking from our perspective is to see how easily he could incorporate contemporary evolutionary thought within this broader discursive framework and how smoothly Darwin's own ideas could be fitted and assessed within it.

Its fusion of classical and contemporary themes is striking. For instance, al-Jisr repeatedly described God as the only true final cause in nature. Alluding to al-Ghazali and other Ashʿari thinkers, al-Jisr counted them among those who offered irrefutable "rational proofs" of the existence of God.[51] He took a skeptical line with regard to our ability to comprehend the sensate world through mere observation and experimentation alone:

this, argued the Sufi shaykh, could not reveal the true causes of natural phenomena.[52] Nevertheless, al-Jisr also emphasized reasoning from nature as a way of uncovering divine truths. He took the view that a rational, and even nomothetic, view of nature could be in line with faith and that certain evolutionary ideas themselves even could be said to be compatible with the Qur'an. This was the real source of the novelty of his approach.

However, he also emphasized the marvels or wonders of nature in line with the classical tropes of Muslim naturalist writings. Hence, his cataloging of evidence from the celestial, plant, and animal worlds (lists of "the proofs for the existence of God"—the stars, oceans, caves, mountains, minerals, plants, and animals, from the microscopic to the human) spanned some forty pages alone and constituted the evidence of God's plan in nature.[53] Paley's watchmaker was cited too.[54] Indeed, one of the ironies of the introduction of Darwin into Arabic is that it led so many to embrace the ideas of Paley, the Anglican natural theologian, sooner than, or alongside, those of the biologist himself. Paley was even updated by al-Jisr using the exact same argument in reference to a steam engine (*alat bukhariya*).[55]

Beirut and the Battle against Materialism

Whatever the classical debts, the *Risala* must be understood as a product of the time al-Jisr spent in Beirut. The intellectual milieu he found there, and in particular the intense debate over materialism that raged through the 1880s, left their mark, and much of the content and style of al-Jisr's arguments overlaps with those of his Protestant and Catholic contemporaries. An avid reader of modern Arabic scientific, literary, and even missionary journals while in Beirut, al-Jisr used the Syrian Protestant College's library in order to follow the latest debates in natural science; and he certainly read *Al-Muqtataf*, from whose writings on evolution he directly borrowed. He also seems to have read another periodical edited by the college's leading antimaterialist, Ibrahim al-Hurani, an important conduit for Christian argumentation that drew on Catholic, as well as Protestant, objections to the new impiety. An excursus is therefore needed to sketch this broader antimaterialist background to the new Muslim theology, for a great deal of the impetus behind the natural theology that coursed through the *Risala* was derived from his readings of these writers.

Much of this can be traced back to *Al-Muqtataf* itself, particularly during its "Beirut phase." While not a combatant in the materialism wars, the journal—like many Victorian popular-science journals in England and America—found that natural theology provided an effective means of transmitting natural historical knowledge.[56] In the hands of Yaʿqub Sarruf and Faris Nimr reports of newly discovered species and taxonomic orders and other news were often presented as an exercise in natural theology, drawing on long-standing classical Arabic topoi such as the wonders of creation (*aja'ib* or *ghara'ib al-makhluqat*) and on the cultivation of an ecumenical reverence for divine wisdom (*hikma*) in nature through God's providential design and the operation of natural laws. These works brought together discursive traditions on the marvelous with the new discoveries and disciplines of modern natural science.[57]

In *Al-Muqtataf*'s early issues, it was also quite common to encounter arguments about how God's providential dispensation was revealed in natural laws and the Chain of Being. The journal's very first account of Darwin's theory, written by Rizq'allah al-Birbari in 1876, dismissed the theory of natural selection on these grounds. Al-Birbari, like so many of his generation, appealed to a conception of natural law—ordained and fixed by divine decree and will—as the clinching evidence for the rejection of natural selection.[58] Bishara Zilzal, another instructor at the Syrian Protestant College, advanced similar arguments. He too championed the concept of a fixed Chain of Being and made an appeal to man's divinely ordained preeminence in nature. "Man," he wrote, "is the most noble of animals, the highest form designed by God in the most perfect fashion."[59] "Perfection," "divine wisdom," and "order in nature"—as evidenced through natural laws—thus became part of the critical terminology emerging through discussions of modern natural history.

Of course, the editors of the journal themselves were more positive about Darwin, but they too referred to natural theological precursors in their discussions. Sarruf, for instance, drew the analogy between an evolutionary theory of the "chain of descent" and the much older view of the "Great Chain of Being": "Ancient Greek philosophers themselves took such ideas from Aristotle, seeing all forms as linked and tied to one another in a chain of descent." Medieval Arabic philosophers, he also pointed out, had adopted Aristotle's concept but added to it the idea of a Creator and a temporal view of "progress toward perfection" (*taraqqa ila al-kamal*). Yet he also suggested that the aim of modern scientists was in fact more or less the same as theirs: "to demonstrate that man has reached his state gradually through such a progress toward perfection."[60] Sometimes the editors even tried to use arguments from design to bolster their pro-evolution position. For instance, in 1885, they cite Bishop Frederick Temple's extension of Paley in which he claimed that evolution "leaves the argument for an Intelligent Creator and Governor of the world stronger than it was before."[61]

For the editors, Christian natural theology mostly provided them with a way of reassuring their readers (and, initially, employers) that the sciences they were helping to popularize were compatible with faith. In keeping with this, they typically ended many of their articles with the invocation "praise be to God, the Wise, the Knowing." Yet there was often something perfunctory about this, and in the end, Sarruf and Nimr were not really concerned to present a full-fledged natural theology of their own. "It would be tiresome," they wrote, "to present even a selection of the proofs of design." One can only "exclaim with the Psalmist, 'How manifold are thy works, O Lord! In wisdom hast thou made them all!'"[62]

They were, however, crucially involved in finding natural theology textbooks for the Syrian Protestant College. Sarruf was said to have translated a book on design in 1877 that would certainly have been one of the earliest works of this kind to make contemporary Christian natural theological arguments available in Arabic. Indeed, it was as much thanks to these textbooks as to their journal that the new vein of natural historical theologies spread so rapidly throughout the Arab provinces and beyond. These textbooks made their way into both Christian and Muslim schools: they were, importantly, used in the Madrasa al-Sultaniya, where al-Jisr taught.

One of these was a textbook on natural history written by the staunch antimaterialist Ibrahim al-Hurani, Shumayyil's main antagonist at the college. His 1883 book on the wonders of nature, *Certain Signs in the Wonders of the Heavens and Earth*, had strong overtones of natural theology and classical revivalism. A riposte to Shumayyil, Hurani's treatise was organized according to a classical division of the three kingdoms of nature—this too was reminiscent of medieval Arabic zoological and botanical compendia—and covered such things as the organization and movement of oceans and mountains, the world of plants, and numerous animal species (presented in no particular order). Yet it was also the work of a rationalist: unlike older works in this genre, for instance, it eliminated any discussion of angels or *jinn*. Hurani argues that it is thanks to the modern European sciences and their novel technological apparatus that we have come to appreciate the true range and vast wonders of the world. New technologies like the microscope and the telescope were making visible that which was previously invisible in nature and thus helped to critically expand our knowledge of God's infinite wisdom in his handiwork. As al-Jisr would later, Hurani begins his discussion with a long lesson on the visible and the invisible in the natural world, or what he called the sciences of the hidden and the manifest: *ʿilm al-ghayb wa al-khifaʿ* and *ʿilm al-shihada wa al-zahur*.[63] As this language was itself an allusion to a key concept in the classical tradition of

Arabic (and particularly Sufi) esoteric writings on the nature of divine being and the manifest and latent, it is not surprising that it would resonate with al-Jisr and other Muslim readers.

As the title of his work suggests, Hurani's emphasis was, above all, on the marvels of creation. His entries placed a strong emphasis on the novel or unexpected, typically left unexplained, as if to stress God's majestic power. But the emphasis on the wondrous was intended to do more than merely entertain. As mentioned, the genre was an old one, stretching back to medieval times. But in the nineteenth century, what began as a broadly ecumenical attempt to sanctify the new European natural sciences in the local linguistic and theological idiom of the day eventually took on a more defensive tone. Faced with the onslaught of Shumayyil's materialism, many of Beirut's Christian intellectuals redoubled their efforts to affirm their faith through natural theology. Warning of the increasing threat from a "school of unbelief" that denied the role of God in nature, the Jesuit mission press issued *The Book of the Magnificent Garden of Natural History* in 1881.[64] This work discussed the latest findings in natural history in a very similar fashion to Hurani.

In this sense, communal lines, while bitterly contested among Beirut's educational institutions when it came to competing for political favor and pupils, seemed much less important intellectually. The Syrian Protestant College's own weekly journal, *Al-Nashra al-Usbuʿiya* (College weekly news), which Hurani edited, serialized an older work by one of the best-known Syrian Catholic writers, Fransis Fath Allah al-Marrash, whose treatise, *Shahadat al-tabiʿa fi wujud Allah wa-al-shariʿa* (Nature's testimonies to the existence of God and divine law) was perhaps the most sustained and serious antimaterialist polemic then available in Arabic. Marrash himself, perhaps reflecting his French orientation, does not refer to Darwin and only hinted at ideas of evolution; nevertheless, his text was useful ammunition for missionaries and shaykhs alike in their struggle with evolutionary materialism. Natural theology was thus clearly a defense that crossed sectarian lines, and to demonstrate the point, a brief consideration of Marrash's thought will allow us to evaluate al-Jisr's own debt to the broader antimaterialist campaign.

Nature's Testimony

We know all too little about the life of Fransis Fath Allah al-Marrash whose treatise, *Nature's Testimony*, was perhaps the first work of its kind in Arabic to update classical themes of theologies of nature.[65] Born in Aleppo in 1836 to a Melkite family, Marrash studied medicine privately before moving to Paris to continue his training.[66] The city clearly impressed him; in *Rihlat Baris* (A journey to Paris), an account of his travels that he published in Beirut in 1867 shortly after he returned to Syria, he described it enthusiastically as the center of enlightenment, civilization, and belles-lettres.[67] Along with its libraries, museums, universities, and the city's Exposition universelle of 1867, Marrash was also impressed with the freedom and breadth of French intellectual life. It seemed to him to prove that liberty to express oneself "by word of mouth or by the pen" led to cultural revival—or what he called a *nahda* in Arabic. This, he declared, was the secret to the West's advance over the Arab East. But the idea of a revival held specific religious overtones for him. Along with their love of knowledge and industry, Marrash also extolled how the French hold "both this world and religion [*al-din wa-al-dunya*] as the greatest of their concerns."[68] Inspired by these experiences and seeking to promote a similar renaissance in Arabic-speaking lands, Marrash went on to publish a wide variety of works before his untimely death in Aleppo in 1873.[69]

His emphasis on the spiritual roots of reasoned discourse was underscored by the very title of his treatise on natural theology: it was nature's "testimony," or *shahada*, for the "existence of God and divine law." At the very beginning of his text, Marrash presents the famous watchmaker analogy, precisely as al-Jisr would twenty years later.[70] Like most works of the kind, Marrash's arguments from design covered such themes as God's providential control over nature and its inhabitants, faith in natural laws, and, of course, proofs of the existence of God.[71] He also arranged his argument from design in a classically deist fashion.[72] Because God created the world in a wise and ordered manner, regular laws in nature proved his existence. How else, he wrote, is one to make sense of magnetism in natural and chemical bodies, laws of repulsion and attraction, and secondary powers in nature such as gravity? Such order in nature could not be the result of mere chance but must be the result of purposeful design, "contrary to what all those 'coincidentalists' [*ittifaqiyin*] claim."[73] Turning to the botanical world, he made much the same point, using evidence of plant respiration, photosynthesis, and reproduction.[74] Drawing on his professional medical knowledge, he did the same with the animal kingdom, citing examples that ranged from descriptions of the mechanical workings of the esophagus and duodenum to the oxygenation of the blood. In the case of humans, he noted how anatomical dissection could demonstrate humanity's standing as the highest creation of all.[75]

This last argument reflected the distinct sense of anxiety that gripped

Marrash. For early natural theologians, man's place in the Chain of Being was unquestioned. But this was not the case by the time Marrash had reached Paris, where materialism had perhaps become more popular than anywhere else in Europe. Modern postrevolutionary France stood as a warning, and in *Nature's Testimony*, Marrash warned of the danger of eliminating religious education and in particular of the risk that research into natural science might lead to disbelief. It had already, according to him, led some of those seeking to unlock the secrets of nature to overlook the limits of their own judgment and deny God. This kind of mistaken interpretation of nature, he thought, was not a new problem, but it was certainly one that worried him greatly in the present.

As a result, he devoted much energy to talking about man's place in the hierarchy of creation. Marrash wrote that man should be contrasted with "lower-ranking animals," and in general, "higher-ranking animals" should similarly be pitted against lower ones, in accordance with what he penned the "chain of descent" (*silsilat al-tanazul*). But he emphasized that the gap between humans and animals was obviously vast.[76] Man stood at the apex of a vast hierarchy, with each class serving that above it: "we come to the conclusion that man is the highest level of living beings since he is served by all." According to Marrash, the rule was that "the one who is served is better than the servant," deploying a classically Aristotelian line.[77] This had fundamental ethical implications, too. For Marrash, that all creation should serve men could be justified only by reference to man's God-given advantage—not the mind but the soul.[78]

Marrash's defense of this point hinted at his anxieties over materialism. "A number of nonbelievers," he wrote, "have claimed that the special properties of the mind are a function of the brain alone, in animals as in men." However, he complained, they went on to prove this "with mere academic explanations" and "much imagination and tongue-wagging."[79] In his defense of the soul, Marrash himself had recourse to the skepticism characteristic of the Catholic apologias current in France in the 1860s. Like many French neo-Thomists, he argued that beyond human knowledge rested matters that no mortal mind could comprehend.[80] Man's knowledge, according to Marrash, was limited in comparison to the huge span of universal knowledge. The human mind could approach such ultimate truths only "by way of certain limited appearances."[81] Marrash urged his readers to put aside questions concerning the mind-body connection and the idea of the soul in particular.[82]

For Marrash, our inability to fully understand the manifest world meant that we should not write off revelation now; it was all the more needed.

The reasonable man appreciated the necessity for faith and God's generosity in guiding us by means of scripture. He defended the Holy Books as compendiums of reasoned truths and as based upon fact, arguing that "many clever scholars" had spent many years of their lives studying them and had confirmed their veracity. "They have inspected every sentence of them precisely, matching prophecies and symbols with future and past events."[83]

And of course there were, for Marrash, other proofs of the true divine authorship of scripture—namely miracles. It was an old argument: precisely because miracles go against nature, they prove God's existence. Is it so inconceivable, he asked, for us to believe that God will sometimes utilize nature to produce something against its laws, as in the case of miracles, to show men his unlimited power? He admitted that religions can spread without miracles. Still, he maintained, "the transgression of natural law alone cannot be taken as evidence against the veracity of these texts."[84]

Beyond mere rationality, then, religion involved intuition if not instinct. If one were to look carefully at all humanity, "in every area and continent, in every time and generation," he wrote, one would find among all peoples "an instinctual tendency to submit to religion, whatever their tribe, barbaric or civilized." All the tribes found to inhabit the Americas, he explained by way of example, had religious laws (*shariʿa*) although these tribes were far from the rest of the inhabited world and untouched by its civilizations. Religion, Marrash felt, was an inherent instinct and fundamental part of human nature, "like the natural instinct to procreate. The latter preserves nature and the former preserves the special attributes of humanity."[85]

Dialogue with a Materialist

Although we cannot prove it, we might count al-Jisr among Marrash's readers: after all, one can easily imagine al-Jisr encountering Marrash's works during al-Jisr's visits to the Syrian Protestant College library or through his reading of the local press. There is little doubt that the Sufi shaykh was as indebted to the arguments of his Catholic predecessor as he was to Hurani and the other instructors at the college. Nevertheless, as becomes clear in the course of the dialogue at the heart of his *Risala*, al-Jisr differed in one important way from Marrash: he was much less interested in the wonders of creation and more interested in its rational ordering and origins. Like Marrash, al-Jisr argued from design; but unlike Marrash, providential lessons were not the centerpiece.

The real thrust of the *Risala* is not so much scientific as epistemological and logical: to extol the superior demonstrative reasoning of a Muslim *ʿalim*, particularly one equipped with familiar syllogistic, dialectical, and theological arguments. The materialists, by contrast, are presented as lacking any rational basis for their position. Denying final causes, they rely on faulty logic to describe "the creation of all varieties of life from matter and motion."[86] Like al-Ghazali in *Tahafut al-falasifa* (A refutation of the philosophers), al-Jisr lays out the materialists' arguments on creation and then proceeds to refute them. Reasoning by syllogisms, his Muslim *ʿalim* triumphs.

The dialogue takes place mainly as a kind of dialectic disputation, and in the course of it, al-Jisr demonstrates the continued richness and power of the *kalam* tradition. For instance, he frequently deploys syllogistic and analogic reasoning (*qiyas*), though he uses the latter more sparingly and often anecdotally. In a typical example, he suggests that a man confronted with a sumptuous feast cannot easily determine whether all the dishes put before him were made all at once or one after the other![87] The notion of a chain of descent, in other words, goes against neither common sense nor logic. The test for al-Jisr is where one comes down on the existence of God. Al-Jisr constantly asks for demonstrable evidence, and in so doing he is basically following the general parameters of post-Avicennian modal logic: indeed, it is Imam al-Razi to whom al-Jisr turns above all, following much of the twelfth-century thinker's own logical formulations and philosophical theology in these matters. He combines this with recourse to classical Ashʿari arguments—following al-Ghazali specifically—on the existence and attributes of God as well as on the nature of matter and the role of causality in nature. We should thus understand al-Jisr's treatise, and his dialogue with the materialist in particular, as an exercise in modern Muslim *kalam*. For al-Jisr, modern evolutionary ideas—and particularly evolutionary materialism—became the grounds for an exercise in metaphysics and philosophical or speculative theology.

He begins by asking his reader to imagine the encounter between a contemporary "atheistic naturalist-materialist" and the "weakest" *ʿalim* among Muhammad's followers.[88] (The point is clear: even the least-trained Muslim scholar could find the flaws in the arguments presented by the former.) The *ʿalim* asks the materialist to explain his *madhhab*, or school of thought, regarding the origin of the world and so forth, while the *ʿalim* promises to then consider these ideas carefully and from there provide rational arguments or logical proofs either for or against these claims. The *ʿalim* then ends his introduction by reminding his disputant that many of these ideas had already been extensively discussed and debated by "our ancient scholars." Indeed, as the dialogue shows throughout, Darwin's Muslim readers—like many of his global readers more generally—did not so much view his ideas as an unprecedented novelty as see them as part and parcel of more long-standing and more familiar cosmological or metaphysical arguments, from which vantage point they were either selectively appropriated or discounted.

In the opening pages, we hear echoes of what would later prove to be the thrust of al-Jisr's antimaterialism. Broadly speaking, there were several key arguments. The first was that many of the postulates of this theory were merely speculative. Even before the dialogue gets under way, al-Jisr voices his misgivings about the basis for the materialists' ideas on the immortality of matter and the origin of life.[89] They say they do not investigate the final cause of matter itself, or even consider the nature of its immortality, yet they seem content to draw inconclusive and unproven arguments from it. He takes issue in particular with the denial of a final cause in nature.[90] For al-Jisr, the denial of a First Cause in nature—both as the final cause of matter and the first cause of life—is itself a logical impossibility.[91] Borrowing from al-Ghazali and others on the impossibility of an infinite regress of finite events, he provides a basic geometrical proof—and offers an accompanying sketch—in favor of this counterargument.[92] Later, he also borrows from classical *kalam* arguments on the nature of matter itself—substance or essence versus accident—to dispute the materialists' ideas on matter, force, and motion.[93]

There is, for al-Jisr, no more damning charge than unsound reasoning. The *ʿalim*'s own presentation of a series of logical proofs regarding God's existence is intended to show how the shariʿa is, by contrast, based on purely logical and demonstrative proofs. And he presents a number of familiar arguments in favor of God's singularity (*tawhid*) through recourse to reasoning about existence and the nature of essences and accidents and of necessity and possibility.[94] Examples of the argument from design—as mentioned before, citing Paley's watch and watchmaker and the steam engine and engineer analogies—also make their way into the text at this point.[95]

Like many of the earlier Arabic works of natural theology that we have examined briefly, al-Jisr's *Risala* gives detailed and numerous examples of the wonders of creation—from oceans, mountains, caves, and deserts to the varieties of minerals and plants, and giving examples from the microscopic complexity of cellular structures to the regeneration of hydras.[96] He also mentions specific organs—in particular, the eye—following a favorite example of previous Christian natural theologians in recounting the extraordinary coordination of the minutiae of its design and its unique adaptability to its function. Faced with these extraordinary designs in nature, al-Jisr concludes that "physiologists should be the greatest believers."[97]

Although al-Jisr added modern physiological, zoological, and botanical knowledge to his list of the proofs in favor of God's existence, for him, only the Qur'an and the shariʿa could be said to ultimately provide certain knowledge of God, nature, and man. He was especially keen to underscore this latter point, arguing repeatedly that man's spiritual nature is the obvious point of distinction between humans and the rest of God's creations.[98] But al-Jisr concludes by arguing something else: namely, that it is religious belief itself that is, in fact, the "most perfect science" and the least subject to doubt or error.[99] This is precisely what his exhaustive recounting of the rational arguments of his *ʿalim* is intended to show. His goal is to disprove the materialist argument that faith (*al-iman*) is not a science at all.

Al-Jisr was thus able to explain why the faith of Muslim students who study the modern natural sciences declines: it is due to the "corruption"—or errors—of their teachers, who are mere sophists and who present illogical and poorly proven arguments to their charges. In fact, if taught properly, these sciences should only strengthen religious convictions.[100] In keeping with his general pedagogic and scholarly interests, al-Jisr repeated the call (*daʿwa*) to "preserve the beliefs of these students" and to safeguard their religious education by "electing faithful instructors" for them, something he argued would benefit Muslims everywhere.[101]

Evolution and the Shariʿa

In the remaining dialectic, al-Jisr developed his critique in light of the rules of Qur'anic exegesis and turned to specific matters of scriptural hermeneutics. He sought to distinguish those cases in which, once confronted with the arguments in favor of evolution, one could be said to be required to adhere to a literal interpretation of scripture from those in which one might be permitted a more flexible interpretation—through *ta'wil*, or analogic reasoning. Once again, we find al-Jisr taking his cue from classical predecessors:

> According to the rules . . . of the Muhammadan shariʿa, it is incumbent that its followers believe in the apparent or plain meaning of the successive and well-known scriptural texts unless their apparent meaning is contradicted by decisive and rational evidence that necessitates their interpretation [*ta'wil*]. Therefore, the belief of those who follow Muhammad, peace be upon him,

> should be that God created each species on earth independently from the others, and not according to evolution and by deriving one from another, even though he is capable of both of these two possibilities. And whether species were created all at once or gradually through the natural laws that God laid out is a point that is not addressed in their shariʿa so as to help them determine between the two options. And according to the aforementioned rules of their shariʿa, they are not permitted to deviate from the apparent meaning toward that which disagrees with it and toward [a belief in] evolution and the derivation of species from other species, as the materialists claim, because it is contradictory to the apparent meaning of the . . . passages [on creation] because there is no concrete evidence that would necessitate their reinterpretation. The evidence that you evolutionists mention in your books is speculative and conjectural. . . . Yes, if there was decisive, rational evidence in opposition to the apparent meaning of these passages, [Muslim theologians] would have to interpret them so as to reconcile them with these proofs and according to those aforementioned rules.[102]

In short, al-Jisr took the view that as yet the theory of evolution was not so much wrong as insufficiently proven. For after outlining his approach to Qurʾanic interpretation, he also put forward the claim that there were some points that a Muslim could concede, both on rational or logical grounds and on purely scriptural ones. Al-Jisr even went on to say that given the general parameters of scriptural interpretation, there was nothing to prevent a Muslim in future from accepting evolution wholeheartedly, given that certain conditions and terms were met. Above all, it would require that much more certain proof, and proper empirical evidence in particular, be established. Yet he also stressed that the Muslim should hold to both such a theory and an unshaken belief in the final claim that there is "no creator but God the Exalted." This would obviously bar materialism but not a nonmaterialist evolutionary view. With such conditions, al-Jisr proposed that one would therefore not be prevented from taking the Qurʾan's "meaning beyond the apparent one and from applying to it what the decisive evidence indicates regarding evolution." Indeed, in that case, one would still "be considered among the people of the Islamic religion," while still having recourse to "the process of deduction from all creation to the existence of God and the absoluteness of his power, knowledge, and wisdom."[103]

Indeed, al-Jisr did not even object to every aspect of evolutionary theory. On the question of what distinguished man from animals, he reached some surprising conclusions. The shariʿa was not incompatible, according to him, with the idea that man's intellect differed only in quantity but not in kind from that of animals. Referring to the medieval philosophers al-Razi and al-Mawardi, he argued that it was one of those "unknown things" that the shariʿa did not explain. Where the materialists erred was in denying that reason had been created by God.[104] But on the issue of the creation of man, he was uncompromising: Islam simply could not accept the idea that man had evolved from other species. Speaking to the materialist, he insisted that "your proofs on this score are mere assumptions and such a view . . . is simply imaginary . . . [and] completely against our beliefs." He insisted that man had been created by God and cited the relevant Qurʾanic verses in support. These were at first sight unpromising because they mentioned different origins: water, clay or earth, dust, and "a clot of congealed blood." But what al-Jisr did was to read these as possible stages in a quasi-evolutionary schema specific to man alone. According to him, Adam had emerged from clay, then a sperm and an egg, then from these a clot, and finally a child. The whole process revealed God's power to transform matter. Yet there was absolutely no evidence, he insisted, that God had used this power to derive man from other creatures.

On other issues, such as the time taken to create the earth, he took a different stance. Referring to the traditional account of the formation of the heavens and the earth in just six days, he argued that some degree of interpretive latitude was possible if not necessary. Were these six days like our days, he asked rhetorically? How could they be when there were as yet no stars or sun to allow for measurement? Or did scripture mean "day" in the wider sense used when referring, for instance, to the "final days of judgment" (*ayam al-akhira*)? Some scholars, like Ibn ʿAbbas, had said that "these days were like one thousand of our years; others that they were like fifty thousand." Nevertheless, they all agreed that God created the heavens and earth "with his infinite wisdom."[105]

As we saw earlier, much of the earlier theological discussion around this issue had been bound up with Aristotelian categories, many of which had been discussed and expanded upon by generations of Arab philosophers and theologians. For the latter in particular, the notion of the temporality (*huduth*) of the universe was critical. Such arguments proved appealing to al-Jisr. First, he himself moved smoothly from "the argument from the possibility [*imkan*] of the universe to the existence of a necessary being [*wajib al-wujud*]" and therefore of a Creator. Second, there was the "argument from the possibility of the qualities of the universe to the necessity of a Determinant of the form, characteristics, and locus of bodies composing it, who is not himself a body." Third came the "argument from the temporality of substances and bodies to the existence of a Maker thereof." And finally, there was the "argument from the temporality of qualities of the universe to the existence of an intelligent designer who disposes things according to his will and power." (Indeed, Razi's critique of *tasalsala*, understood in the classical vocabulary as an argument for a *regressus ad infinitum*, likely connects to al-Jisr's discussion of a first cause in these passages.)[106]

After outlining his views on creation and existence and discussing them in relation to Qur'anic hermeneutics, al-Jisr devotes the rest of the dialogue to examining the specifics of the theory of evolution in greater detail. He begins by summarizing the laws of evolution. Here, he clearly echoes earlier popular-science works, particularly those published in *Al-Muqtataf*. He also adds to his discussion of evolutionary laws the use of geological evidence and the arguments on vestiges and the extinction of species. Although he does not mention Darwin by name, this account clearly follows the outline of the *Origin of Species* given in *al-Muqtataf*'s summaries, and indeed, some of al-Jisr's passages are taken nearly verbatim from the periodical.

But he adds his own interpretation. He starts by describing the four laws of evolution (variation, inheritance, struggle for life, and natural selection) as a mixture of common sense and speculation. On inheritance and variation, for instance, he has his *ʿalim* claim that there is in these "nothing new," adding (on the question of variation and in the commonsense style that marked the *ʿalim*'s rhetorical counterarguments throughout) that if God had made all creatures look the same, "a man would not recognize his wife, or son, or horse." These two laws he regarded as "general laws for all creatures of the earth." In the end, he claimed that none of the first three laws could be said to go against scripture in any way—hence, there was no need for scriptural exegesis with regard to these matters. On the question of natural selection, he argued that if this were merely "the result of the three previously mentioned laws," one could also find it to be within the fold of orthodox Islam, provided that it was ultimately the result of God's handiwork. Reconciling rational and traditional proofs in these matters, as he put it, offered a Muslim believer little difficulty.[107]

In this way, al-Jisr sought, above all, to emphasize the rationality of Islam and highlighted its demand for evidentiary certainty in matters bearing on scriptural interpretation. He chided the evolutionists for their "long and boring claims" and for their books, which contained "only a few real facts that are endlessly repeated and toyed with." On the geological evidence, for instance, he described most of their claims as "mere guesswork" or *zann*, a term of art connoting uncertain or unproven postulates.[108]

Hence, fossil forms did not in themselves prove evolution, he argues, since extinct species could still have been created independently and for a purpose that "God may have felt had come to an end." Similarly skeptical of the extent of human knowledge, he disputed the claims made about rudimentary organs, saying that "they may have a wisdom which is hidden from you." For al-Jisr, God's omnipotence thus meant that he could even have allowed species to evolve if he so wished. A snake, for instance, may have been a lizard, and when men started to kill them because they felt they were dangerous, the lizards began to hide in caves, thereby losing their eyes and their legs and sliding along the ground. We should note too that he also referred to medieval exegetes who argued that God had punished the snake, "which used to be attractive and had legs," for its actions in the Garden of Eden. (This interest in transmogrification had long interested Muslim exegetes, who attempted to similarly interpret verses in the Qur'an that make reference to the transformation of sinful nations or peoples into apes and swine.) In al-Jisr's telling, this suggested a kind of evolutionary adaptation in which one species turned into another, "and this change descended to its branches of offspring by inheritance until it became the form we find today." However, much of the time, God had simply chosen to terminate the life of one species and create another: "God eliminated lower levels and then created higher forms, each one created independently, not evolved one from the other." He created higher forms because the "lower ones were no longer suitable for their environments." Again, however, he stressed that at each point they were created—in the sense of being guided by a divine hand—and not evolved.[109] The progression of species was thus still about the perfection of species for al-Jisr.

The New *Kalam*

It is important to bear in mind that the *Risala* as a whole invokes "science" as only part of a larger exercise in *'ilm al-kalam,* or speculative theology. (The Urdu translation of al-Jisr's work bore the subtitle *Jadid ilm-i kalam* and thus explicitly demonstrated this connection.) *Kalam,* meaning literally "word" or "speech," had a long history. It was originally deployed in Arabic translations of Greek philosophy as a rendition of the term *logos* and hence carried over the latter's allied meanings of "word" "reason," and "argument." As the "sciences of the word," *'ilm al-kalam* traditionally included connotations of conversation, discussion, controversy, and defense.[110] Al-Farabi regarded *'ilm al-kalam* as "a science which enables a man to procure the victory of the dogmas and actions laid down by the legislators of the religion, and to refute all opinions contradicting them."

Proponents of *kalam* (*al-mutakallimun*), from the tenth century AD onward, were to define it similarly: *kalam*, explained one such manual, "is the science which is concerned with firmly establishing religious beliefs [*aqa'id*] by adducing proofs and banishing doubts."[111] It was also sometimes used interchangeably with *ʿilm usul al-din*, as the science of the basic postulates of religion, though unlike *fiqh* (or jurisprudence), this was meant to be a purely rational exercise in establishing the basic tenets and veracity of the shariʿa. In its formal lexicon, then, it came to refer, above all, to reasoning, but reasoning in a very specific sense—namely, conceptual and theoretical reasoning with particular and proscribed rules, conditions, and procedures that produce knowledge whose truth can be said to lie beyond any reasonable doubt. Hence, its more typical translation was the "speculative" or "rational science." In particular, *kalam* had strong connotations of disputation, particularly scholastic disputation (*jadal* or *al-munazara*). As part of a rational or logically regulated discourse, many of the *kalam* writings took the form of dialectical disputations between various schools of thought: al-Jisr's own use of it—in the form of this dialogue between an *ʿalim* and an "atheistic naturalist-materialist"—was thus itself very much an expression of this tradition.[112]

But it was also a product of its time. For the *Risala* was a defense of the faith not so much against errors within the *umma* as against powerful misconceptions about Islam abroad. Since the outbreak of the Near Eastern Crisis of 1876, in particular, with its upsurge of anti-Muslim polemics in the European press, there had been a revival of apologetic works in the Ottoman press.[113] As the British took over Egypt and became more interested in control of eastern Africa and the routes to India, debate erupted in England, widely reported in the Arabic press, about the potential role of Islam and Christianity in "civilizing" Africa and the relative civilizational merits of the two faiths.

The *Risala* in fact begins with al-Jisr describing his reaction to accounts in the Syrian papers of the sensational views of a canon of York, Isaac Taylor, according to which, as al-Jisr put it, Taylor attempted to prove the kinship between the two faiths and provided reasons why Islam was more suited than Christianity to "civilize uncivilized nations" and to "tame the character" of the Africans. In October 1887 Taylor suggested that the British should work with the Arabs in East Africa (this was only a few years after the Mahdi had defeated General Gordon in the Sudan), which elicited enormous opposition. As the "scramble for Africa" intensified, missionaries and their supporters allied with older anti-Muslim propagandists to attack Islam and pour scorn on its civilizing credentials: the slave trade, polygamy, and the supposedly static and unchanging nature of Islamic thought itself were invoked by detractors.[114]

Al-Jisr wanted to rebut all of this; and after reading about Taylor, he allied himself with the Yorkshire canon against his many opponents in England and elsewhere in Europe. He reports with delight the emergence, too, of other signs of interest in Islam in England, such as the establishment of the Woking mosque and the emergence of Arabic and Muslim publications there. Asserting not merely an Ottoman but a global mission, al-Jisr claimed that he had been originally moved to compose the *Risala* to foster these positive developments in the very heart of the new imperial power.

Yet the *Risala* formed part of the "new theology" in another and more specific sense: for as the bulk of the work shows, it was in fact intended to meet the challenges to faith posed by the modern sciences. Al-Jisr was not alone in his consideration of Islam's (versus Christianity's) ability to accommodate science. Both Sayyid Ahmad Khan and his opponent Jamal al-Din al-Afghani, for example, had also seized upon this point and called for a revitalized theology in Islam for similar reasons: "Today we need, as in former days, a modern *ʿilm al-kalam* by which we either render futile the tenets of modern sciences or make them doubtful or bring them into harmony with the doctrines of Islam," wrote Khan.[115] Later Muslim intellectuals in India, including such well-known figures as Shibli Nuʿmani (1857–1914) and Muhammad Iqbal (1876–1938), would make the same plea. Ottoman officials and educators in Istanbul also took up the call and argued much more directly than al-Jisr for the revitalization of *kalam*. Izmirli Ismaʿil Hakki (1868–1946), a theologian and author of *Yeni ilm-i kalam* (The new theology), would continue in much the same vein as al-Jisr for another two decades: "Today our formal theology [*resmi ilm-i kalam*] has lost its scientific basis," Hakki wrote. "It should be changed in conformity with new philosophical theories and expanded in accordance with the needs of the age." By incorporating the ideas of modern evolutionists and materialists in his theological works, al-Jisr was thus blazing a trail many later theologians would follow.[116]

The Politics of Reason

For al-Jisr, the shariʿa and the laws of nature could not conflict, as both were subjects of truth and both were susceptible to rational demonstration. The *ʿulama*—and not those naturalist-scientists he took aim at in his book—were thus the only true guardians of reason. Al-Jisr's work was an attempt to demonstrate their vitality and their ability to guide society, effortlessly combining training in technical and theological and ethical subjects. Like other Muslim reformers at that time, al-Jisr articulated the position of a declining scholarly class that was essentially hoping to reassert its power and prestige.[117]

But al-Jisr's bet on the Ottoman state was a risky enterprise. The Syrian scholar's close connection to the sultan may have initially benefited him from his position as head of the Khalwatiya order, for Abdülhamid was then actively supporting Sufi networks as a means of consolidating his power and as a means of gaining a more populist following, particularly in Arab lands. But this connection probably provided al-Jisr with more trouble than security at the court: it brought him into conflict with the head of the rival Rifaʿiya Sufi order in Syria, Abu al-Huda al-Sayyidi, an Arab ally much more important to the sultan.[118]

Tiring of court intrigues, al-Jisr returned to Tripoli by steamer in the summer of 1892. It was said that the cream of Tripolitan society, Christian and Muslim, welcomed him at the port, and that Rashid Rida recited poetry in his honor.[119] Al-Jisr preserved close connections with the Porte, received several more invitations to visit, and benefited from imperial patronage by being granted a number of positions by the Ministry of Avkaf. In 1893 he obtained the right to teach at the madrasa of a Mosque in his hometown.[120] There al-Jisr remained dedicated to education and the cause of letters. He founded a small weekly journal, *Jaridat al-Tarablus* (Tripoli news), which he helped to edit for a decade, an unusual step for an *ʿalim* at that time.[121]

Al-Jisr's problems at court were bound up with Abdülhamid's abiding suspicions about Syrian protonationalism. In Istanbul, Beirut had a bad reputation as a center of liberal thought. And with the challenge to the Ottoman claim to the caliphate by the Zaydi imam of Yemen (raising new Ottoman fears of an Arab challenge to Turkish rule) and the rise of Wahhabism in Najd, Muslim reformers in the Arab provinces were treated with reserve by Ottoman officials. All this helps explain al-Jisr's cautious and pro-Hamidian approach to reform. However, he did also argue that the empire should genuinely reform the Arab provinces, calling for a transregional rail network and improved management and tighter administrative control over the Hijaz. His *Jaridat al-Tarablus* was closed down on several occasions by the authorities, despite the fact that he generally took a loyalist line and advocated keeping the caliphate in Ottoman hands.

It is not surprising that eventually al-Jisr's approach came to seem too cautious for some reformers. Leaving Ottoman Beirut for British-occupied Cairo, Muslim *ʿulama* as well as Christian intellectuals found a vantage point from which to criticize the empire, the Hamidian conception of modernism, and its apologists. By the end of the century, it was considerably harder to place much faith in the sultan as the leader of a Muslim revival than it had been a decade earlier. Al-Jisr's former student Rashid Rida, who followed ʿAbduh to Cairo in 1897, was the first to seriously criticize him, and in 1899 a fierce exchange of views took place between Rida and al-Jisr in the press over the responsibilities of the contemporary *ʿulama*. ʿAbduh and Rida had founded a new journal, the reforming *Al-Manar*, to publicize ʿAbduh's ideas, and this made a much greater impact than al-Jisr's small and often interrupted paper. Wounded by its criticisms, al-Jisr took issue with certain so-called reformers who had assailed the *ʿulama* for failing in their duties. He not only deplored their aggressive tone but felt that their criticisms were unjustified. These *ʿulama* were just as conscientious but they faced certain material constraints. Unable to appeal to a supportive state, al-Jisr argued for the old charitable model: it was up to Arab notables to provide the funds for welfare organizations, libraries, and schools, and this they were failing to do. In addition, there was the question of the *ʿulama*'s livelihood: if they were not to plunge their families into hardship, there were limits to how critical they could be of the authorities.

Such a line was anathema to Rida. He hit back in *Al-Manar*, accusing al-Jisr of hypocrisy and criticizing his vision of the *ʿulama*. Instead of demanding lavish and expensive ceremonies, they should be useful to the people, clear and simple in their writings, and courageous in their politics. Another article, this time anonymous, accused the *ʿulama* of turning away from the Qurʾanic injunction to "command what is right and forbid what is wrong." (Al-Jisr was not mentioned by name.) Instead, they had turned into sycophants, publishing works to curry favor with tyrants.[122]

By this time Rida, and his mentor ʿAbduh, had begun to advocate a much more direct and uncompromising approach to the reconciliation of Islam and science. Al-Jisr was more cautious; in the classical mode, he had after all also advocated a skeptical position at times, arguing that there were limits to reasoning. He saw no need for the kind of radical rationalism, for instance, that would require one to deny the existence of those things which lay beyond the senses or beyond reason itself, such as the supernatural or the miraculous: on the contrary, he had defended miracles as evidence of "God's unlimited power." (In his *Risala*, he offers a full list of the Prophet's own miraculous acts, not only the splitting of the moon but also speaking to trees and dumb animals and making water flow from between his fingers. And we should remember, too, that al-Jisr's own father had acquired a reputation as a noted miracle worker.)[123] This was a very different line from that later taken by ʿAbduh and Rida, who were keen to highlight the absolute rationality of Islam and to distance it from the "superstitions" of folk practices. This would lead them to argue for a rather different kind of reconciliation between reason and revelation and, as we will see in the next chapter, to a rather different engagement with the religious and political powers of their time.

FIVE

Darwin and the Mufti

In the summer of 1903, the British traveler, essayist, and activist Wilfrid Scawen Blunt visited the eight-three-year-old Herbert Spencer at his home in Brighton. The two, united in their outspoken anti-imperialism, had long been in correspondence, and Blunt often sent the philosopher pheasants and partridges from his West Sussex estate. They had met for the first time four years earlier, and when they met again that August in the new century, it was for another introduction: Blunt wanted Spencer to meet an Egyptian admirer of his, the grand mufti of Egypt, Muhammad ʿAbduh.[1] ʿAbduh was just one of Spencer's many admirers around the world. Indian sociologists, Confucian thinkers, and Japanese reformers were drawn to the philosopher's unified vision of a science of the cosmos, the state, and society.[2] ʿAbduh had been a fan of Spencer's *On Education: Intellectual, Moral and Physical* in particular, his most widely translated book; ʿAbduh himself tried his hand at translating it. The mufti also referred to Spencer repeatedly in his lectures on the exegesis of the Qurʾan (*tafsir*). Indeed, on one occasion he described Spencer as "the greatest living philosopher."[3]

At the meeting between Spencer and ʿAbduh, Blunt was to "act as interpreter." Without one, he feared "much exchange of ideas is impossible" since Spencer "understood little more French than is needed for traveling purposes." Spencer would die just three months later, and no mention of the encounter is to be found in his writings. Blunt, however, wrote vividly about it in his diaries.[4] The aging philosopher had been in a frail condition. Nevertheless, during their morning meeting, the bedridden Spencer offered Blunt and the mufti his views on politics and prophesied the future decline of Europe. Their host lamented the disappearance of "right," denounced the Transvaal war as "an outrage on humanity," and spoke of a coming "reign of *force* in the world" when "there will be again a general war for mastery, when every kind of brutality will be practiced."[5]

After lunch, Spencer turned to matters of philosophy, and he asked the mufti "whether it was true that thought in the East was developing on the same lines as the thought of Europe." ʿAbduh replied (according to Blunt) that "what the East was learning from the West was the evil rather than the good, but that still the best and most enlightened of thought was the same."[6] This opaque answer left Spencer dissatisfied:

> "To go to the bottom of things," said Spencer, "I suppose that the conception of the underlying force of the world, what you call Allah and we call God, is not very different?" In his reply the Mufti made a distinction which struck Spencer as new. "We believe," said Abdu, "that God is a *Being*, not a *Person*." Spencer was pleased at this, but said the distinction was rather difficult to grasp. "At any rate," he said, "it is clear that you are Agnostics of the same kind as our agnosticism in Europe."[7]

We can only surmise what Egypt's grand mufti would have made of being called an agnostic, for the conversation seems to have ended there. Spencer, who had recently had a stroke, was only allowed a few minutes' talk at a time. But Blunt pursued the matter with ʿAbduh as they walked back to the station:

> I questioned the Mufti more closely on the point. *I.* "Do you believe that God has consciousness, that he knows that you exist and I exist, and is not such knowledge personality?" *The Mufti.* "He knows." *I.* "If he knows, he knows that you are good and I am bad." He agreed. *I.* "And he is pleased with you and displeased with me?" *The Mufti.* "He approves and disapproves." *I.* "And he approves to-day because your actions are good, and he disapproves to-morrow because your actions have become bad. Is not this change from approval to disapproval characteristic of personality? How then has God no person?" *The Mufti.* "God knows all things at all times; to him there is no to-day and no to-morrow, and therefore in him is no change; His is an eternal unchanging consciousness of all things. This I call Being, not Personality." *I.* "And Matter? Is not Matter eternal, too, or did God create it? If he created it he made a change?" *The Mufti.* "Matter, too, is eternal as God is eternal." Here evidently is the foundation of Abdu's thought, and we agreed that our ideas are the same.[8]

But the eternity of matter perhaps implied a more specifically materialistic doctrine for Blunt than it did for ʿAbduh, who often spoke against any form of "pure materialism." Blunt, by contrast, spoke of materialism in highly positive tones in his diaries as elsewhere.

In Arabic, the statement *"in al-madda azaliya aydan kama Allah azali"* (matter is eternal as God is eternal) would have struck Muslim readers at the time as a rather familiar one: it certainly did not have the kind of materialistic implications that it might have had for the likes of Blunt, particularly in the light of the spread of philosophical naturalism after Darwin.

Blunt's recollections were later published in Arabic and the dialogue was excerpted and republished as part of a biography of ʿAbduh in the 1930s by Muhammad Rashid Rida, whom ʿAbduh once described as the "transcriber of his thoughts." (Rida compiled and personally published much of the work written under ʿAbduh's name.) But for Rida writing later, the implications of his mentor's encounter with Spencer were quite different from those that led the mufti to visit the English writer in the first place. The Middle East between the world wars was a hotbed of anticolonial foment, and presenting the late grand mufti as the admirer of a British writer—one not quite so eminent as he had been a generation before—was a risky move.

Rida's version, however, gave no cause for worry, and reading the Arabic version against the original, one notices a number of interesting omissions and additions. Whereas in Blunt's recollections, Spencer's lamentations had concerned imperialism and the Europeans' love of war, in Rida's version, he also identified the rise of *materialism* (*al-maddiya*) as the cause of England's—and Europe's—moral decline. Rida's version continues this conversation and it has ʿAbduh reckoning that ultimately "truth and goodness will triumph over materialistic ideas." Spencer agrees and says that this "materialistic tendency will run its course still in Europe." When the Englishman explains to the mufti that "truth for Europeans now is power," ʿAbduh replies that the fascination with power has also gripped Easterners, who imitate Europeans in this. This exchange about truth and power was absent in Blunt's rendition.[9]

Other differences further alter the conversation, guiding it into familiar theological channels:

SPENCER: What do Islam's *ʿulama* say of the Creator—does he exist in or outside the world?

AL-IMAM: The Companions say that God is above all things and beyond the world. The *mutikallimun* [theologians] say that he is neither in nor outside the world. And the Sufis, who profess a unity in all existence, say that all the

> manifestations of the world are a manifestation of his existence. And we believe that God is a *Being* and not a *Personality*.[10]

Blunt's account was thus shown by Rida to fit easily within a standard explication of Muslim theological and cosmological thought, touching on themes that would have seemed familiar: the eternity of matter and God's transcendence and being in the world. Not surprisingly, no mention is made of agnosticism in the Arabic. Rather, at the end, Spencer is quoted as saying "In any case it is clear that you are profound thinkers, like the profound thinkers among us Europeans," Rida's version substitutes "profound thinkers" for Blunt's "agnostics."[11]

ʿAbduh's encounter with Spencer—through its various retellings and afterlives—brings to light the way the British occupation of Egypt and the rise of anticolonial nationalism complicated an earlier generation's enthusiasm for "Western philosophy" and science. Despite Rida's careful editing, ʿAbduh's candid admiration for both Spencer and Darwin had actually worked against him in his own lifetime. As the encounter with Spencer suggests, discussions of evolution also formed part of a broader debate over such things as the role of religion in the progress of civilizations and the relationship between God and nature. These debates threw into relief questions concerning the very role of science, its status as knowledge, and its suitability for transmission among the learned classes of Islam, or the *ʿulama*.

Islam and Modernism

ʿAbduh is still known today as an exponent of "Muslim modernism."[12] The term itself is worth pondering: it was coined after ʿAbduh's death on the basis of an analogy with "Catholic modernism," which Pope Pius X condemned in 1907 as a "synthesis of all heresies" and which was used synonymously with "religious liberalism."[13] Catholic modernists were those who spoke frequently of the reconciliation of theology and science. Indeed, this reconciliation provide a common theme across the various "religious modernisms" as it often served as a key vehicle for the articulation of what ultimately came to count as a new approach to theological doctrine.[14]

ʿAbduh, like other nineteenth-century reformers, did indeed appeal frequently to "science," which he described in what one might roughly identify as modernist terms. His ideas on the subject underpinned his thoughts more generally both on theology and on religious educational reform. Emphasizing rationalism and pragmatism, he helped to construct new exegetical methods and concerns and was among the first of the *ʿulama* to advocate a reconciliation of modern scientific knowledge with religious texts. For example, he helped to popularize the idea—later to spawn an entire and very long-lived genre of interpretation—that one could find evidence of contemporary scientific principles and discoveries (including evolution) in the Qurʾan. Indeed, in his debates with Christian critics and European Orientalists, ʿAbduh frequently made the argument that science and Islam were uniquely compatible and pointed, in contrast, to the evidence of Christianity's own relatively uneasy historical relation with science. For ʿAbduh, modern science and Islam were essentially moving along the same path, that of the advance of civilization.

Yet "religious modernism" is perhaps not the best way to describe what ʿAbduh saw himself as engaging in. In the first place, his vision of science was (as for so many thinkers treated in this book) rather eclectic. Like others of his generation, he drew on the emerging consensus that science was merely the uncovering of the "true principles," or laws, of nature, which he allied to final causes (and divine laws), an approach drawn as much from past Arabic philosophical and exegetical works as it was from contemporary views.

Second, ʿAbduh was primarily concerned with the fate of the Muslim *umma*, not with modernism, and his ideas on civilization and even the "West" cannot be separated from this, particularly as his project of reform was critically couched in an older language of *islah* and *tajdid* (Muslim communal reform and renewal).[15] He drew extensively on Ashʿari themes and borrowed from al-Ghazali and Ibn Taymiya (among others) when discussing contemporary popular practices and beliefs. Struck by his rationalism, which underpinned his advocacy of institutional reform, many of his contemporary critics (urged on or even paid by either the khedive or the sultan) condemned him as a "Wahhabi." ʿAbduh was a harsh critic not only of contemporary popular practices such as the visiting of saints' shrines and the belief in intercession but even of the competence of Muslim jurists, who, he argued, demonstrated a frighteningly wide-ranging ignorance of the true principles of Islam: the *ʿulama*, who in his view should be the real educators of the community, did woefully little for the standing of Muslim civilization in the present. It was this conception of modern civilizational decline that both enraged contemporary *ʿulama* and endeared him to others.[16] For ʿAbduh, civilizational progress was to be ensured by the unification of science and religion—or, more specifically, science and Islam—but it was the latter's conjoined emphasis on reasoning, free thinking, and free pursuit of knowledge that served as the real model for the acquisition of truth. Rida reports ʿAbduh as arguing that despite the advance of the modern sciences and technical arts, they have not provided for man's spiritual needs and that the Arabs can learn from the malaise of Europe. ʿAbduh concludes: "The dilemma of these philosophers (i.e., Spencer) regarding Europe's current state is that it has demonstrated a weakness in its condition [despite the advance of the sciences there] . . . and so what is the remedy? . . . The return to religion . . ."[17] ʿAbduh even cites science in support of this enterprise: science, he claimed in another work, has discovered that there is "more beyond material things."[18] (And, indeed, Spencer's concept of the Unknown further helped him to reinforce this transcendentalism.)

Finally, another problem with the concept of religious modernism is that it dismisses ʿAbduh's critics as "traditionalists." In fact, tradition was a court of appeal for both sides, and the term belies a richer understanding of the very meaning of a Muslim discursive tradition itself, not to mention that it is quite inaccurate to write off opposition to ʿAbduh as merely "traditional" or "conservative" (the latter term often used incorrectly and interchangeably with the former).[19] Such labels also ignore the anticolonial allegiances that brought together religious opponents with either the khedive or the sultan and against the new reformer. Indeed, ʿAbduh was variously lauded as a religious modernist, a rationalist, and a daring reformer, and condemned as a British stooge, a slavish follower of "Spencer's philosophy," and a dangerous adherent of "Darwin's theology" (*madhhab Darwin*).

In considering in more detail ʿAbduh's relationship with modern evolutionary thought, it is important to bear in mind that ʿAbduh's engagement was unquestionably more rhetorical than practical. Unlike the *Muqtataf* group and some of the other figures covered in this book, ʿAbduh was not concerned primarily with the new subjects of contemporary science. Rather, his interest was in their potential contribution to a theological and pedagogic regeneration in the Muslim world. He argued that modern subjects, such as the natural sciences, should make their way into the *ʿulama*'s training curriculum, and he often referred to Darwin and other evolutionists in his lectures and works, but he never spent much time outlining the specificities of the theory of evolution by natural selection or detailing his view on their purported mechanisms and proofs, for instance.

Unlike Afghani, whom ʿAbduh once counted as his most important spiritual and political mentor, his references to Darwin and to the theory of evolution were, on the whole, very positive. By the time ʿAbduh began to bring up the names of Darwin and Spencer in his speeches and lectures, Darwin's reputation had clearly spread, particularly in Egypt through journals like *Al-Muqtataf* and *Al-Hilal*, which ʿAbduh read and admired. To label Darwin a mere infidel and propagator of dangerous materialist ideas, as Afghani did in his lectures against the Indian naturalists of the 1870s, seemed beside the point for the mufti by the early 1900s. What counted when one listed evolution as a universal, natural law was that one also saw it as a divinely ordered law. Any proposition of science held to be true had to be at once good science and good faith.

Like Afghani, ʿAbduh likely never read Darwin's works himself. Nevertheless, tracing how ʿAbduh interpreted Darwin in this indirect, and at times indistinct, fashion sheds light on the ways in which modern ideas of evolution crept into broader discussions of theology, politics, and pedagogy in Muslim lands. As ʿAbduh's discussions of evolution show, he drew on diverse schools of thought, from the traditional schools (*madhahab*) of Sunni law to "Darwin's school" (*madhhab Darwin*). Reading ʿAbduh's works, one can just as easily find him citing contemporary European scientists on the organization of ants as the pre-ʿAsharite ideas of early Muslim theologians.[20]

This eclecticism was partly an expression of his rationalist viewpoint. ʿAbduh was highly suspicious of blind dogma; he often criticized his co-*ʿulama* for customary adherence (*taqlid*) to past rulings in juridical and other matters. ʿAbduh cast a wide net over what counted as part of orthodox Muslim thought—as evidenced by his casual inclusion of the ideas of early Muslim rationalist theologians, the Muʾtazila, alongside those of Ibn Taymiya and by his easy reconciliation with Ibn Rushd on issues of creation and causality while still counting himself an adherent of al-Ghazali. Nevertheless, ʿAbduh was also interested in forging a new sense of orthodoxy, hence his emphasis on such things as the purification of Islam from contemporary heterodox and popular Sufi practices, such as shrine worship and prayers for saints' intercession.

ʿAbduh's eclectic range of hermeneutic references thus show him to be at once embedded in a Muslim discursive tradition while simultaneously fostering new genealogies and outlining new styles of reasoning. He deployed a host of novel arguments: from ideas of the evolution of natural morality in human societies and a new interest in the study of comparative religions, borrowed from the likes of Spencer and Ibn Khaldun, to the construction of historical metanarratives around the rise and fall of a Muslim "Golden Age." If this book has been largely concerned with how contemporary ideas of science came to take hold of the imagination of Arabic readers after Darwin, then, as this chapter shows, this process also helped to create the conditions for a parallel and equally innovative transformation of the notion of religion itself. The chapter looks, among other things, at the rise of a specific notion of "Islam" as an abstract, universal entity, albeit with a distinct historical trajectory and "social" coherence. In the language of the day, it was the view of Islam as a "civilization." As with many civilizational discourses in the late nineteenth and early twentieth centuries, this meant above all discussing the value of science and the specific place of knowledge both within it and against the general run of universal history.

From Exile to Mufti

Twenty years before the visit to Spencer, Blunt and ʿAbduh had dealt with one another in very different circumstances. In the aftermath of the ʿUrabi uprising, an English lawyer described how he

> paid a first visit to Mr. Blunt's personal acquaintance Sheikh Muhamed Abdu. The label on his cell door described him as "journalji" or journalist. At first sight I was a little disappointed. The ex-editor of the Egyptian official journal was a small spare dark-complexioned man entirely dressed in white and wearing a white skull cap. His eyes were intensely black and piercing, and he wore a carefully clipped black beard. The only thing in his cell was a brass ewer, a very humble mattress, and a *koran*. The enthusiasm of the "three Colonels" seemed to be almost wholly wanting in Sheikh Abdu, and even after reading Mr. Blunt's letter to him which we brought with us he seemed to hesitate. He was evidently under the influence of the fear of despair, which had entered into his soul. It was difficult for a moment to recognise in him the patriotic writer and the most militant speaker at the meetings of the council which ruled in Cairo during the sixty days' war.[21]

Muhammad ʿAbduh had been very much involved with Egyptian political life from the start, and politics deeply affected both the formation and the reception of ʿAbduh's ideas during his lifetime. Born in 1849 to a family of *fellahin* (peasants) in Lower Egypt, he was trained in a *kuttab* (elementary school) till the age of thirteen, after which he went to the Ahmadiya Mosque. Unhappy there, he ran away to his uncle, whom he later credited with reinstilling a love of learning in him. He then went to Al-Azhar in 1866, where he was increasingly drawn to politics, journalism, and mystical theology. Much of his time there was spent on his own path of self-education: his experiences at both the Ahmadiya Mosque and Al-Azhar would later form the basis for his critique of the contemporary Muslim curricula. He received his certification in 1877, after which he taught at Dar al-ʿUlum (also known as the Teachers' College) and the Khedival School of Languages, was editor of the official state paper, *Al-Waqaʿi al-Misriya*, and even worked as government censor for a while. Yet ʿAbduh was exiled in 1882 by the newly appointed Khedive Tawfiq because he had been deeply involved in the ʿUrabi revolt—a popular uprising against foreign intervention in Egypt's debt crisis and a revolt that marked the start of British military occupation and administrative rule. In the early 1880s, a coalition of junior officers, merchants, artisans, and intelligentsia had banded together under the leadership of the Egyptian military official Ahmad ʿUrabi Pasha. They claimed they were aiming to secure "the liberties of the Egyptian people," and they also demanded the dismissal of the entire khedival ministry in order to convoke a national, democratic parliament. The revolt led to violence, and the British intervened in support of the royal palace. Although they claimed they had originally conceived of a brief military occupation, British presence in the country would last considerably longer, and they would continue to preside over Egypt's affairs for some sixty years.[22]

ʿAbduh's own career thus coincided with the start of a long and controversial period in Egyptian history. It was his involvement with his spiritual mentor, Jamal al-Din al-Afghani, in particular, and with the various literary salons, political activities, and organizations around the revolutionary thinker that helped to introduce ʿAbduh to a number of key ideas and practices that would later shape his own path.[23] Together with his mentor, he had initially supported Tawfiq's succession to the throne. But this proved to be an unhappy allegiance: the latter, upon taking power, had exiled Afghani to Paris and dismissed ʿAbduh from his teaching post at the Dar al-ʿUlum. ʿAbduh wrote openly for the cause of national reform and took an active role in the ʿUrabists' cause.

After the British invaded Egypt, he was exiled, going first to Beirut, then to Paris to meet up with Afghani, and finally to Beirut again.[24] The British occupation led Afghani to renew his efforts at mobilizing a unified, pan-Islamic front against European imperialism before he ended up a virtual prisoner at the court of Sultan Abdülhamid in Istanbul. But ʿAbduh took another path altogether, one that would see him despair of the Ottomans as an anti-imperialist force and reconcile himself to working with the British as the lesser of two evils.

Back in Beirut, ʿAbduh participated fully in the extraordinary intellectual developments of the early 1880s. He delivered lectures on theology at the Madrasa al-Sultaniya, where he met Husayn al-Jisr and other like-minded Syrian notables. He also completed his translation of Afghani's *Refutation of the Materialists*, no doubt spurred by Shibli Shumayyil's publication of his translation of Büchner's commentaries on Darwin and the broader antimaterialist campaign that took hold of the city while he was there. It was also in Beirut that ʿAbduh first appealed for a pardon from British and Egyptian officials and composed two memoranda—one addressed to Ottoman officials, the other to Egyptian ones—on educational reform. Still looking to Istanbul, ʿAbduh presented the former in 1886 to the *wali* of Beirut, writing of the dangerous spread of foreign missionary schools and the need for national, Ottoman, and religious schools for Muslim students in particular. ʿAbduh's memorandum was very similar to the educational-reform ideas of al-Jisr, with whom ʿAbduh corresponded on the subject; but ʿAbduh and al-Jisr were not alone in their views: as we saw, by the 1880s, there was a growing interest in educational reform throughout the empire.[25] ʿAbduh, like al-Jisr, saw the need for the *ʿulama* to embrace the methods and subjects of modern science as a means not only to ward off the inroads of foreign missionaries but also to safeguard modern Muslims against excessive imitation (*taqlid*) of Europeans—or Westernization (*taghrib*). At first, ʿAbduh seems to have been hopeful about pursuing educational reform under the aegis of the Ottoman state. But it did not take long for ʿAbduh to become disillusioned about the likelihood of reform under Ottoman rule: according to Rida's biography, "Ottoman policemen" seized ʿAbduh's lectures on the theology of unity, revoked his reforms at the Madrasa al-Sultaniya, and shut it down.

In his memorandum to Egyptian officials, sent just a year after he made his proposals to the *wali* of Beirut, ʿAbduh continued to complain of the lack of religious instruction in civil schools—though the problem there seemed to him not so much the presence of foreign missionary educators as the woeful training of Egyptian educators themselves. He recommended the addition of more religious instruction in the general curriculum, but at the same time he warned that religious pedagogues in Muslim schools were themselves in desperate need of reform. He suggested a complete reorganization of Cairo's mosque-university, Al-Azhar. Although earlier attempts at change had failed, he assured the Egyptian government—and British colonial officials alike—that success was possible if reforms were gradual and undertaken by a competent administrator such as himself.[26] With Afghani, in Egypt and Paris, ʿAbduh had been something of a radical. But after his brief stay in Beirut and his return to Cairo—and with his rise in government service there in particular—his approach became more cautious, emphasizing the need for gradual breaks over revolutionary ones. Pedagogy replaced politics as the way forward.

Pardoned by the British, ʿAbduh went back to Egypt soon after the

Madrasa al-Sultaniya was shut down by Ottoman authorities, hoping to engineer a transformation of pedagogy in his homeland. Shortly after his return in 1889, ʿAbduh began his rapid ascent in the state bureaucracy. He was appointed judge in the civil court at Benha, then at Zagazig, then at the chief (ʿAbdin) court in Cairo before being promoted to the post of counselor at the Court of Appeals in 1891. On the accession of Khedive ʿAbbas II (1892–1914), ʿAbduh was appointed to the newly formed 1895 Administrative Council (Majlis al-Idara) of Al-Azhar, established specifically to oversee reforms. ʿAbduh immediately immersed himself in this task, hoping to reorganize the college-mosque's bureaucracy and expand its curriculum. He met with broad resistance from within the ranks of Al-Azhar itself while simultaneously enjoying a certain amount of political protection from none other than the British. Indeed, in 1899, he was appointed to the post of grand mufti, with Lord Cromer's backing.[27]

Even before he was appointed to the post, ʿAbduh had begun to call for religious reform. In 1897 he published the lectures on theology that he had delivered at the Madrasa al-Sultaniya as *Risalat al-tawhid* (Theology of unity), a work that helped to set out his basic theological ideas and program.[28] While mufti, ʿAbduh continued to publish prolifically to promote his views on religious reform and revivalism in Muslim lands: in 1900 he published a treatise on the reform of shariʿa courts in Egypt, *Takrir fi islah al-mahakim al-shariʿiya* (Shariʿa court reforms and regulations); that same year he published his retort to the French statesman and historian Gabriel Hanotaux's writings on Islam; and in 1902 he wrote a series of articles on "science and civilization" in Islam and Christianity (published as a book, *Al-Islam wa-al-Nasraniya maʿa al-ʿilm wa-al-madaniya* (Science and civilization in Islam and Christianity), after his death in 1905). Perhaps most important of all, he collaborated with his student and disciple Muhammad Rashid Rida to publish a popularizing new work of *tafsir*, or Qur'anic exegesis, later known as the *Tafsir al-Manar*.[29] A former pupil of Husayn al-Jisr, Rida was an astute and effective popularizer, and it was he who founded the well-known Muslim journal *Al-Manar* (The lighthouse), where ʿAbduh's exegesis first appeared in serial form and which did the most to help consolidate ʿAbduh's reputation.[30]

As this brief sketch suggests, ʿAbduh's career developed in large part thanks to the rise of new media and organizations, such as newspapers, publishing houses, and modern libraries and schools as well as salons and associations. ʿAbduh's own ideas took shape in these forums, and they affected his writing: for instance, he repeatedly mentions the need for an easy, didactic style that can override the narrow scholastic and classical lexical concerns of contemporary theologians. Utilizing the power of the printed word from the start, ʿAbduh's campaigns in favor of nationalists in Egypt during the ʿUrabi crisis had demonstrated the political effectiveness of the press. As a government official in Egypt, he utilized it in much the same way. But the press was a double-edged sword and his years of greatest professional and public success were also his most contested.

Natural and Divine Law

The consciousness of writing for a new public pervades ʿAbduh's early work, notably the *Risalat al-tawhid* (Theology of unity), which was written in the early 1880s in Beirut and introduces many of the core themes that would recur in his later writings. Delivered in a style that would come to mark the new, more popular didactic approach to theology, it was free of much of the specialized and arcane vocabulary of scholastic theology, had unusually few references to other commentators, and made a conscious attempt to speak directly to the concerns of the educated Muslim reader.

Chief among these concerns was the relationship of natural knowledge to faith. "The study of creation is necessarily salutary in a practical way and illuminates for the soul the way to knowledge of Him whose are these traces," wrote ʿAbduh, drawing on lectures he had delivered as a teacher at the Madrasa al-Sultaniya.[31] Like al-Jisr (who taught at the madrasa when ʿAbduh was there), ʿAbduh argued that the study of "the manifestations of the universe" could ultimately bolster faith, and again like al-Jisr, he thought that contradictions between science and scripture were impossible: "Contradictory views of the universe are part of the conflict between truth and error," he wrote. Like so many others of the time, ʿAbduh upheld that true science could never contradict true religion; yet unlike many, he insisted simultaneously on the rationalist position that true faith could not contravene the laws of logic.[32]

Also unlike al-Jisr, ʿAbduh spent little time over the challenge posed by contemporary natural philosophy. And while ʿAbduh kept up with contemporary Arabic writings on science, his own view of science was multifarious: he drew on past Muslim notions and categories of knowledge, but he also appealed to new disciplines and discoveries that he saw as underlying Europe's material and technological progress. We know ʿAbduh read *Al-Muqtataf* with great enthusiasm: indeed, the young ʿAbduh, when a government official for the censorship bureau, was among the state officials who had enthusiastically greeted the Syrian editors during their brief visit to Cairo in the early 1880s. Not surprisingly, therefore, his discussion of "science" reflected prevailing ideas: he emphasized the use of free thought and independent reasoning and stressed the value of the accumulation of facts and the systematization of natural—and social—regularities, counting them as among the latest contributions to knowledge offered by the modern sciences. Indeed, he referred to "science" broadly as any systematic natural inquiry concerned chiefly with the discovery of those absolute and inviolable natural laws that governed the universe—laws of nature he typically referred to as *sunan*, a term more usually applied to religious laws.[33]

Interpreting the Qur'anic verse "Such hath been the custom [*sunna*] of God, and no change can thou find in it," in an article entitled "Al-haqq, al-batil wa al-qiwa" (Truth, falsehood and power), ʿAbduh defined *sunna* in terms of what we might now call natural law. The universe has laws (*sunan*), evidences of which are to be found everywhere in nature, from the composition of precious stones and other rocks and the development of plants and animals to the composition and disposition of natural bodies. Quoting other verses from the Qur'an on creation—such as "God created the sun and moon to lighten our paths and organized their cycles so we could count the years, and this he did for those who desire to learn and follow the path of truth"—ʿAbduh concluded that he who has knowledge of the order (*nizam*) of creation and of existence (*wujud*) "knows the truth" (in other words, the divine truth) of the laws of life.[34]

Like al-Jisr before him, ʿAbduh was here also building on older Muslim conceptions of nature—and of knowledge of it. To know nature, one had to first uncover the nature of knowledge itself: here *ʿilm* was a broad category, at once a technical term (for moderns and ancients) and an attribute of God, and therefore in essence a divine, spiritual power. The latter point in particular was one that ʿAbduh made at length in his article "The Meanings of *ʿIlm*," where he cites a number of Qur'anic verses to elucidate its meaning.[35]

ʿAbduh's notion of "law," or *sunna*—as in his use of the term *sunan al-tabiʿa* (natural laws)—was also fundamentally built on classical arguments over the problem of causation. This can be seen, for instance, in his treatment of the problem of efficient causes in nature—those that could be said to operate independently of God's immediate intervention. His debate with Farah Antun over Ibn Rushd centered precisely on this point. Their debate, which took place in the Arabic press in the early years of the twentieth century (and discussed at greater length below), saw ʿAbduh take a largely pragmatic stance on questions of causation in nature. For ʿAbduh, causation in nature was akin to predestination (*qada*ʾ), or what the Ashʿarites termed "necessity." But his conception of it left plenty of room for freedom of will, while leaving open the question of efficient causes.

ʿAbduh was fundamentally a pragmatist: on questions of ethics, he essentially appealed to conscience and to everyday experience to make the point that man was free in his actions despite God's omniscience. The same was true of cause and effect in nature. Even as such causation—the essence of natural law—could be said to be merely a reformulation of divine law, they existed independently to some degree.[36] ʿAbduh's reconciliation shows, therefore, that a subtle range of positions were available to Muslim theologians when addressing questions of causation and creation: just as he left room for free will in ethics, there was room for causation and natural laws to exist free of the will of the divinity.[37] ʿAbduh thus cleaved to a notion of "law" that sidestepped the problem of the iron necessity of cause and effect.[38]

Of course, between a positive conception of law and the idea of *sunan*, there were obvious points of tension and difference, but the fact that the latter could also be made to fit with certain positivist critiques of the limits of inductive reasoning served as grist to his mill. The issue of miracles raised this issue inescapably for him. If nature could be said to operate according to regular laws, accessible through rational or scientific investigation, what of those events that might be said to transcend these, such as in the accounts of miracles in the Qurʾan? The question of miracles (*ayat*, which also means "signs" or "verses," as in the verses of the Qurʾan, there implying the miracle of revelation itself) made this dilemma even more apparent. ʿAbduh resolved this by allying the term to his understanding of "God's laws": in other words, he left room for a view of the miraculous as that sphere of events which was ultimately unknowable to the human mind.

His discussion of miracles was therefore another way to get at the nature—and limits—of scientific laws themselves. Such laws, he argued, did not constitute a fundamental ontological reality: they were merely deductions from observed regularities. His ontological skepticism thus allowed him to clarify the distinction between scientific laws and the *sunan* of God. For while there might exist exceptions to the former in the shape of miracles (though ʿAbduh was quite cautious in what he counted as a miracle itself), there were no exceptions to God's laws, whose ultimate truth or reality was fundamentally unknowable to humans. As ʿAbduh would repeatedly argue, the human mind simply cannot know the essence of things.[39] The appeal to nominalism permitted this dual—and yet ultimately unified—view of *sunan*: at once natural (and therefore inviolable and attainable) and divine (and therefore ordained and unknowable). It is not surprising that ʿAbduh thus found the synthetic philosophy of Herbert Spencer, with its own conception of the Unknown, so congenial.

Adam and Evolution

Among the natural laws, ʿAbduh counted the "laws of evolution," which he referred to time and again in his exegesis of the Qurʾan. ʿAbduh had little difficulty reconciling modern principles of evolution with revelation.[40] Interpreting the Qurʾan's Surat al-Baqara, for example, ʿAbduh construed the verse "Were it not for the restraint of one by means of the other, imposed on men by God, verily, the earth would be full of mischief" to be "one of the general principles of natural history, and this is what in this age the learned refer to as the struggle for survival." He added that they say that war is "natural to man" because it is one aspect of this principle.[41]

ʿAbduh in fact at times seemed to go out of his way to demonstrate the potential compatibility of the Qurʾan with contemporary ideas of evolution. He did so primarily by appealing to a broad range of flexible interpretive strategies and hermeneutic traditions of scriptural exegesis in Islam. Take, for example, his treatment of the verse "O men! Fear your Lord who has created you of one soul [*nafs*] . . . and from this many men and women." The reference to "one soul," he claimed, did not refer specifically to Adam. Since all peoples were being addressed, many of whom know nothing about Adam and Eve, how could such a particular reference be intended? "God has left this matter . . . indefinite, so let us leave it in its indefiniteness." The reference to "one soul" should be taken to refer to the concept of humanity itself, and from this he made what, to some, might seem a surprisingly bold judgment: "Men are created from one soul, which is humanity . . . and all men are brothers in humanity . . . which is why it matters little if they claim their father is Adam or a monkey or something else."[42]

ʿAbduh interpreted the notion of man's descent with considerable flexibility. He saw the notion of descent as an expression of the movement of matter (in some general sense) from mineral to animal and then to human. He concluded from this that man could thus be seen as having a dual and yet unified nature: at once possessing an animal nature (body) and a uniquely human one (soul).[43]

Both ʿAbduh and *Al-Manar* were sometimes criticized for these views.[44] Yet they were not deflected. For ʿAbduh (and for Rida) it mattered little if evolutionists claimed man descended from brutes or not—and, as they were careful to explain on numerous occasions, they neither condemned nor endorsed that idea. And while Darwin himself was referred to by *Al-Manar* as that "celebrated natural scientist," his theories, they often repeated, were in truth "merely hypothetical."[45]

Yet ʿAbduh also took care not to draw heretical opinions from a theory of evolution. "Now I disagree with what some philosophers have said regarding Ibn Rushd . . . that he stood against religion by claiming the soul does not remain after the death of the body, arguing that only the spirit of the species [*arwah al-anwaʿ*] remains." These philosophers misunderstood what Ibn Rushd meant. Ibn Rushd, like Aristotle and others, was merely claiming that while individuals come and go, live and die, species do not: "they remain (even as they evolve)."[46]

For ʿAbduh, the crux lay in the social meanings one could discern from Darwin's theory. Evolution was, above all, a process of "social development"; and like so many writers of the time, his interest in social laws was another example of the way in which new conceptions and categories of knowledge were infiltrating Arabic discourses. "The *ʿulama* [scientists] all agree about one of the greatest social laws [*sunnat al-ijtimaʿ*], and that is what is known as natural selection." These "social laws" were also absolute natural laws, though in this case they governed the operations of "society" (*al-haʾya al-ijtimaʿiya* or *al-mujtamaʿ*) (an increasingly popular nineteenth-century neologism in Arabic, as in English and French).[47]

Yet his description of the laws of societies, their rise, stagnation, and fall, could also easily fit within more long-standing traditions of Muslim thought. He was particularly interested in Ibn Khaldun, whom he considered a kind of forefather of modern sociological knowledge. When still a young man teaching at the Dar al-ʿUlum in 1878, ʿAbduh had delivered lectures on Ibn Khaldun and the evolution of human society. He even made the argument that the thirteenth-century Arabic thinker shared a number of critical social theories with Auguste Comte.

ʿAbduh argued that humans have special laws by which their individual lives and their collective life were governed, including their "strength and weakness, wealth and poverty, might and humiliation, domination and subjection, health and illness, life and death." Such laws determine mankind's well-being: only those who know them can "achieve complete happiness." In similar fashion they determine the conduct of nations. In the end, ʿAbduh made clear from where he thought these laws ultimately derived: only those who "live according to the laws of God" continue to be the ones who hold the "widest dominion of all the nations."[48]

Science and Scripture

For ʿAbduh Islam had "set free the human intellect," allowing it to follow nature's established course by "arousing the mind to consider creation and the various signs of God's power in nature." Through an investigation of such laws, therefore, man could also attain knowledge of God.[49] The Qur'an, he emphasized, does not "restrict the mind" with respect to these things in any way: verses that "motivate consideration of the signs of God in nature" total a third, if not half, of all the verses of the Qur'an he claimed.[50] Behind all this lay an old theme: the idea that "God has sent down two books," the works and word of God, or *al-kawn wa al-Qurʾan*. The latter guides men to investigate the former, ʿAbduh wrote, "following the paths of science."[51]

So convinced was ʿAbduh of the harmony between science and scripture that he championed a new kind of Qurʾanic exegesis—one that revolved around scripture's scientific veracity—to meet the "needs of the time." His *Tafsir al-Manar* was one of the earliest works of serial *tafsir* to appear in print and became one of the best known. In it, ʿAbduh interpreted scriptural references in light of the "latest findings in science and medicine." On the possibility of *jinn* (spirits), for example, ʿAbduh wrote: "The *ʿulama* say that the *jinn* are living bodies that cannot be seen. *Al-Manar* has said more than once that it is permissible to say that those minute living bodies made known today through the microscope and called 'microbes' are possibly a species of *jinn*." The *Tafsir al-Manar* made a considerable number of such references on topics ranging from the nebular hypothesis, somnambulism, and embryology to the telephone, radio, airplane, and microscope. In this way, ʿAbduh reiterated the view that there could be no conflict between the findings of modern science and the Qur'an: "We Muslims are fortunately under no necessity of disputing with science or the findings of medicine regarding the correction of a few traditional interpretations. For the Qurʾan itself is too elevated in character to be in opposition to science."[52]

ʿAbduh consistently claimed that Islam in particular was a "friend of science" (or knowledge) and a stimulus for research into the secrets of nature. He connected this compatibility to the particular rationality of the faith: the Qurʾan consistently urges the use of reason in matters of belief. Hence, the very methodology of "science" can be seen as an essential part of the Muslim injunction of rational inquiry. He argued that *al-ʿaql wa-al-naql*—reason and revelation—were uniquely united in Islam and claimed that those fellow *ʿulama* who assert that science and reason should have nothing to do with religion were "ignorant of the true principles of Islam."[53]

Against the claim that faith required only "submission" and not rational investigation, ʿAbduh asked how reason could be denied its right when it is the "judge of all evidence," reaching the truth of all things in order to "know what is divinely given." He concluded: "If this claim were to be allowed, religion would not be a means whereby man could be guided." Religion was a necessary accompaniment to reason. Reason alone could not lead men to happiness, whereas "religion is a general sense by which to discover the things that elude reason among the means to happiness."[54]

ʿAbduh's conception of reason was further tied to his philosophy and history of religions and to his views on the evolution of natural morality. According to him, religion evolved from a primitive state of belief to monotheism, with the latter also progressing over time: first there was Judaism, a religion that appealed to the senses and relied on evidence of God's absolute power through recourse to miracles; then came Christianity, a religion that appealed to sentiment and to the heart, emphasizing the mysteries of nature and God and preaching a message of love (though this would slowly be corrupted over time); and, then finally, there was Islam, which, according to ʿAbduh, brought together sense, feeling, and reason. "Islam reconciled reason with nature, and, recognizing neither master nor mysteries, freed minds from the tutelage of authority and brought man through his faculties closer to God."[55]

The importance of rationality as a means of acquiring knowledge of God in Islam was a common leitmotif for ʿAbduh: most of his writings stressed the need for reason to accompany revelation. So much importance did he assign rationality that in cases of a disagreement between reason and revelation, he argued that conclusions that have been definitively arrived at by reason were to be preferred. For in such cases, there were only two possibilities open to scholars of the tradition: either one acknowledged the genuineness of scripture while confessing man's inability to understand it (thereby "resigning the matter to God" and his supreme, if unknowable, knowledge), or one interpreted the revealed text in such a way that it could agree with what reason had established. Clearly, he favored the latter approach.[56] Indeed, the Qurʾanic esteem for rational judgment, "together with the use of parables in the allegorical or ambiguous passages in the revealed text," is precisely what "forbids us to be slavishly credulous."[57]

Reason thus marked the limits of literalism, and ʿAbduh was careful not to deny any literal meanings of scripture except where reason demanded it. Writing on the story of Noah and the Flood, for instance, he advised faith in its literal meaning unless positive, rational evidence showed that the literal meaning was not the one intended by the text. But arriving at such a conclusion, he added, is a matter that requires "deep study," as well as extensive knowledge of such things as geology and stratigraphy." Making reliable judgments thus required both the traditional hermeneutic skills and the rational sciences. Whereas whoever "speculates irrationally" or without certain knowledge is "reckless," and he should "not be heard nor permitted to propagate his ignorance."[58]

To be sure, ʿAbduh felt ultimately that the mission of science differed from that of revelation. Al-Ghazali, whom ʿAbduh was particularly fond of, was an influence here. "It does not belong to the office of the messengers to be instructors and teachers of craft. They do not deal with the lessons of history nor with the analysis of astronomical worlds and the diverse theories of stars in their courses," wrote ʿAbduh, echoing the medieval philosopher's views on the role and function of the *ʿulama*. "Outside their province are the storehouses of the earth, the dimensions of the world's length and breadth, the sciences that study plants in their growth and animals in their quest to survive. All of these and more belong wholly with the means to material acquisition and well-being and are within those gifts of comprehension whereby God has willed that men be directed." Although revelation, he wrote, sometimes deals with matters of craft or science—such as "the movements in the heavens and the fashion of the earth"—these were "intended only to direct attention to the creator's wisdom or to depth in the apprehension of His mysteries and marvels." In any case, he cautioned, "a long time is necessary for the esoteric meaning to become understandable among the masses."[59]

Similarly, in an interpretation of several verses in Surat al-Baqara that refer to thunder and lightning, ʿAbduh claimed that the "truth about thunder, lightning, and storm clouds, and the reasons for their occurrence," is not among the subjects investigated by the Qurʾan because these belong to the "sciences of nature and of the atmosphere," which men acquire by their own exertions and not through revelation. After all, he pointed out, knowledge of the universe waxes and wanes among peoples and changes over time. People once believed that lightning was caused by material bodies, but now it is attributed to electricity.[60]

ʿAbduh himself was thus one of the first to apply *tafsir ʿilmi* ("scientific exegesis") to the study of the Qurʾan. His only predecessor in this was Muhammad Ibn Ahmad al-Iskandarani (d. 1889), a onetime Ottoman naval officer, Damascene physician, and lay theologian from Syria who published the three-volume work *Unveiling the Luminous Secrets of the Qurʾan and Their Relation to the Heavens, Earth, Animals, Plants, and Minerals* in 1880 and *The Book of the Divine Secrets of Plants, Minerals, and Animal Characteristics* two years later.[61] Both works emphasized the correlation between scriptural and modern scientific (particularly medical) claims.[62] For the physician al-Iskandarani, nature's secrets—equated with the Lord's divine order—were riddles for men to uncover, and both science and the Qur'an were there to help them.[63] This was not an altogether new theme. For generations of Muslim commentators and exegetes, nature—as well as references to the natural world in the Qur'an—had long been regarded as a body of divine signs. Indeed, the idea of God's "two books" was a long-standing theme and many writers in the past had sought similarly to demonstrate how Islam could easily be reconciled with what we might now call "science." Sometimes such demonstrations could be found within the genre of *tafsir* or Qurʾanic interpretation or commentary itself.[64]

What made ʿAbduh's (and al-Iskandarani's) works novel was that they were at the intersection of a number of genres: they were in part broad exegetical reflections and moral guides and in part popular-science compendiums and pedagogical exhortations. As such, they marked the beginnings of the transformation of an enduring scholastic (and primarily philological) tradition of *tafsir*—in the minds of its later critics, corrupted—into a new kind of popular-science literature. By the turn of the century this approach would constitute a full-blown element of Qurʾanic commentary and would be known by admirers and critics alike as *tafsir ʿilmi*, or scientific exegesis of the Qur'an.

ʿAbduh's own *Tafsir al-Qurʾan* was far more influential than al-Iskandarani's work in helping spur an interest in such exegetical approaches to the Qur'an, particularly after his death in 1905. Serialized initially in the journal *Al-Manar*, this pioneering *tafsir* was widely read by intellectuals, educators, journalists, and civil servants—in other words, by newly educated readers who were emerging in the Muslim world at the end of the nineteenth century. Many later exegetes cited it in their own works and followed its style and structure. Like the *Tafsir al-Manar*, many of these works were written not for the traditional scholarly classes but for the new mass readership.

Later works in this vein included those of Muhammad Farid Wajdi and Tantawi al-Jawhari.[65] Like ʿAbduh, these exegetes sought to demonstrate in exhaustive detail that science was in full harmony with scripture. Science, they felt, and its laws of nature and society could only reinforce the divine laws of Islam. Islam, Wajdi wrote, "is in total agreement with the intellect and science, and it is in total agreement with the fixed laws of nature," for it is "a religion that inadvertently benefited from natural science, without the experts realizing it."[66] As for ʿAbduh, this emphasis on the harmony of scripture and science also helped pave the way for a new kind of theology.

Science and *Kalam*

The emphasis on proving Islam's compatibility with reason and science that featured prominently in ʿAbduh's writings was characteristic of the movement throughout the Muslim world toward encouraging the emergence of a "new theology," or what Ottoman scholars increasingly took to calling *yeni ilmi-i kalam*. This was the learned counterpart to the rise of the *tafsir ʿilmi*: if the latter emphasized themes of the "miraculous Qurʾan" and of unlocking the wonders of nature, the move to *kalam* focused on theological reasoning and the tradition of speculative theology, or *ʿilm al-kalam*. As we saw in the previous chapter, many appealed for a revitalization of *kalam*. ʿAbduh was at the forefront of this trend in Egypt.

The need to revitalize *ʿilm al-kalam* was particularly necessary, "especially in these days," when the faith was under attack from critics of Islam, wrote ʿAbduh in 1877. He argued that reasoned proofs of God's existence would prove to be Islam's strongest bulwark against critics and doubters. Such a theology was also urgently needed to address current "spiritual ailments"—and to ensure that Islam would be free of intellectual "stagnation" (*jumud*), a charge, as we will see later, that he was to frequently raise against the majority of the ʿulama, particularly those at Al-Azhar.[67] ʿAbduh thus saw a revitalization of *kalam*, through the use of reasoning (*nazar wa-istidlal*), dialectics (*jadal wa-munazara*), and logic (*mantiq*) as the best means to preserve the faith in modern times.

His strong views on the subject were evident in the *Risalat al-Tawhid*, as well as in another Beirut work—his introduction and commentary on al-Sawi's treatise on logic, *Basaʾir al-Nasiriya*. He borrowed extensively from Ibn Rushd, Ibn Sina, Aristotle, and contemporary French thinkers to argue for the need for logic as a necessary accompaniment to morality and faith itself.[68] In particular, ʿAbduh felt that the revitalization of a rationalist theology was needed to respond to changing methods of proof and verification. With the rising popularity of the modern sciences and the emphasis they placed on inductive methods of observation, proof, and experimentation came the decline of classical and particularly syllogistic logic in arguments of faith. ʿAbduh wanted to wed the two worlds. He appealed with an increasing sense of urgency to the *ʿulama* to revitalize their study of the sciences of scholastic theology (*ʿilm al-kalam*) and to teach "new subjects" (*al-ʿulum al-haditha*), including the modern and applied arts and sciences, in their traditional schools, especially Al-Azhar. In practice, he wrote, the *ʿulama* of Al-Azhar concerned themselves with neither the scholastic nor the contemporary sciences and paid no attention to the fact that "we are in a new world." Indeed, they treated *ʿilm al-kalam* with the highest misgivings, so that if they discovered a student among them studying the subject, they declared: "What a scandal! And they say, 'How can you study the incorrect sciences, so that you come under suspicion?'"[69]

ʿAbduh, like Afghani before him, urged the *ʿulama* to revitalize the study of *kalam* not only to defend Islam against critics but also to pave the way for them to enjoy the "benefits" of the new knowledge acquired by the modern sciences more generally. *Kalam* was the very means and method for the foundation of any rational investigation: for ʿAbduh, as we will see later with his educational-reform policy, the foundations of classical Muslim epistemology would help to make sense of modern subjects themselves.

ʿAbduh regarded the modern sciences (and technologies) as the basic source of Western advancement. In this too he was following in Afghani's footsteps. Science "everywhere manifests its greatness and power," wrote Afghani in his "Lecture on Teaching and Learning," delivered in Calcutta in 1882. "Ignorance had no alternative to prostrating humbly before science." It is evident "that all wealth and riches are the result of science." "There are no riches in the world without science and there is no wealth in the world other than science." He also attributed Europe's ever-growing imperial and colonial expansion to science. "We see no reason," echoed ʿAbduh, speaking of the relative strength of Western over Eastern nations, "for their progress to wealth and power except the advancement of education and the sciences among them." Therefore, Muslims should "endeavor with all our might to spread these sciences in our country."[70]

Orientalism and Islam

Discussions of science and Islam thus raised the issue—at once historical and political—of the "greatness and power" of the Christian West versus the Muslim East. After the Near Eastern Crisis of 1876, many European writers questioned whether Islam had a "civilization" at all: some pointed to the "brutal suppression" of the Bulgarian uprising and argued that the Turks were nothing but "barbarians pure and simple," predicting that the "soul-killing, emasculating and polygamous institutions" of "Mahometanism" were sure to "pass away before the advance of Western civilization."[71]

The French invasion of Tunis in 1881 and the British occupation of Egypt the following year exacerbated the sense of a "vast moral and political chasm" between Christianity and Islam, and this became an increasingly popular motif in English and French journals. This led some to argue either for the need for the imperial suppression of all forms of "Mahometan despotism and superstition" (as many pointed out, Queen Victoria ruled over more Muslims in India than the Ottoman sultan did in his empire) or to an eventual "clash of civilizations."[72] From the early 1880s onward, a number of key Orientalists, such as Ernst Renan, and colonial commentators weighed in on the subject. Renan's writings on Islam and science, Gabriel Hanotaux's discussion of Europe and Islam, and Lord Cromer's depiction of Islam in his colossal 1908 work *Modern Egypt* all charged that Islam was a fundamental impediment to modern forms of knowledge, philosophy, and science—and hence the prime cause of the decline of Muslim polities.

Naturally, such disparaging views attracted a great deal of attention among Muslim writers, leading, among other things, to rebuttal: it is in this process that we witness the emergence of a fundamental reconceptualization of the notion of Islam itself. For the first time, Muslim theologians began to speak of Islam as a kind of "civilization," pointing to the past grandeur of its political and intellectual accomplishments by way of illustration and to the contributions that these achievements made to the development of "Western civilization" itself: indeed, authors such as Gustave Le Bon, who made similar kinds of arguments, proved popular in Arabic for this reason. It was in this context that the rescripting of ideas of "science" for "Islamic civilization" offered important rhetorical reinforcement.[73]

Afghani was among the first to encourage this reevaluation. In 1883, when both he and ʿAbduh were in Paris, he famously engaged Ernst Renan in a debate over the subject. Renan had delivered a lecture at the Sorbonne, "L'islamisme et la science," in which he described the relation of Islam and science as one of opposition and mutual distrust.[74] "This eminent philosopher," Afghani wrote in his reply, in the *Journal des Débats*, "applied himself to proving that the Muslim religion was by its very essence opposed to the development of science, and that the Arab people, by their very nature, do not like either metaphysical sciences or philosophy."[75] Afghani's response made the point that such an opposition to science could in no way be seen as the fault of Islam or the Arabs but was rather the responsibility of those governments and nations that at various times had propagated the faith. He distinguished between Islam and Muslims—or between the "religion itself" and "the character, manners and aptitudes" of those who held the faith.[76] Yet Afghani's argument was also somewhat ambiguous, for he also seemed to suggest that all religious dogma was to oppose philosophy and, hence, the sciences: "religions, by whatever names they are called, all resemble each other" in this respect. A careful reading of the text shows how he limited this opposition to dogma (versus "free enquiry"); regardless, it is alleged that he later objected to the idea of translating his reply into Arabic.[77]

This discussion set the stage for later exchanges. ʿAbduh's response to Hanotaux in 1900 made a similar distinction between Islam and Muslims: "Muslims today are not Muslims," he wrote. Victims of imitation (*taqlid*) and stagnation (*jummud*), they have strayed from the true path of the faithful (this theme of decline would soon come to dominate his writings). Hanotaux, who had been the French minister of foreign affairs, wrote to present Europe, and particularly republican Christian France, as the rightful rulers of Muslim peoples. His conclusion was that France, as the carrier of the spirit of modern civilization itself, was the better overlord of Muslim populations, and even the better preserver of their traditions, than any previous Muslim rule had been. He also presented Arab Muslims (much as Renan did) as basically unchanging, eternally locked in a kind of "Semitic mentality." Whereas Aryan Christianity gave men free wills, saw God as fashioned after man, and hence gave rise to humanist ideals, Islam fostered superstition, encouraged fatalism, and "required men's complete submission to their God."[78]

ʿAbduh, who by then was grand mufti of Egypt, rebutted both the racial typography and its religious implications. After all, he pointed out, was not Aryan civilization itself born in India, the birthplace of idolatry? And had not the Phoenician and Aramaic peoples contributed to the vitality of the Semitic race, particularly in the arts and sciences? As for the ethical divides supposedly at stake, ʿAbduh rejoined that Hanotaux misrepresented the role of predestination and free will in Islam. After all, he pointed out, predestination could be said to be attacked in at least forty-six verses of the Qurʾan.[79]

ʿAbduh's chief defense, however, was to refer to the grandeur of early Islam as a "civilization," and he ended his response with just such an homage. He argued that it was Islam which transformed "simple nomadic Bedouin tribes" into a great and powerful nation, creating from such "barbarians" a new "civilized peoples" while simultaneously giving them a taste for science, art, and industry. With Afghani, ʿAbduh therefore highlighted an argument of civilizational development and progress. This was a point that ʿAbduh himself had long been interested in: after all, while teaching at the Dar al-ʿUlum in the late 1870s, he gave a series of lectures on the "philosophy of society and history," concentrating in particular on "the rise and fall of civilizations" and drawing from both Ibn Khaldun's *Prolegomena* and a recent translation of François Guizot's *History of Civilization in Europe*.[80]

But ʿAbduh placed a very different kind of emphasis on questions of science and civilization in Islam than did Afghani, who seemed to suggest that all religions had been locked in a kind of historic battle between dogma and philosophical enquiry. ʿAbduh argued that Islam alone set free human reason, fostered a spirit of tolerance toward other faiths, and "spread light and truth wherever it penetrated," giving the world innumerable scientific treasures and, from the start, "inoculating its civilization from barbarous Europe!"[81]

ʿAbduh's interest in civilizational progress was thus tied to a new language of human evolution. In turn, this helped to create a new historical narrative around the idea of the rise and fall of Muslim civilization. Of course, as mentioned before, ʿAbduh also added to this a more traditional conception of *islah*, or reform—the need for perpetual moral guidance of the community of believers. For ʿAbduh, true scientific and civilizational progress was predicated above all on moral virtuousness: as history has demonstrated, the pursuit of true knowledge was supported and sponsored by just rulers and states and undertaken by a virtuous moral community of knowledge seekers who valued the use of reason and developed their faculties to bring them both this-worldly and otherworldly success. He believed that the current state of Europe showed that the excessive pursuit of science alone, while ignoring the necessary role of religion in society, led to civilizational malaise. The search for an authentic knowledge tradition and its attendant sociomoral order thus formed the backdrop to the modern interest in the Muslim classical past and played a significant role in discussions of the "Arab Renaissance" (Nahda) of the time.

For other Arab evolutionists, however, this view of science, civilization, and religion was itself part of the problem and the real cause of Muslim decline: the only way to true progress was to follow the course of post-Enlightenment Europe and to separate science and religion entirely. A debate published in the Arabic press in the early twentieth century between the mufti and one such critic shows how different claims about the relation between science and faith would further shape ʿAbduh's own historical narratives of science and civilization in Islam.

Science and Civilizations

Shortly after his reply to Hanotaux, ʿAbduh once again engaged in a highly publicized debate on the historical significance and status of Islam. This time the subject was the medieval philosopher Ibn Rushd, or Averroës, and "the destruction of philosophy in Islam." ʿAbduh published his response in *Al-Manar*, and some of these articles were later republished by Rashid

Rida after ʿAbduh's death in 1905. He also published a series of articles on science in Islam and Christianity more generally that were similarly collected and published by Rida under the title *Al-Islam wa-al-Nasraniya maʿa al-ʿilm wa al-madaniya* (Science and civilization in Islam and Christianity). But on this occasion, ʿAbduh's defense of Islam was a response, not to an Orientalist or European colonial critic, but to the Arab writer Farah Antun. This exchange shows once again how ʿAbduh positioned himself at the very center of arguments concerning both the proper boundaries and aims of science and the relation of science to "civilization" in Christianity and Islam.

Antun was another of those missionary-trained Syrian Christian émigrés who resided in Cairo, and his journal *Al-Jamiʿa* aimed, like *Al-Muqtataf* (the editors of which Antun was close friends with), to popularize the latest scientific lore, though in a much more sporadic and superficial fashion. Between 1902 and 1903 he wrote a series of articles chronicling the life of the twelfth-century philosopher Ibn Rushd that drew heavily on Ernest Renan's *Averroès el l'averroïsme.* (Antun was in fact something of a devotee of Renan's, and he later published a translation of his *Vie de Jésus.*)[82]

Antun's choice of the figure of Ibn Rushd was certainly not accidental. It marked the beginning of a renewed interest in questions of reason and revelation and coincided with the revival of Arabic treatises on the philosopher, who was one among a number of key thinkers to feature prominently in the resurgent discourse of a golden age of Arabic philosophy and science.[83] And although there is no proof one way or the other, it is hard to imagine that in choosing to discuss the predicament of Ibn Rushd, he was not deliberately setting out to criticize by analogy the views of Grand Mufti ʿAbduh himself.[84] For the saga of Ibn Rushd, wrote Antun in his introduction to the later compilation *Ibn Rushd wa-falsafatuhu* (Ibn Rushd and his philosophy), demonstrated the dangers of mixing philosophy and science with religion. After all, Ibn Rushd was an accomplished theologian, jurisprudent, and philosopher from Cordoba, and although once in favor with Caliph Mansur, he was later exiled and denounced by orthodox critics and statesmen alike for promoting a rationalist philosophy of faith, such as that found in his infamous reply to al-Ghazali's *Tahafut al-falasifa* (The incoherence of the philosophers) entitled *Tahafut al-tahafut* (The incoherence of the incoherence). His persecution, Antun felt, was a warning against those who tried to bring together science and religion, which, like reason and revelation, were best left to their separate jurisdictions. Did not Caliph Mansur issue an edict warning men of God's eternal damnation for all who dared to claim truth could be found by unaided reason? Did not the caliph then order that all books on logic and metaphysics were to be burned?[85]

But Antun also turned to Ibn Rushd for another reason: his defense of causality. Ibn Rushd's arguments were delivered in response to al-Ghazali's own rejection of cause and effect, which he had argued, in a fashion one might now associate with modern empiricists, was nothing but the sequence or conjunction of events. For al-Ghazali, God alone was to be assigned as the agent of any so-called effect. Yet for Ibn Rushd, this amounted to nothing but the rejection of secondary causes altogether, which he felt went against the causal efficacy of things (although he also assigned God as the final cause of all events); for Antun, it was this (albeit modified) view of the necessity of causality that lay behind his own view of science itself.[86]

So while Antun admired Ibn Rushd, he also had his reservations. In fact, he wrote of Ibn Rushd's life and times precisely to illustrate these reservations—or the inherent threat, as he saw it, posed by religion to both philosophy and science. Like Renan, Antun was interested in the relation of philosophy and prophecy and what he saw as the inevitable conflict between the two. Ibn Rushd, wrote Antun, attempted to overcome this conflict, claiming prophecy to be a kind of truth made accessible to the religious masses yet attainable through contemplation.[87] His fatal error, Antun thought, lay in his attempts to uncover metaphysical truths with philosophical tools: the only purpose of religion was to exhort men to virtue. Just as the spirit of the age and the demands of modern times require that we separate spiritual and civil authority, he argued (following Renan), then so too should we separate questions of science from questions of religion. Science, he felt, concerned worldly matters (the realm of philosophers and kings, not priests); whereas religion concerned purely "otherworldly" matters (souls and salvation, not systems of knowledge). In fact, whenever religion has attempted to meddle into the affairs of science, disaster has ensued, the only relation between them being one of conflict. The best means of resolving this conflict, he recommended, is to keep the two apart.[88]

In true positivist fashion, Antun insisted that science and religion were different beasts—one an ontology of belief, a sacred and private sentiment of divine exhortation to do good and avoid evil; the other an epistemology of reasoned proofs and calculated observations of the natural world. They entailed different methods of argument and proof, one based on reason and the other solely on faith. It was only by separating the two, Antun felt, that men could advance in their worldly affairs and partake of the fruits of modern civilization. "Men of sense in every community and religion of the

East," Antun wrote, "who have seen the danger of mingling the world with religion in an age like ours, have come to demand that their religion be placed on one side in a sacred and honored place, so that they will be able . . . to flow with the tide of the new European civilization . . . to compete with those who belong to it." Europe's victory over the world and its modern advancement in civilization and science were the result of this separation of religion from worldly affairs: in this sense, Christianity, which had allowed this separation to take place, was in his view superior to Islam.[89]

Unhappy both with Antun's narrative of Ibn Rushd and with the conclusions he drew from it, ʿAbduh published a series of responses between 1902 and 1904, and an acrimonious and occasionally personal dispute ensued. On the whole, ʿAbduh rejected Antun's reading of Ibn Rushd, and at times he hinted at Antun's poor knowledge of Muslim philosophy, and at others he denounced his reading of Christian theology and history, too. Above all, he wrote to counter Antun's claim that science and religion or, for that matter, religion and politics should be separated.[90]

ʿAbduh began by quickly discounting Antun's reading of Ibn Rushd on such issues as causation and creation. The mufti began his rebuttal by offering Antun a quick summary of ancient Greek thought on the question of the creation of the universe, concluding from it that Ibn Rushd was clearly a follower of Aristotle, who adhered to a notion of a Prime Mover in the universe: this meant that the philosopher could thus in no way be counted as a materialist, as *Al-Jamiʿa* claimed. While Antun took Ibn Rushd's notion of a self-moving nature and his acknowledgment of secondary causes as evidence of his heterodox and materialist views, ʿAbduh contended the opposite: Ibn Rushd, he concluded, did not go beyond his *milla*, or religious adherence, and he was certainly no materialist. Rather, his discussions of a Prime Mover or First Cause in nature easily disproved such an assessment.[91]

Next, he turned to Antun's conception of Muslim—versus Christian—civilization. This was something the mufti particularly objected to, and hence this topic occupied the bulk of his response. Two points in particular seemed debatable to him. One was that history showed a greater antagonism toward science in Islam than in Christianity. The other was the idea that civilizational progress in Europe was due to the separation of spiritual and temporal authorities in Christianity: "we read of how spiritual and temporal authority in Islam are linked through the rule of religious laws [*bi-hukm al-shariʿa*], for the ruler is also the caliph. As a result, it is argued that Islam is a less liberal religion than Christianity, for Christianity separates church and state—a great division, it is said, which has made it easy for Christianity to move toward real civilization and progress. In a phrase, 'Give to Caesar what is Caesar's and to God what is his.'"[92] But science, he objected, "is not like Caesar." Rather, he defined it as an attribute of God (in a way that was true to the older understanding of *ʿilm*) and hence something that should fall under the authority of spiritual power.

Drawing on his by now familiar rationalist principles, ʿAbduh then turned the tables and undertook his own comparison between the two faiths, to the detriment of Christianity. Christianity, for him, was a poor friend to science. In the first place, it was fundamentally irrational, and the New Testament (a missionary translation of which he quoted at length) made belief in Christ's miraculous powers a test of faith. This legacy, for ʿAbduh, lay at the heart of Christianity's adversity to science. After all, to induce faith through the preternatural rather than the rational undermined belief in the natural laws uncovered by "men of science."[93] Other "principles of Christianity" that he criticized as opposed to the "spirit of science" were temporal: they included such things as the absolute power assigned to rulers and the clergy and the idealization of worldly abnegation.[94]

Intriguingly, ʿAbduh distinguished between Catholicism and Protestantism: most of his strictures he applied to the former. He castigated the Inquisition for its ill-treatment of scientists and pointed to the persecution of Galileo, the banning of scientific organizations, the burning of science books, and the control over and censorship of printing.[95] In a section he entitled "The Protestant Reformation," on the other hand, he described approvingly the many reformers who appealed to a "return of Christianity to its true state." As he depicted it, they had called for freedom of thought, the limitation of the clergy's worldly power, the abolition of confessions, and the prohibition of the worship of icons. Above all, they had called for reason to stand behind declarations of faith. It was in this way, he suggested, that "science appeared in the West."[96]

All of this came by way of a defense of Ibn Rushd and the legacies of a rationalist Islam. Against the views of Antun (and Renan), ʿAbduh drew upon his wide reading of not only traditional classics familiar to any educated *ʿalim* but also European and American commentators. Guizot's emphasis on the importance of the Reformation for civilizational progress, in particular, was deployed in support of ʿAbduh's own reformist impulses. Of course, he also had Muslim sources for such a conception, as did his mentor Afghani, and both of them drew on eighteenth-century Muslim reformers. It was no accident, for example, that many of the Azhari *ʿulama* termed ʿAbduh a Wahhabi in dismissing his appeals for a return to Islam's "true state." (In fact, as we will see later, ʿAbduh was at the time widely denounced, including by khedival spies, as both a Wahhabi and a mimic of European intellectual fashions.)[97]

Another French authority he invoked was Gustave Le Bon, whom he had gotten to know personally in France. Le Bon's social theories exerted a considerable influence upon at least two generations of Egyptian intellectuals, and ʿAbduh mined his 1889 *Les premières civilisations* for arguments in favor of Arab contributions to the progress of humanity, such as being the first to "show the world how to avoid any contradictions between freedom of thought and religious principles."[98] Le Bon also highlighted the contrast between the caliphs' openness toward scientists and Catholic intolerance. Underlining the moral of his own educational-reform program, ʿAbduh concluded by pointing out that while people might be confused about whether the madrasas were mosques or schools, they had always valued freedom of thought and taught all subjects, from grammar to astronomy: "Science became the life of *adab* [culture] among the Arabs, a spiritual nourishment, the foundation of industry, and a means to advance the state of mankind. Not a single student of all Europe can deny the role the Arabs played in advancing Europe from ignorance to science, teaching them the importance of observation and experimentation."[99]

On questions concerning the broader historical development of the sciences, ʿAbduh's most important source was the American John William Draper, whose works on the struggle between the church and intellectual freedom in European history had an immediate impact upon the reconceptualization of the relation of science and religion in the Anglo-American world (particularly his best-selling 1874 *History of the Conflict between Religion and Science*). It had been at a talk by Draper in 1860 that Huxley and Bishop Wilberforce had had their famous clash. And it was Draper who helped to popularize the idea that science and faith were at war, and that evolution was their final battleground. His ideas were disseminated in *Al-Muqtataf* as we saw in the introduction and embraced by numerous Arab intellectuals: ʿAbduh hailed him as "one of the biggest philosophers among the Americans."[100] Draper provided further authority for Islam's open-minded attitude to scientific inquiry under its early caliphs.[101] Using Draper, ʿAbduh argued that Arab contributions to the historical development of science exceeded those of Europe itself: after all, he pointed out, Europeans had remained in the shadow of the ancient Greeks for over ten centuries, whereas the Arabs had built upon and surpassed them. While Europeans claimed that Bacon was the "father of modern scientific methods of observation and experimentation," leading to a "scientific revolution," these methods, he thought, had been readily accepted among the Arabs from at least the second century AH (eighth century AD).[102]

According to ʿAbduh, Arabs had devised highly accurate astronomical charts and come close to discovering gravity. Above all, they prefigured a theory of evolution (*taraqa al-kaʾinat*). Citing Draper once again, he even argued that such theories went further than Darwin's since they encompassed both living and nonliving things, such as metals. ʿAbduh cited the twelfth-century natural philosopher al-Khazini in support of these ideas: "Khazini knew that gold was a body, which reached its final form from other metals, from the transition from lead to tin to copper and then silver. Philosophers of today recognize that man too reached his state gradually, and not that he was first an ox, then a donkey, then a horse, then a monkey, and then a man."[103] But here as elsewhere in his writings, ʿAbduh was simply repeating, more or less word for word, a genealogy already identified by the editors of *Al-Muqtataf*, in this case from their 1882 article "Madhhab Darwin." In a striking intellectual feedback loop, characteristic of this entire story, a leading Muslim reformer was thus drawing on American intellectuals and a chain of science popularizers stretching from New York to Beirut in order to reconfigure Islam's connection to its own medieval past.

In a favorable review of ʿAbduh's book *Science and Civilization in Christianity in Islam*, *Al-Muqtataf* itself clarified its views on the mufti's thought. "We have no doubt that religion plays a role in civilization," the editors stated, adding that ʿAbduh's message was essentially one that recommended the search for Eastern civilizational progress through the unification of science and religion—and particularly Islam. Their one complaint was that ʿAbduh downplayed the contribution of the original sources of civilization: ancient Greece and Rome. It is true that ʿAbduh was little interested in tracing the origin of human civilization in the way that *Al-Muqtataf* helped to popularize—and he was certainly less interested in Western classical antiquity than the journal's editors were. Yet both the editors and the mufti clearly shared a certain conception of history regarding the rise and fall of Muslim civilization itself. As we saw in previous chapters, this conception easily fit in with the editors' views on theocracies as the cause of stagnation and decline in modern times. At first view, ʿAbduh's narrative of rise and fall appears to take on a similar form, but the differences turn out to be important—if *Al-Muqtataf* aspired to a regeneration of the Arabs, ʿAbduh himself was more anxious about Islam.

Emphasizing Islam's golden age only served to highlight for ʿAbduh the desperate decline of Muslim civilization in the present. "If you look at Egyptians or Easterners today," he bemoaned, "you will see how the situation has changed!"[104] Outlining the causes of the stagnation of the Muslim *umma*, he lists the deterioration of the Arabic language; the decline of social order among the faithful; the obfuscation of religious law (*shariʿa*); the corruption of public and private, state and missionary, schools throughout the Arab East; and finally, the refusal of the majority of the *ʿulama* to teach "new subjects" (or "new sciences") in their schools.[105] ʿAbduh felt that such stagnation (*jumud*) was an "illness" that "can be cured." It will by now be evident where he thought such a cure might lie: in a return to the foundational principles of the Qurʾan and the conjoining of faith with reason.[106]

It was here that ʿAbduh drew very different conclusions from many of the European commentators upon whom he had relied. In his view, the Reformation was ultimately a failure and Europe's progress in the sciences only partially a success since it had led to a break between science and faith. Draper's conflict was one he regarded with dismay. It was the sectarian divisions that had emerged in Europe with the Reformation and the consequent religious wars that, in his view, led to more positivistic, if not materialistic, views of science. Science emerged as the solution to the problem of social order in post-Reformation Europe. Yet, particularly after the French Revolution, this had in fact produced a philosophy that "deliberately limited religious thought."[107]

History had thus left Islam the only religion capable of laying the foundations for a reconciliation since history showed that, when the religious sciences flourished among Muslims, so did the worldly sciences.[108] The irony was that Muslims themselves had forgotten this, along with the insights of their spiritual forefathers. He lamented the fact that, among Muslims today, one could not find the books of al-ʿAshari, Abi Mansur al-Maturdi, Abu Bakr al-Baklani, al-Tibarani, al-Ghazali, and so forth. He complained about the decline of the religious sciences and the *ʿulama*'s reliance on opinion rather than proof. This had led students from the Arab East to leave for Europe to study the modern sciences, putting their faith in jeopardy in the process.[109] What was needed was for reformers to promote the advancement of science and learning among Muslims—in short, to spread a new kind of reformation throughout the Muslim lands. This, he was confident, would eventually take place: "I do not think that this current phase will last much longer, certainly not as long as it lasted for the Christians, . . . who for some 1,000 years were buried in ignorance."[110]

But in the short term, things looked less rosy. He lamented the reluctance of Muslims to change their ways and their refusal to reconsider the meanings of *taqlid* and *ijtihad* or to teach the sciences in their schools. In short, they had buried themselves in "a pit of conformism." In the face of any prospect of change, they merely exclaim: "We are used to what our fathers laid out and we follow their path!" Islam itself was not the issue; the trouble was the conservatism and corruption of its legislators (*fuqaha*) and scholars (*ʿulama*).[111]

As we will see, the majority of the *ʿulama* did not respond well to these attacks against them. Antun's own response to this crushing and lengthy criticism was bitter, and he took full advantage of ʿAbduh's harsh indictment of the *ʿulama*. He wrote several acrimonious rebuttals in *Al-Jamiʿa*, claiming that ʿAbduh had misrepresented not only Christianity but also Islam—accusing him at one point of "insulting the greatest Muslim scholars of Egypt." An unorthodox coalition of ʿAbduh's enemies—Antun, the prosocialist Syrian disciple of Renan, on the one hand, and the traditional *ʿulama* on the other—thus took shape. Antun went so far as to suggest that a council at Al-Azhar be asked to arbitrate the debates between himself and the grand mufti, which suggests he felt that he had supporters within the ranks of Al-Azhar. In 1906 Antun even claimed to have received a letter "with the stamp of Al-Azhar on it" praising *Al-Jamiʿa* and condemning ʿAbduh.[112] Antun left Egypt shortly after and emigrated permanently to the United States. But the *ʿulama* in the college-mosque of Al-Azhar remained and became ʿAbduh's most determined critics, as we will see below.

On Education

The roots of ʿAbduh's conflicts with the *ʿulama* at Al-Azhar can be traced to his views on pedagogy—perhaps the single most important concern of his public life. After ʿAbduh was expelled by Tawfiq from his teaching post at Dar al-ʿUlum, he continued to appeal to the need for educational reform through the press and as a member of the Superior Council in the Department of Education.[113] In 1880 he had been appointed to the Publications Department of the Ministry of the Interior and become editor-in-chief of the official paper, *Al-Waqaʾi al-Misriya*. Under his control, education and pedagogy emerged as chief concerns of the paper.

From these articles we can catch glimpses of themes that ʿAbduh would later discuss at greater length.[114] Many of his early journalistic writings concentrated on the importance of education for both religious faith and national prosperity.[115] Competition from Europe was an ever-present incentive. "Do they not know by what means our neighbors have reached the highest level of greatness?" he asked rhetorically of his country's educators and recommended that they "support the army of knowledge. To all these questions, I say, they do not know [the answers]. I think our contact with European nations for many years now has taught us these causes of weakness and the meaning of strength."[116]

Yet he did not intend for the European style of education to become a model. In fact, ʿAbduh worried about the way in which previous Egyptian administrators had sent their brightest students abroad. And even before he had spent time in Beirut and seen the Syrian Protestant College at close quarters, he had been deeply concerned about the threat posed by missionary schools in Muslim lands. While at the college he had drawn up proposals for educational reform in which he highlighted this threat for the Ottoman governor of Beirut. To the Shaykh al-Islam, or head *qadi,* in Istanbul, he was even more outspoken, referring to the missionary educators as "foreign devils."[117]

ʿAbduh's own prescription for educational reform was framed within a Muslim context and emphasized the benefits of his recommendations in countering what he called the declining "spiritual health" of Egypt. He saw teachers as "physicians of the soul." But he parted company sharply with the *ʿulama*'s "rote learning" and dismally outdated scholasticism, urging them instead to promote the value of individual reasoning and argumentation. In an article he wrote about his own teaching experiences at the Madrasa al-Sultaniya in Beirut, we read the following: "I fitted my various lectures to the different classes concerned. To the first class, I lectured in a readily and comprehensible style. Where the art of discussion was not familiar, I used to start with the making of premises and went through to the conclusions, with no concern other than the validity of the proofs." He was thus willing to dispense with the usual discussions of the various arguments and disagreements that distinguished the differing schools of thought in Islam. "I did not mind if in the process, I diverged from generally accepted arguments. I made only remote allusions to controversial matters, of a kind that possibly only the initiated could have handled." In this way, he "kept closely to the original authorities without impugning their successors' views," while he "avoided the disputations of the schools of law with their barren wrangles."[118]

ʿAbduh's rejection of traditional methods of instruction went back to his own experiences as a schoolboy in Tanta, in Lower Egypt. Later, he felt that he had studied there without "understanding a single thing, because of the harmful character of the method of instruction"[119]—his teachers' use of technical terms of grammar and jurisprudence in particular. Discouraged, ʿAbduh eventually ran away from the school and hid with relatives for about three months, determined to devote himself to farming rather than scholarship, until he was compelled by his father to return to his studies.

In his view, the ʿ*ulama* at Al-Azhar were no better and he was unsparing in his criticism of them—calling them half-wits who "inflict their calamity upon the public" and accusing them of "spreading ignorance among the ignorant."[120] Their inadequacy had led to the decline of well-reasoned Muslim scholarship: the majority of ʿ*ulama* merely embarking upon a path of "blind imitation," or *taqlid*, following dogma with little or no use of reason, and he emphasized the need "to liberate [men's] minds from the chains of imitation."[121]

ʿAbduh's emphasis on the need for education to move away from a rigid and dogmatic curriculum and from outmoded scholastic methods—as well as his stress on the moral and intellectual cultivation of the mind—put him in line with a number of other educational theorists of the time. Above all, it led him to the sage of Brighton. When ʿAbduh read Herbert Spencer's *Education: Intellectual, Moral and Physical*, which we know he attempted to translate into Arabic, he found an unlikely ally in the Victorian popularizer.

Spencer had published this book in 1861, and it proved to be his most popular and most widely translated work. In it, Spencer complains about the tyranny of the classical curriculum in words very similar to ʿAbduh's: "We are guilty of something like a platitude when we say that throughout his after-career a boy, in nine cases out of ten, applies his Latin and Greek to no practical purposes. . . . If we inquire what the real motive is for giving boys a classical education, we find it to be simply conformity to public opinion."[122] Spencer insisted that education's sole aim was to fulfill the mind's natural potential. Schooling should therefore mimic the natural unfolding or cultivation of the mind and foster the power of reasoning, the higher intellectual and moral faculties, and a self-perpetuating and pleasurable curiosity in the nature of things. Such an education, Spencer thought, would lead to self-cultivated and self-sufficient national subjects, thereby contributing to the general progress of civilization. Building upon Enlightenment theories of education, with their emphasis on human perfectibility, the cultivation of the mind, and the fostering of moral rectitude, Spencer added a contemporary fascination with the evolution of mind. For many after Spencer, from the American John Dewey to Muhammad ʿAbduh himself, ideas of evolution would similarly creep into their educational theories.

ʿAbduh found many echoes of his own thought in Spencer. He too saw education as a moral as much as an intellectual enterprise, a concept he would also have derived from canonical Muslim models of education on moral and character development: the publication of classical works such as Ibn Miskawayh's *Tahdhib al-akhlaq* (The refinement of character; republished extensively in Egypt in the 1880s and 1890s) would no doubt have reinforced this connection. Spencer's lamentations on narrow, dogmatic adherence to authority and classicism would also have struck a chord with the mufti. And like Spencer, ʿAbduh thought good pedagogy was good politics: questioning authority on scholarly matters set an example for those who wished to question political dogmatism and autocracies.

There was also the question of the family. For Spencer, care of the family was the key foundation for good society and therefore for good government. The rational management of the family was the starting point for Spencer's chapter "Moral Education": "To prepare the young for the duties of life is tacitly admitted by all to be the end which parents and schoolmasters have in view." But he criticized the current state of affairs: "While many years are spent by a boy gaining knowledge, of which the chief value is that it constitutes 'the education of a gentleman'; and while many years are spent by a girl in those decorative acquirements which fit her for evening parties; not an hour is spent by either of them in preparation for that gravest of all responsibilities—the management of the family."[123] Spencer thought that education should give children the skills to understand right and wrong and that a proper moral development would lead them to acquire an innate knowledge of the natural consequences of both.

For many Egyptian intellectuals at the time, the family was proving to be a similarly key institutional site for discussions of state reform, social progress, and national and communal prosperity. One approach was sociological and technocratic. This proved appealing in particular to those who saw in the new social sciences an expert language of communal (or "social") development that was critical for the fate of the state. Qasim Amin's controversial 1899 *Tahrir al-marʾa* (The liberation of women) proved a good example of this approach and of the new fascination with the idea of the family—with implications for that novel nineteenth-century construction "society" and even the state itself. Amin argued that "any country concerned with its interests should be concerned with the structure of its families, for the family is the foundation of a country." And since the mother was the foundation of the family, attending to women's education and moral rectitude was imperative. The West, he argued, had long realized this: "No Westerner," he wrote, "is ignorant of women's status in society or her important role in the family." He followed up this assertion with an eclectic set of quotes from Simmel, Schiller, Rousseau, and Fénelon. Muslims, however, ignored the importance of women in society. Indeed, Amin even argued that the "decline of Muslims"—like that "of the Indians; the Chinese and all the other inhabitants of the East"—was due to the poor structure of its families.[124]

All this took place through the intensifying debate over *tarbiya,* a term that referred to both child-rearing and general education. Also in 1899, *Al-Manar,* under ʿAbduh and Rashid Rida's joint editorship, published a series of articles on *tarbiya.* These emphasized the need for mothers to adopt a more scientific method of child-rearing from birth. Echoing Spencer, Rida argued that punishment and coercive methods of parenting did little to teach children the true principles of right and wrong. He too emphasized the need for rational management and disciplined routine in the home and the importance of women's education for the national domestic order and for the production of healthy, hygienic, well-behaved, moral, and productive subjects.[125]

ʿAbduh himself had a particular interest in the family from the perspective of the law. His 1899 memorandum on the reform of the shariʿa courts noted that many Egyptians turned to the courts to sort out their private, domestic affairs. This meant enormous responsibilities on the shoulders of the *ʿulama.* Instructing them on principles of good parenting, proper relations between spouses, and the right management of their family affairs was, therefore, imperative—and so was rationalizing the legal code on these issues. Safeguarding the family, in short, was not just a private matter but one that affected the public good and that affected the public role of the *ʿulama* in particular and had immediate implications for the religious legal code of the state.[126]

When ʿAbduh met Spencer in 1904, he praised him as one of the "greatest philosophers" of the time, and it is clear that the English writer's ideas on education—as on other matters—greatly impressed the mufti. But there were points of difference as well as similarity. Both emphasized the need for the intellectual, moral, and physical fostering of their charges (though the latter duty was discussed less by ʿAbduh). But up close, their proposed curricula looked quite different. Spencer emphasized a curriculum of rational education that gave instruction in modern languages, methods of reasoning, and science. Religion was not much of an issue for Spencer. He did cite Huxley to the effect that both true science and true religion serve the search for truth—and the success of one depended upon the success of the other—a formulation that ʿAbduh was also fond of.[127] But Spencer gave little attention to the practicalities of religious instruction as such or to its formal place in the curriculum.

ʿAbduh's primary concern was with the rationalization of a Muslim curriculum, which, in his view, required two things. In the first place, it meant replacing what he had termed the "barren wrangles" of the "disputations of the schools of law" with an understanding of religion in accordance with the ways of the *salaf* (forefathers), "before the appearance of disputes." To this end—and in line with the development of a very particular historical narrative—he advocated a return to the original sources of Islam. Second, it required the introduction of new subjects, especially science, into Muslim schools. Religion and science, he felt, need not be separated. Indeed, the introduction of the latter was crucial, although he was conscious that in wanting this fusion he was in a minority: "I deviated from the two great groups of the community, the students of religion and the students of modern disciplines."[128] ʿAbduh was particularly worried about the opposition of the *ʿulama* of Al-Azhar, whom he felt would reject the validity of "any other science but their own."[129] The long and frustrating experience of attempts at reform at Al-Azhar throughout the second half of the nineteenth century gave him ample grounds for concern. But what ʿAbduh neglected to see was that his very discussion of religion—and his emphasis on its place as one of many *subjects* for school curricula—would transform its very meaning and thence the status of those formerly charged with its communal roles, namely the *ʿulama*.[130] It is scarcely surprising, then, that the opposition to his ideas only grew over time. It was at Al-Azhar that ʿAbduh would attempt to put his educational theories into practice, but it was there, too, that the backlash to his reforming efforts would place him in ever greater tension with other members of the *ʿulama*.

Reform and Backlash

Since as far back as the mid-nineteenth century, Al-Azhar had been suffering from a deteriorating infrastructure, made worse by a tremendous influx of students, especially after the Crimean War, when students flocked there to avoid military conscription. At the same time, there was little central control of its organization. An *ʿalim* of Al-Azhar was usually assigned a column in the central court, where he sat on a mat or chair, with students sitting about him in a circle. There was no formal curriculum and no formal program of study: students merely found a shaykh whose lectures they wished to attend. There was no real organization of subjects taught, no registration procedures, and no examinations.[131] Competition for a column, and even for seating space, was stiff. Thousands of students crammed into the narrow, unpartitioned courtyard (a space of about one hundred square meters), scrambling to listen to one of the dozens of teachers. At the beginning of the nineteenth century, there were forty to sixty teachers and perhaps three thousand students. By the early 1850s, the latter number had risen to over seven thousand.[132] Student fracases erupted periodically as a result of overcrowding.[133]

For the government, this was a bureaucratic nightmare. This was after all the age of national educational reforms worldwide. A concerted effort to transform Al-Azhar had begun under Khedive Isma'il (1863–79)—when 'Abduh himself was a student there. Among the proposals suggested was the introduction of new subjects such as geometry, physics, music, and history. Much of this met with opposition but the government was more successful in challenging the senior *'ulama*'s autonomy and rank.[134]

The battle between the Azhari *'ulama* and state-appointed reformers continued throughout the remainder of the century. In 1872 another series of reforms was issued.[135] The government now stipulated the subjects for examination, thereby creating a set curriculum for the first time but also, in the process, undermining efforts to introduce the natural and physical sciences.[136]

To judge from the numerous reform laws passed over the next fifteen years, it seems clear that the gap between law and implementation was significant.[137] Government rulings on examinations, curricula, and registration procedures were poorly received by the *'ulama* and students alike. But by far the greatest controversy concerned the introduction of "new subjects" (*al-'ulum al-haditha*), particularly the sciences. In 1888 Khedive Tawfiq commissioned a *qadi*, Muhammad Bayram, to issue a fatwa on the permissibility of introducing modern scientific subjects, including the mathematical and natural sciences, into the curriculum of Al-Azhar. Bayram then sought the opinions of the shaykh of Al-Azhar, Shams al-Din al-Inbabi, and Grand Mufti Muhammad al-Banna.[138] Shaykh al-Inbabi ruled, "It is right to study the mathematical sciences, like arithmetic and geometry as well as geography, because they do not contradict truth. Anything contributed by them to spiritual endeavor is needed, just as medicine is necessary." Alchemy and astrology were not permitted, but chemistry and astronomy were. In general, the natural sciences were, so long as study of these sciences was pursued in accordance with shari'a and not from the point of view of metaphysics. "There is no harm," he decreed, "in teaching mathematical subjects, just as other instrumental subjects are taught. The same is true for the natural sciences and chemical sciences, so long as they are taught in a manner that is not understood to oppose the legal tradition, as with the rest of the rational subjects, such as logic, discursive theology, and argumentation." Although al-Inbabi concluded that understanding these subjects was essential for the standing of the *ʿulama*, little changed, and the only texts introduced included an eighteenth-century treatise on astronomy by Hasan al-Jabarti (father of the well-known ʿAbd al-Rahman al-Jabarti) and a medieval text on mathematics by Ibn Haytham.[139]

The government's frustration with this state of affairs led to the development of an alternative teacher-training system, which began in 1872 with the establishment of Dar al-ʿUlum (Teachers' College)—where ʿAbduh himself taught briefly in the late 1870s.[140] It had originally been intended solely to provide language instruction, but its curriculum soon expanded to include arithmetic, geography, natural science, chemistry, physics, and calligraphy, as well as Hanafi law and Qurʾanic exegesis.[141] By the late 1880s, the khedive even considered using it as a training ground for judges—one of the traditional areas of jurisdiction of the *ʿulama* at Al-Azhar. But protests from Al-Azhar forced him to back down. Nevertheless, this gave an indication of where the government might go in future if further reform proved equally unsuccessful.[142]

A vivid picture of the state of Al-Azhar at this time—with its problems seemingly unchanged—emerges from an article that appeared in 1893 in the nationalist weekly *Al-Muʾayyad*.[143] It was written by ʿAbd Allah Nadim, known as the "orator" of the ʿUrabi revolt. Described by Wilfrid Blunt as "a hot-headed young man of genius," he was by then a well-known writer.[144] As a graduate of Al-Azhar, he was able to draw on his own personal experience. "If we examine the state of this great university today," he wrote, it would be found to be in utter chaos and plagued by disease: "We find large numbers of students with no medical attention or control, such that diseases spread rapidly among them . . . [and they are] living in dirty quarters, with no servants to clean the dormitories while they are busy reading and attending classes. As a result of all this, their dormitories are filthy; one finds in them bits of food strewn over the floors, mingled with dirt dragged in by their shoes." Because "students learn neither ethics nor house management," they remain in this squalid state, "their sweaty and fetid bodies attracting dirt," suffering from a lack of proper hygiene out of poverty, ignorance, and lack of proper supervision.[145]

He went on to complain of the students' lack of discipline and poor attendance. Students "moved from shaykh to shaykh" in a disorganized manner, learning hardly anything, while the shaykhs themselves were overstretched and insufficiently specialized.[146] Above all, he criticized the curriculum. Students "know nothing of the progress of civilizations, current affairs, or even wars, they have no knowledge of industry, agriculture, or the basics of trade, and they never research any of the new inventions that surround us—as though they lived in a dark cave, where none other than those as wretched as himself lived."[147]

Echoing ʿAbduh in his critique of the clergy, Nadim saw the decline of the sciences at Al-Azhar as the cause of its degeneration. In a familiar formulation, he drew attention to the past glories of Islamic civilization.[148] He bemoaned the fact that the *ʿulama*, who had once been at the forefront of scientific knowledge, now taught only grammar and jurisprudence. As a result, "we do not find these sciences except through foreigners." He made it clear that he was not advocating turning Al-Azhar into a European-style university: "we do not want Al-Azhar to graduate doctors, chemists, engineers, physicists, and veterinarians." But what was essential, he thought, was that its students become familiar with the latest subjects of modern science.[149]

Like many nationalists of the time, he saw reform and intervention by the state as the only way forward.[150] And indeed the government was as keen as ever to oblige. Taking advantage of student unhappiness with the head of Al-Azhar in 1894, the khedive forced him to resign and then set up a new Administrative Council—one of whose members was Muhammad ʿAbduh.[151] When the young ʿAbbas had become khedive two years earlier, ʿAbduh had anticipated a powerful new supporter for the cause of reform, and it now looked as though he might be in a position to execute many of his ideas.[152] As he later wrote in an article entitled "The Need for Religious Reform at Al-Azhar," he hoped "to arm the *ʿulama*" with genuine knowledge (or "science") to allow them to save the nation from the quagmire of either inherited beliefs or innovative practices, to move, in other words, away from either excessive *taqlid* [imitation of one's own past traditions] or *taghrib* [imitation of 'foreign ways']. He wanted to initiate an indigenous revival, in short, "a scientific movement . . . such as those that came to be during the flourishing epochs of al-Ma'mun and others."[153]

With ʿAbduh in the lead, the Administrative Council at Al-Azhar immediately tried to introduce reforms. Once again, they attempted to expand the curriculum and to standardize examination procedures and, effectively, to place all aspects of Al-Azhar's administration in their own hands.[154] There were many reasons why these measures met with resistance. In the first place, the members of the Administrative Council were regarded as government employees, and traditional arguments of institutional autonomy were advanced against their meddling. Azhari hatred of state intervention intensified when government forces and police clashed with students in June 1896 over an outbreak of cholera at the Syrian *riwaq* (dormitory). Students, afraid to relinquish one of their colleagues to be quarantined at a hospital because "he who was taken off there did not return to smell the fresh air," clashed with government, police, and British officials, the confrontation eventually leading to several student injuries and one death.[155] The event drew wide public attention to the institution and confirmed in the minds of many its reactionary nature. "A riot of a somewhat serious character has just occurred at the mosque and University of El Azhar," wrote Lord Cromer. "The character of the riot is not political, and it is due entirely to the dislike of sanitary measures felt by the most ignorant and fanatical classes."[156]

To be sure, student fears of the hospital were not unfounded. After all, two other students who previously fell ill had recently died while in quarantine there (the students had pleaded to leave the ill student in quarantine at Al-Azhar instead). Government bureaucrats and science enthusiasts, such as the founders of *Al-Muqtataf*, however, blamed the students for their backward and irrational fears and their lack of respect for hygiene. "Rational people," wrote *Al-Muqtataf*, "have agreed that had the government not used determination in this incident, the epidemic would have spread among the people in every place."[157] As a result of the riot, state intervention and supervision only increased. Twelve of the ringleaders were prosecuted, sixty others deported, and the Syrian *riwaq* was shut for a year.[158]

Curricular reform was again another source of conflict. The attempted introduction of new subjects, particularly the modern sciences, raised the question of what Al-Azhar's mission really was. Some reformers—more extreme than ʿAbduh—argued that it should function as a university rather than as a madrasa-mosque. *Al-Muqtataf*, for instance, compared the paucity of subjects, as they saw it, taught at Al-Azhar with what European universities such as Oxford, Cambridge, the Sorbonne, and the University of Paris offered. "Among those large universities," they wrote, "that resemble Al-Azhar as a long-standing school is the University of Paris, which was founded in 1213. . . . But it now has more students than Al-Azhar, . . . and offers more subjects." Many of these universities, they pointed out, "taught religious studies along with the sciences," and they argued that the two need not be taught separately.[159] Indeed, as they saw it, only through the introduction of new subjects in the arts and sciences could Al-Azhar fulfill its ambitions to reform. As it happened, this view suited British policy, concerned as it was with avoiding the expense entailed in establishing an entirely new institution of higher education.[160]

But it was precisely this prospect that worried some critics who condemned the proposed curricular reforms as fundamentally altering Al-Azhar's chief role as a madrasa. Hasan Husni al-Tuwayrani, in the journal *Al-Nil,* for instance, insisted that Al-Azhar was specifically a religious school, as many of the *awqaf* (charitable religious) endowments stipulated that their funds be used exclusively for religious studies. The introduction of these new sciences (*al-ʿulum al-haditha*), he wrote, was therefore out of place. The expansion of the curriculum to include extrareligious subjects was in his view a legal issue to be decided by *ijmaʿ* (consensus of the community) and the *ʿulama* themselves, not "according to the opinions of the newspapers . . . [or] the literati."[161] (The 1895 codes had attempted to get around the argument that the *awqaf* did not support nonreligious study by classifying the new subjects as *wasaʾil,* or auxiliary means to an end.) A contributor to the Azhari journal *Al-Islam* made similar arguments in 1895 and questioned the proposed reforms' usefulness in addressing the problems then plaguing the madrasa-mosque: the addition of new subjects could hardly help with problems of overcrowding, poor methods of instruction, and corruption.[162]

Critics and Conservatives

As attacks on the reform program increased, they became increasingly personalized and directed against Muhammad ʿAbduh, particularly after his appointment as grand mufti in 1899. His brash and outspoken criticisms of Al-Azhar rankled the other senior ʿulama. Shaykh al-Bukhayri challenged ʿAbduh with the following: "Do you not know that you are an Azhari and yet you have ascended to where you are on the stairs to knowledge and have become a unique scholar?" To which ʿAbduh replied: "If I have a portion of knowledge, as you mention, I got it through ten years of sweeping the dirt of Al-Azhar from my brain, and to this day it is not as clean as I would like." When the shaykh told ʿAbduh, "We teach as we have been taught," ʿAbduh replied, "And this is just what I fear."[163]

It was not only his senior colleagues who were suspicious of him. Although he would later be hailed by nationalists and Azhari *ʿulama* alike (not to mention historians) as a daring reformer and champion of the faithful, he was hardly considered so by many students at the time. "When Shaykh Muhammad ʿAbduh came to reform Al-Azhar," wrote Taha ʿAbd al-

Bakhi Surur, "I was one of the younger students. Our professors, may God forgive them, used to constantly criticize the shaykh in our presence and represent him as being dangerous for religion and for the religious." As a consequence, Surur continued, "our young minds were influenced by this, and I used to flee from encountering the professor, for the sake of my religion and to flee from listening to his lessons, even though he was a friend of my father's."[164]

It is often asserted that ʿAbduh's opponents were reactionary and backward looking.[165] He and Rashid Rida, his main disciple and spokesman, routinely denounced ʿAbduh's critics as "intellectually frozen" or stagnant, as "fundamentalist," or as "conservative." In the view of hagiographers and revisionists, ʿAbduh's was a solitary voice of reason and "liberalism," standing against a tide of overwhelming reactionary opposition. "The Conservative opposition," wrote his admirer Blunt in 1896, "was too strong for him."[166]

This view assumes that in the battle over reform at Al-Azhar, the lines were sharply drawn between "conservatives" and "traditionalists," on the one hand, and "reformers" and "modernists," on the other. But this view misreads both ʿAbduh and his critics. First of all, it would be a vast oversimplification to claim that ʿAbduh's critics were merely upholding a narrow orthodoxy against ʿAbduh's own supposed modernist tendencies: the term "conservative" here obscures the fact that there was a contest over the very conception of tradition itself. ʿAbduh saw himself as a *mujtahid*—a reformer whose real ambition was to return the faithful to the purity of Islam's original mission, free of the perversions and innovations accrued over the years. He regarded himself, in fact, as the real traditionalist—restoring the *sunna* (tradition) through a reform and renewal of the faith (*al-islah wa al-tajdid fi al-din*).[167] But his version of tradition, as we will see, was not commonly accepted, and many *ʿulama* suspected ʿAbduh of heterodoxy, especially as his efforts to forge a new theology were bound up and informed by his interest in philosophy and contemporary scientific thought.[168] When ʿAbduh published his *Risalat al-tawhid* in 1897, his detractors accused him of espousing Muʿtazila doctrines—advancing the idea that the Qurʾan was created (and hence historical and not eternally existent). ʿAbduh's own views on the matter were actually somewhat ambiguous but rumors of heresy quickly spread.[169]

Indeed, accusations that ʿAbduh was a Muʿtazilite went back a long way. As a student at Al-Azhar, he had clashed with the head jurist of the Maliki school, Shaykh Muhammad ʿIllish. ʿIllish had chastised him for reading texts that had not been assigned—ʿAbduh was said to be reading the twelfth-century al-Nasif and the fourteenth-century al-Taftazani, two important scholastic philosophers who engaged with Muʿtazilite ideas.[170] This encounter was later portrayed by the ʿAbduh camp as the beginning of the *ʿulama*'s persecution of ʿAbduh.[171] "It was too much for ʿIllish," Rida later wrote, recounting the incident, that a student would study a book like that, "which not even the eldest shaykhs were permitted to teach."[172] According to Rida, ʿIllish felt that this demonstrated ʿAbduh's Muʿtazilite leanings—a serious accusation for ʿIllish, since he was to denounce the Muʿtazila as heretics in one of his fatwas.[173]

What had shocked ʿIllish in particular was ʿAbduh's irreverent attitude to the practice of *taqlid*. This concept in Muslim jurisprudence implied that one should follow a conventional legal opinion or reading unless there were overriding reasons not to. ʿIllish had accused ʿAbduh of abandoning the orthodox Ashʿari school (*madhhab*) of legal opinions by his choice of extracurricular readings. ʿAbduh's response—or so Rida's later admiring account had it—was that he was not an "imitator," neither following blindly the *taqlid* of the Ashʿari nor the *taqlid* of the Muʿtazila. His inclination was to "abandon *taqlid* altogether and use only evidence." Rida tells us that ʿIllish, at that time fifty years older than ʿAbduh, was so taken aback by this reply that he snatched the young ʿAbduh's turban from his head, after which ʿAbduh, fearing physical attack by ʿIllish's supporters, removed himself from the mosque and studied at home.[174]

ʿIllish's reaction has to be understood in the light of earlier criticisms of conformism at Al-Azhar that had been made by the followers of Muhammad ʿAbd al Wahhab and the Maghribi Muhammad al-Sanusi (1791–1859), among others. Many of these reforming sects of the eighteenth and nineteenth centuries, known as the "Muhammadiya," as ʿIllish himself referred to them, favored independent legal thinking, *ijtihad*, over *taqlid*.[175] In response, ʿIllish insisted that *taqlid* was at the very heart of the Sunni *ahl al-sunna wa-al-jamaʿa*, the principles of law that bound together the "community of the orthodox and faithful."[176] "It is not permissible," ʿIllish stated categorically, "for a layman [*ʿammi*] to abandon *taqlid* of the four imams and adopt rulings from the Qur'an and *hadith*." He thought it was doubtful that anyone could claim they possessed a level of knowledge which surpassed that of the imams themselves in order to legitimately claim the rank of independent *ijtihad* (*al-ijtihad al-mutlaq*). For ʿIllish the important point was that *taqlid* had developed within the Maliki *madhhab* as a means of ensuring communal consensus; the individual rulings of an independent *mujtahid* could not match this process for legitimacy.[177]

Beyond their divergent views on the meaning of *taqlid* lay a more radical divergence over the public role of the *ʿalim*. Both ʿIllish and ʿAbduh were nationalists—ʿIllish indeed had been such a strong supporter of the ʿUrabi movement that he was imprisoned by the British as a result (dying in a prison hospital in 1882).[178] Before this he had been a hugely popular figure at Al-Azhar and his lectures had attracted hundreds of students, and he had been centrally involved in earlier reform efforts there in the 1850s.[179] Like many of the so-called conservatives written off by ʿAbduh's followers, he had in fact taken seriously the need to improve facilities, raise stipends, and reduce overcrowding. But the differences between the two men were substantial. In the first place, ʿIllish mistrusted the role of the state and wanted reform to come from within the ranks of the *ʿulama*: he had made this clear in opposing initiatives undertaken under Khedive Ismaʿil.[180] ʿAbduh, a civil servant from the time he was appointed to the Press Office as censor, was a believer in state intervention. He was also much more open to the use of the press for religious proselytism and public debate. Whereas ʿIllish's rulings were only published after his death, ʿAbduh was publishing his opinions, exegeses, and commentaries—not to mention his analyses of public life—from the start.[181]

And then there was the question of science. ʿIllish's fatwas covered not only matters of ethics and practical life but also issues of what would now be regarded as natural science. Thus, he opined on such matters as the nature of the heavens, the sun and moon, the layers of the earth, and the shape of the sky.[182] Believing that such matters naturally lay within the purview of any educated *ʿalim*, he was critical of the government's policy of sending students to Europe and dismissive of European scientific learning: "It is known that the Christians know nothing at all of the shariʿa sciences or their tools, and that most of their sciences derive from weaving, weighing, and cupping, and these are among the lowest trades among the Muslims."[183]

Well after ʿIllish's death, such attitudes remained widespread: in 1905, a member of the *ʿulama* criticized "ʿAbduh's program," which included the introduction of such things as the teaching of mathematics and geography, as tending to "corrupt morality and to discourtesy."[184] The *ʿulama* at Al-Azhar continued to voice their anxieties about reform programs that aimed to "transform education so as to include all subjects and their methods," while making the school open to Muslims and non-Muslims alike.[185] ʿAbduh's critics thus saw his enthusiasm for European science and his penchant for "foreign ideas" as of a piece with his overall heterodoxy and his seeming indifference to mainstream theological consensus. Above all, they connected his pedagogical reformism with his suspect political allegiances.

The British Connection

Memories of the encounter with ʿIllish followed ʿAbduh after his election to the Administrative Council in 1895, and most Azhari senior *ʿulama* still regarded him with suspicion. Indeed, ʿAbduh's uneasy relation with the *ʿulama* lasted throughout most of his career. Yet, despite all this, ʿAbduh was soon elected to the highest judicial post in the country, grand mufti, by Khedive ʿAbbas in 1899. He had also hoped to be elected shaykh of Al-Azhar, as he felt the latter was the more prestigious position. The khedive, however, was unwilling to place so much power in his hands and tried to reassure him, reminding him that he left to the mufti "three agencies, the shariʿa courts, the *awqaf*, and Muslim education, and that by reforming them he would be able to reform the entire nation."[186] Members of the *ʿulama* objected on legal grounds to the appointment: the post required rulings according to the Hanafi *madhhab* (which had been the official legal school of the state bureaucracy since 1856), and ʿAbduh, a Maliki, was, as they saw it, not competent to deliver these. ʿAbduh's subsequent agreement to issue rulings according to Hanafi codes merely confirmed suspicions that he was theologically unsound and lacked the proper legal orthodoxy, since such "*madhhab* crossing" was usually regarded unfavorably.[187] (ʿIllish, ironically, had earlier issued a fatwa on the dangers of this: "A mufti should have only one *madhhab* that he thinks is correct and prefers above all others.")[188] After ʿAbduh's appointment, the khedive continued to receive complaints from Azhari shaykhs, contesting ʿAbduh's performance as mufti, alleging that "he has no understanding of the rules of the *madhhab* and that his fatwas are full of mistakes."[189]

The khedive, however, was willing to ignore these concerns because he hoped that ʿAbduh's friendly relations with the British might be used to his advantage.[190] However, one of the main reasons for ʿAbduh's willingness to work closely with the British was that, like many former supporters of ʿUrabi, he had a low opinion of the Egyptian ruling family and regarded them as responsible for many of the country's misfortunes. ʿAbduh's opponents at Al-Azhar reported back to the palace that he was "a neo-ʿUrabite" and that he intended to monopolize power and run the country.[191] It is therefore not surprising that the relations between the khedive and ʿAbduh quickly became strained.

For his part, Lord Cromer was thoroughly pleased with the appointment. He was happy to see, as he put it, the office filled by one so "animated with liberal sentiments" and hoped that ʿAbduh might "aid the course of reform," particularly in the shariʿa courts, which had previously been the main sticking point with the *ʿulama*.[192] Throughout his tenure as grand mufti, the British consistently consulted him on a range of issues.[193] Lord Cromer himself remained ʿAbduh's most stalwart ally and paid him a tribute on his death in 1905. Describing this "remarkable figure," he noted that "his profound knowledge of Islamic law and his liberal and enlightened views rendered his advice and cooperation of the utmost value." His premature demise was "a great loss to Egypt."[194]

Yet one wonders whether ʿAbduh's close connection with the British really worked to his advantage. On the key issue of educational reform, Lord Cromer was lukewarm at best and always conscious of fiscal constraints. In the 1903 budget, for instance, education was allotted less than 1 percent of total expenditures, a figure that—incredibly—represented an increase from previous years. Lord Cromer was frequently criticized for his administration's neglect of education, but he doubted that the state of Egyptian finances—or intellectual potential—allowed any more to be done. In his mind, the priorities in education lay in reforming the village schools and training natives for civil service positions—a classic late-nineteenth-century colonial position. At heart, Lord Cromer was skeptical of the possibility of real change in a country whose people, as he wrote back to London, "are heavily weighted by their leaden creed, and by the institutions which cluster around the Koran." In particular, he thought that it would be impossible to teach science in Arabic.[195]

His reluctance to reform higher education in particular was reinforced by—and in turn reinforced—the growing mood of anticolonialism. "Evidences of a very considerable change in the attitudes of the natives towards the European are to hand on every side," observed a British consular official in Cairo in September 1897. There was a growing sympathy for the Ottoman sultan, Abdülhamid, and a new emphasis on his role as caliph, strengthened in particular by the victory of the Ottoman army over the Greeks in Thessaly.[196] Added to this was the unsubtle proselytizing of British missionaries, which alarmed public opinion and Lord Cromer himself. In his words, it "stirs up the latent fires of Mahommedan fanaticism and renders Moslems generally more than ever disposed to look askance at the work of the non-proselytising reformer." The dangers for the British-inspired reform effort were spelled out by the consul general: "For years past, I have been endeavouring, with the occasional and very feeble help of a few Moslem reformers to effect gradually some changes in such Moslem institutions as the El-Azhar University and the Mehkemeah Sheraeh. It is uphill work. . . . The least suspicion that conversion to Christianity rather than internal Moslem reform is the object which it is really sought to attain, greatly increases the difficulty, already sufficiently great, which lie in the path of the reformer."[197] These trends, of course, increased hostility not only toward Cromer but toward those associated with his policies. Khedival spies, for instance, reported with dismay seeing Azhari students engaging Protestant missionaries in theological debate and could not help associating this with ʿAbduh and his reported propensity for reading the Bible. Any contacts between Azhari officials and the British also carried the risk of attracting public disfavor.[198]

The khedive himself quickly became ʿAbduh's most powerful enemy. In his *Diaries*, Blunt often described the mufti as working quietly from behind the scenes to protect the public good of Egypt. On one occasion, for instance, when describing the "latest doings" of the khedive, the mufti complained to the Englishman that the Egyptian ruler seemed to have abandoned himself to the pursuit of wealth and ever greater capital speculation. The khedive, it seems, had wanted to take monies out of *awqaf* funds, but ʿAbduh, as mufti, "has a veto on such transactions and he will not allow any swindling of the public purse." ʿAbduh, meanwhile, reported that though the khedive was "ostentatiously friendly" with him, "Abdu knows he is intriguing to get him deprived of the Muftiship. The Khedive will not tolerate anyone who does not fear him."[199] And again: "The Mufti is still being worried by the Khedive," wrote Blunt in December 1903, "who has been putting pressure on a number of Sheykhs of the Azhar to denounce him on grounds of impiety in the form of an *ardahal* (petition)."[200]

In fact, the khedive was running an extensive spy system at Al-Azhar, one of whose main tasks seemed to be to follow the mufti's activities and report the most scurrilous rumors.[201] In the end, however, the khedive's intrigues were of little avail. When in January 1904 he appealed to Lord Cromer to dismiss ʿAbduh, the consul general refused, claiming he would not agree to ʿAbduh's resignation so long as he himself remained in office.[202]

This evident dependence on British patronage led ʿAbduh to be widely suspected of pro-British sympathies, and his alliance with the British set him at odds at times with both the khedive and the sultan. The former was worried about ʿAbduh's attempts to use the connection against him, while the latter may have worried that ʿAbduh was colluding with the British to "proclaim the Arabian Caliphate."[203] Yet ʿAbduh remained consistently opposed to the British occupation of Egypt. Indeed, in his legal rulings and published writings on local affairs in Egypt, ʿAbduh often took a rather critical stance of British policy. In the 1899 cigarette factory workers' strike in Cairo, for instance, he published a fatwa that offered a critique of nonintervention and the free-enterprise state. ʿAbduh argued that both government intervention in the economy and government's role as arbitrator between labor and capital were consistent with principles of good Muslim governance. ʿAbduh's fatwa came at a time when debates over the social welfare state and classical liberal principles of individualism, competition, and laissez-faire were gaining new critics in Egypt, as outside it. The rise of labor unions and other organizations and socialist ideas more generally were coming to be associated in the public eye in Egypt with a critique of the British. ʿAbduh's fatwa therefore would have been widely interpreted in this vein too.[204] In other contexts, the mufti denounced imperialism and the lust for power in Europe more generally—which he typically connected to the vogue for "materialism"—when speaking of the British presence in Egypt.[205]

Yet, right up until his death in 1905, satirists, print pundits, and *ʿulama* banded together against him, accusing him of pandering to foreign interests and imbibing all things European. Indeed, throughout his tenure as mufti he was subjected to an unprecedented onslaught of lampoons and satires. "Never in his life," ran one skit in Muhammad Tawfiq's colloquial journal, the satirical *Humarat Monyati*, "has the *mufti* visited the *hijaz*, but there is no country of the Franks that he has not visited once or twice or three times!"[206] Ibrahim al-Muwilhi, an ardent pan-Islamist and former associate of ʿAbduh's, similarly attacked the mufti. In his pro-Hamidian Egyptian paper, *Misbah al-Sharq*, a caustic article on the mufti's European travels warned against those Egyptians who spent their summers in Europe, adding that it was they who were "spooning out *taqlid* [imitation] of the Westerners."[207] Thus was ʿAbduh's criticism of the unimaginativeness of the *ʿulama* turned against him. ʿAbduh also faced harsh criticism from the journalist and political agitator Mustafa Kamil. A man backed both by Sultan Abdülhamid and by the khedive, who was reported to be financing Kamil's journal, *Al-Liwa*, Kamil made inflammatory speeches denouncing those "traitors" who worked with the British.[208]

The ʿ*Ulama* and the State

In the correspondence of spies, informers, civil servants, and petitioners sent to Khedive ʿAbbas from the early 1880s onward, we gain a vivid picture of the political associations attached to ʿAbduh's name. Take the undated report from khedival agent "294" on a riot at Al-Azhar that brought in the police. This riot had taken place when people began to protest against a certain Shaykh Rashid (Rida?), who had been loudly calling on people to stop praying for divine intercession at the shrine of Husayn Mosque. The agent wrote that Shaykh Rashid was supported by ʿAbduh and that their *madhhab* was that of the Wahhabis. And he reported on another incident, this time at the Hamidiya Mosque, that occurred when Shaykh Rashid attempted to preach against saint worship, causing a mob of angry supplicants to hit him until the police intervened. Agent 294 ended by complaining that if the *madhhab* should in fact spread throughout Egypt, it would be more harmful than the spread of Protestant missionaries there.[209] Another letter sent to the khedive shows the writer complaining of the journal published by Rashid Rida, *Al-Manar*, and of ʿAbduh's ideas on religion as published in it, asking the khedive what happened to the oath ʿAbduh was asked to take to swear off Wahhabism.[210] In fact, much of the intelligence on the mufti that made its way to the khedive assumed an association between ʿAbduh and Wahhabism. This was a subject that held particular interest for Khedive ʿAbbas, who had himself asked a number of scholars to accuse ʿAbduh of being a Wahhabi after Lord Cromer blocked plans to bring about ʿAbduh's resignation.[211]

The eighteenth-century reformer ʿAbd al-Wahhab had been a staunch critic of Sufi shrine worship and advocated a strict return to the thought and practices of the *salaf*, the forefathers, or orthodox founders, of the faith. He had, by ʿAbduh's day, however, also acquired a reputation as not only a radical opponent of the Ottomans but also as a heterodox Muslim thinker, and the early twentieth century saw continued conflicts between Wahhabi dynastic leaders and Ottoman forces. (Rashid Rida's later support for Ibn Saʿud in the 1920s would only have enhanced this association between his and ʿAbduh's thought and the Wahhabis.)

And however strange it may seem today, when Wahhabism has come to acquire very different connotations, there were indeed some similarities between the thought of ʿAbduh and of ʿAbd al-Wahhab. Both authors focused on contemporary Sufi and other superstitious practices as a cause of moral—and even political—decline. Both addressed the issue of *tawhid*, and each published treatises on the subject. But while ʿAbduh's *Risalat al-tawhid* was concerned with a revitalized theology, ʿAbd al-Wahhab's *Kitab al-tawhid* was more limited and confined itself to a discussion of examples of contemporary transgressions of the principle of *tawhid* in orthodox Sunni Islam—from shrine worship and prayers for saintly intercession to sorcery, soothsaying, and quotidian slips like "cursing the wind."

Sufism, so common throughout the Ottoman Empire and so entrenched in its daily life, had posed through its heterodoxy a crucial challenge for many religious thinkers from the eighteenth century onward. ʿAbduh himself had early on engaged earnestly with Sufi thought—an interest he may have maintained throughout his life.[212] Yet, later in life, the mufti emerged as a vocal and extremely harsh critic of contemporary Egyptian Sufi practices. He was particularly disdainful of musical *dhikr*, the meditative and often instrumental incantations that were sometimes accompanied by trance dancing and chanting, and he disliked the practice of visiting shrines and other public rituals, which he classed as leading to "social malaise." In 1904 he even issued a fatwa on the question of *tawassul*, or intercession by prophets and saints, a belief that ʿAbduh roundly denounced.

ʿAbduh's arguments in support of the right to *ijtihad* and his critique of *taqlid* carried similar associations. *Ijtihad* as both a legal and a hermeneutic concept in Muslim theology had long been cited in the creation of new orthodoxies—as was the case with ʿAbd al-Wahhab, Ahmad al-Sanusi, Shah Wali Allah, and others who attempted to oppose the current political and legal order through an appeal to the right of *ijtihad*. And there was a specific Ottoman dimension behind their counterassertions: after all, the official Ottoman code was Hanafi, which had largely been adopted for its view on the caliphate, which emphasized abilities and not the lineage of the ruler.[213] Even the Tanzimat reforms did not challenge this official legal code. The very notion of *talfiq*, or the fusion of different legal opinions and schools, could therefore be seen as an implicit critique of the legal order of the empire itself. Like the argument against *talfiq*, the rejection of *ijtihad* was also connected to the contemporary Ottoman norms surrounding *taqlid*, which was as much a legal category as a hermeneutic method.[214]

Hence ʿAbduh's critics allied themselves as easily with the sultan as with the khedive—or indeed with both. Yusuf al-Nabhani, an Ottoman official from Nablus who spent time in Istanbul, Latakia, Jerusalem, and Beirut before ending up in Cairo, offers an instructive case. He was initially closely connected to the sultan—having spent several years in his court—and was connected to a complex Sufi network in the empire through Abu al-Huda al-Sayyidi, whom we encountered briefly in the last chapter as head of the Rifaʿiya order in Syria. Al-Nabhani promoted complete allegiance to the sultan as the lawgiver and therefore only legitimate *mujaddid* (renewer) and as the caliph of all Muslims. He praised the sultan's care of scholars, Sufis, and the poor, mentioning in particular the attention given to the tombs of the prophets and saints and the building of new mosques and schools throughout the empire. A longtime resident of Beirut, and familiar with the mission schools, al-Nabhani was highly suspicious of proponents of Western-style education: it was their teaching of skeptical subjects like the natural sciences and foreign languages and their lack of guidance for proper ethical conduct that in his view threatened to corrupt Muslim youth.[215]

On all these counts, ʿAbduh's approach attracted his ire. He opposed ʿAbduh's attacks on Sufi popular devotion and his desire to incorporate the modern sciences into traditional Muslim curricula. Al-Nabhani's critique that *ijtihad* was akin to imitating Protestants was also no compliment to ʿAbduh.[216] Al-Nabhani argued that the door of *ijtihad* had been "closed hundreds of years ago" and that no one would now claim *ijtihad* "unless he is insane in his mind and religion." Strict adherence to one of the four legal traditions of Muslim thought was required. He opposed *talfiq* for the same reason.

Al-Nabhani was a professional polemicist at a time when monarchs and Western policy-makers alike were willing to pay for rhetorical ammunition. In 1902 he wrote a book that may have had the aim of undermining Wahhabi influence, particularly after their takeover of Najd. It centered on a critique of Ibn Taymiyya and the Wahhabi's views on saints' tombs and intercession. Around 1909 he wrote another polemic attacking Afghani, ʿAbduh, Rida, and Wahhabism all in one fell swoop. Following the Young Turk Revolution and the deposition of the Ottoman sultan Abdülhamid in 1908, al-Nabhani found patronage in Egypt, where Khedive ʿAbbas asked him to continue his attack against the reformers in return for a salary from the department of Egyptian Endowments.[217]

Al-Nabhani accused ʿAbduh of more than theological error. He castigated him for befriending and collaborating with the British and imitating the Protestants. He attacked his moral character and denounced him as a hypocrite in strong terms, continuing in the same vein when describing his disciple Rashid Rida, whom he described as the "Prophet's enemy" and a "perpetuator of evil" (he also referred to dreams he had of Rida's impending perdition).

Al-Nabhani's critiques echoed those voiced earlier within Egypt, particularly once ʿAbduh's reputation for controversy was cemented by a series of fatwas issued between 1899 and 1901. In a ruling commissioned by the khedive, when the khedive was setting up the Egyptian National Bank and the Post Office Savings Bank, ʿAbduh argued that interest (*ribaʿ*) was permissible (claiming the Qurʾanic prohibition on usury applied only to increases on the principal of deferred debts).[218] This fatwa was considered scandalous by the *ʿulama* and journalists alike. Muhammad Tawfiq, in his *Humarat Monyati*, for instance, ironically described ʿAbduh as a champion of reform: yet what was this "reform," he asked, but following "the will of foreigners at the expense of the people." Was ʿAbduh not conniving against the people by risking *awqaf* monies in order to collect interest for the banks of the Europeans?[219]

ʿAbduh's rulings only added to his already-infamous reputation for heresy: "Was it not ʿAbduh who issued a fatwa permitting usury, although it is prohibited by the shariʿa, saying that it was of no concern if it was to increase revenue or aid development? These are matters the *kharaja* [heretics] and the *muʿtazila* would know about, no doubt. . . . If you asked him to perform a marriage contract, he would write the contract according to the *madhhab* of Darwin."[220] ʿAbduh's association with the British, which damned him in the eyes of the *ʿulama*, could scarcely be disentangled from his suspect theological inclinations and his interest in contemporary European science. His openness to the latter provided a ready topic for slander.

An episode in 1904 made this clear, when a recent graduate of Al-Azhar, Shaykh Muhammad al-Zawahri, published a biting critique of his former madrasa-mosque in a book entitled *Al-ʿIlm wa-al-ʿulama*. "I do not think," wrote al-Zawahri, "that there exists in the world any system so defective as that which exists presently at the noble Al-Azhar." When the pro-Ottoman journal *Al-Jawaʾib al-Misriya* asked a senior shaykh at Al-Azhar, shaykh ʿAbd al-Rahman al-Shirbini, what he thought of these criticisms, al-Shirbini claimed that Al-Azhar's decline was the fault of the Azhar reform movement itself—the fault of those who would "turn the mosque into a school of philosophy and literature to put out Islam's light." Students "duped by Spencer's philosophy," he wrote, reserved only scorn for their imams. True reform, he continued, should include only "the protection of students' health and vigilance over their comfort, providing them with good food, not principles of philosophy or the new high sciences."[221]

ʿAbduh, who just over a year ago had traveled to England to pay his respects to the philosopher, was incensed by the offhand remark, which was clearly directed at him. "Where did he get the name of Spencer?" he wrote. "What student mentioned him? Which of Spencer's principles has entered Al-Azhar? And what does the shaykh mean by it exactly?" It is true, as we have seen, that ʿAbduh was deeply interested in Spencer's ideas, and in 1904 *Al-Muqtataf*—which had long shown its own interest in Spencer—published several lengthy articles on his philosophy and ideas of education. This was almost certainly linked in al-Shirbini's mind with ʿAbduh's own predilection for the British philosopher.[222] It was, however, clearly a travesty of ʿAbduh's views to reduce them to a mere imitation of the Englishman's. Yet it is a striking illustration of the mufti's isolation that only *Al-Muqattam*—*Al-Muqtataf*'s sister publication—would publish ʿAbduh's response to the shaykh. This simply damned ʿAbduh further in the eyes of his critics. For Sarruf and Nimr (the editors of both *Al-Muqattam* and *Al-Muqtataf*) were themselves by then regarded as pro-British lackeys, and their fondness for ʿAbduh, whom they termed a fellow "seeker of reform," did him no favors.[223]

ʿAbduh's Legacies

It was for all these reasons that in his own lifetime, ʿAbduh's ideas were not able to fully capture the imagination of those he so desperately hoped to inspire and reform. ʿAbduh himself became increasingly disillusioned, not only with the bureaucracy but also with the penny-pinching British administration itself. In a letter he drafted shortly before his death in 1905, ʿAbduh complained that "the instruction given by the Egyptian government . . . hardly enables a man to acquire the means for earning a living wage." The aim of the state, he complained, seemed to be "to spread education as little as possible among the people." He too, it seems, finally shared the growing sense that Lord Cromer's policies had aimed at depriving Egypt of an advanced educational system: "Egyptians are persuaded that those who direct their public affairs are not doing all they can to raise the moral and intellectual level of the rising generation." According to ʿAbduh, the result of Egyptian educational development was that the state had promoted vocational training at the expense of scholarly and, indeed, moral values. "True," he wrote, "we possess judges, lawyers, physicians, and engineers more or less capable of exercising their professions; but among the educated classes, one looks in vain for the investigator, the thinker, the philosopher, the scholar, the man in fact of open mind, fine spirit, generous sentiments, whose life is devoted to the ideal."[224]

By the end of his life, ʿAbduh had come to despair of any possibility of real reform at Al-Azhar. "The continuation of Al-Azhar in this age in its current decrepit condition," he wrote, "is impossible: either it shall revive or its ruin will become total. Indeed, I did my best to revive it, but the many adversities I have encountered have cast me into despair."[225] ʿAbduh was by then considering establishing a modern theological school of his own—a project interrupted by his death in 1905 but later realized by Rashid Rida in 1912.[226] ʿAbduh died (of cancer) in the summer of 1905, having only recently resigned from Al-Azhar's Adminstrative Council and having failed to realize his pedagogical ambitions while worn out by the press campaigns and intrigues against him.[227]

One of the prime consequences of the failure to reform Al-Azhar was the emergence of the first modern university in Egypt. There had been much discussion at that time of the need for this, despite the lack of official British enthusiasm. Proponents ranged from the Armenian bureaucrat Yaʿqub Artin to Jurji Zaydan, founder of *Al-Hilal* and a fellow scientific emissary of Sarruf and Nimr, and nationalist-revolutionaries such as Mustafa Kamil and Saʿad Zaghlul. Of course, everyone envisioned something a little different for the school, just as each of their plans differed from ʿAbduh's (who had planned to establish a new religious school of higher education, not a secular one). As early as 1894, Artin was calling for a professional school for technocrats. In 1900 and 1906 respectively, Kamil and Zaghlul called for a national university. Also in 1900 Zaydan launched a campaign in favor of a university modeled after his own alma mater, the Syrian Protestant College.[228]

Buoyed by this groundswell of opinion, the first "Egyptian University" (later known as Cairo University) began operations as a private institution in December 1908—"opening its doors to every seeker of knowledge, regardless of religion." Social evolutionary justifications were part of its raison d'être. A May 1908 fundraising speech for the university began with the following announcement: "It is the law of civilization nowadays for nations to mix freely with one another, most of whom are advanced, educated, civilized, and armed with weapons for the struggle for life [*al-jihad al-hayawi*]. Does one want to remain unarmed until the natural law that the strong eat the weak is fulfilled? Hence, everyone tells you: Learn . . . if you want to remain safe in this age of social struggle [*al-jihad al-ijtimaʿi*]. This can be accomplished only through this university, in whose cause we are gathered here today." The well-known nationalist intellectual Qasim Amin echoed these sentiments: "The main reason for the rise and decline of nations is their way of training and educating." What was needed, read an official university statement, "was to seek knowledge for the sake of knowledge, as was the case in those early days of Islamic civilization and as is now the case among the nations of the West and Japan." It was imperative, therefore, to sponsor those "fields of knowledge which are now neglected in Egypt, like the arts and humanities, history, and the higher sciences—subjects which elevate the individual and his community and make a nation great among nations." Key to the new institution was that it was to be "a university which has no religion, only science [*ʿilm*]."[229] Only three years after his death, ʿAbduh's approach had been replaced by one in which religion and science were seen as embodying separate missions and where science constituted the true universalism.

Yet ʿAbduh's legacy was far-ranging: his vision of religious reform gained him admirers in Egypt and well beyond it, and many later exegetes, religious commentators, and scholars throughout North Africa and Southeast Asia would turn to ʿAbduh's ideas to reinforce their own theological views. Thanks largely to *Al-Manar* and to the dissemination of ʿAbduh's ideas in the press, his ideas traveled the globe and gained him a reputation as one of the most influential Muslim thinkers in the modern world. But that reputation was the product of unexpected associations and surprising appropriations.

At Al-Azhar itself, many of the reforms that ʿAbduh had attempted to initiate were gradually and enthusiastically implemented in the interwar era, acquiring a number of vocal adherents in both the upper ranks of Al-Azhar and government ministries: many later rectors of Al-Azhar and ministers of education would refer to ʿAbduh's educational reform program to drive their own vision for supervision and administrative change there.[230] But this occurred only as the state's jurisdiction over Al-Azhar grew and certainly did not free the *ʿulama* from the forces of government control as ʿAbduh had in fact hoped. In the end, Al-Azhar did indeed come to teach the modern alongside the traditional sciences, and it became one of many higher educational degree-granting institutions of Egypt subject to government funding and regulation. Later rectors Mustafa al-Maraghi (1928–29 and 1935–45) and Mahmud Shaltut (1958–63) would count themselves as ʿAbduh's spiritual, intellectual, and even institutional heirs: they built upon many of his theological and exegetical themes, while a number of the reforms they carried out at Al-Azhar had first been suggested by ʿAbduh.[231]

In Egypt at large, meanwhile, the image of ʿAbduh as a British stooge and object of satirical ridicule was forgotten and replaced by a more orthodox and nationalist one. The process started almost as soon as he died. But it was encouraged by his disciple Rashid Rida, whose growing anti-Ottomanism, anti-British views, and later search for an Arab caliphate reflected his increasingly pan-Arab sentiments and came to replace the earlier pan-Islamist ideology that had helped to form his and ʿAbduh's own early thought. For Rida, establishing a political and religious genealogy from Afghani to ʿAbduh stressed the orthodox faith in unity of the *umma* and ʿAbduh's anticolonial credentials. Yet for Rida it was the Arab state, and not the Muslim world at large, that would come to occupy the horizons of this vision. As part of this struggle, Rida created the image of a national hero for an increasingly anticolonial era.[232] It helped that some of ʿAbduh's other students and disciples—foremost among them the leader of the 1919 revolution, Saʿad Zaghlul—played a prominent role in the nationalist movements of the following years. The result was that later theologians as well as Azhari officials would all argue for reform in his name, although it was not until Gamal Abdel Nasser in 1961 that the modern sciences would be fully integrated into the curriculum at Al-Azhar.

ʿAbduh himself had always been driven by anticolonial sentiment, and his project for a new theology was motivated from the start by the desire to help Muslims withstand European imperialism. But unlike his mentor Afghani, he was willing to gamble on a temporary alliance with the colonial British state against both the palace and the old-guard *ʿulama*. He believed, after all, that one should learn from Europe, but while he admired European science and technology, he wished to avoid the Reformation divorce between knowledge and belief. By demonstrating its rationalist premises and its encouragement of scientific learning, Islam could succeed where Europe had gone wrong. It was for this reason that he spent so much of his later career critiquing Christianity's historical discomfort with science.

This approach was to prove immensely popular with several generations. Many later scholars tried to show that Islam was uniquely compatible with the modern age, and they turned to the Qurʾan to demonstrate its prefiguring of scientific discoveries much as ʿAbduh had done in discussing the struggle for life and other evolutionary principles.[233] Indeed, an entire new genre of "scientific exegesis" of the Qurʾan—the so-called *tafsir ʿilmi*—emerged, and many of its practitioners referred to ʿAbduh's *Tafsir al-Manar* as inspiration.

Yet as the *tafsir ʿilmi* proliferated, ʿAbduh's attitude toward evolution was abandoned. He and Rashid Rida had argued earlier that evolutionary principles could be accommodated with Qurʾanic authority through the use of analogic interpretations of the scriptural text. Rida, in 1905, had insisted in defense of what he had called "a rationalist view of religion" that verses referring to the creation of man need not be taken literally.[234] But it was precisely on this issue of the creation of man that later science exegetes would increasingly come to diverge from this reading. At the same time, ʿAbduh's Spencerian readings of social progress through a reform of the self and his emphasis on education as the means of creating a new citizenry were displaced by more interventionist and collectivist creeds. Thus, paradoxically, as ʿAbduh's reputation grew, many of the views he had actually espoused were left behind.

SIX

Evolutionary Socialism

In 1903, when Farah Antun clashed with Grand Mufti Muhammad ʿAbduh over Ibn Rushd, the exchange was carried out in Antun's journal *Al-Jamiʿa* (The collective).[1] Antun had founded the journal after his move to Cairo in 1897; by 1903 it was widely known across the Arab world for its collectivist approach to contemporary "social problems." The same year, Antun elaborated on his ideas in a short novella. Entitled *Al-din wa-al-ʿilm wa-al-mal aw al-mudun al-thalath* (Religion, science, and wealth: The three cities), this blend of social realism and apocalyptic utopianism was essentially a reflection on political reform and the value of technocratic expertise. In the manner of tales (*riwayat*) "like Tolstoy's and Zola's," it took aim at "the Social Question" (*al-masʾalla al-ijtimaʿiya*) and proposed, via the neologism *al-ishtirakiya* (socialism), what was still at that time a relatively novel solution in the Arab world.[2]

In the novella, the story's two protagonists, the young Halim and his friend Sadiq, journey to "three cities": Wealth, Religion, and Science. Along the way, they witness violent political conflict and social disorder. The city of Wealth is the most divided of the three and is quickly embroiled in a revolutionary struggle between workers and capitalists that eventually draws in members of the other two cities. Agitating for workers' rights and social reform, a few revolutionary laborers cite Marx while calling for radical change. Among their demands are the abolition of private property, an equal share in profits for capitalists and workers, and a general revolution against the capitalist mode of production. Their employers counter these demands with classic liberal arguments in favor of laissez-faire, individualism, and the virtues of competition and struggle. Members of the "community of religion" (*ahl al-din*) come to the defense of the workers but ultimately advise them to direct their gaze beyond material concerns altogether. Members of the "community of science," or the "intellectuals" (*ahl al-ʿilm*), prove to be the true intermediaries, responding effectively to both liberal capitalists and ascetic preachers and helping to devise an economically viable, equitable, and, above all, rational solution. They point out that while it is true that struggle is an omnipresent fact of nature, it does not have to be so in society—particularly in "advanced societies"—and they propose a compromise: eight-hour workdays with a minimum wage, restrictions on female and child labor, and the establishment of pensions, hospitals, and educational programs guided by scientific principles and technocratic expertise. Those unable to find employment in the private sector are guaranteed jobs in government factories and farms.

In the story, extremists reject the plan and a few rabble-rousing workers take up arms, leading the novel to an incendiary and apocalyptic finale. The cities are razed, and Halim and Sadiq, disgusted by the class warfare and fanaticism they have witnessed, call for their own utopian republic founded on the shared virtues of equality and fraternity and guided by a moderate, welfarist, and "socialist" elite of the kind called for earlier by the intellectuals' spokesman, or *shaykh al-ʿulama*.[3]

As Antun suggests at the start, the revolutionary finale conveyed precisely the kind of social disorder and political unrest that he saw as currently threatening Europe. Yet the problem of inequality between rich and poor was universal. Nevertheless, few people, he states in his introduction, were addressing its implications in contemporary Arabic discourses. Hence, even as Antun ultimately presented such class warfare and ideological conflicts as a quintessentially European problem, he nevertheless found echoes looming over Egypt's own future. Similarly, he clearly saw parallels between his imagined technocratic resolution of Europe's ills and a model that could be applied closer to home.

Conflict between rich and poor was the crux of the problem, yet its solution could be entrusted neither to the propertied classes nor to the revolutionaries, and certainly not to religious leaders. Like so many of his contemporaries, Antun emphasized his preference for reform over revolutionary change, but this reform was to be entrusted to scientists or rational intellectuals (*ahl al-ʿilm*): for it was *their* kind of knowledge (*ʿilm*), based on scientific expertise, that would win the day, not that of the old-guard *ʿulama*. It was this, too, that gave them their special claim to both political and social authority: for as Antun himself states, humanity has no need for guidance other than the judicious exercise of its own powers of reasoning. Clearly, Antun saw the *ahl al-ʿilm* as the only truly enlightened guides to reason: they alone might lead people out of the darkness of their own irrationality.[4] For Antun, it was science that was to serve as the intermediary between religion and capital.

Antun's fictional account was a mediation on several fronts, and it straddled genres in more ways than one. For example, its reference to the "three cities" harkened back to long-standing Arabic discussions of just or virtuous and unjust or erring city-states. This topic can be traced back to the tenth-century Muslim philosopher Abu Nasr al-Farabi's Neoplatonic reworking of *The Republic* in his *Mabadi ara' ahl al-madinat al-fadila* (Principles of the people of the virtuous cities): among the "ignorant cities" he warned against was the "city of meanness," whose sole aim was "the acquisition of wealth and riches, not in order to enjoy something else which can be got through wealth, but because they regard wealth as the sole aim of life."[5] But like discussions of the ideal society elsewhere at this time, the very concept of utopia had undergone radical transformations: still rendered in Arabic as *al-madina al-fadila* or else simply transliterated as *al-utuwbiya*, modern discussions of socialism, Marxism, and other forms of radicalism looked above all to science and technology, and not statesmanship or even divine guidance, as the true means of establishing the perfect collective state. It was thanks to human reason, economic equality, and social justice that this might be achieved, and not to humanity's realization of the Active Intellect, its comprehension of the principles of divine existence, or even in the perfect ordering of the souls of citizens.

Like so many of the other modern variants on utopia that were developed from Saint-Simon to Proudhon, Owen, and Marx, Arabic versions were similarly based on the idea that the golden age of humanity, as Saint-Simon once famously put it, was ahead and not behind us. Science would be its new religion, and technique its new magic.[6] Indeed, as Antun's own fictional *shaykh al-ʿulama* announced in his response to the "call of the religionists" (*daʿawat ahl al-din*), religion and science might diverge in their methods, but their aims were the same: concord or perfect harmony between people.[7]

Antun saw his technocratic utopia and the "city of God" as committed to the same principles. He even has his preacher citing early Christian and Muslim communities as models for "true socialism" (*al-ishtirakiya al-haqiqaya*). And indeed in certain circles, the early history of the Islamic caliphates was being similarly reread at just this time. Just as Christian and other forms of religious socialism gaining ground in Europe and elsewhere in the 1890s looked back to the communalism of earlier times, so Muslim anti-individualists made a similar historicizing move.[8] For example, Jamal al-Din al-Afghani's argument, much as Antun's cleric's, relied on the idea that both the Qur'an and the early Muslim *umma* followed the principles of equity and material partnership. Afghani cited the social conscience of the early Muslim *umma* and the first four caliphs, manifested in their fair distribution of war spoils and the modesty of their rulers. (In keeping with historical narratives of this kind, this social conscience was then said to have been abandoned by the profligate Umayyads.) God's legislation itself rested on universal principles: indeed, "an examination of the human potential for good or evil will reach the same conclusion." It was the greed and arrogance of the wealthy today—particularly those fueled by Western capital and industry—that made "the working classes resort to socialism, they feel deprived." Afghani's conclusion—not wholly unlike Antun's—was that this would "work up to a grave situation in the West, which will not leave the East unaffected." Thus, while Afghani saw socialism as part of the divine and therefore universal principle of the good, he also saw its Western variant as fraught with risk, threatening the East: "Socialism, like other ideologies is likely to go to extremes (the best thing, of course, is moderation)." Many, like both Antun and Afghani, saw Western socialism as subject to extremism, one that had the potential, as the latter put it, "to end up in bloodshed, the massacre of innocents, a devastation that will benefit no one." Indeed, this is precisely what the finale of Antun's story suggests.[9]

Antun's views on socialism were, as his novella shows, rather different, but in the fight for the rights of labor he had been willing, in his role as a supporter of the workers' movement in Egypt, to solicit the support of the *ʿulama*. Before his dispute with ʿAbduh, for instance, Antun had asked the grand mufti to issue a fatwa backing the cigarette workers' strike in 1899. The close of the nineteenth century in Egypt marked the beginnings of labor union formation and worker mobilization, and this strike was one of the earliest and most important of its kind. It lasted three months and mobilized about thirty thousand workers, involving colonial officials, government bureaucrats, and the police. In the end, the police—with explicit British backing and command—intervened on behalf of factory owners, setting off widespread public debate. Antun was highly critical of Lord Cromer's policy—one in which "all the Government has to do is to keep order in case of any threatened disturbance." "I do not think it is the business," wrote Lord Cromer in one of his letters, "of any Government official to interfere between employers and labourers."[10] *Al-Muqattam* served as a British mouthpiece through which the consul general stressed the value of free enterprise. Antun accused him of turning a blind eye to capitalist exploitation and criticized the laissez-faire protection of "free enterprise" and the "individualist" implications of his stance.

ʿAbduh's fatwa, for its part, addressed the legitimacy of government intervention from the point of Islamic jurisprudence and history more specifically. Like Antun, it also viewed the state as a necessary arbiter between labor and capital. And like Afghani, it emphasized early Islamic communities as examples of just governance and material moderation—the Aristotelian golden mean, or *mizan*, between liberality and parsimony or, in contemporary terms, between the extremes of capitalist free enterprise and revolutionary labor radicalism.[11] The fatwa was reminiscent of Afghani's earlier discussion, but it took the argument further by explicitly rejecting both free-enterprise state economic policy and "possessive individualism" (*istifrad*).[12]

In real life, then, Antun had turned to ʿAbduh for arbitration. Yet in his novella, only a few years later, he has the rational planner, not the cleric, acting as society's guide. Divine justice might well be on his side, but the call for rational economic planning and government reforms would be derived ultimately from modern principles of science and technical training, whose true value lay precisely in their potential for resolving vexing social problems through a new technocratic order. Further grounding Antun's optimism in science was his faith in the "school of evolution and progress." Indeed, many of his writings expressed the view that the laws of nature offered the basis for understanding the evolution of states and societies. Like other proponents of natural law and social development, he used the idea to bolster ethical arguments and to draw a collectivist moral—hence his emphasis on the rise of altruistic ethics among "advanced societies."

The 1903 novella reflected the declining hold of the individualism, utilitarian ethics, and laissez-faire approach to state and society shared by other early Arab evolutionists. Mutual moral development (as much as national collectivism) became the new mainstream reading of social evolution. It was for this reason that Antun was so keen to publicize the writings of Alfred Russel Wallace, and only a few months after Wallace's article "Man's Place in the Universe" appeared in the *Fortnightly Review*, Antun published an enthusiastic summary entitled "The Heavens and Earth (*al-samaʿ wa-al-ard*) and Man." Wallace argued that the new astronomical findings on nebulae and clusters could equally support the view that the "universe is a manifestation of Mind" and support the cause of those who believe that "we ourselves are its sole and sufficient result." Antun classed this as a "wonderful opinion": it restored "mankind's greatness and the splendor of his humanity" by arguing that humans are the highest spiritual creation.[13] Antun was certainly not alone in turning to Wallace in this way. The latter's view of the intervention of "the unseen universe of Spirit" in the natural world—as in the case of the creation of life, consciousness, and humanity—won him many supporters among Arabic readers. His interest in socialism won him more. In fact, the first full-length history of socialism published in Egypt in 1915 began with a quotation from Wallace on its frontispiece.[14]

Socialism proved to be as much about social evolution as anything else. Since at least the 1870s, *Al-Muqtataf* and other periodicals had been presenting the notion of "society" as an organism. This, too, had classical roots; classical Arabic political theory was itself deeply organicist.[15] Now, however, this acquired a new valence: as an object of political intervention, society came to be defined as an abstract entity subject to its own laws and to rational regulation. It was the scientific social reformer who emerged as the physician of the state and as the real intermediary between workers and capitalists—counseling moderation rather than the extremism of capitalism, on the one hand, and Marxist revolution, on the other.

Darwin, not Marx

Marx and radicalism were fused in these discussions, yet Marx's actual ideas were clearly unfamiliar to most and often the product of mere hearsay at best. This was true in Antun's work: the novella gets Marx's name wrong, and many of the principles ascribed to him could have equally applied to any of a number of radical socialist or anarchist thinkers. Yet whether generated from Marxism or anarchism, revolutionary thought in general proved much less popular among the fin-de-siècle Egyptian intelligentsia than among the union activists spread throughout the eastern Mediterranean. For many Arab intellectuals, as elsewhere in the colonial world at this time, socialism was an idealist as much as a materialist reformist enterprise. Rather than prescribe the inevitability of revolution, they were more inclined to reach for technocratic solutions, albeit ones that also counseled a new emphasis on social cooperation and political collectivism. Indeed, for many Arab intellectuals, as with many colonial intellectuals elsewhere, socialism was at first tied more strongly to evolution and to Darwin than to communism or to Marx, although few have reflected on this genealogy. It is no accident, then, that many early writers on socialism in Arabic such as Shumayyil and his protégé Salama Musa were more familiar with Darwin than they were with Marx (*Das Kapital* was not even translated into Arabic until 1947).[16]

This is hardly surprising. As historians have recently underscored, Marxism was only one variant of socialist thought and, before 1914, a fairly insignificant one in many places outside eastern Europe. Other socialist and labor movements arose dramatically in the 1880s and 1890s: Egypt was no exception. As a "utopian wave of working class hope" gathered force, radical thought found expression in a number of diverse forums: theaters dramatized political assassinations, workers' schools distributed anarchist literature, and the press carried out a vibrant debate which was often read in salons and coffee shops.[17]

With foreign investment and industry increasingly transforming capital, class, and property structures under British rule, socialism proved an increasingly attractive anticolonial and nationalist political platform. Yet it was not until 1917 that references to Marx became commonplace.[18] The fact that Marx was little discussed among radical circles in the eastern Mediterranean may be one reason why, until recently, historians of the region have tended to assume that socialist thought was of relatively little consequence until the First World War.

The tensions between nationalism, colonialism, and internationalism may have been another reason for the neglect of early socialist thought in the region. Historians of socialism in Egypt in particular have often suggested a conflict between nationalist and internationalist agendas and used this to explain the relatively minor role of revolutionary, Marxist or Leninist radical politics in Egypt. But more recent work has shown how radical politics and ideas in the region were tied to growing global and local networks of capital, labor, and information that themselves accelerated the processes of imperialism and migration in the region and brought together a diverse range of actors and forces, both intellectual and political.[19] Many of the early discussions of socialism also took place at a time when political parties were first being formed in Egypt, and so the development of socialist ideas can scarcely be separated from the rise of nationalist politics.[20] The power of evolutionary socialism lay precisely in its ability to bring together an emphasis on national development and a growing international critique of Western capitalist and imperial expansion outside Europe. For the Nahdawi elite, eschewing a dialectical and revolutionary view of political change was, by the early twentieth century, no mere epiphenomenon of competing global vectors, then, nor was it simply the natural outgrowth of previous ideas about nature and society. Rather, it was a synthesis of a diverse number of competing ideologies that came through the recasting of an older vision of the natural moral economy (and the abandonment of classical liberal moral and political thought).

After all, discussions of socialism raised larger ethical and social problems. Like its rival ideology, liberalism, or what many early Arab socialists simply called individualism, evolutionary socialism appealed to nature and its proper moral economy. It was invoked to popularize any of a number of collective social reforms as much as socialism proper. For most of those who wrote about it, the problems of political economy were social and ultimately moral ones. Of course, classical liberal political theory had also been couched as a moral argument—this was one reason why early *laissez-faire* Arabic readers of Darwin went to such pains to emphasize the morality of individualism and utilitarian ethics. Yet the growing disillusion with liberal individualism transformed the wider reading of evolution's basic tenets by the late 1890s and early 1900s. Worker utopianism, global radical networks, and, later, the Russian revolutions convinced people like Shibli Shumayyil, who died the year the Bolsheviks gained power, that the true moral lesson of evolution was the rise of the mutualism of scientific socialism.[21] Socialism itself was therefore almost always described as an outcome of the natural evolution of states and societies—one reason why so many of the intellectuals examined here promoted an organicist notion of the state. It was this faith in the organic development of societies and the moral authority of nature that gave birth to the "soft" radicalism of evolutionary socialism.

These trends called forth a new Darwin for Arabic readers. In place of self-preservation and individual cultivation in the "struggle for life," themes that had been stressed by the first generation of evolutionary popularizers, Darwin's name was increasingly invoked in reference to the "law of mutual aid" and the power of collectivism. The emphasis on competition and laissez-faire individualism by early Arab evolutionists soon gave way to a new focus on social cooperation. This was especially true in the first decades of the twentieth century and through the writings of a rising generation of evolutionary socialists. Even journals like *Al-Muqtataf* and *Al-Hilal* that did so much to help promote the value of individualism and competition, and to link them with their view of evolutionary progress, began to change their tune. By the turn of the century they no longer took the view that competition and struggle were sacrosanct laws of nature: indeed, it was just around this time that they began to publish scores of articles on animal cooperation, mutual parasitism, and symbiosis.[22]

The idea that nature provided models for collective advancement—even at a mortal cost to the individual—could be found in Darwin's own writings, particularly in his *Descent of Man*. There, Darwin singles out these forms of self-sacrifice for the sake of group advancement as the key causal mechanism for moral progression.[23] Meanwhile, the writings of other naturalists at this time, such as John Lubbock, on the newly dubbed "social animals," such as bees and ants, further encouraged this trend. The fact that the Qur'an itself has chapters on these (Surat al-Nahl and Surat al-Naml) only seemed to underscore these connections. Indeed, this was a point that the well-known physician, writer, and avid beekeeper Ahmad Zaki Abu Shadi sometimes made in his journal on apiculture, *Mamlakat al-nahl* (The bee kingdom).[24] It was in this context, too, that Kropotkin proved a decisive influence. His 1902 book *Mutual Aid* begins, after all, with two chapters on mutualism among animals, where references to bees and ants also abound.

At the same time, popular science journals also began to write differently about socialism itself, taking a much more positive view of its potential. Indeed, many early Arab evolutionary socialists first tried out their ideas in their pages.[25] And like the British Fabians and other social reformers who took up the banner of democratic socialism in the early twentieth century, these Arab evolutionary socialists turned to arguments about the role of collectivism and "mutual aid" in nature to help run a public campaign for a new social and political order in Egypt.

Collectivism and Darwin were further tied after 1914. Once again Kropotkin proved a crucial influence, even as many ignored or sidestepped his actual anarchist agenda. Spencer too was conscripted into this new emphasis on collectivism, and the latter's individualism and laissez-faire view of the state were often overlooked or reworked to fit a more centralized, cooperative, and evolutionary model of government. Spencer's liberal utilitarianism, for instance, was adapted to fit the concerns of moral collectivism, and his law of equal freedom was frequently cited to argue for a strong state-led effort to induce the proper development of the national body. This effort was led by a new generation of *effendiya*—that same professional elite whose ties to government and to the civil service did so much to consolidate a new politics based on evolutionary and gradualist principles.[26]

Against the view that evolution necessarily implied either a pitiless tale of individuals locked in struggle or a crassly materialist view of the state, many of these thinkers showed how science could point the way toward a communitarian morality as the basis for a twinned material and spiritual progress of the nation. This also bolstered their credibility in their rebuttal of criticisms that socialism was essentially an irreligious or immoral creed: in fact, early proponents like the schoolteacher Mustafa al-Mansuri even went to some lengths to establish its compatibility with belief in God. Yet even for al-Mansuri, it was no longer God's vice-regents who had the best claim on virtue or on governmental expertise but the new "men of science." If ʿAbduh had worried that divorcing science from religion would lead to moral crisis, for these evolutionary socialists, science itself would provide the basis for a new social morality.

The Meanings of Socialism

Arabic translations of the term "socialism" itself contained as many if not more ambiguities as their European counterparts. Although Ottoman Arab encounters with left-wing European ideas had occurred as early as the 1848 revolutions, references to them were rare until the Franco-Prussian War and the 1871 Paris Commune. In that year, Butrus al-Bustani, in his Beirut paper *Al-Jinan*, described the Commune as those who "were known as the reds [and] who demand participation [*al-ishtirak*] in wealth." A neologism derived from the Arabic root *sh-r-k*, *ishtirakiya* (socialism), primarily connoted egalitarianism and shared wealth. (It also contained the much more negative connotation of polytheism.) The root itself also alluded both to the pre-Islamic sharing of land in the Arabian peninsula and to *hadith* references to the rights of all to share in the necessities of life. By extension, this came to imply a common division of property: Ahmad Faris Shidyaq's Istanbul paper, *Al-Jawaʾib*, for instance, described the Communards as calling for "sharing the property of those with property." The 1898 edition of Butrus al-Bustani's well-known *Encyclopaedia* deployed the neologism *susialism* as well: a potted summary detailed its European origins and development, highlighted its connections with New Testament teachings, and cited texts from Plato's *Republic* to Thomas More's *Utopia* and Robert Owen's *Addresses*. He also mentioned nihilists (*ʿadamiyun*), anarchists (*fawdawiyun*), and communists (*ʿammiyun*) but did not distinguish between them.[27]

In his description of Halim and Sadiq's utopia, Antun used the term *ishtirakiya*. But in other contexts, he spoke of *al-taʿawan al-ijtimaʿi* (social cooperation), which better described the idea of a cooperative socialist state.[28] Shibli Shumayyil also preferred this term—*ijtimaʿiya* (from the root *j-m-ʿ*, "to gather" or "collect"). Niqula Haddad, another Syrian émigré and collaborator of Antun's, argued similarly: *ijtimaʿiya* should be used instead of *ishtirakiya* since the latter referred specifically to the idea of sharing wealth rather than to social cooperation more broadly. In 1908 *Al-Hilal* echoed this. Specialists complained that the latter term did injustice to the precise meaning of socialism, and some even suggested it be used instead to mean "communism" (which was how ʿAbduh had deployed it in the 1886 Arabic translation of Afghani's *Refutation of the Materialists*, following the latter's lead).

Yet *ijtimaʿiya* never fully caught on, and *ishtirakiya*, despite its emphasis on property, remained the most commonly encountered term for "socialism," perhaps because of its egalitarian tinge.[29] Although *ishtirakiya* was indeed occasionally used for "communism," after the First World War the term *al-shuyuʿiya* (whose root forms possessed a wide range of connotations, from joint property and universal ownership to factionalism or partisanship) gained ground for this instead.[30]

Behind this terminological confusion lay the multiplicity of Arab encounters with European and American political activism. The rise of Egyptian socialist thought simply cannot be understood without appreciating the intensity of its international and cross-regional engagement. This took a variety of forms. Arab students abroad formed radical associations of their own and participated in the left-wing movements they found, associating with French and German socialists, Fabians in London, and anarchists in Lausanne. At the same time, Syrian emigrants to Brazil, Mexico, and the United States reported in the Arabic press on the Left there: Antun himself established an Arab socialist society during his stay in New York, where his partner, Niqula Haddad, had come to know the American labor leader Eugene Debs.

Back home, the rapidly developing Arabic press played a key role in satisfying public curiosity with details of anarchist protests abroad, labor activism, and potted lives of Tolstoy and others. "What are the socialist and nihilist associations that are constantly mentioned in newspapers, and what are their origins?," a couple of readers asked *Al-Hilal* at the end of 1897. Meanwhile, contemporary events quickly inspired poems and dramatic reconstructions: a play commemorating the Spanish anarchist Francisco Ferrer was staged in theaters in Beirut only a few days after his execution, and speeches were delivered on the merits of socialism.[31] Soon numerous articles and debates were carried out in these journals and in countless others.

From Individualism to Collectivism

As Antun's story suggested, socialism was first seen as a foreign problem, associated above all with the ills of industrialization. Egypt's predominantly agrarian economy was commonly regarded as an obstacle to industrial class conflict. Yet as foreign capital and manufacturing took root in

Egypt, labor unions and syndicates were formed. Under the influence of anarcho-syndicalists, workers began to organize cultural and educational associations and to publicize their cause.[32] As the first wave of strikes—in the cigarette factories and on the trams—hit Cairo and Alexandria at the turn of the century, a reader wrote to *Al-Muqtataf* to inquire into the likelihood of further labor activism "as in the West." "There is no possibility of an expanding scope for strikes here," the editors confidently replied.[33]

In the pages of *Al-Muqtataf* and *Al-Hilal*, where some of the earliest discussions of socialism took place, the tone was almost uniformly negative. In an 1890 discussion entitled "The Socialists and Their Corruption," for instance, the editors of *Al-Muqtataf* warned their readers that though socialism "might be thought at first to be useful, as it advocates the sharing of the money of the rich with the poor . . . this sharing is as fleeting as a summer cloud." They concluded bluntly: "socialist principles are harmful however they are presented."[34] Reading socialism against the background of the wave of anarchist terror of the early 1890s, their opposition to revolutionary violence was evident in their disapproving report on the assassination of French president Carnot by an anarchist in 1894.[35] Citing terms such as "struggle for existence" and "survival of the fittest," they attacked the very idea of equality as unnatural. It was also immoral.[36] The rich, they argued in 1889, had acquired their wealth by virtue of their exertion, economy, thrift, and self-reliance. The poor, by contrast, had only their laziness and incompetence to blame. As liberal Spencerians, they foregrounded individual self-interest and self-help over public beneficence, state welfare, or the idea of equality.[37] Their solution to the problem of poverty was thus charity, and the values associated with it. Like many Victorian liberals, *Al-Muqtataf* saw private or individual beneficence as the proper obligation the rich owed to the poor—one that would aid the development of a healthy social body. It was crucial to ensure that no one sunk below a certain minimum standard of living; otherwise, insurrection beckoned. But redistribution could not trump the rights of the individual without causing substantial moral decay and slowing the advance of progress.[38]

Al-Hilal similarly criticized socialism as a violation of the laws of nature even as it clarified the different currents on the radical Left for its readers. Contrary to popular assumptions, it argued in 1897, reformist socialism (or *al-ijtimaʿiya,* which it distinguished from *al-ishtirakiya*) did not call for the overthrow of religion, marriage, or property. Rather, it sought an end to the monopoly of wealth by the rich and the exploitation of workers by capitalists. Nevertheless, even this was both an impractical and an unnatural vision. Socialism, it argued, destroyed the sense of individual initiative and competition, which was a fixed law of nature and led to counterproductive government interference in social affairs.[39]

Through such figures as Ahmad Fathi Zaghlul and Ahmad Lutfi al-Sayyid, this championship of laissez-faire, individual initiative, and competition acquired powerful patronage. Zaghlul, the son of wealthy provincial landowners, joined with Muhammad ʿAbduh and his brother, the future nationalist hero Saʿad Zaghlul, to form the Islamic Charitable Society in 1892. Ahmad was a prolific translator of works by Demolins, Spencer, Bentham, and Le Bon and a member of the political elite: he was one of the figures chosen by the British to hear the infamous 1906 Dinshaway case.[40] Pleading opposite him in that case in defense of the peasants, the lawyer Ahmad Lutfi al-Sayyid played a key role in the founding of the constitutionalist Umma Party—one of the first political parties in Egypt's history. Both men were paternalists who built upon older Arabic ideas about the obligations of the rich toward the poor. They demanded a constitutional monarchy, guided by party political leaders and intellectuals, and remained deeply unconvinced as to the merits of democracy or notions of absolute equality among citizens.[41] Advocates of a liberal system of governance, under elite discretion and control, they were committed to a fundamentally developmentalist and evolutionary worldview. For them, socialism was the enemy of "true liberty."[42] Until his death in 1914, Zaghlul remained a "believer in progress through evolution" and was said to have "detested revolution even in thought." Ironically, a mere five years later, his brother Sa'ad, a far more prominent member of the political elite and organizer of the new nationalist Wafd Party, would make international headlines as leader of the 1919 Egyptian Revolution.[43]

Yet well before 1919, the widely perceived crisis of liberalism in Europe and the evident parliamentary success of reformist socialism there forced even journals such as *Al-Muqtataf* and *Al-Hilal* to begin to portray socialism in a more favorable light. Increasingly, Egyptian intellectuals championed versions of democratic or evolutionary socialism much like those of Antun's intellectuals in the "City of Science." In 1900 Khalil Thabit, a younger member of the *Muqtataf* group and another Syrian Protestant College alumnus, began to write on the theme of "democratic socialism" (though the article was based loosely on an 1885 article from *Contemporary Review*!).[44] Among the first of the journal's contributions on the relationship between individualism and collectivism, Thabit's article describes social evolution as founded on "the exchange of aid" (*tabadul al-musaʿida*), not competition. "Democratic Socialists," it asserted, were "bringing the world many benefits." Survival of the fittest could not justify disparities between rich and poor, concluded Thabit, since the former's wealth was based on exploitation of the laboring classes; what was needed was redistribution of wealth, nationalization of utilities, and free public education.[45]

With time, even the editors of *Al-Muqtataf* came to stress the importance of cooperation and to argue that there were clear limits to the social utility of the law of the struggle for survival. "People originally misunderstood this law," they wrote. Men's "savage morals were tamed" only through the "law of cooperation," leading them "in the way of morals and virtues." Soon, the editors were publishing a number of articles on "struggle and cooperation" and on "altruism" (a neologism in Arabic, as in English).[46]

Individualism had been under attack from a wide range of social critics in Europe, and Arab commentators were clearly aware of this.[47] They were drawing on at least three major influences—the British sociologist L. T. Hobhouse, the American social evolutionist Benjamin Kidd, and the Russian anarchist Piotr Kropotkin. Their ideas were widely discussed in the Arabic press from the early 1900s, and Kidd's major work, *Social Evolution*, was translated in 1913—a very early date when we consider that the major works of neither Darwin nor Marx had been published by then. Kidd's idea that social solidarity was essential in a world of national competition was deeply attractive to many radical nationalists, as was Hobhouse's universal humanitarianism.[48] The result was a fusion in collectivist terms of nationalist and internationalist ideas that was characteristic of much anticolonial thought. Kropotkin's declaration that "mutual aid is as much a law of animal life as mutual struggle" appealed to virtually all the major evolutionary writers in the early twentieth century, from Shibli Shumayyil and Salama Musa to Darwin's translator Isma'il Mazhar.[49] But true to their gradualist inclinations, they read Kropotkin's *Mutual Aid* as a sociological or evolutionary doctrine rather than a political one and mostly ignored his anarchist agenda. Fear of censorship, or worse, might also explain this, given that the young student tried for assassinating the Egyptian prime minister Butrus Pasha Ghali in February 1910 was accused by the police of harboring "anarchist sympathies."[50]

The First World War intensified the turn to collectivism. Many saw the "Great War"—"that most awful of human massacres among the most civilized of men"—as born of the savagery of greed and an untamed lust for might.[51] "We had expected to see civilization . . . swept up by progress, overcoming barbaric inclinations, especially in Europe."[52] Yet at the same time, the war played an important part in intensifying the attack against materialism, which was increasingly equated with vulgar self-interest and a brute ethic of competition. "What would prevent a man from killing anybody who hinders his interests," wrote Sarruf to Shumayyil in 1916, "exactly as he kills lions, wolves, and flies? Why should not a man from Paris or Berlin kill blacks who prevent him from hunting in Africa? Is not the deterrent that keeps strong people from having a free hand with weaker people a moral and not a materialist one?" For many, like Sarruf, the war was the "end product of the materialist philosophy."[53] There was widespread agreement on this point: Rashid Rida echoed it in 1917: "Those who have been infatuated with materialist views have been disproved now." So did Salama Musa in 1919: "Germany drank from materialism until it poisoned her body and then she rushed, at the height of her materialist insanity, into this ugly war."[54]

Of course, the veteran materialist Shumayyil disagreed. The problem revealed by the war was individualism, not materialism. Only the abuse of power by "corrupt individuals" had been responsible for its outbreak. Materialism proper should be thought of as natural philosophy. It taught men the value of collective, as much as individual, interests, as in the "relation of a living organism and its interrelated parts." Shumayyil concluded: "People would change their minds about materialism if they knew it was a philosophy of collective interest."[55]

Yet few people saw materialism this way. Even Musa—otherwise normally an ardent supporter of evolutionary materialism—spoke of the move "from materialism to spiritualism" as part of the "new movement in science."[56] Rida called the war the "War of Civilization" (*al-Harb al-Madaniya*), specifying that he meant an internecine war of "materialist civilization" (*al-madaniya al-maddiya*) that turned its most powerful tool—science—into a weapon of sheer brutality and an instrument to pursue material ends unhindered. In his article "The War of European Civilization Compared with the Arab Conquest of Muslim Civilization," Rida made explicit just what he thought the difference between materialist and nonmaterialist civilization amounted to: the latter, exemplified in the spiritual rule of the early Muslim *umma*, would never have led to the kind of destruction witnessed by the Belgians at the hands of the Germans.[57]

Cooperation had thus quickly become the new lesson of nature, and this shaped the terms in which Darwin's ideas were discussed. Many began to speak of "social selection"—or "group selection"—instead of "natural selection." In this way, they emphasized altruism and mutualism over competition and struggle between individuals.[58] The power of Darwin's (and Spencer's) metaphor of "the struggle for life" was on the wane. In a lecture he gave shortly before his death in 1914, Jurji Zaydan considered how societies that rely upon this precept alone suffer from "reverse selection" (*al-intikhab al-ʿaksi*)—a threat to the very "Future of Humanity." Citing Max Nordau's *Degeneration*, he listed nervous illness, alcoholism, insanity, and suicide among the social problems it created, while Henry Drummond's 1904 *Ascent of Man* gave him the opportunity to emphasize how "love, cooperation, and friendship are also laws of nature."[59]

The war had changed the morals of nature as much as its politics. Woodrow Wilson's speeches were widely cited in the Arabic press; even *Al-Manar* reprinted them in full. The vision of self-determination and international cooperation he outlined in 1918 sounded to many, particularly those outside Europe, like the beginning of a new world order.[60] Yet the Americans' indirect support for British colonialism, the agreements over Palestine, and the inefficacy of the League of Nations to act as mediator for mandated territories soon turned the enthusiasm into bitter disappointment. It seemed "liberation was not intended for the non-Europeans," wrote a disillusioned Rida after the Paris Peace Conference of 1919.[61]

Bitter disillusionment followed the dashing of "the Wilsonian moment" and the collapse of the Egyptian Revolution at the end of the war. Yet it is striking testimony to the enduring power of the evolutionary paradigm that discussions of socialism and collectivism continued to be shaped by it. Dreams of national self-determination dashed, the drastic fall in the standard of living after the war only furthered the sense of dissatisfaction with the political status quo. Strikes and other forms of radical labor activism increased among Egyptian workers during and after the war. Meanwhile, socialism continued to grab readers' attention. In 1920, against a background of nationalist unrest, economic depression, revolution in Russia, and an upsurge in labor activism at home, Niqula Haddad, Antun's onetime collaborator, published a little pamphlet entitled *Socialism*, a broadside against what he called the "Individualist System." He saw "individualism" as the polar opposite of "socialism," and creating his own dialectical naturalism, he saw "struggle" as balanced by its antithetical law "cooperation." Haddad credits Darwin with establishing the former as an irrefutable law of nature. But he added to that the law of cooperation and, citing from the Qur'an, gave this law a different kind of universal sanction. He also argued that the capital invested in large-scale transportation and other public works represented national capital (*mal al-umma*). The tram, for instance, should provide revenue not only for individual lenders. As a public work, it required state oversight and equitable distribution of profits to the populace at large. "Individualism," Haddad concluded, was simply not appropriate for modern civilizations.[62] His position was also fairly moderate; not only did violent revolution receive little coverage, but so too did the idea that socialism would in any way favor workers over capital investors or industrial owners: socialism works toward the betterment of all, he claimed.[63] A combination, therefore, of religious authority, Darwin, and Eugene Debs (the American Socialist leader whom Haddad had gotten to know during a stay in the United States), Haddad's pamphlet was characteristic of the new turn toward evolutionary collectivism.

The Social Body

Shibli Shumayyil, the elder statesman in this group and one of the best-known early Arabic authors to promote evolutionary socialism, similarly avoided revolutionary exhortation and wrote little on the political economy of capitalism itself.[64] For Shumayyil, socialism was more of an umbrella term for a social reform program than a philosophy of revolution: his 1908 proposal for a socialist party (ultimately unrealized) outlined national welfare and educational policies—including the banning of "useless books"—but was notably light on the economic considerations that younger authors such as Haddad highlighted. A 1913 essay, "True Socialism" (*al-ishtirakiya al-sahiha*), was, more than anything, an exercise in clarification if not reassurance.[65] As he had written to a challenger some years earlier, socialism was "a weight which exposed its devotee to the worst suspicions, not only in the eyes of the public but also in the eyes of the elite."[66]

Concerned to counter the charge that socialism fostered vice and laziness, Shumayyil insisted that it was widely misunderstood. His own conception, however, was decidedly idiosyncratic. Most people, he wrote, thought socialism was nothing more than an attempt to "divide money illegally" and to allow "a lazy person to share the money of an industrious one." Shumayyil responded that "true socialism" teaches respect for the proper regulation and reward of labor so that "no injustice be committed against individuals of the society, injury to whom may hurt the whole society." Asked what its aims really were, Shumayyil answered that it was that "the social system . . . be organized in such a way that all people in the society work and become useful, as well as being beneficiaries, each one according to his effort, until society becomes free from unemployment, and from those others who are deceived and who distort and corrupt."[67]

For Shumayyil, ever the evolutionist, science was the key to politics: socialism was not only ancient but more importantly "natural and based on the principles of natural science." Above all, it was the future—an inevitable stage of the progress of nature, essential to the healthy functioning of the collective social body. The relationship between socialism and natural science "is made clear by the parallelism which exists between society and living organisms. The social scientist considers society as a living organism, like other living bodies . . . they are compared and submit to the same laws. . . . Every member, even every living cell in that body, works for the security of all, and the whole works for the security of the part."[68] For the physician Shumayyil, inequity or injustice was a pathology of the social organism, and reform was the medicine.

This organicism explains his ambivalence regarding the role of the state. In a lengthy essay much indebted to Spencer, Shumayyil argued that state action could be either positive (maintaining law and order) or negative (deployed in active interventions that forced citizens to act in certain ways). The ideal state would be one where government interventions of the second kind were kept to a minimum, allowing society to function according to natural laws, which would guarantee harmony and cooperation between individuals.[69] This was a moral economy of nature firmly rooted in natural scientific principles. "Thus, the great do not overlook the rights of the small nor the small ignore those of the great lest both be equally hurt and the general condition of the society become worse. Moreover, such knowledge facilitates the performance of duties with clear and strong convictions. This is what drove me to say that socialism cannot spread in a society without the spread of the principles of natural science."[70]

Progress and Poverty

Europe's industrial and financial expansion affected land ownership in Egypt, as well as labor relations. Private landownership of large estates spread as older Ottoman systems of land tenure were abolished.[71] European oversight and colonial governance reinforced these trends. Khedive Ismaʿil had mortgaged about a tenth of Egypt's arable land as a guarantee for loans he took out in 1870 and 1877, and an international body of controllers, run by the British and French, was set up to manage the properties.[72] When mortgage loan companies were first set up in the 1880s, the majority of their loans went to purchase land rights.[73] The counterpart to this capitalization of the land was that millions of peasants were landless by the time of Egypt's fiscal crisis in the early 1880s, and peasant revolts against high taxation and threats of expropriation played a major role in the ʿUrabi uprising. A small number of proprietors owned immense estates (the minimum collateral required for a loan was usually set at around 50 *faddan*, or acres, and these loans were often used to purchase more land),

and many of the older institutions for the care and management of the poor, such as *waqf* distributions or other forms of elite patronage, were on the decline.[74]

It was against this background that rural poverty began to be discussed as a social problem, whether subject to rational state regulation or the moral conscience of the monied elite. Muhammad ʿAbduh described the condition of Egypt's poor as follows:

> At one time the people were required to pay their annual taxes in the first month of the year; at another they were forced to pay a year in advance. They had no choice. Those who failed to pay were severely beaten, sentenced to life imprisonment, or their property was confiscated. Many could find no way out except by borrowing from the banks, which were then great supporters of the oppressors. Seeing the peoples' need of them, they became greedy. . . . The banker demanded at least ten percent interest per month. . . . Those days were miserable and a calamity for both the people and the government, but brought happiness and prosperity to merchants and intruding foreign bankers, who spread among the people like wolves among sheep. Farmers and other citizens became overloaded with huge debts, which forced them to sell their property or to put it up as security or to abandon it completely. Poverty surrounded them.[75]

Questioning the role of the state, he noted that though these burdens were temporarily lifted through tax reforms and fairer interest rates, such relief did not last long: "The government, when relieved of the burdens of injustice, opened the door to poverty, through which it has entered willingly. . . . It is the door of extravagance, prodigality, exaggeration in luxury and embellishment, and of false pretenses." The greed of government officials put the very nation's wealth at risk: "I have heard that one of the dignitaries in our country has put his fertile, arable land in security for 25,000 pounds. He will pay back in fifty years the amount of 100,000 pounds or more. . . . By such doings, and by their irrational extravagance and spending, our dignitaries prove to be unworthy of wealth. . . . Rather, they like always to be poor and penniless, owning nothing, although they appear wealthy and rich."[76]

ʿAbduh's criticism was directed at the rich, warning them against incurring too much debt and promoting the value of careful savings. Yet, for him, the question of the relation between "rich and poor" was also part and parcel of any consideration of the common good (*maslaha*) as much as self-interest.[77] He thus urged the wealthy to establish charitable or benevolent societies for the poor and to learn in this way how to act for the national interest and to mobilize for the larger good: "a just and free government" would not be established until its dignitaries acted as role models for the citizenry. Muslim notables in Syria, Iraq, and India were making similar arguments.[78]

After 1907, when the fiscal crisis was most strongly felt and after which there was a sharp decline in the standard of living, particularly in rural areas, the problem of landed poverty and reform was approached from a very different angle by the younger generation.[79] Among them was one of the Arab world's first historians of socialism, Mustafa al-Mansuri, a schoolteacher who is little discussed today. In 1915 he published one of the first historical studies of socialism in Arabic (*Tarikh al-madhahab al-ishtirakiya* [History of socialisms]). That same year there was an attempt on the life of Sultan Husayn Kamal, whom the British had installed shortly after the outbreak of the First World War in place of his nephew, Khedive ʿAbbas II. Nationalist organizations had denounced Kamal as a traitor and vowed to kill him, and in the course of police investigations into the botched assassination in Alexandria, the youthful al-Mansuri, considered suspect for his socialist sympathies, was arrested as a suspect and briefly detained before being cleared and released.

Al-Mansuri later returned to publishing. In 1919 he translated Henry George's *Progress and Poverty* and dedicated it to "those who live in misery and toil in drudgery as a result of the inequality of wealth and who see that a more advanced and just system is possible."[80] Al-Mansuri classed George among the visionaries of the age and claimed his greatest achievement had been to show how the poverty of workers was not due to their nature, to Malthusian arguments about the competition for resources, or to an inherent opposition between laborers and capitalists but rather to the "monopoly of the earth"—a consequence of private ownership of the land. Land should be public property. Yet, rather than actually expropriate land from individuals, al-Mansuri followed George's suggestion that the best solution would be to establish a single land tax. He pointed to the victories by the peasantry and poor in Russia and extolled the spread of Bolshevism and socialism, ignoring the Russian revolutionaries' overall communist ideology in favor of their wartime land reform.

Al-Mansuri, as mentioned, became an authority on the history of socialism, publishing his *Tarikh al-madhahab al-ishtirakiya* two years before the Bolsheviks came to power. Influenced by Shumayyil, *Al-Muqtataf*, and *Al-Jamiʿa*, as well as by George, his own sympathies for evolutionary socialism are evident in his introduction to the work.[81] Like Shumayyil, he presents socialism as the solution to the "defects of limitless competition," modifying the harshness of the struggle for existence. Emphasizing Darwin's importance to socialist thought, al-Mansuri nevertheless followed George in arguing that Darwin's Malthusianism was not applicable in the modern world (especially given the availability of cultivable land worldwide).

Tarikh al-madhahab al-ishtirakiya basically provided a comparative survey of socialism in the major European powers. It is not clear if al-Mansuri knew English (most of his citations in the Arabic text are to French works), yet the book appears to follow Thomas Kirkup's highly popular 1892 *History of Socialism*, which ended, like al-Mansuri's history, with a chapter entitled "Socialism and the Evolution Theory." Mansuri's history had chapters on France, England, Germany, and Russia—with the latter mostly discussed in relation to the anarchist movements of the 1890s. He pointed out that different governments had adapted socialism to suit their national needs, and he ultimately drew a democratic moral: it was where democratic regimes had long been in place that one saw the greatest advance in socialist principles.[82]

Al-Mansuri's book also offered the first really systematic coverage of Marx and of the First and Second Internationals in Arabic. Yet his own recommendations for Egypt were firmly reformist, promoting land reforms and progressive taxation and including such things as the guarantee of free and fair elections, the introduction of a minimum wage, the prohibition of polygamy, and the nationalization of public services.[83] If, like Shumayyil, al-Mansuri was particularly concerned with the plight of the poor, whom he described as locked in a "cycle of misery," he was—unlike the older man—unambiguously in favor of unionization and the abolition of certain kinds of private property.[84]

Al-Mansuri was also much more concerned than Shumayyil had been to refute the charge that socialism was against religion. Emphasizing even more than his predecessor socialism's indebtedness to evolutionary thought, al-Mansuri noted that both creeds had been unfairly accused of atheism, and he invoked Wallace to show that evolution was compatible with belief in a Creator. As for socialism, he admitted that it was influenced by materialist principles but insisted that it was not against religion per se. Noting the Christian socialist tradition in Europe, al-Mansuri added that socialism and Islam were equally compatible. But he had (like Antun) little time for "the men of religion." The problem, according to him, lay in theocracy and in the fatalism and unquestioning obedience to the rich and powerful that accompanied it. Once people had believed that slavery was not against "God's justice" and were bought and sold like beasts; only recently had "this disaster" been ended, thanks to "great states like England."[85] The solution was the spread of reason through education and in particular an understanding of economics that would make people think for themselves and see that disparities in wealth were not part of God's law but rather the result of social injustice.

Al-Mansuri invoked Bacon and Darwin as the bearers of enlightenment—the latter for revealing "the natural laws that apply to all creation." He admitted that socialism had come to be identified in the public mind with unbelief (*kufr*) and polytheism. Yet this view was mistaken. He praised "religion," in fairly abstract terms, as a force against "lying, hypocrisy, and injustice," or against vice in general, and saw it as a righteous source of morality (and comfort). He then clarified where he saw the real battle as taking place. Socialists did not combat religion so much as the "men of religion," "who were a disaster and still are." Socialism thus implied putting the latter in their place so that they could not "distort the modern sciences."[86]

Not surprisingly, this viewpoint attracted critics among the Azharis, who especially disliked his support for reform at Al-Azhar.[87] His theology of nature did not leave much room for traditional theologians. And even if many *ʿulama* would also have agreed on the need for greater social reform and care and control of the poor, few of them at this time offered a countervision of religious socialism or collectivist social policy. Indeed, it would not be until the postnationalist revolutions of the 1950s that a vision of "Islamic socialism" would be systematically developed (though we see the roots of this in the 1890s). Until then, most Arab socialists looked to the kind of rational planning and political mediation that were offered by the emerging corps of technicians, experts, and policy-makers, who were hailed by al-Mansuri and Antun as the antidote to the once-powerful *ʿulama*.

An Egyptian Fabian

At the turn of the century, democratic socialism crowded out other radical ideologies of the Left (such as anarchism) in the minds of the Arab intelligentsia. It was gradualist and cooperative, egalitarian and yet governed through technocratic reforms: it allowed for nationalist party politics to support the cause of social justice and reform while still leaving room for bureaucratic and professional oversight by an elite. Some invoked Darwin, and emphasized a collectivist politics based on the law of mutualism and group selection; others looked to America and to the democratic socialism of Debs or the progressive taxation of Henry George. Yet perhaps not surprisingly given the Anglocentric orientation of much Egyptian intellectual life under the British occupation, it was the Fabians and their vision of state-led social reform democratization that dominated Egypt's arguments about socialism, especially after the First World War. A onetime member of the Fabian Society himself and the founder of the first Egyptian Socialist Party, Salama Musa provides the best example of this. Born into a prosperous Coptic family, Musa dedicated his life to writing, and in his numerous works and his brief political activism, we can trace the combination of intellectual and political forces (both domestic and international) that were to affect his generation so differently from previous ones.

Musa was a graduate of the elite British-run Madrasa al-Khidiwiya (Khedival school), which was originally founded in 1863 as a school for the wealthy, bureaucratic, and landed classes of Egypt. Under an English headmaster, its teachers provided students with a bilingual education together with the customary brutality of an English public school and helped to produce many of Egypt's nationalist elite. Among its alumni were the pioneering political leader Mustafa Kamil, Ahmad Lutfi al-Sayyid (the founder of the Umma Party), and the writer and politician Muhammad Husayn Haykal.[88] The school gave Musa a taste for English literature. But he also prided himself on his wide reading and omnivorous literary tastes. A devotee of Nietzsche and Shaw from his time with the Fabian Society in London, Musa would turn later to Marx and Gandhi. Politically, his career spanned the emergence of nationalist party politics in Egypt and the Young Turk Revolution, the collapse of the Ottoman Empire, the Bolshevik Revolution, and the struggle for Egyptian independence.

Musa finished his schooling at a time of mass political ferment. The infamous Dinshaway incident of 1906, when British officials clashed with peasants during a pigeon-shooting expedition in the Delta countryside and subsequently hanged and exiled several of them on trumped-up charges, caused enormous outrage across Egypt. It contributed to an upsurge in anticolonial feeling and also brought the "problem of the peasantry" (*mushkilat al-fallah*) to the forefront of nationalist consciousness. Meanwhile, the fiscal crisis of the following year affected the livelihood of a workforce that had expanded significantly with the influx of foreign capital over the previous decade. Lord Cromer's retirement increased the sense of an era passing. It was at this time that the first political parties were formed, calling for greater democratization, constitutional government, and British withdrawal. Although organized socialism was as yet marginal, care of the laboring poor—from urban factory workers to peasants—emerged as one of politicians' key rallying points. At the same time, students at home and abroad began to mobilize: the first strike at the Cairo Law School took place in 1906, the first large-scale demonstrations in 1908.

Musa traveled to Paris, where as well as learning French, he read the journal *L'Humanité* and became interested in socialist ideas. Then he settled in London, where his law studies took second place to moving in socialist and rationalist circles. He formed part of a small group of about sixty or seventy Egyptian students who would soon become the object of police surveillance. The British, already anxious about Indian student radicalism, were so conscious of the threat of the anticolonial mood of this new generation that the authorities eventually extended their surveillance of Egyptian students to the whole of Europe.[89]

It was during his trip to England that he became fully aware of the colonial predicament and in particular began to see Egypt's plight in relation to India and the empire more generally. Already on the voyage to London, he had been snubbed by British officials returning from India: "treated by them as an Indian, I was a slave and they the masters." It was to this experience that he traced his hatred of British racial prejudice (although this would not prevent him from adopting scientific racism later, as we will see). In 1909 he called on Keir Hardie, having been struck by his denunciation of the violence of British rule in India; Hardie told him that imperialism must be ended, and that socialism would ultimately triumph. As Musa recalled in his memoirs, Hardie advised the Egyptians to concentrate on first getting rid of the British before embarking on social reform.[90]

While in London, Musa joined the Fabian Society.[91] Echoes of Beatrice and Sidney Webb's and Ramsay MacDonald's views of evolutionary socialism, of Graham Wallas's views on property rights, and of George Bernard Shaw's views on the transition to social democracy could be heard in much of Musa's writings after that.[92] Indeed, he continued to cite Shaw for the rest of his life, and Shaw's ideas on good breeding, property, marriage, and the "political need for the Superman" were constant themes in Musa's own works.[93] At about this time Musa started writing on socialism and other topics for the Arabic press. He published several articles in *Al-Muqtataf* —the journal that had made such an overwhelming impression on him in his schooldays—although its editors tried to distance themselves from the radicalism of his views. To one 1910 letter he penned on the "education of the nation," they appended the warning that "if all our men who study in Europe" returned to us with such opinions, "the result would be unrest, revolution, and struggle whose end we do not know."[94] In 1911, back in Cairo, his attempt to start a newspaper with the aim of "spreading Fabian socialism and rationalism" was rebuffed by the British. Musa complained that their action was "an obstacle against our progress," but the authorities held their ground. They had followed his career closely and described the articles he had published in *Al-Muqtataf, Al-Jarida* (an organ of the Umma Party), and *Al-Liwa* (the paper of the Nationalist Party) as "not only violently socialistic but . . . also directed against the government and the occupation." Some of his more inflammatory phrases were quoted, in which he had urged worker solidarity, unionization, and strikes against foreign capitalists in order to oppose the occupation.[95] In the 1910 letter that *Al-Muqtataf* criticized for its radicalism, Musa had attacked the British for spending the bare minimum on schooling for Egyptians. Indeed, Musa accused them of intending them to be nothing but cotton pickers. *Al-Muqtataf*'s reply (not unlike ʿAbduh's!) was that there should be more Islamic benevolent societies like those of the American missionaries and therefore placed the responsibility for education outside the state.[96]

The authorities had for some time been trying to tighten their control of the press; fear of pan-Islamism and intercommunal tensions as well as of attacks on the British and other foreigners in the country had led Eldon Gorst immediately on taking office as consul general in 1907 to recommend increased use of the penal code to "protect the Egyptian Government from seditious criticism." Student radicalism was one of their prime concerns.[97] As a result, it was not until 1914 that Musa was given a license to publish a weekly, *Al-Mustaqbal* (The future). Modeled on the *Fabian Essays*, this venture involved Musa's collaboration with the older Shibli Shumayyil, with whom he had become close.

This vehicle of "extreme modernism, and even futurism," urged Arabs to relinquish "slavish imitation" (*taqlid*) and to look to the future "as though it were a blank page." It saw Europe as a model for thought and expression and the embrace of European ideas as the only means of overcoming imperialism—an ambiguous stance characteristic of that generation of colonial intellectuals. In this journal, Musa clarified his own understanding of socialism, criticizing the "great prophet of socialism" Marx for assuming that small proprietors would inevitably be overwhelmed by the process of industrial concentration. For Musa, this class was essential for the development of local democratic communities, which he saw as a part of socialism itself. In fact, he downplayed Marx's role in the emergence of scientific socialism.[98]

The journal, however, was short-lived, and after only sixteen issues the British censors stopped it. In his final interview with the Directorate of Press Affairs, Musa defiantly proclaimed his socialist credentials, no doubt confirming the authorities' suspicions. Some of Musa's British contacts, notably the veteran anti-imperialist Wilfred Blunt, met earlier that year in London with the Nationalist and Subject Races Committee, after which Blunt and others (including Hobhouse) had written to the khedive deploring the increased use of press laws since 1910 and requesting their relaxation. (It is surely no coincidence that Musa had written to him a few years earlier, from his lodgings in West Hampstead, about translating Blunt's *Secret History*, warning him that publishing this would not be possible under the prevailing press laws.) The effect of the outbreak of the First World War, however, was that the censorship only became tighter.[99]

Musa's commitment to what were in the Egyptian context rather radical socialist views was in any case by this time a matter of public record. After his numerous articles on the subject, he published his first full treatment of it in a brief pamphlet in 1913. In the pamphlet he was anxious to defend socialism against what he felt were common charges of the time: that it was against religion and Darwinian natural selection and that it was revolutionary. But Musa's vision of socialism was staunchly nationalist; he had already parted company with his idol Shaw on these grounds. This form of socialism, in the Egyptian context, downplayed religious difference in favor of national solidarity and offered a vision of a new political community that rejected charges of atheism while nevertheless relegating religion to the margins of politics. Intercommunal relations in Egypt at this time had become increasingly politicized. The first wave of the nationalist movement had targeted the role of foreigners in the country and identified Arab Christians as their agents; the assassination of the Coptic prime minister in 1910 had deepened the alarm. Musa was no doubt sensitive to these trends; and as a member of the new generation of anticolonial intellectuals, he was naturally drawn to a nationalistic version of socialism. In this he differed from socialists of the earlier generation like Shumayyil, who had promoted reform for the Arabs within a decentralized Ottoman framework rather than an exclusively national one. With the rise of the Young Turks and the emergence of nationalism throughout the empire, such a position became increasingly difficult to sustain.

A classic example of the evolutionist view, Musa's pamphlet depicted a rather moderate view of socialism, with room for private property and concerned as much about ethics as about the redistribution of wealth. After a brief historical sketch, it deplored the destructive impact of capitalism on morality, eroding the honor of individuals and their finer feelings. The profit motive, Musa went on with typical imaginative vigor, led men to cheat and forced women into loveless marriages. Following the classic arguments of the British Left, he denounced capitalism's fostering of war and imperialism (referring primarily to the plight of India). Musa also asked how socialism could come to a country such as Egypt and outlined a practical program of reform. Arguing for the need to educate the public and for representative local democratic rule, he avoided any mention of what should happen to the British. What was important for national revival was the dissemination of socialist principles, and their gradual introduction into the government.[100]

Socialism, Musa argued, emphasized gradual change: "The socialist demands that our government make progress . . . by way of deliberate gradation—not by a rapid leap." Much like the Fabians themselves, he largely bypassed Marx and covered him only very briefly in the 1913 essay.[101] Within a few years this would change: in 1921, at the height of Bolshevism's popularity in Egypt, Marx's star shone much more brightly. Much later—after the 1952 revolution and the consequent rise of Arab socialism—Musa would claim that Marx had been one of the key influences on his intellectual trajectory all along.[102] In fact, he had been a confirmed socialist long before he turned to Marxism and always remained true to his Fabian origins.

Fabian influences were evident in another respect as well. Picking up a theme that others had dealt with before him, Musa drew out the political implications of natural selection in a new way. Objecting to arguments that socialism hindered natural selection, he claimed that it simply created equality of opportunity and was not intended "to equalize compensation."[103] Unlike the earlier collectivist evolutionists influenced by Kropotkin, with their emphasis on altruism and anti-individualism, Musa highlighted the virtues of struggle and competition: certain kinds of inequality were natural and even desirable. Thus framed, Musa's socialism led him to advocate a radical program of state-controlled biological and social reform dear to many Fabian hearts: *ʿilm al-ujaniya*, or eugenics.

Eugenics and the State

Although Francis Galton, Darwin's cousin, had coined the term in 1886, it was not until the early 1900s that eugenics really gained ground. Galton himself first advocated these ideas in 1901, the same year that *Al-Muqtataf* began to track the British debate closely. It was then that the editors translated Galton's speech "The Possible Improvement of the Human Breed under the Existing Conditions of Law and Sentiment" to the Royal Anthropological Institute. In 1909, marking the centennial of Darwin's birth, they published a short biographical piece highlighting Galton's scientific standing, presenting him as the preeminent heir of Darwin, and prominently placing his picture on the cover. This was followed up with an admiring obituary upon Galton's death in 1911.[104]

Musa was also following the debate and sending back regular reports to *Al-Muqtataf* while in London. Around the time he was there, eugenics was taking on some of its very first institutional forms: between 1907 and 1911, for instance, the Eugenics Laboratory and the Eugenics Education Society were founded. Ideologically, the movement could look both right and left, though its early incarnations were generally associated with progressive middle-class professional sentiments. This was the version that Shaw and other Fabians had helped to promote—one that was hostile to laissez-faire and stressed the need for a kind of scientific social reform that required direct state action under the tutelage of the new professional or technocratic classes.[105]

Musa took a similar line and between 1909 and 1913 published a number of articles that covered eugenic, among other, themes.[106] He turned to figures like Schopenhauer and Nietzsche to identify the ills of the modern world and to herald the means to overcome them, and to visionaries like Wells to capture a new vision of the future of humanity. The result was an eclectic mix of tempered philosophical pessimism and anticapitalism and eugenicist and progressivist rhetoric. Like the Fabians, he saw the capitalist state as essentially based on dysgenic, inherited wealth and the landed classes as "parasitic." He also advocated greater state control over the population, subject to elite discretion and supervision and following scientific principles. That same year he published "The Coming of the Superman"—initially carried in *Al-Muqtataf* and later reissued as a pamphlet. In this sprawling essay, which owed much to Shaw's reading of Nietzsche, the twenty-five-year-old Musa covered such themes as the relation between man and monkey, the need for sexual equality and women's liberation, and the dangers of marrying off Egyptian daughters to black Africans.[107]

His promotion of a highly racialized version of eugenics was something *Al-Muqtataf*'s editors disagreed with. Although their journal had long dealt with the relationship between race and civilization, they clearly felt that Musa's views on race were much too extreme. In a debate between them on the supposed inferiority of blacks, *Al-Muqtataf* rejected the view and Musa categorically upheld it, claiming that science had proved that blacks' brains were closer to those of apes. The implications of this kind of discussion for Egyptian national identity were far-reaching. Responding to the newly globalized discourse on race, Musa insisted on learning from across the Atlantic: Americans, unlike the Egyptians, demonstrated their love of progress by their attention to racial purity. "The American does not consent to the marriage of his daughter to a Negro because he believes that the Negroes are inferior to Whites, but the Egyptian does that, thus increasing the contamination of Egyptian blood." Musa was thus concerned—from the start—to distinguish Egyptians from Africans. In this lay not only the making of a potentially racialized social policy but the origins of the interwar movement now sometimes dubbed Pharaonism, of which he was to become the key proponent.[108]

To be sure, Musa agreed that there was an important role for social policy more broadly and projected a much more state-centered message than Shumayyil did. "The environment, meaning the surroundings in which an [organism] lives . . . is the creator of organisms," he argued, going on to indicate that in the case of man's environment, this would encompass "government, education, religion, marriage laws, and all that influences man's social life." Nevertheless, sexual selection was key. He stressed that "improving the environment is beneficial to the generation that lives in it, but . . . is not beneficial to the species." What mattered was control of the reproductive process. In this respect, he was highly influenced by the works of Weismann, De Vries, and others who had overturned neo-Lamarckian ideas of inheritance on the basis of Mendelian genetics.[109]

It was for these reasons that Musa outlined a program for the state regulation of marriage and recommended castration for the unfit, as he would do to the very end of his life. Sexual equality—and an economic equality that would ensure a better system of marriages and more effective sexual selection—were centrally important for Musa. He was a fierce critic of the situation of Egyptian women, whom he saw as the victims of social and political oppression. Their long-term degeneration was endangering the nation's future. He deplored the lack of men's attention to the physical and aesthetic condition of their prospective mates and their indifference to issues of racial purity. All this betrayed a deep, underlying anxiety: what he called "the myth of progress" which had led in reality to subservience, superstition, and national retardation. The contrast with the civilizational optimism of *Al-Muqtataf*'s founders could not have been more evident. It was Musa's uneasy sense that society unaided would regress toward some savage state that underpinned his emphasis on strong state intervention.[110]

Bolshevism and the British

As everywhere else, discussions of socialism in Egypt were transformed by the success of the Bolshevik Revolution.[111] The first small cells of pro-

Bolshevik sympathizers were reportedly established almost immediately but do not appear to have fared very well. The British police and intelligence services had stepped up their monitoring of all forms of radicalism and were especially anxious about the infiltration of foreign agents—men like the Italian Pizzuto, who told them when arrested that he had become sympathetic to the Bolsheviks when a prisoner of war in Hungary. (He described Bela Kun as "a great friend of his" and "the most intelligent man he had ever met in his life.") Before he was deported, the otherwise unknown Pizzuto told his interrogators that "unfortunately, the people of Egypt are not yet quite ripe for [Bolshevism]."[112]

But the British were taking no chances. On 2 July 1919 Muhammad Bakhit, the grand mufti, issued a fatwa clumsily denouncing Bolshevism as a version of Zoroastrianism, a "threat to the whole world" and contrary to Islam. His rhetorical assault—alleging everything from adultery and prostitution to murder and theft—did not prove very effective. The British were widely suspected of being the instigators of the fatwa, and indeed, they adopted the fatwa for official use elsewhere in the empire; greeted with enthusiasm by the viceroy in India, it was roundly rejected by the authorities in Iran and Iraq, who pointed out that its attacks on Persians would not go down well among Shiites, not to mention the Parsees. As for Egypt itself, the reaction was uniformly negative. When a British informant asked ʿAbduh's former amanuensis Rashid Rida (whom the informant described as "the only Oriental that can be compared to Machiavel") what he thought, Rida replied that it was a propaganda blunder: he claimed that no one in Egypt had known so much about Bolshevism—"and the newspapers never wrote so much about [it]"—before the publication of the fatwa. This was confirmed by another British intelligence source, who wrote in September 1919 that Bolshevism was "the general topic now in Egyptian circles. . . . News of success or victory by the Bolsheviks seems to produce a pang of joy and content among all classes of Egyptians."[113]

The public may have been curious but it was uncertain what Bolshevism really was. The intelligence report cited above described it as a German movement, guided by the kaiser in exile, while Allied unity fragmented and Wilson returned home to defeat, playing out "the last scene of the American comedy on the stage of the Great War." Pro-Bolshevik pamphlets presented it chiefly as a movement of anticolonial liberation and the wave of the future. One pamphlet denounced the mufti and argued that Bolshevism was not anarchic, irreligious, or immoral, crying, "O Egyptians! Embrace Bolshevism!" On the contrary, God had sent Bolshevism to "purify the world from the pollution of oppression and slavery." The author went on to predict that the Russians, "worshippers of oppression in the past," would in this way turn toward Islam "as if God hath willed that all the people of the earth should become Moslems"; somewhat contradictorily, it also claimed that Bolshevism had no bearing upon religion. After describing how the Germans had allowed Lenin to return to Russia and the subsequent course of the revolution, it went on to call Bolshevism "moderate socialism"—a doctrine according to which "everything will be the property of the nation and . . . no individual would be left to starve." Predicting that the new creed was destined to "overthrow the throne of George and bury underneath it the policy of imperialism, oppression and slavery," it presented it as Egypt's salvation. Another critic greeted Bakhit's fatwa with equal scorn, asking whether the mufti would publish another ruling for the "case of the Great Powers, who destroy the liberties of small nations and take their country, their wealth, their sons and murder them." Musa himself dismissed the fatwa in an article in *Al-Ahali* in August 1919 designed to dispel some of the misconceptions surrounding Bolshevism—in particular, its supposed assault on the institution of marriage and on national sovereignty.[114]

All this was taking place at a time of acute political crisis in Egypt. Hopes that President Wilson would support the cause of national independence at Versailles had just been dashed, and the British repression of Saʿad Zaghlul's Wafd movement—the best-organized nationalist party in the country's history—had increased public anger. There was a massive rise in strikes and labor activism, as well as an upsurge of assassination attempts on British officials. Between 1919 and 1922, Whitehall inched painfully toward the recognition that the protectorate (inaugurated only in 1914) would have to be wound up, and it was in this period of great uncertainty and excitement that Musa became a political activist.

In the summer of 1921 he became involved in discussions to found an Egyptian Socialist Party.[115] By this time the revolutionary wave in Europe had broken: pro-Bolshevik movements had been crushed in Italy, Germany, and Hungary, the Polish-Soviet War had ended in stalemate, and the famine in Russia itself suggested a grim outcome for Bolshevism. In these circumstances, the question of whether Bolshevism would help or threaten the cause of Egyptian independence was inescapable. Niqula Haddad, Antun's collaborator, had published a work the previous year warning about the growing divide between evolutionary and revolutionary strategies. But because Bolshevism was still widely regarded as a pro-German movement, both the Wafd movement and socialist intellectuals like Musa (already of course anti-Marxist) were keen to avoid being tainted by association. Indeed, Musa warned against "that which could be taken as pretexts by those who oppose independence in England: that Egypt has communists and Bolshevists."[116]

Musa had originally formed a study group on socialism, modeled on the Fabian Society, with several other intellectuals—Husni al-ʿUrabi, Antun Marun, Muhammad ʿAbd Allah ʿInan, and Shaykh Safwan Abu Fatah. Although the predominant tendency among them was moderate, they had diverse ideological backgrounds. Musa was a Fabian and al-ʿUrabi was translating Ramsay MacDonald, while ʿInan followed the Second International; Fatah and Marun were further to the Left. In 1921 they met with the veteran activist Joseph Rosenthal, who had been involved for years in trade union and workers' movements. Rosenthal was keen to amalgamate the diverse socialist groups in Cairo and Alexandria, whereas they wanted to learn about the foreign workers' socialist organizations. What concerned the young Egyptians above all was whether joining forces with Rosenthal would serve Egypt's national interests. They found it hard to agree as to whether their organization should turn itself into a party or remain a "scientific society" intended to "transplant and apply the creed [of socialism] to Egyptian conditions." Musa himself worried that "the present time may be the worst for forming this society, for we are in a critical political situation."[117]

The new socialist party was formed that August. Musa had hoped to become its secretary-general and had indeed even signed its "Call for the Egyptian Nation" (which appeared in *Al-Ahram*) as "Secretary-General of the Egyptian Socialist Party." But in fact, many felt that his reputation was too controversial, and the leadership was given instead to Muhammad ʿAbd Allah ʿInan. Yet despite supporting appeals for aid to the famine-stricken Russians, Musa remained true to his reformist and anti-Marxist principles. In an article the same month, he insisted that the new party was moderate in nature and repeated his belief in the essentially evolutionary character of socialism. He was signatory of the new party program: it highlighted liberation from imperialism and other Bolshevik foreign policy themes, alongside a set of much vaguer economic demands and a vision of social reform that was remarkably similar to what he had been calling for for years. Challenged by critics on the right and left to define what made this program socialist, Musa retorted that socialism depended upon industrialization. To avoid the fate of the Russian Revolution, it was necessary to prepare the ground through mobilization and education. As it was, he insisted, it was still too early to predict precisely what form socialism would take in Egypt.[118]

Disputes between the new party's reformist and revolutionary wings soon led to a split, triggered by the debate about whether to join the Third International. Musa had hoped that the party would join the International Working Union of Socialist Parties instead. He had envisaged it as akin to the Fabians, taking an active advisory role in government and supporting a gradual, parliamentary, and moderate form of socialism.[119] Nevertheless, in July 1922, party members voted to join the Third International, and Musa was eliminated from the central committee. He was deeply dismayed, viewing this decision as a violation of their original agreement. He had always believed that socialism could not come to Egypt without the support of the middle classes, and this was one of the fundamental strategic differences between Musa and the more radical Alexandria branch membership. Following his public criticisms of the party's new line, Musa was expelled. He warned that the specter of class warfare hung over Egypt and could only set the country back: better follow the lead of European and American socialist movements than the Russian revolutionaries.[120]

Thanks to British intelligence in Alexandria, we have a detailed description of the way in which the reoriented Socialist Party celebrated the anniversary of the Russian Revolution that November. According to the report, the meeting began with "a choir of ten Jews chanting the International" to great applause, and this was followed by a lecture on the history of the Bolshevik Revolution. The first half of the evening was brought to a close with a tableau vivant in which a veiled young woman, accompanied by two workers, held up an enormous red flag, representing liberation through the revolution. Afterward, Rosenthal talked about the global importance of 1917. He went on to describe Kemalism as "a great step" toward "a free and communist Turkey and Orient." As for Egypt, he said it was the duty of all Egyptians to support the Russian policy toward the country: abolition of the foreign debt, internationalization of the canal, and democracy.[121]

The new Communist Party, however, faced intense pressure from the authorities, for the suppression of communism was one of the few issues upon which the Wafd and the British could agree. The former inherited from the latter the repressive 1923 Law of Association, which together with the amended penal code allowed for the prosecution of those spreading "subversive, anarchist, communist and anti-constitutional ideas." Once in government, Wafd politicians made numerous arrests, deported suspected activists, and banned the Communist Party. The Ministry of the Interior continued to watch over communist activity after the movement was forced to go underground. One government agent warned in 1925, for instance, that the "communist party is reassembling a second time . . . and are now promoting revolutionary ideas against the principles of the national constitution of the Egyptian government . . . and are aiming to change . . . the social structure of Egypt . . . through revolution, force and terrorism." But in fact such fears were exaggerated and the revolutionary Left was marginalized as a political force for the rest of the interwar period.[122]

Salama Musa himself had feared arrest in 1924, and in an article he wrote on the "Socialists and Communists and their history in Egypt," he denounced the party and distanced himself from it, reiterating his close allegiance with Fabian socialism. The parliamentary elections of 1924, which handed an enormous majority to the Wafd, had proved a bitter disappointment. The subsequent crackdown on political dissent and the failure of the party itself to pay any attention to the country's pressing social problems led Musa to abandon organized party activity for the rest of his career and to return, once again, to journalism.

Social Evolution

In Egypt as elsewhere, the interwar era was the heyday of science popularization, and the prolific Musa quickly became the most important figure in this. This role began in 1923 when he was appointed editor of *Al-Hilal*, helping turn it into one of the most widely read journals in the country. A condition of his appointment was that he publish a book each year and the results were quickly evident. Whereas he had published mostly pamphlets and articles before then, he published no fewer than nine books before leaving *Al-Hilal* (in 1929) and at least twenty-five more before his death in 1958. At the same time, he continued to write regularly for the daily and weekly press. In 1929 he left *Al-Hilal* to found his own journal, *Al-Majalla al-Jadida* (The new magazine), which covered everything from politics to popular science, and as a vehicle for literary modernism, carried the first published contribution by the young Naguib Mahfouz. The first issue had articles on such diverse topics as Zionism in Palestine, the Perfect Body, and Ramsay MacDonald. Special subscription deals offered its readers books annually.

One might think of Musa in this period as an Egyptian H. G. Wells, and indeed Wells was an important influence upon him. Like Wells, he held on to his faith in Fabian technocracy, believed in the education of the common man—giving regular weekly seminars at the Cairo YMCA, which sparked much discussion—and emphasized the importance of clear and accessible prose as a vehicle for this. Like Wells, he was a utopian and his futurism affected his attitude to history and to Egyptian nationalism. As literacy spread among the new urban middle classes, the national readership quickly expanded: the total circulation of Arab-language newspapers in Egypt is estimated to have increased from 180,000 in 1928–29 to over half a million two decades later. It was upon this audience that Musa left his imprint.[123]

As with Wells, the popularization of science was at the heart of Musa's mission: as he put it later, spreading "scientific thinking" was important because "the industrial civilization which is now dominant is in fact the civilization of science." The theory of evolution for him, was "the greatest theory to reign over European culture, and to shape the minds of thinkers all over the world today." This was how he introduced his 1925 *Nazariyat al-tatawwur wa-asl al-insan* (The theory of evolution and the origin of man), the first popular book-length treatment of the subject in Arabic. At the start, he paid homage to his predecessors—*Al-Muqtataf*, *Al-Hilal*, and Shibli Shumayyil—but stressed the need for an accessible and simple account of evolutionary thought and its implications.[124] Strikingly, there is no mention of Isma'il Mazhar, the first systematic translator of *Origin of Species*, even though Mazhar's translation of the first five chapters had appeared in 1918, and in 1923 he had published a detailed discussion of evolutionary theory and the controversies surrounding it. But Mazhar's work was written in a more complex style and demanded much more of the reader.

Musa's priority, evident here, was popularization. In his introduction, he reassures the reader that he has avoided complex terminology (including the names of animals unknown in Egypt) and even gives instructions on how best to read the book. He describes evolution as a kind of "scientific Sufism" that connected man to all beings, and he even stressed the word's positive mental impact, contrasting it with the "bad feeling" produced by words like "coup" and "revolution."[125]

The book itself provides a history of evolution that is drawn largely from popularizing British accounts of the time. Although he traced the origins of the theory back to the ancient Greeks and medieval Arabic thought, Musa made it clear that the modern European contribution had been decisive, and he presented the theory as the work of a trio of thinkers: Lamarck, Darwin, and Spencer.

The theory of evolution had revolutionized man's sense of both his past and his future, but it was the latter that Musa emphasized: evolution "is the theory of hope and advancement and the key that will open the previously closed doors of the Unknown."[126] Although he covered the creation of the cosmos and the development of species, his focus was upon the theory's social and cultural implications. Understanding it had allowed men to think for themselves and to question the political order, the nature of human psychology (here he referred to Freud), and gender relations.

He also preserved his old emphasis on eugenic reform and racial science, yet strikingly, there is only a passing reference to socialism. This he sees as Europe's future: the more Europe becomes mechanized, the more it must limit private ownership of property. Describing Egypt as the "origin of all civilization," he stressed that in the present as much as in the past, its agricultural economy constituted an obstacle to democracy.[127]

To what extent, then, did Musa remain an evolutionary socialist? Burned by his flirtation with party activism in the early 1920s and wary of the increasingly repressive authorities, he downplayed his interest in the politics of the Left for several years. He could, of course, afford to take the long view, as he remained convinced that Egypt was not ready for socialism.

Very long, in fact: the book that followed his work on evolution included his own version of a Wellsian utopia. In his *Ahlam al-falasifa* (Dreams of the philosophers), Musa drew a picture of Egypt in the year 3105. Socialism had brought prosperity, and large, efficiently organized collective farms required only one hour's work a day from their members, who could pursue intellectual advancement in their ample leisure time. Additional employment was provided by a factory in each village, but Musa was uncharacteristically vague about the details and said virtually nothing about urban industry. A syndicalist government was in the hands of the professional classes. Religion had been completely reformed: philosophers now led a "religion of humanity," and their temples were decorated with murals depicting the evolution of mankind. The key to this transformation was the long-term implementation of the eugenics program. Politics was no longer the way forward; rather, the foresight of a scientific elite breeding a population of large-headed men guaranteed the nation's future.[128]

In other respects, however, Musa betrayed a new interest in the possibilities of revolutionary change. It is about this time that he appears to have started taking Marx more seriously as a guide to the analysis of history. However, much more important was the impact of current events. He was—like many contemporaries in the late 1920s—impressed by the achievements of the Soviet regime. He was even more enthusiastic about Kemalism in Turkey. For Musa, Mustafa Kemal's creation of the Turkish Republic and his drive to modernize post-Ottoman society provided a model for how the rapid adoption of European culture could facilitate national independence and overcome European imperialism. Kemal's achievements reinforced Musa's Fabian faith in the way decisive leadership could drive the masses forward. "The experience of nations tells us that sometimes the masses must be reformed by force. . . . The Turks realized this, acted accordingly, and have enjoyed great success."[129]

This approach put Musa at odds with those Egyptian intellectuals whose anticolonialism was turning them against what they described as "the West." An Eastern League had been formed in the early 1920s, but Musa was disapproving. He felt it would be a regressive force for the country, and he called instead for radical Westernization and made the argument that the country formed part of the West. There were, he felt, several problems with the Easterners' analysis: Egypt was racially distinct from other peoples in Asia and Africa, and it was a mistake to believe either that the East could be a source of superior spiritual inspiration or that it offered any kind of model for Egypt's future. Attacking the Easterners' interest in Arab and Islamic history, he wrote this off as a fixation with "forefathers." In fact, according to Musa, the Arab-Islamic political inheritance had involved little but despotism and stagnation all the way up to the rule of the khedives and Sultan Abdülhamid. Above all, he asserted that neither Arabism nor Islamicism could provide the Egyptians with a true sense of nationalism.[130]

Pharaonism provided the way forward. Drawing on the theories of the British anthropologist Elliot Smith, Pharaonists like Musa asserted the biological racial continuity linking the modern Egyptians to their ancient forebears. Musa, whose interests in Egyptology dated back to his first visit to France, highlighted ancient Egypt as the originator of civilization and in this way provided a historical rationalization of the country's connection with Western civilization on the path of universal progress. This nationalist reading of Egyptian history—which regarded the Arab conquests and spread of Islam as an unfortunate parenthesis—was combined in Musa's case in the 1920s with an emphatically internationalist orientation that included support for the League of Nations as the precursor of a new world order organized along scientific lines.[131]

Under the impact of the world economic depression, however, Musa moved sharply in the direction of economic nationalism. Inspired by Gandhi's Swadeshi movement in India, Musa went so far as to found a boycott movement—the Society of the Egyptians for the Egyptians—in the summer of 1930. It was enormously successful at first (similar movements, equally inspired by the Indian example, were established elsewhere in the Middle East), attracting ten thousand members in only a few months. Musa had always highlighted the dangers of foreign capital's grip over the country but he had never gone to such extremes before. To this, he added a highly personal critique of the dominance of Syrians in the Egyptian press, having resigned from *Al-Hilal* to found his own paper the previous year. However, his involvement with the boycott was short-lived; his former employers at *Al-Hilal* hit back, not only accusing Musa of heresy and immorality but, more damagingly, publishing private correspondence that cast doubt on his nationalist credentials.[132]

For a time, his economic nationalism also led him to a relatively favorable appraisal of the new National Socialist regime in Germany. He admired how the Nazis led their country out of an economic slump, their success in reducing unemployment, and their eugenics program. But this did not last long: by the end of the 1930s he was disillusioned with their totalitarian tendencies and realized after their invasion of Czechoslovakia that they intended a war of aggression.

Evolution in the Age of Revolution

Musa's interest in social engineering had always formed the background to his interests: in fact, it was this faith in the rule of experts and technocratic professionalism in matters of the state and society that marked the very character of the Nahda in Egypt during most of Musa's lifetime. Like many *effendiya* of his generation, he held on to an ideal of intellectual or cultural reform and renaissance as the path to social advancement and progress. The very idea of a *nahda*, or renaissance, was one that Musa often returned to. In 1935 he wrote a book entitled *What Is the Nahda?* In it, Musa listed the fifteenth-century Renaissance, the seventeenth-century Scientific Revolution, the eighteenth-century French Enlightenment, and the "Darwinian Revolution" of the late nineteenth century (though Musa also emphasized that this last revolution was still ongoing) as marking modern Europe's rise, reformation, and final enlightenment.[133] For Musa, the West was still a model, a conceptual intellectual, social, and political ideal, for the East to emulate. With the increasingly negative connotations attached to the "West," the popularity of this formula would fade. So too would the Egyptian Nahdawi intellectuals. Evolution had been major ideological tenet for them, but the decline of Darwin also helped to seal their fate. Other political developments also moved this along.

Musa's own career demonstrates the point well. He was one of the main popularizers of Darwin in the first half of the twentieth century. He was also part of a new generation of *effendiya* who looked for alternatives to classical liberalism. Yet many associated liberalism with the name of Darwin and the ethos of competition and the individual struggle for life. Socialism, defined in gradualist, reformist, collectivist, and evolutionary terms provided the resolution. Like a number of other early Arab socialists, Musa started out by looking to Fabianism as the model for technocratic social reform. He would be influenced by Fabianism for some time: many of his later writings echoed ideas of "anti-Marxists" such as Sidney Webb and Ramsay MacDonald. Yet he also creatively adapted the ideas of a host of other European thinkers to bolster his political prescriptions: H. G. Wells, George Bernard Shaw, and historians of civilization like Thomas Buckle proved the most influential. He also attached to Darwin's ideas those of Galton and later Marx and Freud. Like so many evolutionary socialists of that generation, he turned to eugenics as a key part of his political and social program, advocating for strong state interventions into the public and private lives of citizens.

Yet, for much of his life, Musa's literary prescriptions proved more popular than his political activities. He was one of the founding members of the short-lived Egyptian Socialist Party in the early 1920s, but that was the final extent of his experiments with party politics. Disappointed by left-wing party mobilization and by a growing split between reformists and revolutionaries, Musa eventually retreated to his more purely intellectual and journalistic endeavors. Yet he continued to promote the idea that national revival was dependent upon a strong and interventionist welfare state. For obvious reasons, his blend of nationalist politics and advocacy for technocratic expertise appealed to a wide range of emerging professional classes in Egypt for much of the interwar years. Even the main party, Saʿad Zaghlul's Wafd, which dominated the early Egyptian political scene, co-opted elements of this social reformist agenda.

While this version of socialism might have been reformist in terms of its social program, it still had some radical political implications. In particular, it was closely identified with the anticolonial nationalism that emerged powerfully in Egypt after 1906 and exploded with the First World War and its bitterly frustrating aftermath. Although it bore the imprint of Fabianism and the British labor movement, it was unmistakably anti-British in its focus and attuned to anti-British currents in India and other imperial possessions. Like many anticolonial socialist movements between the wars, it was both internationalist and nationalist at the same time. By the 1930s a new radical Left in Egypt would emerge, carrying the banner of anticolonial nationalism alongside a new interest in revolutionary and even potentially violent change. This, coupled with the decline of the power of Darwin's ideas, meant that this interwar generation's view of evolutionary socialism would no longer represent the hopes of Egyptian nationalists and notables.

The evolutionary politics of notables had been on the wane since at least the 1930s. The rise of populist and extraparliamentary movements and the general decline in liberalism and its political solutions worked against it.

Yet despite Musa's literary fame, and his own refashioning as a precursor to Arab socialism after Nasser, only a handful of his ideas survived intact after the war. He published an autobiography in the 1950s that presented him as committed to anticolonial nationalism and revolutionary Marxism from the start; the reality was rather different. He was never a supporter of revolutions, and in his autobiography he simply enfolded Egypt's latest revolutionary fervor into a longer history of gradual political change, even as he conceded it might be punctuated by periodic, radical ferment. He turned to Marx fairly late in life, but by then the rise of a new and much more radical Left had made his opinions on the subject moot. Musa's view of socialism was not so much realized by the 1952 coup as repudiated by it.

Over the course of Musa's lifetime, new reading publics—and counterpublics—began to express a growing hostility to ideas of evolution and development. It was Darwin, and not Marx, who had provided Musa with the greatest ammunition throughout much of his career. This meant reading Darwin through the lens of collectivist politics, of course, and wrestling him from previous associations with individualism and liberal laissez-faire policies. Yet this rehabilitation was only partially successful. The final fate of Musa's ideas also mirrored the waning appeal of Darwin after the Second World War. The rise of a revolutionary Left and the shifting geopolitical landscape in the postnational era critically reshaped Darwin's image yet again, and in ways that would prove less salvageable for former enthusiasts.

The period just after the war saw Musa's literary reputation and popular appeal decline rapidly, and he emerged as something of a political outcast. During the war, working as editor of a journal in the Ministry of Social Affairs, he turned to the Left again, reestablishing links with Marxian intellectuals and publishing houses. But his own socialist agenda remained fairly gradualist in nature, which did not fit in well with the rise of the new Marxist and revolutionary Left in Egypt after the war.

Nevertheless, Musa was still subject to some governmental harassment: his journal was shut down, and most of his books were banned by the political police. An article he wrote on the plight of the poor in Egypt—and their subsistence on meager salaries—led local authorities to regard Musa with suspicion, and he was arrested and suspected of antinationalist sentiments. He was also later charged with attempting to overthrow the monarchy and was arrested again in 1947. When the 1952 revolution broke out, he was initially uncertain whether to support it. What seems to have changed his mind was the Suez Canal. This destroyed his residual faith in the British and made him turn to the USSR instead as Egypt's natural tutor along the path to industrialization and social progress.

It was not long before Musa's own writings would portray the Egyptian military coup, and the revolutionary moment more generally, in very different terms. In 1954 Musa published *Kitab al-thawrat* (The book of revolutions). In it, Musa devoted chapters to each of the major revolutions of human history, including the Scientific Revolution, the French, American, and Russian Revolutions, and four chapters on each of the successive Egyptian revolutions. For Musa, so long the gradualist in matters of social and political reform, revolution suddenly took on new meaning.[134]

A few years later, he rejoiced in the nationalization of the Suez Canal and hailed Nasser as Egypt's first genuinely national leader in nearly eight hundred years. He ended his life in a mood of optimism. The revolution now seemed to him a vindication of the ideas he had held all along: technocratic leadership and scientific socialism leading to national revival. In 1952 a technocratic Permanent Council for the Development of National Production was established, and the same year it introduced an agrarian reform law that broke the power of the old landed elite and vastly expanded the cooperative system in the countryside. The 1956 constitution talked about committing the state to a program of wide-ranging social welfare.[135]

All this could look oddly like a realization of Musa's Fabian ideas, and that was how he presented it. However, such an interpretation involved overlooking a number of inconvenient facts. The Suez Canal had never been such a key issue for Musa in the past, and indeed in the 1930s he had not been willing to advocate breaking with the British over it. For a man who had always been scathing about pan-Arabic ideologies, about the interwar Easternist movement, and, above all, about any association with Africa, it was more than a little ironic that he enthusiastically attended the Afro-Asian Peoples' Solidarity Organization conference in Cairo, which was the 1957 follow-up to the first Asian-African Conference, held in Bandung, Indonesia, in 1955. Also in 1957, he claimed that Marx had always been his greatest inspiration and that the only reason he had not confessed this earlier was for fear of being accused of communism. But for someone who had devoted little thought to the details of modes of production and strategies of industrial growth, this claim was also rather unconvincing. In fact, Musa was still committed to the evolutionism that increasingly marked him out as belonging an older generation. In the post–Second World War world, gradualism had come to be associated with the old political elite and with its habit of "taking one's time" (*al-tarayuth*), an excuse trotted out for several decades to stall social change. The era of Darwin's sway over Arabic intellectual life was coming to an end.[136]

SEVEN

Darwin in Translation

... ٩-تولد الحى من الحى أو الحى لايتولد الا من الحى...
... ١٤-الانواع و التنوعات...
...٢٩ - النغولة والانغال...
...٣١ - التناسل الجنسى...
...٣٢ - نكتارين...
...٣٣ - شواذ الخلق...

. . . 9—*tawlad al-hay min al-hay, aw, al-hay la-yatawlad ila min al-hay* [Biogenesis]

. . .

. . . 14—*al-anwaʿ wa-al-tanawaʿat* [Species and Varieties] . . .

. . . 29—*al-naghulat wa-al-anghal* (Hybrids and Hybridism) . . .

. . . 31—*al-tanasul al-jinsi* . . . (Sexual Reproduction) . . .

. . . 32—*naktarin* (Nectarine) . . .

. . . 33—*shawadh al-khalq* (Monstrosities) . . .

—From Ismaʿil Mazhar's 1928 translation of the *Origin of Species by Means of Natural Selection; or, The Preservation of Favoured Races in the Struggle for Life*[1]

In 1918 the first verbatim translation of Darwin's *Origin of Species* appeared in Arabic as *Asl al-anwaʿ wa-nushuʾiha bi-al-intikhab al-tabiʿi*. The title itself was faithful to the original save for one difference: it adds the word evolution (*nushuʾ*)—"the origin of species and their *evolution* by means of natural selection"—as if to highlight for its readers the centrality of this concept for the book even though Darwin himself never uses the word in his *Origin of Species*.[2] The subtitle—*Hifz al-sufuf al-ghaliba fi al-tanahur ʿala al-baqaʾ*—provides another example of this subtle rearrangement of meaning; it substitutes "the preservation of the victorious ranks" for "favoured races," thus introducing the implication of dominion and ascendancy between natural kinds and eschewing other cognates for the modern concept of "race."[3]

Before this edition, only summaries of Darwin's work had been available in Arabic, mostly carried—as we saw—in popular-science journals like *Al-Muqtataf*. Published in Cairo, the first Arabic translation consisted of chapters 1–5 of the sixth edition of *Origin*: the translator described these chapters as the "true core" of the book. Ten years later, he added another four chapters, an expanded introduction, and a new, and quite extensive, glossary.[4]

The glossary, nearly a hundred pages, took up the bulk of the second volume of the 1928 edition and revealed the variety of readerships that translation was intended to reach. It included a number of the terms in Darwin's original glossary, such as "crustaceans," "lepidoptera," "parasite," and "albinism." Yet it also left many more out: specific taxonomic terms of plants and animals and more general concepts like "analogy," "homology," and "correlation."

Take the example of "analogy," in some ways a rather strange omission. Darwin defined it as the resemblances between structures with similarity in function.[5] The Arabic glossary makes no mention of it, despite the fact that Arabic terms for "analogy" have quite an extensive range and genealogy of meanings. In the translation, the term receives only a brief footnote (and only in the 1962 edition): the original English typescript for "Analogous Variations" is accompanied by a strangely unhelpful explanation in Arabic "meaning that which is inferred by the term." It also uses the construction *"al-tahawulat al-naziriya,"* which we could translate as "comparable change," and avoids using terms for analogical reasoning (e.g., *qiyas*) that have philosophical or theological implications.

The glossary was mostly concerned with names: people like Isidore Geoffroy Saint-Hilaire, Lamarck, and Goethe, places like the Galapagos Islands, and exotica like nectarines. It gave potted biographies and brief geographical descriptions; objects like "nectarine" were usually described in relation to more familiar ones, in this case defined as "akin to our domestic apricot."[6] There are also greyhounds, terriers, bulldogs, setter dogs, pointers, and spaniels. Darwin had made passing references to these precisely because they would have been entirely common to his English readers, helping to make his argument for natural selection less disconcerting. It had the opposite effect in Arabic, however; as the glossary makes clear, few Arabic readers understood the differences between English canine pets.

The glossary revealed more fundamental and complex problems for the translator as well. Take the entry for "species and varieties." Darwin himself admits to their fundamentally ambiguous distinction: "I look at the term species as one arbitrarily given . . . it does not essentially differ from the term variety."[7] The translation of "species" contra "varieties" was made even more difficult because the terms used for "species," "variety," and "variation" were all neologisms derived from the same root, *n-ʿ-w*: "By 'varieties' [*tanaʿwuat*], [the author] means a group of individuals of a species [*naʿw*] different from others related to it of its species [*naʿwha*]; hence, it appears as another kind [*naʿw*] from that from which it transformed."[8] The inclusion of entries for "species" and "varieties" suggests the novelty of the concepts and the distinction that the translator wanted to make between these words in Arabic: the term used to translate "species" was the general term for "kind," while the term used to translate "variety" also was used for "variation."[9]

The translator of *Origin* was the Egyptian writer Ismaʿil Mazhar, and while it may seem that he was primarily providing a technical translation of expert terms and ideas, a close reading of his text suggests that he also had a strategy in mind aimed at a more general literate public. Mazhar was keen to domesticate Darwin and his ideas: the very vocabulary he used reflects this, and he often cited medieval verses or texts to justify his choice of words.[10] Mazhar made much of the fact, for instance, that the fundamental concept of evolution was not itself new to Arabic readers. Much more systematically than any other figure discussed in this book, Mazhar placed Darwin's theory of evolution within an expanded vision of both "Arabic science" and "Western science." Like Vedic scholars in India, Confucian writers in Qing China, and other colonial intellectuals of his generation, Mazhar was thus able to act as publicist for modern evolutionary studies while seamlessly reviving older cosmological traditions and in the process transforming their signification.

Mazhar was also more conscious than any other figure discussed here of the limits and possibilities of translation. Indeed, he was as much a professional translator as anything else, and he was hailed in his day for perfecting the art of "summary translation" in particular.[11] By the time of his death in 1962, he had Arabicized many contemporary Anglophone texts, including those of Darwin, Andrew Dickson White, George Sarton, and Edmund Sinott; he wrote numerous books and articles that summarized other English works; and late in life he served as official translator for the Arabic Academy of Language, advising them on technical and scientific translations in particular. He also wrote often on the value of translation itself. In the 1928 introduction to *Origin*, which offered a lengthy meditation on the subject, he spoke of expanding the intellectual horizons of his readers by introducing them to "entire subjects of modern thought," referring to "agnosticism" and even "the Unknown" in this context.[12]

This chapter explores this process of translation as a complex project of intercalating linguistic, conceptual, and historical references and metatexts by focusing on the figure of the translator himself. Viewing Darwin through Mazhar's eyes, we capture the local referents through which Darwin was read. The focus on Mazhar, meanwhile, takes us into the nexus of ideas, places, and people that helped to construct this particular reading of Darwin in translation.

Philosophical Experiments

Ismaʿil Mazhar is curiously little known today, though he was an important figure in the Egyptian intellectual landscape for much of the twentieth century.[13] He was born in 1891 into a family of prominent technocrats and government officials: his paternal grandfather was a chief engineer under Muhammad ʿAli, and his maternal grandfather was the first minister of education in Egypt. Like so many notables and colonial intellectuals of that generation, Mazhar held a lifelong commitment to the power of science and education, both as a universal program and as a means toward individual refinement and national maturity.[14] The young Mazhar attended the Khedival School, graduating a few years after Salama Musa. Like Musa, he had little favorable to say about his schooling. He complained of the effect of its curriculum of "sick books" in particular. Afterward, he took courses on Arabic language and literature at Al-Azhar with Shaykh ʿAli al-Marsafi (a disciple of Muhammad ʿAbduh), but unsatisfied with this, or perhaps with the state of Al-Azhar itself (as later writings of his suggest), he decided to pursue his own course of "self-education" (once again like his older colleague and fellow evolutionist Musa). In a kind of intellectual refashioning, Mazhar read widely in contemporary Western and medieval Arabic science and philosophy, eventually devoting himself to a life of reading, writing, and translating. Mazhar was in many ways quite typical of the new class of *effendiya* in his orientation toward the "West," but he was also the last of that generation which came of age in the 1910s and 1920s to so enthusiastically take on the mantle of scientific positivism, cultural evolution, and social progress under the name of Darwin.

As he wrote in 1928, Mazhar's introduction to evolution had occurred shortly after his graduation from the Khedival School in the summer of 1911. It was then that he had come across an advertisement for Shibli Shumayyil's essays and translation of Ludwig Büchner's lectures on Darwin. Its title immediately caught his eye: *Falsafat al-nushu' wa-al-irtiqa'* (Philosophy of evolution and progress). Though he was relatively new to what he called "literary books" and he "had passed over many advertisements for newly translated books before," this piqued his interest and left him confused. "If people want to know why I translated [*Origin of Species*]," he later wrote, "I have one word to say: 'confusion.'" *Falsafa* and *nushu' wa-al-irtiqa'*, Mazhar explained, were words he little understood at the time: his education had given him a very poor knowledge of things philosophical; the idea of a "philosophy of evolution and progress" was new, appealing, and powerful.[15]

Given the turmoil of the times, Mazhar, like many others, was understandably attracted to such an approach: in 1909, in the wake of the Young Turk Revolution, he had already edited a short-lived student journal, *Al-Sha'ab* (The people). Mazhar read through Shumayyil quickly—in a sleepless and frenzied state according to him: the book and its materialist philosophy had a profound and initially disorienting effect. Shumayyil's, and even more so Büchner's, atheism shook his worldview and led Mazhar to "wander into the desert of materialistic opinions." He had previously studied medieval Arabic texts on ethics, and on the nature and existence of the soul in particular—whether as part of his schoolwork or as extracurricular reading is unclear.[16] Reading these, he had imagined a life after death, where his soul would "soar through the heavens" and immortality would be his "everlasting companion." He now abandoned these ideas: "I moved from the world of conviction to the world of solid materialism and philosophical atheism which that school of evolution and progress had then followed." And while the later Mazhar felt that this had allowed him to "overcome certain unthinking customs [*taqalid*]," it also left him in a "state of utter pessimism [*ilya's*]." He became, in his words, an "Epicurean," regarding himself as nothing more than a material entity with no moral obligations beyond the safeguarding of his own pleasure. Had not the nebular hypothesis (*nazariyat al-sadim*) shown that the universe sprang from chaos and blind chance? Then, why morality? What of the spirit?[17]

Mazhar's descent into a bleak and pessimistic nihilism did not last long, however. For though he "failed" his "first philosophical experiment in life," as he put it, his "confusion" led him also to read widely among those Greek and Arabic philosophers whose writings were enjoying something of a revival. He had already read up on the debates between the Mu'tazila and the Ash'ara (which presumably would have introduced him to centuries' old philosophical debate on the nature of creation, the cosmos, and the soul) and had studied texts by Plato on the "transmigration of souls" (*tanasukh al-arwah*) and the daring works of al-Maʿari, whose verse on the rejection of God he also cited. Print editions of these texts would have been newly available in Arabic, and they no doubt passed through Mazhar's hands. But they acquired new meaning in light of his readings on evolution.

According to his 1928 self-presentation of his education, it was Socrates who saved him. Plato's account of Socrates' trial showed Socrates to be a rationalist and "freethinker" but no materialist—presenting his case for the eternity of the soul and one's lifelong duty therefore to guard its health and happiness. This offered Mazhar a basis for ethics that was both rationalist and spiritually rich at the same time. With his faith in the eternity of the soul restored, Mazhar became, as he put it, "human again." Socrates also offered Mazhar something else: a picture of the lone intellectual who defies common opinion for the sake of truth and the divine. Through Socrates, he finally came to know what *falsafa* (philosophy) meant: a combination of reasoned thought and a metaphysical quest (or the search for what lay behind reality, *wara' al-haqiqa*). Only this would serve as "guidance for the darkness of this-worldly life."[18] This was the path, with its striking combination of rationalism and transcendentalism, that Mazhar would follow for the next fifty years of his life.

A later spiritual mentor served a similar purpose. For Mazhar, Darwin was the quintessential rationalist. He presents him as "a freethinker, who did not limit himself to *taqlid* [imitation] or accept a common opinion before weighing it well with wisdom." It is here—a decade before the more systematic translation of 1928—that we get the clearest picture of Mazhar's own views on the relation of nature to God. Mazhar's 1918 Darwin resembles how Mazhar would eventually come to see himself: a rationalist in search of metaphysical truths and a deist finding God in nature. Mazhar argues that while Darwin himself had largely skeptical tendencies in matters of faith, he had also admitted that "this world was built by and would return to a First Cause." To support this claim, Mazhar cited an 1873 letter that Darwin wrote in answer to a Dutch student on the existence of God. (Mazhar likely encountered this letter in G. T. Bettany's 1887 *The Life of Charles Darwin*.)[19] The letter expressed Darwin's characteristic ambiguity on the issue, an ambiguity Mazhar skated over. In his summary translation of the letter, moreover, Mazhar omits an important line of Darwin's about suffering: "Nor can I overlook the difficulty from the immense amount of suffering throughout the world." Theodicy in general proved a less central concern for Muslim theologies of nature, as we have seen throughout; "the impossibility of conceiving that this grand and wondrous universe, with our conscious selves, arose through chance" was the real point of the letter for Mazhar.

Mazhar concludes by reminding his readers that while it was true that Darwin had been routinely charged with atheism in his lifetime, familiarity with his thought led one to a different conclusion.[20] For Mazhar, belief in a First Cause of "this grand and wondrous universe" was tantamount to belief in a Maker of the universe. This was a subject that would reappear regularly in Mazhar's later writings. Rejecting the materialist postulates that had led to his youthful confusion and despair, Mazhar found a way to hold on to a belief in God while reading Darwin. Mazhar's Darwin was thus a progressive rationalist and a deist seamlessly blending contemporary speculations about nature with past Arabic philosophies of creation.

Evolution and the Arabs

The year after Mazhar's translation was published—with the First World War and the 1919 nationalist uprising in the background—it was reviewed in *Al-Muqtataf*. The editors pointed out that Darwin was far from unfamiliar to most Arabic readers: Shumayyil had translated Büchner and introduced Darwin that way, while they themselves had spent the last forty years reporting on Darwin's theory: "Every time we publish an issue of *Al-Muqtataf* it either has something on Darwinism [*al-madhhab al-Darwini*], or it refers to the existence of species . . . according to natural or sexual selection." They described *Origin of Species* as the book which had had "the greatest impact on our recent scientific revolution" (*al-inqilab al-ʿilmi*).[21] Yet, curiously, it had never been translated into Arabic before. They commended Mazhar in glowing terms for having taken on the job.

But they also raised a significant objection to his attempt to fit Darwin within a longer history of evolutionary thought that emphasized the role of medieval Arabic and Persian thinkers: this simply did not do justice to the novelty of Darwin's theory. Mazhar might have cast Darwin's predecessors as Muslim evolutionists, but the only one to have formulated a theory (and therefore explained the cause) of evolution (i.e., natural selection) was Darwin himself.[22]

Mazhar had indeed constructed a comprehensive genealogical ancestry for Darwin, including the Greeks, the Assyrians, and Renaissance and Enlightenment philosophers. Yet he spent the most time detailing the contribution of what he classed as Arab natural philosophy. Many of these authors were in fact Persian. But his easy use of the term "Arab" demonstrates its range: for Mazhar it was an ecumenical term that referred to these authors' use of Arabic as their lingua franca.

The editors of *Al-Muqtataf*, we might recall, had long resisted this kind of move, though they too engaged in occasional cross-historical references, as did many of their writers and readers. But the growth of nationalist aspirations and party politics in Egypt after the First World War increased the appeal of such intellectual genealogies. Mazhar, a younger man, was among the first to extensively develop this theme vis-à-vis the evolutionary sciences.

His was a novel genealogy, of course. As he explained, while he found only a few formal references to evolution in his research on the "Arab sages," they covered a great many general principles that have since been identified and explained more fully, including inheritance, regression, extinction, and natural selection. Of course, what he counted as evidence of inheritance or natural selection was as much a matter of interpretation as anything else. The point was not so much to show a strict genealogy of ideas, or even to investigate the particularities of the analogy; rather, it was to endorse ideas of modern science in the idiom and tradition of older ones.

Nevertheless, in his revised introduction to the 1928 translation, Mazhar also conceded that it was not until the nineteenth century that a more analytical theory of evolution could be developed, thanks to increased knowledge of the world's flora and fauna and the general advancement of geology and geography. Still, he stood his ground with respect to the idea that the conceptual analogies had stood the test of time: and his second introduction retained the entirety of the original, 1918 discussion of medieval evolutionists.

These connections rode slipshod over discontinuities, to be sure. Take the example of one of the earliest authors discussed by Mazhar: al-Jahiz. This ninth-century philosopher was the subject of an intense revival of interest by Orientalists and Arabists alike in the late nineteenth and early twentieth centuries, and his *Book of Animals* was frequently likened to Darwin's *Origin of Species* by the latter in particular.[23] Composed in the early 800s in ʿAbbasid Baghdad, the book was a collection of personal and anecdotal observations and lore about animals taken from the Qurʾan, the *hadith*, pre-Islamic poetry, proverbs, and, of course, Aristotle's *Generation of Animals*. It covered a wide variety of topics, including the influences of various climates and diets on men, animals, and even plants of different geographical regions; as well as discussions of animal mimicry, intelligence, and social organization. Much of the work emphasized the marvels and morals of creation, and it had strong natural theological undertones.

Indeed, it was not so much a zoological treatise as an exercise in *adab* literature: written in a highly expressive and ornate style, its emphasis was on philosophical and religious edification. Mazhar, like others before him, read it selectively. When discussing the ninth-century philosopher's views on the transformation of species, for instance, he says nothing of al-Jahiz's repeated discussions of *maskh*, or the transformation by God of sinful nations or peoples into pigs, apes, and other "lower creatures."[24] Mazhar merely ends his section "Evolution among the Arabs" by citing al-Jahiz's discussion of the copulation of two different varieties of dung beetles and the production of a new one as evidence of the prefiguring of evolutionary principles. He closes with the ambiguous statement that "these observations were equal to the time spent on them."[25]

Mazhar's discussion of Ibn Miskawayh and Ikhwan al-Safa' did something similar. Both were also the subject of new scholarly attention by European Orientalists and their Arabist counterparts. Ibn Miskawayh, a tenth-century thinker from Isfahan, was enjoying a widespread revival in Muslim lands thanks to the republication of his ethical writings in particular.[26] Mazhar had also cited him as one of the authors that he had encountered in his youthful dabblings with philosophy and ethics. The philosopher's writings on the proper cultivation of manners—which had strong Neoplatonic elements and also contained a number of themes common to the literature on "advice to kings"—were a mixture of the search for a rationalist faith and a call to a faith in divine sanction. His influential *Tahdhib al-akhlaq* (The refinement of character)—first published in Egypt in 1871—had helped Mazhar recover from his youthful flirtation with materialism. Wrote Ibn Miskawayh: "Now, as the soul is a divine, incorporeal faculty, and as it is, at the same time, used for a particular constitution, and tied to it physically and divinely . . . each one of them [body and soul] is dependent upon the other, changing when it changes, becoming healthy when it is healthy, and ill when it is ill."[27] This quest for the perfection of ones' intellect and manner of thought and being was precisely the image of the rationalist thinker and reasoned believer that Mazhar was cultivating for himself.

Yet here Mazhar was interested specifically in Ibn Miskawayh's naturalist observations. Mazhar was among a number of prominent Arabic writers of the time who were popularizing the link between the philosopher's cosmological thought and modern evolutionary theory. Ibn Miskawayh's cosmology outlined a familiar Aristotelian model of creation: from God-created matter there came the various vapors of the cosmos and particularly water, and thence all the major classes of life, the vegetable and animal kingdoms, man, and, then finally, the angels. Various intermediate forms were listed as evidence of the seamless transition from one class to another: coral, with its treelike vegetative qualities; date palms, with their separate sexes and their curious fate of dying when the "head is cut off"; the apes; and finally the prophets or saints. Much in particular was made of the mention of apes as humanity's next of kin in this chain of being. Yet other comparisons were avoided: one might speculate, for instance, on the nature of the differences between the metaphor of a central authority in date palms versus the stress laid on the mobility of insectivorous plants in later English discourses; yet Mazhar glosses over these finer points. Unsurprisingly, he also neglects to mention the ranks of angels, saints, and prophets in Ibn Miskawayh's account.

A similar ladder of beings could be found in the writings of the eleventh-century illuminationists, the group of pseudonymous writers known as Ikhwan al-Safa' (Brethren of Purity).[28] The Ikhwan drew a similar picture of the hierarchy of beings descending from the divine through the heavenly spheres and angels and down to the world of the four elements. Mixing in various degrees with the universal spirit or Soul, these graded elements were then said to have formed the three kingdoms, culminating of course with man.[29] The Ikhwan's cosmology was a chain of beings that linked man and the rest of the created world to the divine.[30]

Mazhar's comparison of the Ikhwan with Darwin avoids the broader moral cosmology and chorography in which these ideas were placed. Yet he nevertheless identified their discussion of the power of *al-hikma al-ilahiyat*, a divine wisdom or spiritual power, with Darwin's "natural selection." Though the names are different, the ends are the same, he concludes: the preservation of beneficial traits in the natural progression of species.

He argued similarly when he compared the ideas of Ibn Khaldun with those of Lamarck. Focusing on the former's discussion of the effect of climate on the moral evolution of peoples, Mazhar connected this with Lamarck's heritability of acquired characteristics.[31] As these examples show, reading Darwin in Arabic was part of a larger metatextual process: it accompanied a revival of interest in classical Arabic texts that served as both the conceptual and even linguistic substratum from which to construct new translations and textual interpretations. They also helped to endorse new disciplines of study and categories of knowledge. The example of Ibn Khaldun here is a good one: alongside Mazhar's Lamarckian reading were other similar revivals of Ibn Khaldun as the forefather of modern sociological or even historical inquiry.[32]

Novel genealogies such as these were proving increasingly popular, and they were intended to show how Arab civilizations of the past had contributed to the universal progress of science. Mazhar himself developed these connections in light of growing interest in classical Arabic literature among lay readers, stimulated by the widespread print reproduction of books of medieval Arabic philosophy. The rise of Orientalist literature on the Arabic "golden age" of philosophy and literature also helped to contribute to this general resurgence. Indeed, many of Mazhar's ideas borrowed from this. His articles in *Al-Muqtataf* on "the history of Arabic thought" followed De Lacy O'Leary's 1922 *Arabic Thought and Its Place in History* in detailing the persistence of the "Greek spirit" in the lands of the Arabs and Persians through their role as translators of Greek texts.[33] In the collection of essays Mazhar republished and expanded upon later, he also relied on O'Leary in his discussion of Jabir Ibn Hayyan, or "Geber," the eighth-century Muslim alchemist whose alchemical and laboratory experiments were said to be a kind of precursor to the modern "scientific method."

Stressing the connections between older, Arabic ideas on evolution in this way was, for Mazhar, a key semiotic, philosophical, and political strategy. Reading Darwin in light of one's own tradition could reveal certain shared values and thoughts, across languages, traditions, and even religions. At the same time, referring his readers to Arab predecessors of Darwin, however distant, offered a means by which modernity's reassuring antiquity could be highlighted, casting universalizing narratives in more familiar, local forms. This, for Mazhar, was the real point of translation.

Translatio Studii

By 1928, when the second, expanded edition of his translation came out, it was not clear that his attempt to domesticate Darwin through genealogy had worked. Indeed, Mazhar complained that his first edition had not been much discussed in the Arabic press (*Al-Muqtataf*'s review was an exception). The bitter outcome of Egypt's bid for national self-determination after the First World War and the beginnings of revolutionary foment, he concluded, had probably diverted people's attention. After a prolific decade of deliberating on the subject, including the publication of two popular studies on evolution and numerous articles on the topic in his own literary journal, Mazhar offered another version of the translation, adding more chapters and expanding his introduction considerably. He spent much of the new introduction discussing the value of the work of translation itself and raised the fundamental question: why bother with translation at all?

He recalled that many people had objected that those who were interested in the subject most likely already had knowledge of English, in which case they were better off reading it in the original; while those who did not know English would probably find the Arabic translation too unfamiliar, too bogged down with strange words and concepts. He himself was not persuaded. Why would someone who knew English not also read the book in Arabic, and why should those who did not know English not also find the Arabic a useful text? Might it not in fact expand their sense of the science and extend their vocabulary? He argued that *Origin of Species* was in fact a "manual for belief," a "science of conviction." Surely this was a subject that could benefit all "students of philosophy," whatever language they happened to know. His translation was thus not intended for an elite of highly educated Arabic readers alone.

And at that moment in the history of the region, he urged a wider consideration. His translation of *Origin*, like all such translations, offered an important way to expand, revive, and develop both the Arabic language itself and the modes of thought possible in it: translation thus introduced new technical terms, refined the language's conceptual vocabulary, expanded its cultural connections, and allowed the development of a modern and yet indigenous scientific discourse. Always ready to construct genealogies of legitimation, Mazhar was among the first Arab authors to stress the explicit importance of the "translation movement" (*al-haraka al-tarjamiya*) for the classical golden age of Arabic thought. Through translations such as these, he argued, another, contemporary renaissance of Arabic thought, or Nahda, would emerge. This was the idea of a *translatio studii* once more, with the added hint of a reversed *translatio imperii*: by transferring "Western science" eastwards, one could also point the way toward transferring the locus of political might and returning the "East" to its former glory. In this way, the primacy of translation in the movement for political and cultural regeneration that had been under way for a century was reaffirmed. Mazhar's harkening back to the original Arabic translation movement added another perhaps more ambiguous dimension in his case: the idea that the political legitimacy that would be conferred in this process would be that of the early Islamic empires. He would return to this later, but for much of the interwar period Mazhar was concerned with issues of translation and language more specifically.

By the time of Mazhar's 1928 edition of *Origin*, a fierce debate was well under way in the Arabic press over the language question in general and over translation strategies in particular. Indeed, Mazhar was a key participant in this and had already worked out his own philosophy of translation, which highlighted the positive contributions made by the medieval translation movement and attributed the stagnation of Arabic thought in the present to its abandonment. Genealogy and evolution formed the dual motifs of his approach. Translations, he argued, demonstrate that language is flexible and subject to the laws of change; without them, the language would simply become a "dead and immobile form." Language had a life of its own, which the expansion of meanings preserved and helped to evolve.[34]

These were, to be sure, not new ideas: the organicist conception of language as subject to the laws of evolution had been propounded by language reformers and particularly science translators for some time. Jurji Zaydan, for instance, had published his *Tarikh al-lugha al-ʿArabiya kaʾin hay* (History of the Arabic language as a living entity) in 1904.[35] But Mazhar tied this idea to his Socratic commitment to freedom of thought: "independent thought" only came with "independent language." The present was thus in his hopeful reading "an age of translation," analogous to an earlier time when translators had facilitated the transmission of ideas, first during the ʿAbbasid era, from west to east and then, during the European Renaissance, in the reverse direction. The intellectual revival of Arab thought meant restoring this mobility of knowledge and helping to eliminate "unstable Eastern thinking."[36]

Mazhar illustrated his point with the terms for "natural selection" (*al-intikhab al-tabiʿi*) and "struggle for life" (*al-tanahur ʿala al-baqaʾ*). The word *intikhab*, he argued, had previously only indicated selection or choice in general; with the new writings on evolution, it therefore acquired a totally new meaning for Arabic readers. (At the time he was writing, it had also come to mean "election"—as in parliamentary or political elections—as well as "selection.") Similarly, the phrase "struggle for life" gave new meaning to old words: he preferred to use *tanahur* here because he thought that this (more than the word *tanazuʿ*, which Shibli Shumayyil used) expressed the sense of a mutual struggle unto death. This substitution had occurred to him when reading a verse of old Arabic poetry that expressed this sense of struggle perfectly; now, he thought, in the context of a science translation, it acquired an exact and technical meaning hitherto unknown. Other words, such as "agnostic" (which he rendered as *al-laʾadri*), "the unconscious" (*al-la-shaʿuri*), and "infinity" (*al-la-nahaya*)—all new compound terms in Arabic—have specific, philosophical meanings whose introduction, he believed, might open up whole new areas of modern thought to their readers.[37]

The older generation had also, of course, been believers in the value of translation and shared Mazhar's optimistic attitude toward linguistic change. But where they and Mazhar differed was in their translation strategies and in their attitude toward neologism in particular. At that time, there were various lexical strategies for constructing these. First, there was the derivation of words (*ishtiqaq*) through analogic reasoning (*qiyas*), such as deriving new words through the extension of trilateral root forms (*al-ishtiqaq al-saghir*), the figurative semantic extension of existing words (*al-wadʿ bi-al-majaz*), the use of older vocabulary to express new meanings (*gharib al-lugha*), and the use of descriptive paraphrases (*al-ishtiqaq al-maʿnawi*). Second, there was the formation of compound words (*naht*); and, finally, the assimilation of foreign words, or Arabicization (*taʿrib*).[38] It was this last, in particular, or the use of Arabicizations, that came under increasingly harsh and sustained criticism. For many, particularly in an era of growing anticolonialism, this represented a corruption of the proper rules of Arabic and, worse still, a sign of Arab political weakness and the dangers of excessive Western importation.

Geopolitics clearly shaped this debate. *Al-Muqtataf*'s long-held view was that there was "no shame in borrowing from the West." Reviewing the various scientific advances its editors had publicized in their first fourteen years, for example, they had reminded their readers in 1890 that "borrowing from the West does not impede the advance of knowledge in the East." Science offered benefits to all: "There is no need to be hindered by our pride," for science belongs to everybody. "The West has borrowed from us when we were once great, and now it is our turn to take from the West." The real shame would be if one were to insist on starting from where Europe did two hundred years ago: "That would be like one who abandons a steam engine for a simple tool."[39]

Not everyone was persuaded even then. One reader had written in to *Al-Muqtataf* as early as 1891 complaining about the editors' excessive use of "foreign words" (*al-kalimat al-aʿjamiya*). This was despite the fact, he thought, that "the noble Arabic language" was one of the "most comprehensive languages" in its ability to grasp a whole host of meanings and expressions. The editors pointed out that the majority of the foreign words used in the journal, like *al-uksujin* (oxygen) and *al-haydrujin* (hydrogen), had no synonyms in Arabic. And even where there were such synonyms, often the imported term was still generally better known—like the archaic *al-murqashiya* versus *al-bismuth*. They also reminded their readers that generations of Arabic scholars, like Ibn Sina and al-Razi, had engaged in this process.[40]

After the First World War, and the Egyptian Revolution of 1919 in particular, the politics of this issue became even more contentious. Mazhar himself clearly preferred the use of simple derivations or descriptive paraphrases and the resuscitation of archaic terms, and he was much more reserved than *Al-Muqtataf* in his use of transliteration. Only in cases where he may have felt the foreign term was either too technical or particularly novel—for example, Acrasieae or amoeba (*al-aqrasiya, al-amibia*)—did he resort to these. Even when he used an Arabicized term, he usually added an alternative description in common Arabic: in this way, he tried to balance novelty with familiarity. For instance, "raccoon" was *al-raqun aw dubb amriki* (literally "*raqun*, or an American bear") and "bulldog" was *al-bulduj aw al-kalb al-ʿajla* ("*bulduj*, or the speedy dog"). At other times he translated a word twice, using an Arabicized term alongside an archaic synonym, such as his translation of "hybrid" as *haybrid* together with the resuscitated *al-anghal*, or "dandelion" as both *al-dandilin* and *al-handiba* (a kind of wild chicory). He saw the necessity for Arabicization but, as he admitted later, felt that it was the "laziest" of strategies.[41]

Like *Al-Muqtataf* and Tahtawi before him (and, indeed, like Darwin himself), Mazhar relied on glossaries. As we have seen, unfamiliar names such as Tierra del Fuego were usually transliterated directly: Stratford as Astafurd, Surrey as Sari, and Tierra del Fuego as Tiara Dalfawiju, to which he added *jazaʾir ard al-nar*, "peninsula of the land of fire," a direct translation from the Spanish. Similarly the Galapagos Islands were rendered as Al-Ghalabaghus or *jazaʾir al-sulahfa*, "tortoise island," which Mazhar explained—here turning to English—as "named from . . . the Spanish name for tortoise." Among the plant and animal kinds he described were *Spongilla fluviatilis* (*isfinj al-miya al-ʿazba*, or freshwater sponge) and dicotyledons (*dhawat al-falqatayn*, or a two-halved being), which had both the Latin term in the Latin alphabet and a descriptive Arabic rendition of the creature alongside.

The glossaries not only permitted Mazhar to elucidate unfamiliar objects or terms but also allowed him to make connections necessary for his readers but undreamed of by the original author. Mazhar's inclusion of "monstrosities" and "hybrids and hybridism" (*shawadh al-khalq* and *al-naghulat wa-al-anghal*) conveyed the findings of other contemporary scientists and simultaneously catered to an enduring fascination in Arabic literature with the curious and spectacular. When introducing the latter, moreover, an obscure quotation from the fourteenth-century lexicographer Firuzabadi (presumably in reference to some form of hybrid plant) opened the entry. This helped Mazhar justify his translation of the term, even as he assigned it new meaning in Arabic. Largely concerned with the classification and genetics of developmental abnormalities, the entry also made considerable reference to more familiar domestic crossovers like the onager (given as the more colloquial *himar al-wahsh*, or wild ass). An entry on biogenesis—the term is neither in Darwin's original glossary nor in the text of *Origin*—allowed Mazhar to engage with the old debate over spontaneous generation: listed as *tawlad al-hay min al-hay, aw, al-hay la-yatawlad ila min al-hay*, it summarized for Arabic readers once again Pasteur's victory over the materialists, helpfully drawing attention to the fundamentally antimaterialist message, as the translator saw it, behind Darwin's theory of natural selection. And as a kind of supplement to the glossary, Mazhar incorporated a number of diagrams and charts that (much like *Al-Muqtataf* once again) he took from various English originals, including his transcription of Darwin's own branching diagram on the divergence of character and his transcription of an animal kingdom classificatory schema (fig. 7.1).

All this begs the question of whom Mazhar was actually writing for. The contrast with Salama Musa's popularizing book on evolution, published only three years earlier, is striking. Perhaps modeling himself on popularizers like H. G. Wells, Musa had reassured his reader at the outset that he would not encounter any technical terminology or unfamiliar place-names or species. He felt no need to provide a glossary, and his language was simple and at times colloquial. Mazhar, however, had a different pedagogic strategy in mind, one that was both longer term and more specialist in its orientation. He wanted both to revive the language and to help create a modern scientific vocabulary in Arabic. In this, he reflected the ambitions of a new generation of Arabic modernists who sought to demonstrate not only the flexibility and potential of Arabic itself but also its connections to their own past traditions of thought.

Transcendental Positivism

Mazhar's interest in translation was clearly not the only reason he decided to translate *Origin*. As he writes in his preface, he felt it was important to present Darwin in response to the by then widely held view that modern evolution was necessarily a materialistic doctrine, and in particular to the radical materialism of Shibli Shumayyil's *Falsafat al-nushu᾿ wa-al-irtiqa᾿*, which had made such a powerful early impression on him and on so many other Arabic readers. Mazhar assigns himself the role of correcting the misperceptions that evolution was a threat to belief in an "immortal soul," to religious ethics, or even to religious faith more generally. The theory of evolution by natural selection was, rather, "far from opposing religion and its laws [the *shariʿa*]."[42]

Yet Mazhar is if anything a study in paradoxes: for even as he never tired of advising his fellow countrymen to avoid the dangers of mixing science

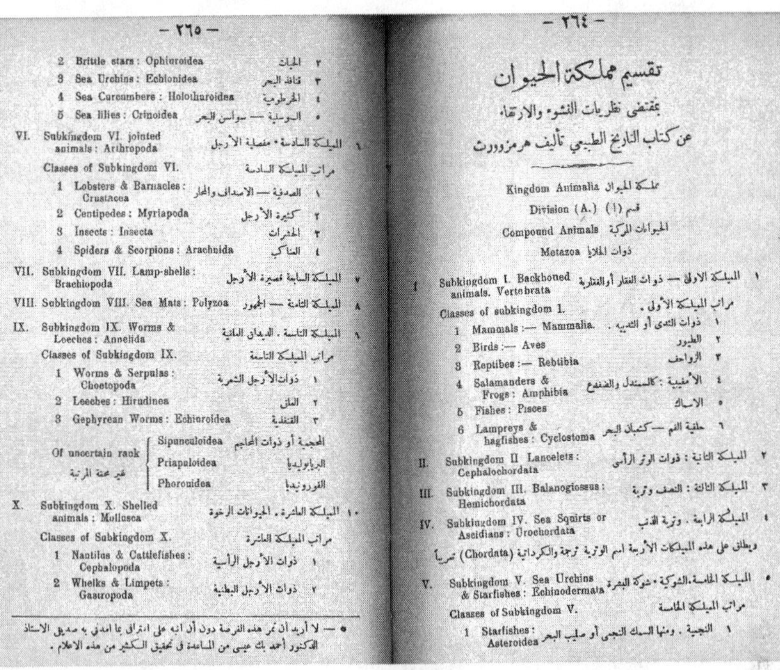

- ٢٦٤ -

تقسيم مملكة الحيوان
بمقتضى نظريات النشوء والارتقاء
عن كتاب التاريخ الطبيعي تأليف هرمزوورث

Kingdom Animalia مملكة الحيوان
Division (A.) قسم (ا)
Compound Animals الحيوانات المركبة
Metazoa ذوات الخلايا

I Subkingdom I. Backboned animals. Vertebrata — ١ الميلكة الاولى — ذوات الفقار أو الفقارية
Classes of subkingdom I. — مراتب الميلكة الأولى .
1 Mammals :— Mammalia. — ١ ذوات الثدى أو الثدييه .
2 Birds :— Aves — ٢ الطيور
3 Reptibes :— Rebtibia — ٣ الزواحف
4 Salamanders & Frogs : Amphibia — ٤ الأمفيبية : كالسمندل والضفدع
5 Fishes : Pisces — ٥ الاسماك
6 Lampreys & hagfishes : Cyclostoma — ٦ حلقية الفم — كثبان البحر
II. Subkingdom II Lancelets : Cephalochordata — ٢ الميلكة الثانية : ذوات الوتر الرأسى
III. Subkingdom III. Balanogiossus : Hemichordata — ٣ الميلكة الثالثة : النصف وترية
IV. Subkingdom IV. Sea Squirts or Ascidians : Urochordata — ٤ الميلكة الرابعة . وترية الذنب
ويطلق على هذه الميلكات الأربعة اسم الوترية ترجمة والكرداتية (Chordata) تعريباً
V. Subkingdom V. Sea Urchins & Starfishes : Echinodermata — ٥ الميلكة الخامسة . الشوكية • شوكة البشرة
Classes of Subkingdom V. — مراتب الميلكة الخامسة
1 Starfishes : Asteroidea — ١ النجمية . ومنها السمك النجمى أو صليب البحر

- ٢٦٥ -

2 Brittle stars : Ophiuroidea — ٢ الحيات
3 Sea Urchins : Echionidea — ٣ قنافذ البحر
4 Sea Curcumbers : Holothuroidea — ٤ الخرطومية
5 Sea lilies : Crinoidea — ٥ السوسنية — سوسن البحر
VI. Subkingdom VI. jointed animals : Arthropoda — ٦ الميلكة السادسة • مفصلية الأرجل
Classes of Subkingdom VI. — مراتب الميلكة السادسة
1 Lobsters & Barnacles : Crustacea — ١ الصدفية — الاصداف والمحار
2 Centipedes : Myriapoda — ٢ كثيرة الأرجل
3 Insects : Insecta — ٣ الحشرات
4 Spiders & Scorpions : Arachnida — ٤ العناكب
VII. Subkingdom VII. Lamp-shells : Brachiopoda — ٧ الميلكة السابعة قصيرة الأرجل
VIII. Subkingdom VIII. Sea Mats : Polyzoa — ٨ الميلكة الثامنة — الجمهور
IX. Subkingdom IX. Worms & Loeches : Annelida — ٩ الميلكة التاسعة . الديدان الحلقية
Classes of Subkingdom IX. — مراتب الميلكة التاسعة
1 Worms & Serpulas : Choetopoda — ١ ذوات الأرجل الشعرية
2 Leeches : Hirudinea — ٢ العلق
3 Gephyrean Worms : Echiuroidea — ٣ القنفذية
Of uncertain rank / غير محققة المرتبة:
Sipunculoidea — الحجمية أو ذوات المحاجم
Priapuloidea — البريابوليديا
Phoronidea — الفورونيديا
X. Subkingdom X. Shelled animals : Mollusca — ١٠ الميلكة العاشرة . الحيوانات الرخوة
Classes of Subkingdom X. — مراتب الميلكة العاشرة
1 Nautilus & Cattlefishes : Cephalopoda — ١ ذوات الأرجل الرأسية
2 Whelks & Limpets : Gastropoda — ٢ ذوات الأرجل البطنية

• — لا أريد أن تمر هذه الفرصة دون أن انبه على اعتراف بما امدني به صديق الاستاذ الدكتور أحمد بك عيسى من المساعدة فى تحقيق الكثير من هذه الاعلام .

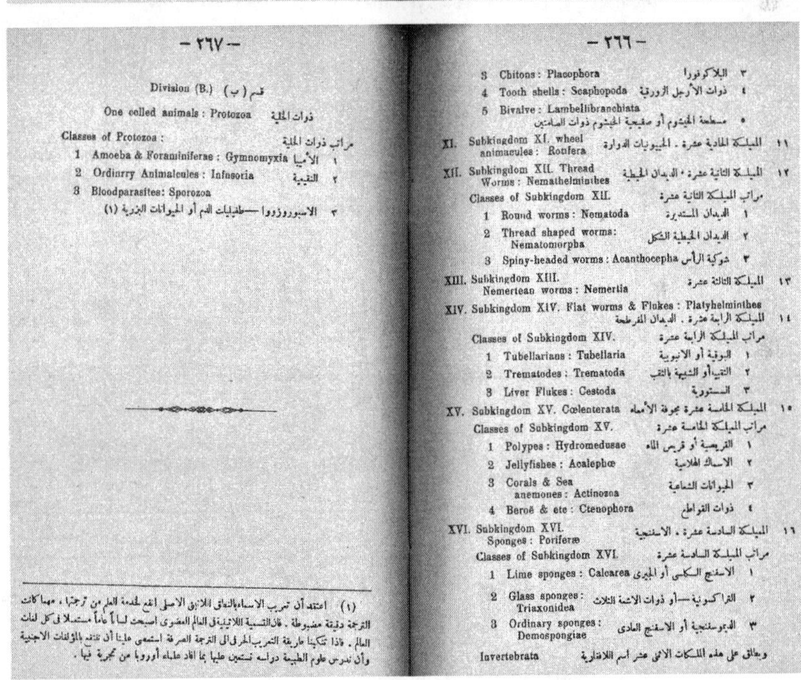

- ٢٦٦ -

3 Chitons : Placophora — ٣ البلاكوفورا
4 Tooth shells : Scaphopoda — ٤ ذوات الأرجل الزورقية
5 Bivalve : Lambellibranchiata — ٥ مسطحة الخيشوم أو صفيحية الخيشوم ذوات الصامتين
XI. Subkingdom XI. wheel animacules : Rotifera — ١١ الميلكة الحادية عشرة . الحييوينات الدوارة
XII. Subkingdom XII. Thread Worms : Nemathelminthes — ١٢ الميلكة الثانية عشرة • الديدان الخيطية
Classes of Subkingdom XII. — مراتب الميلكة الثانية عشرة
1 Round worms : Nematoda — ١ الديدان المستديرة
2 Thread shaped worms : Nematomorpha — ٢ الديدان الخيطية الشكل
3 Spiny-headed worms : Acanthocepha — ٣ شوكية الرأس
XIII. Subkingdom XIII. Nemertean worms : Nemertia — ١٣ الميلكة الثالثة عشرة
XIV. Subkingdom XIV. Flat worms & Flukes : Platyhelminthes — ١٤ الميلكة الرابعة عشرة . الديدان المفرطحة
Classes of Subkingdom XIV. — مراتب الميلكة الرابعة عشرة
1 Tubellarians : Tubellaria — ١ البوقية أو الانبوبية
2 Trematodes : Trematoda — ٢ الثقبيأو الشبيهة بالثقب
3 Liver Flukes : Cestoda — ٣ السستودية
XV. Subkingdom XV. Cœlenterata — ١٥ الميلكة الخامسة عشرة مجوفة الأمعاء
Classes of Subkingdom XV. — مراتب الميلكة الخامسة عشرة
1 Polypes : Hydromedusae — ١ القريصية أو قريص الماء
2 Jellyfishes : Acalephœ — ٢ الاسماك الهلامية
3 Corals & Sea anemones : Actinozoa — ٣ الحيوانات الشعاعية
4 Beroë & etc : Ctenophora — ٤ ذوات القواطع
XVI. Subkingdom XVI. Sponges : Poriferæ — ١٦ الميلكة السادسة عشرة . الاسفنجية
Classes of Subkingdom XVI. — مراتب الميلكة السادسة عشرة
1 Lime sponges : Calcarea — ١ الاسفنج الكلسى أو الجيرى
2 Glass sponges : Triaxonidea — ٢ التراكسونية — أو ذوات الاشعة الثلاث
3 Ordinary sponges : Demospongiae — ٣ الديموسفنجية أو الاسفنج العادى
Invertebrata — ويطلق على هذه الملكات الاثنى عشر اسم اللافقارية

- ٢٦٧ -

Division (B.) قسم (ب)
One celled animals : Protozoa — ذوات الخلية
Classes of Protozoa : — مراتب ذوات الخلية
1 Amoeba & Foraminiferae : Gymnomyxia — ١ الأميبا
2 Ordinrry Animalcules : Infusoria — ٢ النقيعية
3 Bloodparasites: Sporozoa — ٣ الاسبوروزووا — طفيليات الدم أو الحيوانات البزرية (١)

(١) اعتقد أن تعريب الاسماء بالنطق اللاتيني الاصلى انفع لخدمة العلم من ترجمتها ، مهما كانت الترجمة دقيقة مضبوطة . فان التسمية اللاتينية فى العالم العضوى اصبحت لساناً عاماً مستعملا فى كل لغات العالم . فاذا تنكبنا طريقة التعريب الحرفى الى الترجمة الصرفة استعصى علينا أن ننتفع بالمؤلفات الاجنبية وأن ندرس علوم الطبيعة دراسة نستعين عليها بما افاد علماء أوروبا من تجربة فيها .

Fig. 7.1a–b. Diagram from Isma'il Mazhar, *Malqa al-sabil fi madhhab al-nushu' wa al-irtiqa' wa-atharuhu fi al-fikr al-hadith* [Evolution at the crossroads: The theory of evolution and its impact on modern thought] (Cairo: Al-Matba'a al-'Asriya, 1924), 264–67.

with religion, he himself addressed the ways in which Darwin's natural philosophy could also serve as an argument for the existence of God. His own positivist commitments sharpened rather than diminished his sense of the metaphysical: examining the conditions and therefore the limits of knowledge led to an interest in the transcendental itself. In this way, for Mazhar, Darwin brought back rather than expelled one from the garden of creation. To do this, Mazhar armed himself with positivist arguments while simultaneously stressing the limits of human knowledge, as did many of that generation of what we can only call transcendental positivists.[43]

This emerged, for instance, at the very end of his introduction in his discussion of the question of the implications of evolution for the origin of life. Mazhar divided contemporary scientific opinion on the subject into three broad camps. Among the first he classed those who believed in special creation, citing Agassiz and Pasteur and declaring their motto to be "Omne vivum ex vivo." The second included those who, like Hermann Richter, postulated that life in the universe has existed and will exist for all eternity, propagating from place to place by means of living germs in the cosmos; their view can be summarized by the motto "Omne vivum ab aeternitate e cellula."[44] And, finally, the third group included those materialists who professed a belief in spontaneous generation, men like Bastian and Haeckel, whom Arab evolutionists like Shumayyil followed, or so Mazhar claimed. Clearly, Mazhar did not think very highly of any of these schools of thought: he felt it was not the job of science to even pose the question of where life had come from.[45] He added that while we might over time uncover some of the secrets of life, we would never uncover them all, for behind each discovery lay yet another "secret." And here he directed his readers to a verse from the Qur'an (17:85): "And they ask you about the soul. Say: The soul is one of the commands of my Lord, and you are not given aught but a little knowledge of it." For Mazhar, this was a reminder of the limits of human intelligence and these conjectures were thus all merely useless and unfounded disputes—or what he described as, ultimately, purely "arbitrary" (*ma anzala ilahu bihi min sultan*, or more literally, "God hath revealed no legitimation for it").[46]

Mazhar concluded by arguing that the real disagreement lay primarily between materialists (*al-maddiyun*; or those who profess a belief in spontaneous generation despite having no experimental evidence for it) and causalists (*al-ʿillayun*; or those who believe in a first cause: *ʿilla awla*).[47] Once again, his very language harkened to past philosophical categories, in this case reviving an Arabic Aristotelian vocabulary.

Mazhar also pointed out to his readers that though Darwin became increasingly skeptical regarding the question of God's existence—which people had too hastily used to label him an atheist—he had maintained a faith in a first cause and had similarly acknowledged the limits of man's intellect. Mazhar thus presented Darwin as a "causalist" and not a materialist. Indeed, materialism and positivism were de facto incompatible: one could not square a belief in a positive scientific method, or positivism—the only scientific method, if not philosophy, Mazhar recognized as valid—with materialism and its ultimately metaphysical claims about the origin of life. For him, positivism demanded and did not merely accommodate, still less challenge, a transcendental view of life and its origins.

In other respects, Mazhar certainly took a classically positivist view of the proper relation of science to theology. He referred frequently to Comte's philosophy, describing it as the "greatest discovery in human thought."[48] In this he followed *Al-Muqtataf,* which had been among the first to use the term "positivism" in Arabic and was, by 1898, defining it as knowledge "established on reason and its proofs," as distinct from materialism, Epicureanism, and all those schools of thought that "rely on sensation."[49] As this suggests, by the turn of the century, Arab thinkers, as intellectuals elsewhere, were moving gradually away from many of the tenets of classical positivism. Idealism, phenomenalism, and pan-psychism were some of the alternatives that found adherents.[50] They reacted against the exclusion of metaphysics in particular: many wanted to bring it back through an appeal to transcendental logic, reviving traditional logical and cosmological arguments in the process. They shared with positivists of previous generations an interest in the criteria for knowledge-claims, only they saw little tension between the kinds of claims mandated by classical Muslim learning and those of contemporary Western science: increasingly, the point was to bring the two together.

In this way, Arabic writers in the early twentieth century moved toward some kind of transcendental positivism. For many of them, Herbert Spencer's ideas on the limits of knowledge and his transcendental logic proved useful, and Mazhar was no exception. One of the most systematic of the early Arabic transcendental positivists, he took an essentially positive, epistemological approach to metaphysical questions. Later, he would turn more directly to what we might label transcendental materialism and showed a more explicit interest in the merging of matter and mind, as his translation of Edmund Synott's *Life of Matter* demonstrates. It was a version of transcendental materialism that was inspired as much by Neoplatonism as it was by modern hylozoism.

The intellectual space for Mazhar's particular form of transcendentalism was shrinking, however. He might write vociferously against materialism. Yet the "modern school of evolution and progress" was by the interwar years increasingly losing favor, and it was becoming harder and harder to divest ideas of "evolution and progress" of their associations with materialism—and atheism. Mazhar spent much of the 1920s and 1930s trying to battle the current, and he was not helped by books such as Isma'il Adham's highly controversial 1937 *La madha ana mulhid* (Why I am an apostate), which cites such authors as Darwin, Huxley, Haeckel, and Büchner as those who smoothed the author's path to atheism.[51] In fact, both critics and supporters of evolution increasingly identified it with materialism if not atheism, and rationalist reconciliations of evolution and theology appeared passé. Mazhar was perhaps the last of his generation of Arab evolutionists to persist.

Much more common by the 1920s were denunciations of materialism. Some of these were clearly inspired by Afghani; others, by al-Jisr. Take the example of Ibrahim Muhammad Sayhi. He too was part of Mazhar's generation of Arab nationalists attempting to reconstruct an Arab or Muslim "golden age of science" and to reverse the current of civilizational progress from West to East: among his works in this vein are *Science among the Arabs* and *Arab Civilization and Its Impact upon Europe*. But he also wrote a book in 1920 that offered a sharp criticism of evolutionary and hylozoic materialism. The title of the refutation, *Falsafat al-takwin* (which we might translate as "the philosophy of creation"), used a term—*takwin*, meaning "to bring into being"—that was more commonly associated with alchemical treatises on the generation of beings (mineral, plant, or animal, including humans) by elaborate artificial means, mimicking processes in nature and inducing souls to mix with earthly matter. But Sayhi used it here as a cognate for *tawallud*, as in the spontaneous generation of matter, which he described as the basic tenet of the new materialism.

Sayhi charged modern scientific naturalists with "corrupting the country's youth" with their "atheistic and materialistic delusions": "There appeared in the East during the twentieth century a philosophical movement that sprouted from the atheistic seeds of Western scientists. And it traveled to Egypt and elsewhere in the Middle East by some itinerant philosophers, spreading into the souls of youth . . . and entering their hearts without restraint." This, Sayhi complained, happened in particular to those youth who were sent to Europe to further their studies, who then returned home ignorant of the teachings of the truly beneficial sciences.[52]

Sayhi himself cited standard English school textbooks as well as Arabic literature on the subject.[53] Although he did not list Shumayyil's translation of Büchner, he did include several journals that Shumayyil contributed to, including *Al-Muqtataf* and *al-Hilal*, and his descriptions of those who believed in the creation and evolution of matter closely resembled Shumayyil's views, or, rather, those of Shumayyil's avatar as portrayed by Husayn al-Jisr's evolutionary materialist. Indeed, Sayhi was clearly influenced by al-Jisr, whom he cites directly at several points in the text. But he also appealed to such figures as Bacon, Newton, Spencer, and Thompson to arrive at what he classed as the logical and necessary conclusion that the world has a First Cause.[54] These are familiar arguments, as we have seen; indeed al-Jisr was not the only ʿ*alim* to have influenced Sayhi. Sayhi's text is suffused with Muslim cosmological and theological themes, and he borrows from the arguments of the Ashʿara to argue that an infinite regress of temporal events was impossible and hence that a First Cause, Prime Mover, or final creator was logically necessary.

For Sayhi, materialism's loss was clearly Islam's gain: referring and quoting liberally from the Qurʾan, Sayhi repeatedly returned to natural theological arguments and saw science as merely illustrating the truths of revelation: citing the verse on the "formation of the heavens and earth," for instance, he confirmed the truth of contemporary ideas on the formation of gases, nebulae, other heavenly bodies, and the earth. The verse "we have given you but little knowledge" provided the decisive argument for those who would confess to our persistent ignorance of the mystery behind the origin of matter and life. According to him, contemporary materialists, including those who would follow Darwin as "their village chief" (*umda*), missed the true point of science. This was Sayhi's version of a transcendental positivism: antimaterialist and anti-Darwin, it was also increasingly bound up with a new search for the justification of a specifically Muslim metaphysics.

Appealing to a new Muslim cosmogony, Sayhi especially wanted to bring back both metaphysics and classical Muslim cosmological theories into the study of the contemporary natural sciences. This would allow positive natural science a reasoned faith in a Prime Mover while reviving older Arabic philosophical and theological arguments about creation. Clearly, he thought that only the ʿ*ulama*, the teachers and guardians of Islam, were equipped to do this: this was another reason why he complained so bitterly of those youth of the nation who went to Europe to be educated by the wrong kind of ʿ*ulama*.

The following year, another Muslim ʿ*alim* whom Sayhi also cited favorably, Muhammad Farid Wajdi, published his own extensive critique of materialism. This was his four-volume study ʿ*Ala atlal al-madhhab al-maddi*

(On the refutation of materialism). Wajdi had begun his career in publishing after a dream he had in which he saw himself preaching Islam to foreigners, and much of his work was concerned with the defense of Islam in modern times. He wrote against a number of Orientalists, including Hanotaux (as Muhammad ʿAbduh had before him), as well as against prominent Egyptian intellectuals, like Qasim Amin and Taha Husayn, whose views on women and Islam and on early Muslim sources respectively he objected to. (Qasim Amin was himself an early critic of Wajdi's and accused him of mixing arguments of science with those of religion, a strategy that Amin thought had met with failure in the past.) Wajdi naturally dismissed Ismaʿil Adham's defense of atheism. But his main concern was with the dangers of Western materialism, and much of his writings turned ironically to European science and spiritualist studies to make the case against it.[55]

The first volume of *ʿAla atlal al-madhhab al-maddi* offered a potted history of materialist thought, from the ancient Greeks to nineteenth-century France and England, covering such figures as Robinet, Lamarck, and Darwin. There was one very revealing difference from the kind of antimaterialist genealogies constructed earlier by Afghani: this time, Eastern and errant Muslim materialists such as the Mazdaists, Zoroastrians, and the *batiniya* (Ismaʿilis) were left out; materialism was increasingly seen by nationalist intellectuals of this generation as a uniquely "Western disease." Wajdi ends with Darwin and claims that his theory does not support a purely materialistic vision of the world. Then, with a final flourish he claims that a verified citing of an apparition in 1846 America provided decisive proof in favor of the scientific reality of the soul.

The second volume picks up on this theme. Republishing articles and letters that Wajdi had sent to *Al-Muqtataf* earlier in response to the journal's dismissal of mediums as charlatans, he cites numerous books that show the contrary. Experimental psychic research from Europe provides the bulk of the evidence for the rest of the book. The two other volumes carry partial translations of Jean Vinnault's *L'âme est immortelle* (which Wajdi had also previously published in his own journal, *Wajdiyat*) and Camille Flammarion's *La mort et son mystère*. Exploring such supernatural themes as telepathy, hypnosis, séances, and other paranormal phenomena, he argued that modern spiritualist research was providing a final and decisive argument in favor of the existence of the soul, indeed confirmation of the truth of prophecy and revelation itself.

In a kind of reversal of the positivist hierarchy of knowledge, Wajdi argued that a new "spiritual era" was dawning in both the West and the East.

Centuries' old materialistic doctrines had finally been conclusively refuted by spiritualist science, thereby ending the epistemological anarchism (Wajdi linked this with modern communism and atheism) that had seemed so appealing only a generation ago. He saw this new era as the final stage of human knowledge. It followed from both the natural state of man's religious instincts (*fitra*) and the metaphysical, philosophical, and scientific evidence of experience more generally.

He tied this critique of Western materialism to the theory of evolution and to Darwin. He repeatedly complained about the number of Arab evolutionists in his time who continued to follow this school of thought despite the fact that many Western scientists were themselves now rejecting the theory, citing August Weismann's rejection of Darwin's theory of pangenesis in support. He may have even had Mazhar in mind here: as we will see later, Wajdi and Mazhar were at loggerheads on many issues of metaphysics and belief. Wajdi focused his criticism of evolution in particular on the idea of "missing links"—echoing contemporary Christian and other creationist discourses. Yet he also focused on the more general and traditional question of a First Cause. This was reminiscent of Mazhar, but unlike his contemporary, Wajdi used his own particular breed of scientific skepticism to point out the uncertainties of both materialist and evolutionary beliefs, attempting to replace them with his own version of a Muslim empiriospiritism.

These were ideas that Wajdi would continue to return to over the next two decades until his death in 1954. It was revealing that his four-volume work proved an instant hit: three editions appeared in ten years, and it was frequently lauded by others who were similarly in search of a new spiritualism.[56] In the meantime, Wajdi edited the journal *Nur al-Islam* (from 1936), which later became *Majallat al-Azhar* and was a mouthpiece for a new generation of ʿ*ulama* at Al-Azhar and an increasingly vocal critic of the dangers of "Western materialism." There were articles on the corruption of materialist thought, on scientific proof for the existence of the soul, and diatribes against the worship of natural laws. Once again, Wajdi turned to contemporary psychic research, translating the ideas of Ernesto Bozzano, Gustave Geley, and Andre Cresson. In short, Wajdi was part of the new generation of ʿ*ulama*—one that would prove to be increasingly influential over the course of the twentieth century—who saw the combination of metaphysics and modern science itself as providing the best defense of traditional forms of knowledge and belief.[57]

Mazhar had already tried engaging with these antimaterialist critics of evolution. In his 1923 *Madhhab al-nushu' wa-al-irtiqa'* (The theory of evolution and progress) and the expanded 1926 edition, *Malqa al-sabil fi madhhab al-nushu' wa-al-irtiqa'* (On the crossroads of the theory of evolution and progress), he sought to demonstrate that contrary to the common misperception held by many, modern evolution was by no means a materialistic doctrine.[58] He felt that Afghani in particular had misunderstood the term "materialism" and had confused it with the doctrine of the eternity of matter and with the problem of atheism. Dismissing Afghani's accusation that Darwin was a materialist, he doubted openly if Afghani ever actually read any of the British naturalist's writings.

But Mazhar was just as little impressed with the many other Arab authors who professed themselves to be materialists. Public confessions like those of Adham, and popular writings like those of Shumayyil, were in fact interfering with people's understandings of Darwin and of the theory of evolution itself, or so he argued. Doubting if Shumayyil too had ever read Darwin, he charged that the Arab materialist's views on spontaneous generation, in particular, were lacking all scientific evidence. In his attack against religion, Mazhar saw Shumayyil as merely attempting to replace one religion with another, namely the "worship of matter."[59]

Mazhar insisted that the theory of evolution was not concerned with the ultimate nature of things (*ma hayat*), and he spent a considerable amount of time criticizing Shumayyil's use of concepts like "force" and "matter," which he classed as abstract and fundamentally unempirical—and therefore, ultimately, "unknowable."[60] Like Herbert Spencer (a figure much cited in this text), Mazhar felt that a belief in the Unknowable permitted him to sidestep any metaphysical commitment concerning the nature of reality and to avoid subscribing to "materialism, determinism, or to any 'ism' that implied knowledge of reality as it exists apart from consciousness." The true nature of space, time, matter, motion, and force were, in Spencer's words, "absolutely incomprehensible." "Ultimate Scientific Ideas, then, are all representations of realities that cannot be comprehended," and the man of science "truly knows that in its ultimate nature nothing can be known." The "reality existing behind all appearances is, and must ever be," for Spencer, "unknown."[61]

Like *Al-Muqtataf* before him and like many late Victorian popularizers who "busily exaggerated theistic themes" found in Spencer and others[62] (one might also cite John Beattie Crozier, whose *Civilization and Progress* Mazhar summarized in his 1926 work), Mazhar argued that there was room for belief in a deity within a positive philosophy. Indeed, Mazhar followed Crozier by outlining several examples of principles that positive science must believe in though it cannot scientifically verify them: the existence of a world outside ourselves, the persistence of force, and a belief in scientific causation. Crozier's conclusion, which Mazhar followed, was: "If then, the main foundation stones on which our intelligence is built must be believed, although they cannot be known by Science, Comte's first great argument against the existence of a Deity, viz., that He cannot be known by Science, falls to the ground."[63] Mazhar took a similar line regarding the question of ultimate truths and the existence of a "Greater Being."[64]

In the end, however, Mazhar's sympathies were made very clear: he thought that only evolution—and not the doctrine of independent or special creation, or *al-qawl bi-khalq al-mustaqill,* as he put it—could explain and bring together the evidence available from the extant fossil and geological record as well as from the new disciplines of comparative anatomy, embryology, and ethnology. Only evolution could make sense of the distribution, diversification, adaptation, and finally the extinction of flora and fauna. He thus felt that the scientific evidence in favor of the evolution of humans from some lower form—particularly as presented in Darwin's *Descent of Man*—was irrefutable. The point was that this did not threaten religious faith in any way: after all, Wallace, he reassured his readers, believed in this fact and held on to his faith in God.[65]

The Evolution of Belief

For Mazhar, evolution was ultimately about the progress of man's moral or ethical life. He no doubt took Darwin's lead on this, as Darwin argued much the same in his *Descent of Man*. Yet Mazhar also gained inspiration elsewhere. Shumayyil had been keen to promote the idea of the social inventedness of religious systems. Evolution, on this reading, could undermine the foundations of faith by showing its history. Mazhar disagreed. It was not the origin of religion per se that interested Mazhar but the evolution of its social function.

Belief in the soul, argued Mazhar, had had a profound effect on the evolution of groups and the "progress of civil systems." Studying the evolution of religion, he argued, could help to shed light on how specific religious tenets and traditions had helped to advance particular peoples at particular times. It was this social function of religions that one needed therefore to investigate. But this did not involve the attempt to evaluate the truth or error of specific beliefs. Science had no competence to do that. (Hence, one could also argue that it in no way threatened a belief in divine inspiration or even in revelation itself.) Science merely explores "the effect of established beliefs on the creation of social systems"; as for matters of belief, they are best "left to theological scholars and the like." After all, evolutionists "do not concern themselves with the origin of life or man's fate after death."[66] As for the social impact of religion, Mazhar—borrowing from the work of Benjamin Kidd and others—argued that this helped to foster a spirit of cooperation and an ethic of mutual aid and brotherly love, traits he felt were necessary for the proper "social evolution" of any peoples or polity.[67]

Like others of his generation, Mazhar stressed the need for a more collectivist view of evolution. In 1923, for instance, he argued that cooperation—and not competition or struggle—was the greater natural law, and citing Kropotkin on "mutual aid," he claimed that "all animals and primitive and civilized men follow this law."[68] In the later *Malqa al-sabil*, he used the same author's "law of mutual aid" to argue that altruism was the origin of such instincts in animals as care of the young and herding. It was also, he argued, the source of man's highest ideals such as parenthood, brotherhood, and love and thus the very origin of ethics and morality itself. And like Edvard Westermarck and Leonard T. Hobhouse, Mazhar felt that the "highest ethics is that which express the most complete mutual sympathy and the most highly evolved society, that [in] which the efforts of its members are most completely coordinated to common ends."[69] But these same ideals of fraternity and group cohesion, Mazhar warned, could also lead to antagonism and excessive self-preservation (e.g., tribalism or even patriotism). Emphasizing his ecumenical internationalism, he speculated that in future men might one day extend their sympathies to all other men, regardless of race, nation, or origin.[70] Indeed, citing Darwin, Mazhar even left open the possibility of "acquired characteristics" in the further perpetuation and perfection of moral sentiments from one generation to the next.[71]

In 1928 Mazhar published a translation of an article by the biologist and self-proclaimed agnostic J. B. S. Haldane that neatly summarized these views. Mazhar presented what was in fact a very loose translation of Haldane's essay, removing certain references such as the bulk of Haldane's dispute with Dean Inge on eugenics (he was himself highly opposed to eugenicist principles) and highlighting instead Haldane's more general comments on morality and God.[72] Discussing the possible existence of a Comtean "Great Being," he focused in particular on Haldane's views on the possibility of a "super-human" consciousness that was neither "external to man, [nor] even existing apart from human co-operation." Haldane concluded that "if the Great Being is wholly independent of individual men, their well-being must be disregarded in its service." And, added Mazhar, we must consider "that their life is its life . . . and their rights, responsibilities, and happiness are its rights, responsibilities, and happiness."[73] For Mazhar, then, the laws of evolution, applied to the social and intellectual progress of faith, meant that it was conceivable that the progress of knowledge itself might also lead the way toward a new, more rational, and scientifically determined basis for both belief and ethical life itself. But while Mazhar was happy to present the case for a scientific intimation at such transcendental truths, he was less keen to encourage a full program for the reconciliation of science with specific religious beliefs in the Muslim world.

Events in Egypt in the mid-1920s made these issues of more than academic importance and raised more sharply than before the issue of the roles of religion and freedom of expression in public life. The ʿAbd al-Raziq affair, which erupted in 1925 (to be discussed below), brought to light the problem of how to interpret the 1923 constitution's protection of free speech and highlighted the question of the *ʿulama*'s role in public life. These issues drew the attention of many evolutionists: Salama Musa compared the ʿAbd al-Raziq affair with the Scopes trial in the United States. There was another public controversy when the well-known writer and university professor Taha Husayn published a study of pre-Islamic Arab poetry that challenged traditional interpretations of the Qurʾan, describing certain stories in it as myths and arguing for a new kind of literary criticism of scripture. His critics, who deemed him an apostate, demanded his dismissal from the Egyptian University, arguing that the state had an obligation to protect Islam as its official faith.

Mazhar, like many others in his circle, was greatly dismayed by these developments, which many felt called into question the very idea of an Arab Renaissance (Nahda).[74] With these events clearly in mind, Mazhar published an article in 1928 in which he defended "freedom of thought" (*hurriyat al-fikr*), a.k.a. rationalism (Mazhar listed "rationalism" in English beside the Arabic title). He argued that both education and the judiciary should be free from religious interference. But he was equally wary of the politicians: only three years after the Egyptian University had become a state institution, Mazhar saw the Husayn affair as an illustration of the dangers not only of the meddling of the *ʿulama* but also of the involvement of Parliament itself in curricular matters, and quoting Andrew Dickson White he recalled the papacy's attacks on Galileo and Newton. If left unchecked, he warned, such arbitrary meddling in academic affairs would lead to backwardness and oppression. Alarmingly, Parliament had turned against the forces for change and pandered to the ignorant. Mazhar invoked Bertrand Russell's attack on political intolerance and even criticized the ongoing persecution of Bolsheviks ("although this is not to defend them, because they are extremists") as against freedom of thought. Turning to Egypt, he explicitly referred to "prosecutors and shaykhs in our time" who were doing the same thing.[75]

The following year, Mazhar published his highly abridged translation of Andrew Dickson White's *History of the Warfare of Science with Theology in Christendom*. Introducing this work, Mazhar accepted that religious sentiments were both an instinct in men and a necessary feature of social life. As such, like Crozier, he criticized those who tried to argue in favor of a future "natural religion" based solely on the principles of reason. Religion was a form of "reason beyond reason," serving as the glue that tied together individuals and formed the basis of collective life, for the application of reason alone could not lead individuals to act selflessly and for the good of the collective whole: individuals had one rationale, and collectives had another.[76] Progress and *jummud* (stagnation) existed in a dynamic relationship with one another, and if societies were to evolve, they had to do so gradually in order to avoid chaos. In this sense, Mazhar was also reversing the claims of earlier modernists like Muhammad ʿAbduh and the editors of *Al-Muqtataf*, who saw progress and retrogression or stagnation as opposed.

Yet Mazhar also clearly felt that true faith was acquired through reasoned deliberation. This was why he sometimes referred to "philosophy" as that which brought together the contradictory, though necessary, impulses of religion and science. It was reasoned and metaphysical speculation, he thought, rather than mere blind faith or imitation (*taqlid*) that should guide the "enlightened" classes (*al-tabaqat al-mustanira*) in matters of faith. If reason was what distinguished man from the other animals, the willingness of individuals to challenge accepted beliefs was what accounted for progress toward truth.[77]

Confining himself to the first three chapters of White's volume, Mazhar reversed their original order and began with the trial of Galileo and ended with the coming of Darwin as the triumph of evolution over theology. Mazhar began his introduction to White by saying, "Many people say these days that science and religion are opposed to one another." And he agreed that this idea reflected a certain reading of history. Yet over the centuries, neither had managed to eliminate the other, for both were natural "instincts" (*fitrat*) of human social life.[78] There was, he stressed, no struggle between them since each possessed its own specific "intellectual formation." Even if they might sometimes be thought to possess contradictory impulses and divergent methodologies, it was the nature of human social life to balance opposing forces—such as self-interest and altruism—much as it was the nature of the material world to be subject to forces of both repulsion and attraction, for example.

Insofar as conflict existed, it was—as White's book argued—between science and theology, not religion as such.[79] This was because theologians had wanted religion to serve as a kind of educational compendium for all natural and metaphysical knowledge in the world. But the true function of religion, he argued, was guidance (*al-irshad wa-al-huda*), not education. It provided, as he put it, "the truth in but one of its forms," for truth has many forms and could not be constrained by *taqlid*. The implications for the Husayn case—and for Egypt's future—were clear: the universities should be bastions of independent thought, free from interference either by the *ʿulama* or by politicians. Were they to fail in this mission, the results would be felt not only by their students but by the country as a whole, which would be condemned to remain in "stagnation."[80]

East and West

The geopolitical stakes had been transformed in particular by the success of the Turkish Republic under Atatürk; Mazhar, like many of his class, was a keen supporter: "No contemporary, cultured mind," he wrote, "can deny that the [Turkish revolution] contains the elements of truth and power for the general constitution of a reformed mentality in the entire East."[81] In the civilizational terms that were so popular at the time, Kemal's achievement was to demonstrate how a previously "Eastern people" could first emancipate itself from its own yoke and then join "Western civilization."

At the same time, the question of the public role of religion was attracting enormous attention in Egypt. As its new intelligentsia and its *ʿulama* coped with the collapse of Ottoman authority and the slow withdrawal of British control, they debated the place of religion within the nation. Mazhar argued that the Turkish revolution, in separating statecraft from traditional theology, was a step in the right direction in that it overcame the "Asiatic" mentality (*al-ʿaqliya al-asiyawiya*) and introduced a modern, civilized, and European one in its place.[82]

It was when reporting on a recently published book on Mustafa Kemal (written in Turkish) that Mazhar claimed that "the happiest people" in the world are in the West. By contrast, Eastern peoples—whether they follow Buddha, Confucius, Brahma, Abraham, Moses, or Muhammad—have denied themselves the right to happiness in this world since they see life itself as temporary and only life after death as eternal. Mazhar thus identified the reason for Eastern misery as their adherence to religious laws and social-spiritual conformism or imitation (once again, *taqlid*, and again, used outside its original juridical connotation). The difference between East and West, in short, was that the former was still bound to the old bonds of religious authority, whereas the latter had advanced to a positive and, above all, rational state of cultural, social, political, and even moral life.

In 1929 he published *Wathbat al-sharq* (Advance of the East), and subtitled it *A Study on the Contemporary Turkish Mentality Which Is an Example of a Sound Mentality Which the East Should Adopt so as to Follow along the Path of International Civilization*. In it he summarized three books, or what he dubbed "three schools of thought," on recent events in Turkey.[83] Atatürk's reforms were being followed closely in the Egyptian press at this time. The pace of change had been dizzying: the sultanate was abolished in 1922 and the caliphate two years later; the Fez and Sufi orders and lodges were prohibited in 1925, and a new civil code was introduced the following year. Two years later a new Roman alphabet replaced the older Arabic, all religious terms were removed from the constitution, and attempts at reforming religious services and liturgical practices (such as installing "pews" in mosques) followed shortly after that.

Many Egyptian writers were shocked by these measures and troubled by their implications for Egypt itself; others saw in them an opportunity and a model.[84] The question of the caliphate in particular attracted a great deal of controversy. The Turks' abolition of the position was largely condemned, and one consequence was that a debate began, fueled in large part by the palace and sympathetic *ʿulama*, over whether the caliphate should be invested in the Egyptian royal house instead. It was in this context in particular that the 1925 publication of ʿAli ʿAbd al-Raziq's book *Al-Islam wa-usul al-hukm* (Islam and the foundations of governance) became a cause célèbre. Influenced by events in Turkey, ʿAbd al-Raziq argued that the caliphate had never been a part of the early legal foundations of Islam and, indeed, that the shariʿa implied no particular form of politics. He was brought to trial by the Higher Council of the *ʿulama* and dismissed from his position as a *qadi*.

Mazhar, in his analysis of the situation in Turkey, criticized ʿAbd al-Raziq's interpretation. Although he too opposed any revival of the caliphate, he thought a discussion of its origins was beside the point. The caliphate had emerged over time—and here he cites the German Orientalist Carl Becker—to become a recognized institution, but it had ceased to correspond to what he called "the spirit of the times." It had thus become one of the factors retarding the East's potential for progress. If the Muslim East had not followed this path, he thought, it would have advanced as a civilization along European lines rather than remaining, as it were, arrested at the metaphysical stage and mired in an Asiatic mentality.[85]

His notion of an Asiatic mentality was connected to an evolutionary (and Comtean) view of the universal development of the human mind. For Mazhar, the "civilized mind" was the final stage in an evolutionary process, after the "animal mind," the "infantile mind," and then the "savage mind" had developed. These were marked by "instinctive," "theological," "metaphysical," and "positive" mentalities, and each needed to be researched through its appropriate science. So far as the civilized mind was concerned, it could be researched through history, sociology, and social psychology (*ʿulum al-tarikh wa-al-ijtimaʿ wa-ilm al-nafs al-ijtimaʿi*).[86] His eyes set on an Arab future, Mazhar wanted Arabs to abandon the purely metaphysical deliberations of their predecessors and embrace a positivist philosophy, albeit with transcendental overtones, alongside a modern education that would pave the way toward a "scientific future."

Mazhar's evolutionary view of civilization was by no means an uncommon theme of the time. The first few decades of the twentieth century, particularly in Egypt but also in other parts of North Africa and western Asia more generally, witnessed the rise of a renewed interest in questions of "civilizational" or "racial" ascent and decline—both highly ambiguous terms that gained new meanings in the era of high nationalist sentiments and party politicking. The vogue for Gustave Le Bon is a good example of this engagement.[87] Ahmed Fathi Zaglul translated Le Bon's *Lois psychologiques de l'évolution des peuples* as the highly popular *Sirr tatawwur al-umam* (Secret of evolution of nations) in 1913 (a second edition was issued in 1921), and in 1924 Sadiq Rustum translated the section on ancient Egypt of *Les premières civilisations*.[88]

These translations fed the growing concern with questions of national character, racial unity, and collective identity.[89] Mazhar's own perspective was an interesting blend: on the one hand, it was strongly Comtean and it was a constant theme of his that the "civilization of the East" had still not moved beyond Comte's "metaphysical stage" (*al-daraja al-thaniya aw al-uslub al-ghaybi*) and that the Arab world had not yet produced any coherent philosophical school or effective political ideology. On the other hand, he placed Egypt in the "Asiatic" and "Semitic" worlds, eschewing the contemporary debate, popularized by people like Salama Musa, on Pharaonism and civilizational (and racial) connections between ancient Egypt and the West. Mazhar's conception of the "East" was much broader than that of many other Egyptian intellectuals, for it encompassed not only the Arabs and Muslims in general but a broad geography that was east of Europe and included numerous faiths and nations.

This was far from the mainstream of so-called Easternism in interwar Egypt, a broad-ranging movement that during the 1920s represented the principal intellectual alternative to the largely pro-Western elite. The Society of the Eastern Bond had been founded in 1921 to promote contact among "Eastern peoples." In the hands of Rashid Rida and the *Al-Manar* group, in particular, this took on a pan-Islamic dimension. It was also one of the key movements in support of the revival of the caliphate, and to help the society, they revived the memory of Jamal al-Din al-Afghani through lectures and conferences.[90]

Indeed, it was in 1926, the year after Rida had accused ʿAbd al-Raziq of atheism, that Mazhar had singled out Afghani for the criticism described earlier: Afghani's misunderstanding of modern philosophical schools and scientific methods betrayed a deeper problem with his political program. In an article in *Al-Muqtataf*, Mazhar argued that he had merely inherited the metaphysical concerns of his Arab predecessors and committed what for Mazhar was always the cardinal sin: stagnation of thought. "Afghani," he wrote, "learned the ancient methods of scientific thought that the Arabs have held on to since the Middle Ages . . . and his political tendencies are like those of a fossilized skeleton whose body still lives among us." And that, he concluded, was "the general tendency of Arabs in the nineteenth century," for there was "no difference between him and other religious leaders of the time" save for the fact that "he used the power of religion as a means to affect politics." Indeed, Mazhar argued that Afghani's polemical works were positively harmful to the development of scientific thought in the Arab world in that they confused hard, positive facts with metaphysical speculations and uninformed commentary. In short, he was no role model for Eastern revival.[91]

This scathing critique led two of *Al-Muqtataf*'s readers—Amin al-Khuli, the head of the Egyptian legation in Berlin, and Mustafa al-Shihabi, a prominent Muslim scholar in Syria—to send in letters in protest. They defended Afghani in particular—and Arabic thought and civilization more generally. Al-Khuli was particularly incensed by Mazhar's outspoken criticism of Afghani, whom he felt he had badly mischaracterized. As for al-Shihabi, he argued that Arab philosophy was not merely concerned with metaphysical matters. Both men made the now familiar point that Arab thought had actually contributed to the development of Western scientific thought itself. Mazhar was unfazed and stood by his position that Arab philosophy was by and large arrested at the metaphysical stage.[92]

Popularization and Politics

A member of the new *effendiya* class, Mazhar was deeply enmeshed in a network of politicians and journalists throughout the 1920s and 1930s, a particularly volatile period of party factionalism and constitutional debate. He was a member of the writer Mayy Ziadeh's literary salon, which in the 1920s included such well-known public figures as Mahmud ʿAbbas al-Aqqad, Muhammad Husayn Haykal, Ahmad Shawqi, Khalil Mutran, Dawud Barakat, and Hafiz Ibrahim.[93] Together, they constituted a powerful group: Mutran had helped to found the Egyptian National Bank, Bank Misr; Shawqi was a member of the Senate; Ibrahim was, like Barakat, an editor of *Al-Ahram*; Mutran, Shawqi, Ibrahim, and Aqqad were also writers of considerable renown. This was not a group united on party lines, however: Aqqad had been the editor of the leading Wafd Party paper, *Al-Balagh*, while Haykal helped to found the splinter Liberal Constitutional Party in the early 1920s and later served as minister of education for several cabinets.

At this apogee of his career, Mazhar found himself straddling the poles between popularization and elite discretion. He was also part of the generation caught between the rise of the professional, laboratory scientist—whose writings on evolution increasingly made their way to the press and to government schools by the early 1940s—and former science popularizers like *Al-Muqtataf*'s early founders. As we will see later, Mazhar's brand of refined scientific popularization would find itself increasingly eclipsed by the rise of another tradition of populist science writing in Arabic, which would do much more to shape the style of science writing for the rest of the twentieth century.

It was at exactly this time that Mazhar turned his sights on a new effort at disseminating science, one with an avowed political program at heart. In 1927 he founded a monthly magazine, *Al-ʿUsur* (The Times), where he presented the political implications of his ideas and engaged with nationalist politics for the first time. (A few years later he also became a member of the Egyptian Academy for Scientific Culture, which Fuʿad Sarruf and Salama Musa formed in 1930 with the express aim of propagating "scientific culture among the people at large.")[94]

For Mazhar, this new "critical journal of arts, sciences, and politics," published by the *Muqtataf* press, would fulfill five main political and literary objectives: to fight for the independence of the Nile Valley within its natural boundaries; to help to reform the parliamentary system for the sake of the nation; to promote absolute freedom of speech and critical inquiry for the sake of arts and letters; to provide a forum for independent opinions that was free from any specific allegiances; and, finally, to work toward spreading "general knowledge."[95] On this latter point, not everyone was so confident. The editors of *Al-Muqtataf*, despite supporting him, worried that the journal was too much "like any European or American journal of the day." By this they seem to have meant that Mazhar was too engaged with European intellectual debates and overlooked the needs and tastes of local readers. Certainly, the tone of his journal was at times very different from that of the popularizing mode of Salama Musa or Al-Muqtataf itself; he had a tendency to wax philosophical and was at times rather cryptic. Some of this divergence in approach may be gleaned from the repeated criticisms of Musa that appeared in *Al-ʿUsur*: Musa's writing style was written off for its inelegance and colloquialisms, and even the depth of his knowledge of science was sometimes challenged.[96]

Mazhar made a considerable effort to cover a wide variety of topics of general interest. The journal included popular pieces entitled "Man and the Universe" and "The Relation between God and Man"; but it also included rather more demanding discussions, as evidenced by the article entitled "The Philosophy of Relativism, or Relativism and Metaphysics." The journal's contributors included a number of well-known intellectuals, poets, and literati, such as Ahmad Shawqi and ʿAssam al-Din Hifni Nasif. His close friend Ahmad Zaki Abu Shadi was perhaps the most striking example of this new class of "Renaissance men" who brought together science, literature, and the arts for the sake of cultural revival. Abu Shadi, a poet and naturalist, became chair of bacteriology at the University of Alexandria in 1942. Before this, he had also started his own society and journal on beekeeping in Egypt, called *Mamlakat al-Nahl* (The bee kingdom), which was a monthly review of "modern bee culture." He was also a keen translator and often participated in debates, along with Mazhar, on the most appropriate literary style for Arabic translations. Mazhar's journal also published many anonymous and pseudonymous writers, such as Filawpuns (a play on philo-*penser* perhaps), some of whom were probably disguised versions of Mazhar himself—not surprisingly, since the journal was essentially a vehicle for his own particular vision. Even its backer, *Al-Muqtataf*, commented in its review, after the first two issues of the journal had appeared, that "the journal is a picture of its founder's views."[97]

Nearly every issue returned to Mazhar's idées fixes: evolution and science, the need for freedom of expression and thought, the "release from imitation" in the Arab world, and, above all, his deeply positivistic philosophy. He made it clear that he hoped that his journal would serve as a vehicle for diffusing the arts and sciences of Europe—and "Anglo-Saxon culture" in particular—to the Arab world. This was a theme he returned to again and again. But his promotion of social and political reforms along the "Anglo-Saxon" model and his technocratic and elitist ideas were quickly falling out of public favor, while the journal's attacks on what it dubbed the "reactionary" class at Al-Azhar aroused condemnation from the *ʿulama*.

In one article, signed by "an Azhari freethinker" who may very well have been Mazhar himself, the journal described the mosque-college as still arrested in the "mentality of the Middle Ages," in which adherence to the letter of the law replaced the exercise of reason.[98] We should note that at exactly this time a former student of ʿAbduh's, Shaykh Mustafa al-Maraghi, was attempting to push through a set of reforms similar to those his mentor had proposed and meeting with no less hostility than ʿAbduh had. The article had a pessimistic prognosis. Acknowledging the effort at pedagogic reform, based on the "new elements" that ʿAbduh had begun, it also reminded readers of what had happened to ʿAbduh himself: ʿAbduh's educational reform efforts made him an unbeliever (*kafir*) in the eyes of some Azharis, who predicted he would end up in the "seventh level of hell."

Accepting, as Mazhar had always done, that religion and rationalism both had their place, the author was nevertheless gloomy about Al-Azhar's tendency to veer toward reactionary politics. He described it as excessively involved with political life and willing over the centuries to work with whoever was in power—emirs, Napoleon, the khedives, and Saʿad Zaghlul. As for today: "the treasury department has become their God." After Interior Minister Ismaʿil Sidqi Pasha, whom the *ʿulama* at Al-Azhar had denounced as an atheist, had helped to raise their salaries, the *ʿulama* suddenly classed him as a defender of the faith! Only a complete change of mentality, the article concluded, could save Al-Azhar.[99]

Mazhar's journal was highly critical of the *ʿulama*'s outmoded outlook and especially of the increasingly popular genre of *iʿjaz al-Qurʾan*, or the literature on the "miraculousness of the Qurʾan." There was an old tradition of works that emphasized the linguistic or rhetorical inimitability of the book; but in the twentieth century, many authors began to expand this by pointing to interpretations (*tafsir*) that foreshadowed contemporary scientific discoveries and theories. Known as *tafsir ʿilmi*, "scientific exegesis," this was a very different kind of science popularization from that of previous generations and genres. Of course, not all of the *ʿulama* approved of the new trend: men like Rashid Rida and later, the young Sayyid Qutb, felt that there were dangers in tying the eternal Qurʾan to temporal scientific findings. But despite this, the genre had a keen readership from the late 1920s onward.[100] One early example was Mustafa Sadiq al-Rafiʿi's 1926 *Iʿjaz al-Qurʾan*. Mazhar himself reviewed it after the author sent it to him. He used the review to deconstruct the very composition of the work and to concentrate on the genealogy of the idea of the miraculous and of miracles and their place in religious faith. He prefaced his remarks with the statement that he felt he must publicize these ideas for the sake of "attaining the truth" through the "open expression of opinions," even though he thought some people might find such views upsetting.[101] It was characteristic of Mazhar's approach—for reasons of real or self-censorship—that instead of naming his precise target, he widened the discussion to a comparative and larger genealogical analysis. He offered a potted history of accounts of miracles in the "world religions"—a concept that had also recently gained ground in Arabic[102]—and commented on the revealing similarities between accounts of miracles across many faiths, such as the virgin birth of Krishna and of Christ.[103] This also included a long and laudatory digression on the rise of deism in Europe. These tales, he said, emerged as a means of persuading people of the truth of their faith, but because they generated conviction not through reason but through a kind of awe or power, their effect could be compared to the "power of tyranny in politics."

As a rationalist, however, he was categorical: religion should not espouse the idea of miracles. Its main purpose should be confined to defining the relationship between an individual and God; miracles were an irrelevance.[104] He also made one more daring comment: the miraculousness of the Qur'an involved a "relative" rather than an "absolute miracle." The latter went against natural laws—an impossibility that would only harm a true and reasoned faith in God. The Qur'an, however, could be said to be a relative linguistic miracle in much the same way that Dante's *Divine Comedy* was a miracle of the Latin languages (*ahl al-latiniya*).

Mazhar would make similarly outspoken remarks about other attempts by Azhari *ʿulama* to popularize science. When *Nur al-Islam* (Light of Islam)—one of the first journals to be published at Al-Azhar—appeared, the first editor had a keen interest in science and Mazhar was invited to work with him. But their relationship soon turned sour.[105] Indeed, the very first pages of the first issue of the Azhari journal contained a warning against certain journalists—and particularly some of the *ahl al-ʿilm* (men of science)—who were said to be "spreading falsehoods" and whose activities were deemed to be against the best interest of the nation. An article by Shaykh Yusuf al-Dijwi, a noted scholar and spiritualist, spoke in equally harsh terms against people who incorrectly "used science against religion."[106] These atheists, Dijwi claimed, had tricked people into thinking they were speaking of proven, scientific matters when in fact they only trafficked in hypotheses and not certainties.

Mazhar took all this personally and cited a verse from the Qurʾan against what he felt to be an appalling "lack of manners." But he was not slow to offer a more substantive reply. "It seems the good Shaykh al-Dij-wi," retorted Mazhar in an article entitled "Faith in God," "does not know the difference between science, knowledge, and ethics [*al-ʿilm, al-maʿarifa wa-al-adab*]"—pointing out the profound reorientations that the concept of *ʿilm* and allied forms of knowledge had long been undergoing. He noted derisively that though al-Dijwi seemed to know only Arabic, the two men were scarcely speaking the same language. Whereas al-Dijwi offered an argument from within the tradition, dense with Qurʾanic allusions, Mazhar's Arabic was full of allusions to Spinoza, Kant, Voltaire, Comte, Spencer, and Crozier.[107] One wonders what kind of dialogue this really was. Mazhar argued loosely from Kant that one cannot attain knowledge of God as a real or actual existence in and of itself, any more than one can attain knowledge of the external world beyond the mind's comprehension. And he cautioned against confusing scientific with ultimate causes: science concerns itself only with the chain of relations (*tasalsala*) between manifest appearances; therefore, it devises natural laws that are not to be mistaken for actual or real causes. The shaykh's own professed faith in God, he concluded, was therefore really no different from the views, albeit opposing, expressed by materialists: they were both simply relying on assumptions about a First Cause.[108]

If Mazhar was unsympathetic to the direction of the new journal at the start, he would become far more so. In the last issue of Mazhar's own journal, he gave space to an attack on him by Muhammad Farid Wajdi, who would later become the second editor of *Nur al-Islam*. Wajdi's attacks on materialism and his turn toward spiritualism and trends in contemporary psychology were all utilized to bolster the by-now-familiar argument that Islam was the only real "religion of reason" and "a friend of true science." Nothing could have been further from Mazhar's own views, but it was characteristic of the man that he opened his journal to Wajdi's criticisms of him. Wajdi addressed some of Mazhar's charges head-on. He wondered how the editor could express such confused opinions on the conflict between religion and science. Did he not know that Islam was a religion of reason and science and, as Le Bon and others had shown, had contributed to the progress of science? And from the theological viewpoint, Muslims were free to accept anything established with absolute proof by science—even if it required interpretation as a result of an apparent contradiction with the

Qurʾan or *sunna*. Unchanging adherence to one interpretation at all costs, he wrote, and blind imitation in matters of faith were strictly forbidden; Islam requires that men be free in their thoughts and follow reasoned proofs in all matters. But it also advises caution, for as the Qurʾan claimed, man "hath but little knowledge." Scientific knowledge itself, moreover, waxes and wanes as expert opinions change. Therefore, Muslims must accept only those things that are established with absolute certainty.[109]

We can imagine what Mazhar would have replied to this; after all, he had been combating such arguments for some years. On this occasion, however, he was not given the chance. In 1930 critics—some say Azhari students—burned down his press, and the fierce attacks the journal suffered even led one of its contributors, Ibrahim al-Haddad, to leave Cairo for Beirut, where he started a splinter journal—*Al-Duhur* (The ages).[110] Yet it was not only Mazhar's ideas on religion that attracted such controversy: his political ideas too were fiercely criticized.

Mazhar had a rather ambiguous relation to politics. He had wanted his journal to be independent of factional politics and proscribed political opinion. At times, he also seemed to believe that too much political interference into questions of social and educational reform was detrimental to the national cause. Yet, simultaneously, Mazhar was also very much part of the new era of party politics in Egypt. He admired Saʿad Zaghlul in particular: the very first volume of his journal carried a eulogy of Zaghlul, whom it contrasted favorably with Mussolini and De Riviera because he had ruled with popular approval rather than by force. But when he came to write about politics, it was through the filter of natural science, and his analyses remained abstract and couched in the language of evolution. With little interest expressed in the here and now, his gradualism—so characteristic of Arab evolutionists (and socialists, as we saw)—allowed him to engage in ideas of reform while eschewing direct confrontation.[111]

In 1927 he published a slim book entitled *Al-ishtirakiya taʿwuqa al-nuʿa al-insani* (Socialism retards human progress).[112] Borrowing from Benjamin Kidd, Spencer, and others, Mazhar laid out his argument against Bolshevism, which he regarded as an offshoot of socialism, and against sudden, revolutionary change in general. While he believed in change as the motor of progress, he also thought that radical or sudden change led to severe social unrest. Revolutions, like the French and the Bolshevik, did not bring long-lasting change; they simply replaced one tyranny with another: after all, the French returned to despotism under Napoleon, and the system of government that the Russian Revolution brought in was unlikely to last more than a couple of generations.

Once again, translation allowed Mazhar to transfer the lessons of Europe to his home, for this was a warning against revolution in general. The basic problem he saw as the "strangling of nature" by tyrannical governments that ignored the need for gradual social reform; it was their obstinacy—like that of the *ancien regime*—that created the conditions for revolution just as the attempt to suppress trade unions would. At the same time, the principle of absolute equality that socialists preached was also an impossible fiction, and in its attempt to annihilate religions, it eliminated one of the most essential forces for social harmony and cooperation. (This was a very different reading of Benjamin Kidd and Kropotkin from that of the evolutionary socialists.) Socialism was bound to fail, he predicted, so long as it stood against the principles and laws of nature.

He talks about capitalism and economics in equally ambiguous terms. Once again, everything is analyzed in terms of its relationship to "nature." Thus, the principles of capital were themselves based on nature, he thought, for "natural capital" distinguishes men from one another physically and intellectually just as finance capital supposedly does. He accepted that capitalism was one of the causes of "evil in the world"—but not the only one. The real problem lay in human nature: until humanity learned to follow the "laws of the universe," evil would persist and social ills would tempt men to revolution. Mazhar's main recommendation was to educate Egypt's population in the natural sciences—hardly a comprehensive program for political action.

It comes, therefore, as something of a surprise that in 1929 Mazhar became the first to propose an agrarian party in Egypt, publishing a manifesto for this in his journal and later as a pamphlet: "A Project for the Founding of an Agrarian Party in Egypt" was dedicated to the then head of the Wafd Party, Mustafa al-Nahas—whom Mazhar described as "the follower of the great and late leader Saʿad Zaghlul and the inherent protector of true democracy."[113]

Mazhar's Agrarian Party proposal was his most ambitious political project but he conceived of it as being as much a social movement as a political party—"a pillar for the social life of Egypt." (Indeed, he suggested that it should form part of the Wafd Party itself.) Politics, he argued, was susceptible to rapid and sometimes fickle change, whereas society was controlled by the gradual and long-lasting laws of social evolution. Agrarian reforms, moreover, were the most enduring of all social reforms—a view he had long espoused in *Al-ʿUsur*. In fact, as he himself emphasized in the pamphlet, he wanted only to recommend certain "social principles" established "scientifically" and on the basis of "natural laws." He was happy to leave all national and international "political objectives" to the Wafd. His interest in Egyptian politics was thus motivated primarily by his concern for general social reforms along "social scientific" and evolutionary lines and by an increased sense that the political elite needed to act to avert an internal crisis.[114]

Mazhar saw Egypt facing a Malthusian dilemma as the country's population increased and as poverty in the countryside worsened. But, through agrarian reforms based on sound, scientific principles, it could be saved from the chaos and anarchy that afflicted the West and from its susceptibility to Bolshevism. Such reforms, he believed, could lead to social progress according to the "Anglo-Saxon" model, which the recent electoral success of the Labour Party in England seemed to presage. Among the reform measures he recommended were promoting equality for all through the spread of general education; extending the smallholding class and limiting the large landowners; improving the health and housing of the *fellahin* and helping them express their rights and demand greater "independence"; improving the standard of agricultural products, particularly cotton, through the introduction of scientific farming principles and procedures; and, finally, encouraging emigration to agricultural colonial settlements in the Sudan in order to protect Egypt's political and economic rights there and over the waters of the Nile. Alongside the agricultural colonization of the Sudan, Mazhar advocated a pacifist foreign policy of placing the Egyptian armed forces (once established) under the control of the League of Nations, as an example for other nations to follow. The program was internationalist and yet deeply nationalist along familiar, majority-party lines; it was intended to give voice to the *fellahin*, yet it in effect excluded any thought of popular participation; the whole attitude toward the peasantry was deeply paternalistic, and practical advice was confined to teaching them how to breed better chickens. One does not sense that political mobilization was what Mazhar had in mind.

But even this limited association with organized party politics soon led Mazhar into trouble. The details are sketchy, but shortly after the publication of these proposals, Mazhar was arrested and briefly detained on the charge that he was a communist. This was likely the result of the machinations of the opposing Liberal Constitutional Party, who accused him of having ties to Moscow and used his dedication to Nahas as evidence that the Wafd Party had a secret pro-communist agenda. Later that same year, Mazhar wrote that he regretted linking his program to the Wafd Party, as he had failed to see that this would allow the British and other parties in Egypt to accuse it of communist sympathies.[115] His arrest, coupled with the burning of his press, led Mazhar to abandon his political and literary ambitions. He closed down his journal and, after 1930, retreated once again to his country home.

Evolutionary Scripts

In many ways Mazhar was caught between competing genres and shifting generations. His attempt at science popularization alienated the new class of popularizing *ʿulama* and the vocal counterpublic that they gave rise to at precisely the time that Mazhar was offering his own program in popularization. And whereas he hoped that translation would allow him to mediate and then ultimately bridge the worlds of the East and the West, to most Arabic readers in his time he seemed caught uncomfortably between them. His national internationalism and a peasant party agenda both populist and elitist—not to mention the nuances of his own transcendental positivism—only added to the confusion.

Mazhar would return to journalism in very changed circumstances. In 1945 Fuʿad Sarruf invited him to become the editor of *Al-Muqtataf*, and there he continued to proselytize for evolution and for the sciences in general. He had always been close to the editors of the journal and had earlier written a highly respectful obituary on the death of Sarruf's father, one of the founding editors. Yet with the end of the Second World War, Mazhar's political outlook also appears to have altered, and the few articles that he published in *Al-Muqtataf* expressed not only his shaken faith in the kind of universal humanism that had characterized his thought between the wars but also a new assertiveness about the Arabs and their civilization. He now embraced the idea of Arab solidarity that he had previously rejected and talked in ever more glowing terms about their past civilizational achievements and their contribution to the modern world.

The reasons for this shift in outlook remain obscure. But it modified his views on a subject he had always been interested in: the future of the Arabic language and the construction of a technical and scientific vocabulary. After his withdrawal from political debate, Mazhar had worked on the translation of scientific terminology under the aegis of the Royal Arabic Language Academy.[116] There he supervised the compilation of dictionaries and composed encyclopedia entries. He also worked on translations commissioned by the Franklin Book Program—a postwar US government–funded initiative to fight communism in the Middle East through the translation and dissemination of American literature: it was for them that he translated

Edmund Sinnott's *The Biology of the Spirit* (*Hayat al-ruh fi diwa' al-ʿilm*) and George Sarton's *The History of Science and the New Humanism* (*Tarikh al-ʿilm wa-al-insaniya al-jadida*).[117]

Mazhar also began to write extensively on the problem of modernizing Arabic and the creation of technical terminology; he advocated the reform, expansion, and modernization of the language so as to accommodate new works, words, and concepts in the modern sciences. He was involved in the Royal Arabic Language Academy of Cairo, which had been founded in 1932 to standardize and modernize written Arabic.[118] The standardization of translations and the creation of specialized and scientific vocabularies were among the academy's primary tasks, and all the linguistic strategies for creating new terms—*ishtiqaq, naht, taʿrib*—were considered at length by its various committees. They contributed little, however, to the effective creation of technical vocabulary, and language reform got bogged down in what looks like a parody of Egyptian bureaucracy. This was due, in large part, to the circumstances of the academy's formation. Fearful that the academy might otherwise foment radical political change in the guise of language reform, Fuad I and his cabinet had been careful to select members who they thought would be most obedient to the royal court and whose attitude toward language change was not too "revolutionary."[119] There was little consensus over translation strategies and a great deal of infighting over abstract discussions on the nature of linguistic reform.

Mazhar criticized those members of the academy who held an overly conservative approach to linguistic derivation.[120] Other well-known language reformers of the time made similar criticisms. Salama Musa, a long-time advocate for language change, mocked the academy for its penchant for creating clumsy alternatives to already-established everyday terms for new technologies. "The nation needs useful books on the subject of the *atumubil, talifun* and *radiyofun* and doesn't need to be told that these are actually called *sayyara, misarra* and *midhya*."[121]

Nevertheless, as his Arabist orientation emerged after the war, Mazhar himself became more conservative. In his discussion in *Tajdid al-ʿArabiya* (Revival of Arabic), for example, which was published in 1950, he dealt extensively with the problem of Arabicization and criticized the school of thought that preferred to take, as he put it, the "short route." He admitted there was a necessity for it at times. But strikingly, the future translator of George Sarton insisted that those who supported such wholesale transliterations were operating under the mistaken impression that the language of science was universal: in fact, the language of botany and zoology was "universal" only in the Indo-Germanic languages (*al-lughat al-*

indujarmaniya) and not in the Semitic ones. These transliterations could never be said to be universal coinages: indeed, they were entirely foreign in Arabic. Rather than translate a word once in Greek, once in Latin, and once in an Indo-Germanic language, he argued, why not simply create a proper Arabic scientific language, for the language was certainly well equipped to accommodate this.[122]

Yet as a vehicle for elite control and linguistic reform, the academy effected little change in linguistic usage. Although it would go on to produce dozens of dictionaries and compiled lists of translated technical terms in the arts and sciences—a feat dwarfing the individual efforts of Mazhar and others and one that Mazhar himself participated in—it achieved little control over their proper or consistent usage. Just six years after its founding, Muhammad Husayn Haykal, the minister of education, expressed his concern over its failure to influence the way in which authors constructed new terminology in Arabic. They continued to translate in an idiosyncratic and haphazard manner, he alleged, and he called on Arabic writers to adopt the recommendations of the academy's councils in the name of national unity (to little avail).[123] Mazhar himself called for official language regulation and suggested the creation of a *dustur*, or constitutional regulation, for the use of Arabicized terms in particular.[124] Yet, in the end, no one could agree on the major questions the academy had been set up to tackle. What the academy—much like Mazhar himself—failed to realize ultimately was that language was as much a matter of usage and convention as it was of standardization and elite control. Outflanking such official efforts was the unregulated impact of mass journalism, print capitalism, and changing forms and norms of expertise. Once again, Mazhar had found himself chasing an unrealizable ideal.

The year 1952 was the end of an era in more ways than one: *Al-Muqtataf* finally closed its doors, only a few years after Mazhar stepped down from its editorship, and this coincided, of course, with the fall of the monarchy and the officers' coup that was to pave the way for Nasser's rise to power. Mazhar now tried to adapt his ideas to the realities of the Cold War. Close to the new regime through his association with its newspaper, *Al-Jumhuriya*, he published books and articles that combined his older anticommunism and evolutionism with a new stress not only on Arab solidarity but more broadly on the unifying power of Islam.

In 1958 he published *Al-din fi zill al-shuyuʿiya* (Religion in the shadow of communism)—an attack on "materialist, imperialist, Marxist, and Leftist critics," from whom he wished to rescue Islam, and from whose aggression he wished to preserve Arab unity. It was just at this time that the regime's stress on a "Third Way" was acquiring an anticommunist inflection. The formation of the United Arab Republic and Nasser's fears of communist takeover in Syria explain why communism was coming to be seen as a threat to Arab solidarity.[125] Behind all the talk of West versus East, communism, democracy, missiles, satellites, and nuclear submarines, Mazhar discerned that the fundamental problem facing Egypt was how to chart its course in a newly divided world. It was a final opportunity for his tendency to synthesize. Egypt, as he put it, straddled three continents and could not abdicate from history: it indeed had a duty as the "greatest power in the Middle East" to direct the "Arab-Islamic world" in the right direction. Reminding his readers of the crimes of Western imperialism, on the one hand, and of communist aggression, on the other, he turned to Islam as the guide and "that which will raise us." He warned in particular against the communists' attempt to stifle religion and strongly criticized the outcome of having tried to realize Marx's ideas: in Russia not even the peasants had benefited.[126]

Three years later, he returned to the same subject. In *Al-Islam la al-shuyuʿiya* (Islam, not communism), Mazhar critiqued communists and certain socialists for their attempts to suppress religious ethics—and, crucially, identities—with a crude ideology of equality. His thoughts on Islam, which he argued offered a more proper basis for communal politics and ethics, he now developed at greater length. He pointed to the concept of "mutual responsibility" (*al-takaful al-ishtirakiya*) and offered the history of the first Muslims as the exemplary model for a true spirit of human collectivism. More than ever before, he presented Arab identity as stemming from Islam, which he now characterized as a "religion and a state" (*din wa-dawla*), and he even described Muhammad ʿAbduh as "our imam." He depicted Islam, in its original state, as a religion of freedom and racial and religious tolerance. Kropotkin's notion of "mutual aid" was now overlaid with a Muslim concept of "mutual responsibility," the only way of escaping the extremes of capitalism and communism—"individual despotism," on the one hand, and "collective despotism," on the other.[127] But the evolutionary perspective remained: it shaped his view of the rise and decline of civilizations, and it helped explain the rise and fall of political philosophies. The contrast with the language of someone like the reformist head of Al-Azhar, Shaykh Mahmud Shaltut, is striking: both men present Islam as the only true socialism, but Shaltut's constant references to the Qurʾan are largely absent in Mazhar, who continued to talk about society and individuals in organicist terms, about social parasites, and about the importance of human norms conforming with natural law. His gradualism also survived:

"Marxist revolutions are nothing but rushing the cycles of time. There is nothing natural in them. They are incomplete experiments."[128]

Mazhar's faith in an evolutionary ethic had allowed him to adapt his political critique over his lifetime—from staunch anti-Easternism in the 1920s to a Muslim Third Way thirty years later. But he had never seen his faith in evolution or in science generally as in conflict with religion. Even at the end of his life, he continued to be interested in the kind of metaphysical evolutionism that he had spent much of his life trying to define in the face of several generations of materialist challenges. The very last translations he produced before he died in 1962 made this clear. Edmund Sinnott's *Biology of the Spirit* focused on bridging the world of matter and of spirit. George Sarton's *Science and the New Humanism,* which came out in the year of Mazhar's death, emphasized the unity of humanity under the banner of science. It was in a cosmic vision of evolution, with its potential to speak to the grand questions of human social and political life, that he thus found the most powerful lens to focus upon matters of politics and faith. For Mazhar, as for so many around the world, this was the real value of viewing Darwin in translation.

Afterword

In the late 1990s, a court case in Egypt brought Darwin into the headlines once again. It revolved around a book published by the Egyptian religious scholar and Cairo University professor ʿAbd al-Sabur Shahin. Entitled *Abi Adam* (My father Adam), it offers a novel approach to the question of man's origins, in the light of both modern science and the scriptural evidence of the Qurʾan. Shahin begins the book by lamenting the antiquated state of contemporary Islamic exegesis. Despite recent advances in carbon dating and paleochronology, as well as in disciplines such as biology, geology, and anthropology, he complains, the majority of Qurʾanic exegetes are still borrowing from the tales of the ancients, and particularly from what he considered the outlandish and unorthodox tales known as the Israʿiliyat.[1] By contrast, Shahin's own account makes ample use of contemporary scientific findings—particularly paleontological remains. But it is also a striking blending of gnosticism with a rationalist, modernist approach to Qurʾanic hermeneutics. In the beginning, he tells us, starting in a fairly orthodox fashion, there was nothing; then God said, "Be!," and thus he created time and place (*al-zaman wa al-makan*).[2] Basing his evidence entirely on a linguistic analysis of the verses of the Qurʾan that deal with creation, he then describes the "creation of humanity" as having followed various stages: the first creation occurred when Adam was fashioned from clay; the creature was then shaped into a human form; but the "spiritual creation" of humanity was completed only when God breathed his spirit into it. The main novelty of the argument here is that the two words in these verses that have traditionally been taken to refer to "humanity"—*bashar* and *insan*—actually have, according to Shahin, very different meanings. The former (*bashar*), he argues, does not refer to what most people would understand as "human" today. Rather, Shahin interprets the term to refer to all hominids, and he classes the evidence of remains of early anthropoids as part of this broader rubric. The latter, *al-insan*, he understands as the only real definition of humanity proper, for which he provides the more technical definition of *Homo sapiens*. (He claims that this distinction is present only in Arabic: no other language of the world has an equivalent to the word *bashar*, which he defines as any other species of the genus *Homo*.)[3]

Reading the various verses in the Qur'an that refer either to *bashar* or to *insan*, Shahin provides a novel interpretation of the verses on human creation. It was through a gradual—but, of course, divinely guided—progression that the evolution from *bashar*, sentient hominids, to *insan*, our modern humanity, took place. Corroborating, even if rather generally, "current anthropological, geological" evidence, he offers his *tafsir* as one possible way to bring together the different forms of evidence available to students of scripture and of science.

Yet Shahin was keen to separate his account from the views of Darwin and other modern evolutionists. In a section entitled "Scientific Theories," he claims that the "majority of scientists" now reject Darwin's ideas. This is a commonplace refrain in discussions of Darwin in Arabic now, and one that likely has the criticisms of contemporary creationists (many of whom *are* scientists) in mind. He concludes his discussion by siding with the "victorious" views of the proponents of a "fixed creation" against Darwin's supposedly defeated view of a gradual "evolution from some lower form."

How to categorize such a presentation? Shahin's account was not a strictly evolutionary one, to be sure. Yet by the standards of contemporary creationism, Shahin's exegesis might count as "liberal." To some of his critics, it was perhaps even radical. This is what one of his colleagues at Cairo University, Shaykh Yusuf al-Badri, argued when he initiated a controversial court case against Shahin for his allegedly pro-evolutionary stance. Relying on the so-called *hisba* law, al-Badri brought Shahin to trial on a charge of blasphemy.

As if further to emphasize the difficulty of categorizing contemporary opinion on the question of evolution—especially in the context of a legal order that has allowed for a kind of bidding war on the line between orthodoxy and heterodoxy—Shahin himself had earlier taken advantage of the religious courts and been seen altogether differently as the epitome of a Muslim conservative, if not "fundamentalist" (itself a late-twentieth-century neologism in Arabic). For Shahin had previously been one of the main parties responsible for declaring another Cairo University colleague an apostate. This was Nasr Hamid Abu Zayd, a figure much better known than Shahin in circles outside Egypt and an Islamic studies scholar who has written extensively on the need for a new, literary hermeneutics of the Qur'an. In 1995, thanks to the case that Shahin helped bring, Zayd was declared an apostate, and he went into exile in the Netherlands, where he taught. The case gathered so much attention that it also eventually led to a revision in legal procedures in the Personal Status Courts, not to mention considerable debate over issues of law and hermeneutics and the nature of secular critique.[4] As for the latter case, al-Badri lost, and Shahin escaped being branded an apostate in his own right. But his triumph was not complete. His book was added to the censored list of banned reading material, as it was deemed to be too complex for the untutored reader.[5]

In contemporary Egypt it is thus perfectly possible, as in the case of Shahin, to be both in favor of some variant of evolutionism and a zealous defender of religious orthodoxy, to be pro-science and anti-Darwin. The battle lines so easily discerned by Western commentators thus continue to be blurred in practice. Novel in some respects, Shahin's work brings together a number of elements that have appeared throughout this book: it is both a fairly traditional, linguistic exegetical reading and at the same time a modern one with respect to its extensive invocation of contemporary scientific findings; it emphasizes the transcendental nature of creation but offers its own rationalist reconciliation; and, finally, it sees no incompatibility between the authority of science and the word of God. In this sense, then, it testifies to the continuing strength of an overarching framework that has informed readings of Darwin in Arabic from the very beginning.

In the thirty-some years between Mazhar's death and Shahin's publication, writings on Darwin have proliferated. This more contemporary discourse has, like the one covered by this book, been critically shaped by the massive global production of literature on evolution, in this case by a host of writings on "creationism," "intelligent design," and other neologisms fostered by those who see themselves as opposed to a theory of evolution in one or more critical ways. It lies well beyond the scope of this book to consider this episode in any great depth. Yet the Shahin anecdote is instructive, and it reminds us that contemporary discussions of evolution in Arabic, even when labeled as part of a new "creationism," show as much diversity and flexibility of views now as they did nearly 150 years ago.

By and large, it is true that since the 1970s, in particular, references to Darwin in Arabic have expressed a more skeptical—if not outright hostile—view of the man and his theory. One book published in Baghdad recently coined him "the accomplice of the devil."[6] Highly colorful street pamphlets in Arabic that make a similar argument are today widely distributed: the current, sad state of affairs in the world (defined variously as moral decadence, rampant consumerism, or unrestrained warfare) is the result of the materialism spawned by Darwin's theory of evolution.

Of course, laudatory, as well as critical, works on Darwin in Arabic may be found throughout this period. The rapid proliferation of universities along with the institutionalization of laboratory science across the region at precisely this time encouraged academic discussions of evolution in Arabic. New journals of science also appeared.[7] The teaching of evolution was—and, in places like Egypt, is still—part and parcel of secondary and higher education in Arabic (although most universities will teach these subjects in English, the de facto language of science across the world today). When the newly renovated Alexandria library held a conference on Darwin as part of the 2009 bicentennial celebrations, most reporters—both within and outside Egypt—reported on the relative staidness of the affair, with none of the expected criticisms or protests occurring.[8]

It would also be wrong to assume that the literature on the subject that hit the Arabic bookshelves after the 1970s took a common or, for that matter, a necessarily critical line. Materialism remains the bugbear, as it did for Shumayyil's critics a century earlier, and thus, the majority of works single out the materialist interpretation of evolution as the real culprit, while happily conceding much other ground to modern evolutionary principles. Many of the approaches and tropes outlined in earlier chapters remain popular—the denial that evolution contradicts established belief, for instance, and the insistence that earlier Arabic philosophers, and indeed the Qurʾan itself, anticipated the so-called discoveries of the nineteenth century.[9]

In some ways, therefore, the arguments presented in the recent Arabic literature on the subject are not new: there are echoes of the creative intellectual genealogies of *Al-Muqtataf*, Shumayyil, Mazhar, Afghani's critique of materialism, the exegetical techniques used by al-Jisr to circumscribe the implications of the theory for scriptural hermeneutics, and the *tafsir ʿilmi* tradition from which many such works sprang. What was genuinely new, however, in the 1970s was the emergence of creationism proper as a new kind of global discourse. As a glance at the bibliographic references in these works reveals, they borrow extensively from American and European creationist literature. Like their counterparts elsewhere, these discourses mobilized new media, using the television and the Internet to play the kind of role the press had played a century earlier.[10] In a story replete with ironies—one that begins with American missionaries bringing Darwin to Arabic readers—these intertwining bibliographies provide a fitting finale.

That contemporary discussions of Darwin should be flanked by Christian creationist rhetoric and references is hardly surprising. Early science popularizers like the *Muqtataf* group, Husayn al-Jisr, and even Muhammad Rashid Rida had also borrowed from the works of Christian exegetes and commentators on evolution in particular.[11] Nor is the appetite for books on evolution unprecedented, as I hope to have shown here. It has merely changed in focus and style.

This book opened with Andrew Dickson White's visit to Cairo in 1888, when he was surprised to be greeted by Arabic readers who shared his optimism about the progress of science and its role in the shaping of the modern world. But when nearly four decades later Isma'il Mazhar translated White's *History of the Warfare of Science with Theology in Christendom,* the audience for this kind of literature was being rapidly replaced by others. Mazhar's technocratic and evolutionary vision of social progress was already losing ground to other forms of radical nationalism, whether the religious nationalism of the Ittihad Party and other Azhari *'ulama* whom he debated in the pages of his journal or of those "artificial revolutionaries," Bolsheviks and Communists alike, whom he also critiqued. In intellectual and even literary terms, Mazhar became more and more isolated. By the late 1920s, even rationalism—a capacious affiliation shared by almost all covered in this book—would lose its appeal. Over time, a growing interest in the ontological, phenomenological, and even mystical roots of religious experience would make the claims of a more purely positivist faith in rationalism seem obsolete. And with this turn to the spiritual, a new kind of science popularizer would also emerge.

It was Mazhar's critics who were the voices of the future. Men like Muhammad Farid Wajdi used Al-Azhar's journal (of which he assumed the editorship) not only to intensify the crusade against materialism but also to demonstrate that the spiritual dimension of life itself could be scientifically demonstrated. This new genre of popular science criticized positivism as much as materialism and replaced the earlier interest in the "evolution and progress" of the world with a more spiritualist interest in the creation of the cosmos, which was part of a new vogue for writings on "*takwin al-kawn,*" a kind of theistic cosmogony that drew upon medieval, Neoplatonic, and modern empiricism.

The authority of science had not waned, but its message and value had changed. In Wajdi's own words: "Ever since we devoted ourselves to the service of Islam, it has been our custom to familiarize ourselves with the natural sciences and with Western philosophy, realizing that the relation between our culture and that of the West imposes on us the duty of understanding [these]."[12] Yet Wajdi also foresaw the end of Western power carrying its corollary: the need for a new spiritual dawn in the Arab East.[13] His journal's attitude to evolutionary thought itself was tempered and ambivalent, sometimes suggesting that it had been prefigured by the Arabs, at others decrying it as materialist and even—in 1939—attributing to it the outbreak of the Second World War.[14]

There was also a new type of popularizing Qur'anic exegesis known as *tafsir ʿilmi*, or "scientific exegesis," that enjoyed a wide-ranging appeal.[15] Its origins lay in the Qur'anic commentaries published by Muhammad ʿAbduh and his disciple Rashid Rida in the latter's journal *Al-Manar* from the early 1900s. ʿAbduh and, even more so, Rida had conceived of the *tafsir* as a kind of platform for their broader political and pedagogical aims. But they had not by any means focused exclusively on discussions of science. The new scientific exegesis was very different, both more tightly focused in its attention to science and looser and more informal in its readings of the Qur'an.

In the interwar period the leading example of this new melding of science and piety was Tantawi Jawhari's *Al-jawahir fi tafsir al-Qur'an* (Gems of the exegesis of the munificent Qur'an). His *tafsir* was published in twenty-six volumes between 1923 and 1935, and it proved incredibly popular, going through numerous editions and translations. Where ʿAbduh's and Rida's *Tafsir al-Manar* had stressed reason as the path to faith, Jawhari inculcated in his reader a sense of the awe and wonder of God in nature as the basis of a new spiritual communion, and he used the authority of modern science to bolster his own particular brand of natural supernaturalism. His exegesis centered on discussions of modern science and offered an interpretation of all Qur'anic verses with potential cosmological significance. It was also aimed at a general readership, and the language was informal and anecdotal—very different in spirit from that of the classically trained *ʿulama*—and full of biographical and bibliographic digressions. Issues of jurisprudence and philosophy were eschewed in favor of an eclectic array of references that set established exegetes alongside such figures as the English naturalist John Lubbock (of whom Jawhari was particularly fond) and the Bengali biologist Jagadish Chandra Bose.

It was Jawhari's single-minded emphasis on the harmony of science and scripture that really set his exegesis apart. Nearly every sura interpretation led to a reflection on the wonders of creation in nature—as uncovered in modern science and anticipated in the Qur'an. The Qur'an in his reading had anticipated the findings of everything from physics and astronomy to botany, zoology, and spiritualism. When offering his exegesis of the opening verse of the Qur'an, and considering its use of the word *hamd* (praise) in particular, he spoke of the wondrous insects in the natural histories of Henri Milne-Edwards, for example. He repeatedly drew attention to the beauty, perfection, and majesty (*al-jamal, al-kamal wa-al-jalal*) of God's work in nature and offered a far more ecstatic and mystical interpretation than ʿAbduh and Rida did.

This was in many ways a reworking of the natural theology traditions beloved of an earlier generation of science popularizers. And indeed Jawhari himself represented a fusion of the Azhari tradition (he had attended Al-Azhar in 1877) and the new state-education system. He had taught at various elite secondary schools—including the Khedival School, around the time that the young Salama Musa and Ismaʿil Mazhar graduated—as well as Dar al-ʿUlum (Teachers' Training College) and the newly founded Egyptian University. Enormously prolific, the kind of popularization he believed in had very different political connotations than the popular works published by the editors of *Al-Muqtataf* and by Salama Musa. Far less Anglophile than them, he had started out as a member of Mustafa Kamil's Nationalist Party before gravitating toward the Muslim Brotherhood, whose journal he helped to edit for a time. He also participated in setting up the International Organization of Islamic Brotherhood as well as in a number of international educational and other schemes aimed at unifying Muslim communities around what he considered to be their "heartland": Cairo and, especially, Al-Azhar.

The influence of both nineteenth-century positivism and the recently revived tradition of Arabic rationalism was thus declining by the turn of the century. This did not mean that there was a rise of irrationalism, of course, but rather it indicated a turn toward a more trenchant transcendentalism. For Jawhari, science was not merely a method for revealing nature's mechanisms and laws but a path toward a mystical appreciation of creation and its Creator. It was, in short, a means of "enchanting" the world. "All that exists is love and beauty," he wrote, adding that were it not for the act of "mutual love" that God created between hydrogen and oxygen, for example, we would not have water.[16] Many of his writings are suffused with Neoplatonic concepts and terminology.[17] The obsession with "existence"—and, in general, the rising concern with matters of ontology rather than epistemology—took hold of theological discourse at this time, the very era when *al-wujudiya,* or existentialism, began to be debated in Arabic. A growing interest in the nature of religious experience in particular also helped to spur a reawakening of interest in Sufi and other transcendent philosophies.

Jawhari was typical of his era in seeing no contradiction between his commitments to naturalism and supernaturalism. Spiritualism itself was best described as a science, and he frequently cited European mesmerists, psychic researchers, and hypnotists to prove this.[18] For much of his life, Jawhari was active in a spiritualist society in Cairo, Jamʿiyat al-Ahram al-Ruhiya, held regular séances at his house, and also organized Sufi *dhikr* sessions (Jawhari, in the society's journal, would urge the confluence of Sufism and modern spiritist or psychic research). Thus, on the one hand, Jawhari publicized his contact with Al-Ruh al-Aʿzam, the Supreme Spirit or Spirit Guide also known as Silver Birch.[19] (Birch had also been popularized by the London-born Maurice Barbanell, who acted as his medium and published Birch's teachings in *Psychic News.*)[20] On the other, he drew—as his references to Sufi devotional practices suggested—from an older Islamic eschatological literature for his ideas about communion between the living and the dead.[21] Much of his terminology on this—and even some of the titles he used for his works—stretched back to al-Razi, Ibn Qayyim, and al-Ghazali.

Synthesis and unification—intellectual, religious, and political—were critical goals for Jawhari. He had begun his *tafsir* while teaching at Dar al-ʿUlum in the early 1900s, and one of his primary motivating factors had been a desire to unite secular subjects of study with religious ones. Jawhari, like ʿAbduh, worried greatly about the growing bifurcation of the worlds of science and Muslim pedagogy. He was disappointed with those students who had gone to Europe to study the natural sciences: rather than returning to Egypt to reconcile those subjects with the knowledge of creation outlined in the Qurʾan, they went instead into worldly professions. The *ʿulama,* he complained, did much the same thing in their own way. Rather than study the cosmos as a whole, they became overly "satisfied with their positions" and concerned themselves with jurisprudence to the detriment of a broader approach to knowledge. His *tafsir* was intended to rectify this. Science could bring together Muslims as a moral community and unite them again in the face of political fragmentation and sectarian strife or theological disputes. His exegesis aimed to "open hearts and guide nations" and "lift the curtains from the eyes of Muslims." He wanted to show people, both East and West, that the Qurʾan was a guide for all, calling upon them to study both the "lower" and "higher" worlds. His emphasis on *tawhid*, or unity, was thus threefold: it encompassed not only strength in the unity of the faithful but also the unity of knowledge and the unity of God's word and work.[22]

In the struggle for the public understanding of science in the interwar Arab world, Jawhari's brand of popularization won out over *Al-Muqtataf*'s. His *tafsir* was rapidly followed by others in the "miraculous Qur'an" tradition (*i'jaz al-Qur'an*): the older focus on the rhetorical inimitability of the Qur'an was now supplemented by a new concentration on the "scientific miracles" to be found in the text, and each new international scientific and technological breakthrough triggered new interpretations in this spirit. The launching of the two *Sputnik* satellites in 1957, for instance, was reported in detail and involved discussions among a number of high religious officials, including the grand mufti of Jerusalem: eventually, several *tafsir* in the Jawharian vein even appeared with an official imprimatur, testimony both to their undeniable popularity and to the state's desire to catch up. Thus, in 1961, the High Council on Islamic Affairs brought together scientists and religious scholars to produce a compendium of all possible scientific evidence in the Qur'an.

Jawhari's approach certainly had its critics. Mazhar likened it to a "contagion." But Jawhari was criticized from very different quarters as well. The new genre was too unconventional for many of the traditional Azhari rank and file, and bans were placed on his work. Rashid Rida was among the first to target those who tried to derive contemporary scientific meanings from verses of the Qur'an. Another of 'Abduh's protégés, Mahmud Shaltut (later the rector of Al-Azhar), thought that Jawhari made the error of applying temporal knowledge to "eternal truths" and thereby detracted from the "true purpose of revelation." One of the misuses of *tafsir* that he warned against was the "derivation of modern knowledge" from the Qur'an. If such an interpretive strategy commits to a given scientific theory, what happens to the Qur'an's authority? Such *tafsir*, he concluded, attempt to speak of "hidden matters" of which only God has ultimate knowledge, reviving once again the notion of *al-ghayb*, or the Unknown, as among the attributes of the divine itself.[23]

There was even criticism of this kind from within the ranks of the Muslim Brotherhood with which Jawhari was for a time associated. In fact, Sayyid Qutb took much the same line as Rida and Shaltut in his highly influential thirty-volume *tafsir* of the Qur'an. Because he sought to highlight the eternal verity of the "divine order of life," Qutb cautioned against "engaging in futile debates." Repeating an old theme, he argued that human knowledge—including temporal and inconstant knowledge of the natural world—was necessarily imperfect and inferior to the eternal truth of God. Hence, Qutb disapproved of efforts to attribute scientific meaning to verses of the Qurʾan. "The Qurʾan was never intended to be a book on astronomy, chemistry, or medicine, as some of its admirers and detractors try to make it." It was best to avoid anything that detracted from an "understanding of the nature of the Qurʾan and of its function and scope," which was concern with "the human soul and the state and direction of the human condition." The approach of those who attempted to find evidence of scientific facts and theories was "naïve and counterproductive" and involved "fantastic methods" of Qurʾanic interpretation. For these facts and hypotheses were, after all, merely derived from "human empirical methods" and subject to change. Thus, referring to Darwin's theory of evolution, he insisted that it was pointless and indeed wrong to try to find references to evolution in the Qurʾan: "To begin with, the theory is not conclusive, and within a century it has undergone several amendments and changes that have made it almost unrecognizable. . . . The Qurʾanic statement ["We (God) created man from an extraction of clay"] establishes the origin of man without giving any details of the process itself. It does not aim at more than that and carries no other connotations or meanings."[24]

Qutb was keen to separate the empirical and yet ultimately hypothetical results of science from the eternal truth of revelation. This betrayed a fundamentally different attitude toward science from that of a generation earlier. The rationalist reconciliation of science and religion no longer had the appeal that it did for so many of the colonial intellectuals and literary elite covered in this book. Since at least the 1930s, new popularizers—and new publics—had emerged.

These shifts in the focus of readings and translations of Darwin in Arabic were underpinned by profound socioeconomic shifts. In Egypt, for example, illiteracy decreased from 94 percent in 1897 to 77 percent fifty years later—still high but a significant drop nonetheless. Access to the state-run educational system was rapidly broadening: student numbers in higher education rose from three thousand in 1926 to ten times that number by 1950, and there were nearly two million students at all levels of schooling. Circulation figures for newspapers rose from an estimated total circulation of 24,000 in 1881 to 360,000 for weeklies and 200,000 for dailies in 1947. The cities, also growing fast, were filling up not only with these newly literate graduates but also with workers coming in from the countryside: the numbers in industry alone trebled between 1937 and 1947.[25]

This newly urbanized, literate society was growing ever more disillusioned with the performance of the political elite that had guided the country since independence. The monarchy was more unpopular than ever, and the continued presence of the British after formal independence (their domination of Egypt was starkly revealed in the crisis of 1942) was a source of resentment. Meanwhile, student protests and signs of the resurgence of union mobilization also showed that the old elites—including the "generation of 1919"— were losing control.[26] After 1945 there was a sharp rise in political violence, but police repression did nothing to stem the numbers flooding into the Marxist Left and into the Muslim Brotherhood, whose organization boasted half a million members by the end of the 1940s. In these movements, the old Westernized *effendiya* class was still prominent, especially among their leadership, but their composition was much more diverse, and their attitude toward politics more aggressive than that of the notables who had guided the country's fortunes for the first decades of the century. With the old elite accused of time-wasting and gradualism, evolutionary solutions to the country's socioeconomic problems seemed less and less persuasive. To be associated with the British, as *Al-Muqtataf* was now, was to be doubly condemned as both antinational and out of date.

Even the imperialism of the 1880s had not prevented critics of colonialism from signing on to a notion of "progress and development"—either national, civilizational, or spiritual—that shared a common telos with contemporary European and American discourses. After the 1930s, ideological conflict was sharpened in the region, and within a decade the old liberal evolutionism was replaced by neoliberal American-style modernization or Soviet communism and, later, by the Third Way offered either by new Arab socialists or by other revolutionary alternatives like the Muslim Brotherhood. *Al-Muqtataf* shut down in 1952, shortly after the army officers' coup, and leading members of the second generation of science popularizers such as Musa and Mazhar passed away. Even as professional scientists and scientific institutions flourished, then the public appetite for popular science was moving in new directions.

NOTES

INTRODUCTION

1. Andrew Dickson White, *Autobiography* (New York: Century, 1922), 2:433.
2. The serialization in *Popular Science Monthly* was followed two decades later by the author's two-volume book of the same title. Andrew Dickson White, *History of the Warfare of Science with Theology in Christendom*, 2 vols. (New York: D. Appleton, 1896).
3. White, *Autobiography*, 2:436.
4. The well-connected White visited a number of Egyptian officials, including Artin Pasha, the minister of public instruction, Riaz Pasha, the prime minister, and even Khedive Tawfiq himself. A noted educator, among the founders of Cornell University and first president of the American Historical Association, White also visited the "great technical school of Cairo," which he described as "under the charge of an eminent French engineer . . . training admirably a considerable number of Egyptians in various arts applied to industry" (ibid., 2:435). Around the same time that White founded Cornell, his "old Yale friend" Henry Jessup helped found a similarly nondenominational Christian school of higher education in Beirut, the Syrian Protestant College.
5. Ibid., 2:436.
6. Fiske helped White acquire a vast collection of books and historical documents. He was himself a veteran traveler to Egypt: he composed a short work on navigating the Egyptian postal service for foreigners and another on colloquial Egyptian Arabic, which even made its way into local debates on language reform.
7. George Burr, "Sketch of Andrew Dickson White," *Popular Science Monthly*, February 1896, 556.
8. White was "much struck" by his host's interest in freemasonry in particular: Nimr described it to him as his principal hope for "bringing Christians and Mohammedans together under the same roof for mutual help, with the feeling that they were children of the same God" (White, *Autobiography*, 2:437). Among the rational dictates it supported, Nimr listed the "censure of oppressive use of authority" and "respect by all individuals for proper laws governing society." Others included "dedication to improved interpersonal relations"; "compassion for the weak"; "deemphasis on the presumed value of personal contact with people of high rank"; and "honor for women, 'who should not be vexed in their social intercourse.'"

From Makarius's Masonic journal, *Al-Lata'if*, cited by Bryan Cannon, "Nineteenth Century Arabic Writings on Women and Society: The Interim Role of the Masonic Press in Cairo (*al-Lataif*, 1885–1895)," *International Journal of Middle East Studies* 17 (1985): 466.

9. White, *Autobiography*, 2:436–37.
10. This will be discussed further in chapter 1.
11. White's chapter on "Diabolism and Hysteria" was translated, for instance, in *Al-Muqtataf* 25 (1900): 281.
12. Burr, "Sketch of Andrew Dickson White."
13. Two authors who helped shape how Darwin was read in Arabic in particular did travel there: Thomas Henry Huxley and Herbert Spencer. Huxley only later mentions this visit, however, when referring to fellow Anglicans: hoping to shock them, he compared them unfavorably to the irate "Mohammedans" he encountered when he, an "infidel Englishman," tried to enter Al-Azhar, or what he misspelled as the "Hazar Mosque," the grand Sunni madrasa-mosque college in Cairo. T. H. Huxley, "Agnosticism," in *Collected Essays* (Cambridge: Cambridge University Press, 2011), 5:234. Herbert Spencer's only discussion of his voyage to Egypt in his diaries was that he had to cut short his trip down the Nile due to "dyspepsia and morbid fancies." Herbert Spencer, *Autobiography* (New York: D. Appleton, 1904), 2: chap. 53. Spencer himself often entertained visitors voyaging in the opposite direction. Among those he met at his home in Brighton was Egypt's grand mufti Muhammad ʿAbduh.
14. It was no coincidence that in the Chinese, Arabic, and Japanese cases, translations of Darwin, Huxley, or Spencer were followed by translations of key texts in political thought, including those of Adam Smith, John Stuart Mill, and, later, Marx and Kropotkin. Some would understand this in terms of the appeal of "social Darwinism." I prefer to avoid this term altogether: a pejorative in its original late nineteenth-century construction, it also assumes that there was a divide between using Darwin to understand "society" and using him to understand "nature." Yet for much of the period under discussion here, such a distinction seemed to make little sense. See Donald C. Bellomy, "'Social Darwinism' Revisited," *Perspectives in American History* 1 (1984): 1–129; James Moore, "Socializing Darwin: Historiography and the Fortunes of a Phrase," in *Science as Politics*, ed. Les Levidow (London: Free Association Books, 1986), 38–80; Paul Crook, "Social Darwinism: The Concept," *History of European Ideas* 22 (1996): 261–74.
15. For more on this approach, see Patrick O'Neil, *Polyglot Joyce: Fictions of Translation* (Toronto: University of Toronto Press, 2005). My thanks to Carol Gluck for this reference.
16. Paul de Man, "'Conclusions' on Walter Benjamin's 'The Task of the Translator,'" *Yale French Studies* 69 (1985): 35.
17. For more on the history of reading, see Robert Darnton, "History of Reading," in *New Perspectives on Historical Writing*, ed. Peter Burke (Cambridge: Cambridge University Press, 1991); Robert Darnton, "What Is the History of Books Revisited," *Modern Intellectual History* 4 (2007): 495–508; Roger Chartier, *The Order of Books* (Stanford, CA: Stanford University Press, 1992), esp. "Community of Readers"; Michel de Certeau, *The Practice of Everyday Life* (Berkeley and Los Angeles: University of California Press, 1984), esp. "Reading as Poaching"; Wolfgang Iser, *Act of Reading* (Baltimore, MD: Johns Hopkins University Press, 1978); Wolfgang Iser, *Range of Interpretation* (New York: Columbia University Press, 2000); Susan Suleiman and Inge

Crosman, eds., *The Reader in the Text: Essays on Audience and Interpretation* (Princeton, NJ: Princeton University Press, 1980); David Finkelstein and Alistair McCleery, *Book History Reader* (London: Routledge, 2006); Alberto Manguel, *A History of Reading* (London: Flamingo, 1994).

18. See Roy MacLeod and Philip F. Rehbock, eds., *Nature in Its Greatest Extent: Western Science in the Pacific* (Honolulu: University of Hawaii Press, 1988); Marwa Elshakry, "Global Darwin: Eastern Enchantment," *Nature* 461 (2009): 1200–1201; James Secord, "Global Darwin," in *Darwin*, ed. William Brown and Andrew Fabian (Cambridge: Cambridge University Press, 2010), 31–57. For more on the global enterprise behind the HMS *Beagle* itself, see also Mark Graham, "'The Enchanter's Wand': Charles Darwin, Foreign Missions, and the Voyage of H.M.S. *Beagle*," *Journal of Religious History* 31 (2007): 131–50; and Janet Browne, "Biogeography and Empire," in *Cultures of Natural History: From Curiosity to Crisis*, ed. Nicholas Jardine, James Secord, and Emma Spary (Cambridge: Cambridge University Press, 1996), 305–21.
19. Duncan Bell, *The Idea of Greater Britain* (Princeton, NJ: Princeton University Press, 2007), 85. See also Nicholas Daly, *Literature, Technology and Modernity, 1860–2000* (Cambridge: Cambridge University Press, 2004); and Wolfgang Schivelbusch, *The Railway Journey: The Industrialization of Time and Space in the Nineteenth Century* (Berkeley and Los Angeles: University of California Press, 1986).
20. See, e.g., the example of Yen Fu's translation of Huxley's *Evolution and Ethics* discussed by James Pusey, *China and Charles Darwin* (Cambridge, MA: Harvard University Press), 1983; and the chapter on "Buddhism and Science" in Gerard Godart, "Darwin in Japan: Evolutionary Theory and Japan's Modernity (1820–1970)" (PhD diss., University of Chicago, 2009).
21. See, e.g., Pratik Chakrabarti, *Western Science in Modern India: Metropolitan Methods, Colonial Practices* (New Delhi: Permanent Black, 2004), 175; C. Mackenzie Brown, "Western Roots of Avataric Evolutionism in Colonial India," *Zygon* 42 (2007): 423–48; David Gosling, "Darwin and the Hindu Tradition: 'Does What Goes Around Come Around?,'" *Zygon* 46 (2011): 345–69.
22. Ironically, the editors themselves were less keen to draw such a genealogy, as they considered Darwin's theory of evolution to be uniquely modern. See chapter 1.
23. These articles were later published as a book in 1905. ʿAbduh also repeatedly wove references to Darwin and to ideas of evolution into lectures he delivered on the exegesis of the Qurʾan. See chapter 5.
24. The locus classicus of this view is Albert Hourani, *Arabic Thought in the Liberal Age* (Cambridge: Cambridge University Press, 1962).
25. They were also tied together by literary salons, like that of Mayy Ziadeh. See Rose Ghorayeb, "May Ziadeh (1886–1941)," *Signs* 5 (1979): 375–82.
26. This is discussed later in the introduction and in the afterword. See also Samah Selim, "The People's Entertainments: Translation, Popular Fiction, and the Nahdah in Egypt," in *Other Renaissances: A New Approach to World Literature*, ed. Brenda Deen Schildgen, Gang Zhou, and Sander Gilman (New York: Palgrave Macmillan, 2007), 35–58; and Ziad Fahmy, *Ordinary Egyptians: Creating the Modern Nation through Popular Culture* (Stanford, CA: Stanford University Press, 2011). See the discussion in the section "Publics and Counterpublics" below for more on this.
27. So does the literature on the subject. For the classic, early Cold War modernization theory view, see George Basalla's "The Spread of Western Science: A Three-Stage Model Describes the Introduction of Modern Science into Any Non-European Na-

tion," *Science* 156 (1967): 611–22. Basalla's own argument closely mirrors Walt Rostow's classic Cold War (and also tripartite) scheme in *The Stages of Economic Growth: A Non-Communist Manifesto* and has since served as the subject of much critique. See Roy MacLeod, "On Visiting the Moving Metropolis: Reflections on the Architecture of Imperial Science," in *Scientific Colonialism: A Cross-cultural Comparison*, ed. Nathan Reingold and Marc Rothenberg (Washington, DC: Smithsonian Institution Press, 1987), 217–50; S. Irfan Habib and Dhruv Raina, eds., introduction to their *Social History of Science in Colonial India* (Oxford: Oxford University Press, 2007), xx. For more recent works, see Kapil Raj, *Relocating Modern Science: Circulation and the Construction of Knowledge in South Asia and Europe, 1650–1900* (New York: Palgrave Macmillan, 2007); Mark Harrison, "Decentering 'Colonial' Science,'" *Metascience* 16 (2007): 543–47; Andrew Cummingham and Perry Williams, "De-centering the 'Big Picture': The Origins of Modern Science and the Modern Origins of Science," *British Journal for the History of Science* 26 (1993): 407–32; and, for a review of the literature, James McClellan III, "Science and Empire Studies and Postcolonial Studies: A Report from the Contact Zone," in *Entangled Knowledge: Scientific Discourses and Cultural Difference*, ed. Klaus Hock and Gesa Mackenthun (Munich: Deutsche Nationalbibliothek, 2012). While these later works emphasize mutual flows or exchanges of knowledge in the construction of European science, others have also emphasized the need to consider notions of the universalization of science from different local perspectives. For more on rethinking the global history and historiography of science, see the essays in Simon Schaffer, Lissa Roberts, Kapil Raj, and James Delbourgo, eds., *The Brokered World: Go-Betweens and Global Intelligence, 1770–1820* (Sagamore Beach, MA: Science History Publications, 2009); essays in Jürgen Renn, ed., *The Globalization of Knowledge in History* (Berlin: Edition Open Access); essays by Shruti Kapila, Niel Safier, Sujit Sivasundaram, and Helen Tilley in the Focus section "Global Histories of Science," *Isis* 101 (2010): 95–158; and Fa-ti Fan, "The Global Turn in the History of Science," *East Asian Science, Technology and Society* 6 (2012): 249–58.

28. See Raj, *Relocating Modern Science*.
29. I have written about this subject elsewhere; see my "When Science Became Western," *Isis* 101 (2010): 98–109.
30. Qasim Amin, *Tahrir al-mar'a* (Cairo, 1899), translated as *The Liberation of Women* by Samiha Sidhom Peterson (Cairo: American University in Cairo Press, 2000), 62–63.
31. Ibn Khaldun's much-cited fourteenth-century work *Al-muqaddima* discusses civilization only in general terms. Particular dynastic Muslim civilizations rose and fell, to be sure, but the idea that an Arab or Muslim civilization rose and then fell from a "golden age" summit did not form part of his analysis. For instance, he outlines the various conditions for its development, including such things as climate and abundance or scarcity of food. He also describes the various civilizations that were formed by Arabs and that counted as dynastic "Muslim civilization" (such as "Muslim civilization in Spain"), but his use of the term referred primarily to urban settlement and habitation.
32. See Karam Rizk and Dominique Avon, *La Nahda: Réveils de la pensée en langue arabe, approches, perspectives; Colloque organisé à l'USEK les 28–29 octobre 2004* (Jounieh, Lebanon: Université Saint Esprit de Kaslik, 2009).
33. The Nahda remains one of the most enduring concepts in modern Arab historiogra-

phy. For more recent, critical works, see Stephen Sheehi, "Toward a Critical Theory of al-Nahdah: Epistemology, Ideology and Capital," *Journal of Arabic Literature* 43 (2012): 269–98; and Shaden Tageldin, "Proxidistant Reading: Toward a Critical Pedagogy of the Nahdah in U.S. Comparative Literary Studies," *Journal of Arabic Literature* 43 (2012): 227–68.

34. For more on this, see Thomas Philipp, "Approaches to History in the Work of Jurji Zaydan," *Asian and African Studies* 9 (1973): 63–86, esp. 68–74.

35. In his histories of the Arabs and of Islam (published in the early 1900s) and in his later writings and lectures on history (particularly those he composed but did not deliver for the newly founded Cairo University in 1910), Zaydan turned from the subject of universal history to Arab and particularly Islamic history. Zaydan was forced from his appointment as lecturer in Islamic history before he could present these thanks to controversy over his views and appointment. See ibid., 71–72; and for a translation of one of these, first published in *Al-Hilal* in 1910, see Anne-Laure Dupont, "L'histoire de l'Islam au regard des autres histories: Un article de Gurgi Zaydan (1861–1914)," *Arabica* 4 (1996): 486–93.

36. See his *Tarikh al-tamaddun al-Islam* (History of Islamic civilization), published between 1902 and 1906, and his *Al-Arab qabl al-Islam* (The Arabs before Islam), published in 1908, in particular.

37. Carl Schorske, *Thinking with History: Explorations in the Passage to Modernism* (Princeton, NJ: Princeton University Press, 1999).

38. Rifaʿa Rafiʿ al-Tahtawi, *Takhlis al-ibriz fi talkhis Bariz* [Purification of ore in Parisian lore], translated by Daniel Newman as *An Imam in Paris: Al-Tahtawi's Visit to France (1826–1831)* (London: Saqi, 2004), 102. Tahtawi also subscribes to a classic Enlightenment view of man's original character in the "state of nature": "Originally, man was simple and devoid of adornments; he existed in a purely natural state, and knew only instincts. Some people acquired knowledge . . . uncovered by chance, accident, inspiration or revelation" (ibid., 101). He speaks of three human groups: "wild savages;" "uncivilized barbarians;" and those "who are cultured, refined, sedentarized [*tahaddur*], civilized [*tamaddun*] and have attained the highest degree of urbanization [*tamassur*]." In some ways he is borrowing from Ibn Khaldun. Yet he also lists Egypt and "the Franks" as Europeans, or those whom he sees as having political institutions, sciences, industries, law, and trade; and he also places the French among the "highest rank" thanks to their advances in mathematics, natural science, and metaphysics, adding a contemporary angle to the conception. Indeed, it was thanks to science in particular—which he connects to their "versatility and inventions of warfare"—that they surpassed the Arabs themselves. Ibid., 102–5. He uses both *madaniya* and *hadara* to refer to civilization. *Madaniya* more specifically connotes material culture or technical culture; whereas *hadara* refers to any urban or civil settlement; hence, discussions of the former often dwell upon such things as the advancement of learning or arts and industry; the latter upon civil as much as moral orders; but the distinction was often fuzzy. Other terms like "culture" and "heritage" were also being critically updated and drawn into such discussions. For more on the concept of civilization in Arabic, see Joseph Massad, *Desiring Arabs* (Chicago: University of Chicago Press, 2005), 1–29, 51–57.

39. J. S. Mill, "Civilisation" (1836), in *The Collected Works of John Stuart Mill*, ed. John M. Robson (London: Routledge and Kegan Paul, 1977).

40. This substitution furnishes an interesting example of the "seductions of translation"

for many Arab colonial intellectuals at this time: using *umma* or *umam* instead of "race" or "races" transformed a particular colonial or highly racialized conception into a familiar dialogue in Arabic about moral reform and civilizational progress. For more on this, see Samah Selim, "Languages of Civilization: Nation, Translation and the Politics of Race in Colonial Egypt," *Translator* 15 (2009): 139–57; on translation, see Lawrence Venuti, *The Translator's Invisibility: A History of Translation* (London: Routledge, 1995); and Shaden Tageldin, *Disarming Words: Empire and the Seductions of Translation in Egypt* (Berkeley and Los Angeles: University of California Press, 2011).

41. François Guizot, *History of Civilization in Europe* (London: David Bogue, 1846). On Jamal al-Din al-Afghani's reading of Guizot, see Margaret Kohn, "Afghani on Empire, Islam and Civilization," *Political Theory* 37 (2009): 398–422.
42. See Damian Howard, *Being Human in Islam: The Impact of the Evolutionary Worldview* (New York: Routledge, 2011), for more on this.
43. Discussed in chapter 7.
44. Al-Jisr himself borrows from Anglican natural theology—citing Paley's famous watchmaker analogy and others of his own in similar fashion. This he likely encountered through his eclectic readings in Arabic: a frequenter of the Syrian Protestant College library, his ideas echo those of Catholic natural philosophies, the Protestant College's weekly paper and textbooks, and *Al-Muqtataf* itself.
45. Yevgeniya Frolova, "The Problem of Knowledge and Belief, Science and Religion in the Arab Thought," in *Muslim Philosophy in Soviet Studies*, ed. Mariëtta Tigranovna Stepaniants (New Delhi: Sterling Publishers, 1988), 118.
46. Ibid., 117.
47. See ibid., 122; Mary Baine Cambell, Lorraine Daston, Arnold Davidson, John Forrester and Simon Goldhill, "Enlightenment Now: Concluding Reflections on Knowledge and Belief," *Common Knowledge* 13 (2007): 429–50.
48. Frolova, "Problem," 123.
49. Al-Ghazali, *The Alchemy of Happiness*, www.sacred-texts.com/isl/tah/index.htm.
50. Compare with the turn to transcendentalism among Bengali positivists. See Geraldine Hancock Forbes, *Positivism in Bengal: A Case Study in the Transmission and Assimilation of an Ideology* (Calcutta: Minerva Associates, 1975).
51. The rise of new forms of scriptural criticism, comparative religious studies, and ideas about the evolution of faith were equally a part of this story. For more on the intellectual background to this story in Europe, and new understandings of "world religions," see Tomoko Masuzawa, *The Invention of World Religions; or, How European Universalism Was Preserved in the Language of Pluralism* (Chicago: University of Chicago Press, 2005).
52. D. B. Macdonald, "ʿIlm," *Encyclopaedia of Islam*, http://referenceworks.brillonline.com; Jean Jolivet, "Classifications of the Sciences," in *Encyclopedia of the History of Arabic Science*, ed. Roshdi Rashed (London: Routledge, 1996), 1008–25; A. I. Sabra, "The Appropriation and Subsequent Naturalization of Greek Science in Medieval Islam," *History of Science* 25 (1987): 223–43. See also Frolova, "Problem." For interesting parallels, see Nicolas Standaert, "The Classification of Sciences and the Jesuit Mission in Late Ming China," in *Linked Faiths: Essays on Chinese Religions and Traditional Culture*, ed. Jan De Meyer and Peter Engelfriet (Leiden: Brill, 2000), 287–317; David Reynolds, "Redrawing China's Intellectual Map: Images of Science in Nineteenth Century China," *Late Imperial China* 12 (1991): 27–61.
53. Tahtawi wrote: "Know that the French divide human knowledge into two sections,

the sciences and the arts. The sciences consist of verified facts by established proofs, whereas art is the knowledge of techniques. . . . But with us the sciences and arts are often the same thing." Indeed, much of what the French took for "science," Tahtawi would identify with "craft," and he argued that normally "a distinction is made based only on whether an art is an independent science or serves as a tool for another." Tahtawi, *Takhlis al-ibriz fi talkhis Bariz* (Cairo: Bulaq, 1834), 331.

54. This is a subject that has begun to interest a number of scholars. See, e.g., Mohammed Abattouy, Jürgen Renn, and Paul Weinig, "Transmission as Transformation: The Translation Movements in the Medieval East and West in a Comparative Perspective," *Science in Context* 14 (2001): 1–12; Jamil Ragep and Salley Ragep, eds., *Tradition, Transmission, Transformation* (Leiden: Brill, 1996); Robert Wisnovsky, *Avicenna's Metaphysics in Context* (Ithaca, NY: Cornell University Press, 2003). See also Scott Montgomery, *Science in Translation: Movements of Knowledge through Cultures and Time* (Chicago: University of Chicago Press, 2000). For recent and illuminating studies on the modern history of science and translation, see esp. Georgia Petrou, "Translation Studies and the History of Science: The Greek Textbooks of the 18th Century," *Science and Education* 15 (2006): 823–40; Michael Dodson, 'Translating Science, Translating Empire: The Power of Language in Colonial North India," *Comparative Studies in Society and History* 45 (2005): 809–35; Nicolaas Rupke, "Translation Studies in the History of Science: The Example of Vestiges," *British Journal for the History of Science* 33 (2000): 209–22; Roger Hart, "Translating the Untranslatable: From Copula to Incommensurable Worlds," in *Tokens of Exchange: The Problem of Translation in Global Circulations*, ed. Lydia Liu (Durham, NC: Duke University Press, 2000), 45–73; David Wright, "The Translation of Modern Western Science in 19th-Century China, 1840–1895," *Isis* 89 (1998): 653–73. See also Sundar Sarukkai, *Translating the World: Science and Language* (Lanham, MD: University Press of America, 2002). For general works on translation and translation studies, see Reinhart Schulte and John Biguenet, eds., *Theories of Translation: An Anthology of Essays from Dryden to Derrida* (Chicago: University of Chicago Press, 1992); Lawrence Venuti, *The Translation Studies Reader* (New York: Routledge, 2000); and *Übersetzung: Ein internationales Handbuch zur Übersetzungsforschung* (Berlin: W. de Gruyter, 2004).

55. See exceptions cited above. I have also written on this subject elsewhere, see my "Knowledge in Motion: The Politics of Modern Science Translations in Arabic," *Isis* 99 (2008): 701–30.

56. Charles Darwin, *On the Origin of Species by Means of Natural Selection; or, the Preservation of Favoured Races in the Struggle for Life* (London: J. Murray, 1859), chap. 2.

57. Hasan Husayn, *Fasl al-maqal fi falsafat al-nushu' wa al-irtiqa'* [The definitive treatise on the philosophy of evolution and progress] (Cairo: n.p., 1924); translation of Ernst Haeckel, *Last Words on Evolution: A Popular Retrospect and Summary*, trans. Joseph McCabe (London: A. Owen, 1906).

58. According to Stetkevych, the verb *tatawwur* ("to develop" or "evolve"), a Form V derivation that made possible the more recent Form II derivative *tawwara*, does not in fact constitute a modern derivation and was used by Arabic scholars as early as the tenth century (and later by Ibn Khaldun). Its modern meaning—if not form—however, is novel: it originally meant "to disguise oneself," from the noun *taur*, "time" (synonym of *tara*, "state," "condition," "manner"). See Jaroslav Stetkevych, *The Modern Arabic Literary Language: Lexical and Stylistic Developments* (Chicago: University of Chicago Press, 1970), 40–42. See also *Majallat al-Majma' al-'ilmi al-'Arabi*

28 (1953): 502–3; and Reinhart Dozy, *Supplément aux dictionnaires arabes* (Leiden: Brill, 1881), 2:66.

59. On this latter point, Zilzal included such things as their freedom from religious zealotry and blind imitation (*taqlid*) of past theological or juridical traditions, alongside their technological advancements and their intrepid explorations of new lands, including flora and fauna, around the world. See Bishara Zilzal, *Tanwir al-adhhan* [The evolution of mind] (Alexandria: n.p., 1879), 4–15.
60. On the former, see Zilzal, op. cot.; on the latter, M. A. Helmy, "Notes on the Reception of Darwinism in Some Islamic Countries," in *Science in Islamic Civilisation*, ed. Ekmeleddin Ihsanoglu and Geza Gunergun (Istanbul: IRCICA, 2000).
61. Husayn al-Marsafi, *Risalat al-kalim al-thaman* [Discourse on eight words] (Cairo: n.p., 1881), 85.
62. Abduh, like many others at the time, used ideas of evolution as extended metaphors for the development and natural progression of selves, families, and states. For more on this, see chapter 5. See also Samira Haj, *Reconfiguring Islamic Tradition: Reform, Rationality, and Modernity* (Stanford, CA: Stanford University Press, 2009); and Talal Asad, *Formations of the Secular: Christianity, Islam, Modernity* (Stanford, CA: Stanford University Press, 2003), for more on ʿAbduh's view of shariʿa and family law.
63. See Abdelslam Maghraoui, *Liberalism without Democracy: Nationhood and Citizenship in Egypt, 1922–36* (Durham, NC: Duke University Press, 2006); and Walid Kazziha, "The Jaridah-Ummah Group and Egyptian Politics," *Middle Eastern Studies* 13 (1997): 373–85.
64. Mona Russell, *Creating the New Egyptian Woman: Consumerism, Education and National Identity, 1863–1922* (New York: Palgrave, 2004), 51
65. Albert Hourani, "Ottoman Reform and the Politics of Notables," in *The Modern Middle East*, ed. Albert Hourani, Phillip Khoury, and Mary C. Wilson (London: I. B. Tauris, 2004), 83–109.
66. British literacy rates from Carlton J. Hayes, *A Generation of Materialism: 1871–1900* (New York: Harper, 1941), 174–75. Egyptian literacy rates from Ami Ayalon, *The Press in the Arab Middle East: A History* (New York: Oxford University Press, 1995), 144.
67. Jonathan Rose, *The Intellectual Life of the British Working Classes* (New Haven, CT: Yale University Press, 2001).
68. Craig Calhoun, "Introduction: Habermas and the Public Sphere," in *Habermas and the Public Sphere*, ed. Craig Calhoun (Cambridge, MA: Massachusetts Institute of Technology Press, 1992), 9.
69. Cited in Leah Price, "Reading: The State of the Discipline," *Book History* 7 (2004): 306.
70. For an interesting reflection on this, see Brinkley Messick, *Calligraphic State: Textual Domination and History in a Muslim Society* (Berkeley and Los Angeles: University of California Press, 1993).
71. See *Al-Muqtataf*'s article on subscribers' professions: "Haquq al-suhuf wa al-mushtarikun fiha" [The rights of journals and their subscribers], *Al-Muqtataf* 28 (1903): 666–68.
72. See Michael Allan, "The Limits of Secular Criticism: Reflections on Literary Reading in a Colonial Frame," *Townsend Center for the Humanities Newsletter*, February 2007.
73. Charles Hirschkind, *The Ethical Soundscape: Cassette Sermons and Islamic Counterpublics* (New York: Columbia University Press, 2006), 67–74, quotation on 106. See

also Armando Salvatore, *The Public Sphere: Liberal Modernity, Catholicism, Islam* (New York: Palgrave Macmillan, 2007).

74. See the afterword for more on this.
75. Nancy Fraser, "Rethinking the Public Sphere: A Contribution to the Critique of Actually Existing Democracy," in Calhoun, *Habermas and the Public Sphere*, 123.

CHAPTER ONE

1. "Al-talighraf" [The telegraph], *Al-Muqtataf* 1 (1876): 24, 126–28, 182–83, 276–78; 2 (1877): 3–4, 25–27.
2. On the introduction of electricity to Syria and Egypt, see Charles Issawi, *Egypt in Revolution* (London: Oxford University Press, 1963), 29; Rudolph Peters, "Religious Attitudes towards Modernization in the Ottoman Empire: A Nineteenth Century Pious Text on Steamships, Factories and the Telegraph," *Die Welt des Islams* 26 (1986): 76–105; Roderic Davison, "The Advent of the Electric Telegraph in the Ottoman Empire," in *Essays in Ottoman and Turkish History, 1774–1923*, ed. Roderic Davison (Austin: University of Texas Press, 1990), 133–65; Yakup Bektas, "The Sultan's Messenger: Cultural Constructions of Ottoman Telegraphy, 1847–1880," *Technology and Culture* 41 (2000): 669–96.
3. Kamil al-Ghazzi, *Nahr al-dhahab fi tarikh halab* [Golden river of Aleppine history] (Aleppo: Al-Matbaʿa al-Maruniya, 1924), cited in Charles Issawi, *The Fertile Crescent, 1800–1914* (New York: Oxford University Press, 1988), 87; and Eugene Rogan, "Instant Communication: The Impact of the Telegraph in Ottoman Syria," in *The Syrian Land: Processes of Integration and Fragmentation; Bilad al-Sham from the 18th to the 20th Century*, ed. Thomas Philipp and Birgit Schaebler (Stuttgart: F. Steiner, 1998), 113–28.
4. "Al-Duktur Yaʿqub Sarruf" [Doctor Yaʿqub Sarruf], *Al-Muqtataf* 71 (1927): 127.
5. "Manzilat Al-Muqtataf" [*Al-Muqtataf*'s standing], *Al-Muqtataf* 7 (1882): 249.
6. The first six chapters were published in 1918, with four more added in 1928 by Ismail Mazhar, and the complete work, also by him, published only in 1964; Ismaʿil Mazhar, *Asl al-anwaʿ wa-nushuʾiha bi-al-intikhab al-tabiʿi wa-hifz al sufuf al-ghalib fi al-tanahur ʿala al-baqaʾ*, vol. 1 (Cairo: Wizarat al-Thaqafa wa al-Irshad al-Qawmi, 1918). See chapter 7.
7. ʿUthman Ghalib, *ʿIlm al-hayawan* [Zoology] (Cairo: n.p., 1886).
8. "Al-Muqaddima" [Introduction], *Al-Muqtataf* 1 (1876): 1.
9. See Nadia Farag, "*Al-Muqtataf*, 1876–1900: A Study of the Influence of Victorian Thought on Modern Arabic Thought" (PhD diss., Oxford University, 1969), appendices 1A–1C.
10. Sarruf and Nimr were also joined by Shahin Makarius, who was largely responsible for the printing and production of the journal. For more on Makarius, see below.
11. "Al-Duktur Yaʿqub Sarruf," *Al-Muqtataf* 71 (1927): 125.
12. *Al-Muqtataf* 2 (1877): on the phonograph, 210; on the radiometer, 98; on whitening teeth, 238; on unwanted hair, 45; on bedbugs, 259; on magic, 28–29; on the zebra, 33–34; on blood transfusion, 39–40; on healthy hair, 44.
13. "Al-Duktur Yaʿqub Sarruf," *Al-Muqtataf* 71 (1927): 126.
14. After 1889, the emblem was discontinued. It was replaced with a nonpictorial title page, displaying only its title and volume number in both Arabic and English—adding the latter with American and European booksellers in mind; Farag, "*Al-Muqtataf*," 62.
15. These were founded, respectively, in 1872, 1845, 1818, 1877, 1872, 1881, 1843,

1828, 1877, 1865, and 1788. *Popular Science Monthly* and *Nineteenth Century* were the two journals Sarruf and Nimr relied upon most. *Popular Science Monthly*, like *Al-Muqtataf* itself, also drew liberally upon other journals, such as *Edinburgh*, *Fortnightly*, *Cornhill Magazine*, *Chambers Journal*, *Spectator*, and *Nature*.

16. For a complete list of Sarruf and Nimr's sources, see Farag, "*Al-Muqtataf*," 119–31.
17. Nevertheless, a full subscription to *Al-Muqtataf* was still well beyond the means of most late Ottoman subjects.
18. On pricing, see Farag, "*Al-Muqtataf*," 87–95. Nadia Farag, in her comprehensive study of the journal, found no evidence on the financial viability of the journal during its early Beirut years. The editors most likely subsidized it, at a loss, from their own personal funds. Indeed, as late as 1887, the editors were chastising their readers for late payments. See the note at the end of the advertisements in *Al-Muqtataf* 11 (1887): 256. This was also the subject of a later article: "Huquq al-suhuf wa-al-mushtarikin fiha" [The rights of journals and their subscribers], *Al-Muqtataf* 28 (1903): 666–68. With the journal's increasing success, and the expansion in the number of pages, however, the cost of the annual subscription to the journal would have proved prohibitively high: Ami Ayalon estimates that the subscription, at one hundred Ottoman *qurush* (or one Ottoman pound) per annum, represented about 10 percent of the average annual income in the Syrian provinces; see Ami Ayalon, "Modern Texts and Their Readers in Late Ottoman Palestine," *Middle Eastern Studies* 38 (2002): 32.
19. Sarruf and Nimr's journal and the various supplementary pamphlets that their press issued can be compared to Victorian periodicals and literature sponsored by the Society for the Diffusion of Useful Knowledge, founded in 1826 by Lord Brougham. On the society, see James Secord, *Victorian Sensation: The Extraordinary Publication, Reception, and Secret Authorship of "Vestiges of the Natural History of Creation"* (Chicago: University of Chicago Press, 2000), 48–51; Aled Jones, *Powers of the Press: Newspapers, Power and the Public in Nineteenth-Century England* (Aldershot: Ashgate, 1996), 105–8; Scott Bennett, "Revolutions in Thought: Serial Publications and the Mass Market for Readings," in *The Victorian Periodical Press: Samplings and Soundings*, ed. Joanne Shatock and Michael Wolff (Leicester: Leicester University Press, 1982), 225–60. This society may have also provided the model for Muhammad Arif Pasha's 1868 Egyptian Society of Knowledge for the Publication of Useful Books (Jamiʿyat al-Maʿarif li-Nashr al-Kuttab al-Nafiʿa); see Timothy Mitchell, *Colonising Egypt* (Cambridge: Cambridge University Press, 1989), 90.
20. For more on Bustani, see below, particularly n. 103.
21. "*Al-Jinan*," *Al-Jinan* 1 (1870): 1. On al-Bustani's commitment to the "diffusion of knowledge through the medium of Arabic," see Butrus Abu-Manneh, "The Christians between Ottomanism and Syrian Nationalism: The Ideas of Butrus al-Bustani," *International Journal of Middle East Studies* 11 (1980): 291; Ami Ayalon, *The Press in the Arab Middle East* (New York: Oxford University Press, 1995), 35–36, 211. Al-Bustani was also the secretary of the Syrian Scientific Society (discussed below). In an 1859 address to the society, he called on the need for newspapers to educate the public (Ayalon, *Press*, 31–32).
22. Martin Hartmann, *The Arabic Press of Egypt* (London: Luzac, 1899); Ahmed Bioud Abdelghani, *3200 Revues et journaux arabes de 1800 à 1965* (Paris: Bibliothèque National, 1969); Ayalon, *Press*, 28–72. Ayalon (*Press*, 50) counted 514 newspapers and 113 journals in Cairo and Alexandria from 1880 to 1908.
23. Population figures from 1884 to 1913, from Charles Issawi, *The Economic History*

of Turkey, 1800–1914 (Chicago: University of Chicago Press, 1980), 17; Donald Quataert, "Population," in *An Economic and Social History of the Ottoman Empire, 1300–1914*, ed. Halil Inalick and Donald Quataert (Cambridge: Cambridge University Press, 1994), 780. Donald Cioeta estimated Syria's literacy rate in the late nineteenth century to have been roughly equivalent to that of Egypt's; Donald Cioeta, "Ottoman Censorship in Lebanon and Syria, 1876–1908," *International Journal of Middle East Studies* 10 (1979): 180.

24. "Al-Muqtataf," *Al-Muqtataf* 8 (1884): 676–79; Ayalon, *Press*, 148, 158; Hartmann, *Arabic Press*, 7.
25. They attracted a wide readership from the start: in the first few years of the journal they received inquiries that ranged from such quotidian questions as how to dye cloth, remove iron rust, and protect eggs from rotting to more technical ones such as questions about the latest printing techniques. See *Al-Muqtataf* 1 (1875): 94, 114, 209, 233.
26. Jurji Zaydan, "Tarikh al-nahda al-ʿilmiya al-akhira" [History of the latest renaissance of knowledge], *Al-Hilal*, 1901, 201–5; Salama Musa, *The Education of Salama Musa*, trans. L. O. Schuman (Leiden: Brill, 1961), 33.
27. *Al-Muqtataf* 8 (1883): 174–75.
28. They also met with many other familiar figures (and government employees) like Riaz Pasha and ʿAli Pasha Mubarak.
29. Farag, "*Al-Muqtataf*," 72–73, 86.
30. ʿAbdin, archive box 233: *Taʿlim: Murasillat*, 1908–9, Egyptian National Archives.
31. "Al-Muqaddima" [Introduction], *Al-Muqtataf* 1 (1876): 1.
32. Jurji Zaydan, *The Autobiography of Jurji Zaydan: Including Four Letters to His Son*, trans. Thomas Philipp (Washington, DC: Three Continents Press, 1990), 37.
33. See, e.g., "Al-muʿaribat," *Al-Muqtataf* 8 (1883): 107–12, 212–16, 294–96, 341–46, 401–2, 467–69.
34. Cited in Ayalon, *Press*, 180, 184 (citing *Al-Muqattam* in 1889). "Tabaqat al-kuttab" [Classes of writers], *Majallat Sarkis* 13 (1910): 253–59, classified Yaʿqub Sarruf's style as "scientific," and Faris Nimr's as "journalistic." See Farag, "*Al-Muqtataf*," 92. Salama Musa (*Education*, 38) describes Sarruf's style as "telegraphic."
35. "Al-madhhab al-Darwini" [The Darwinian school of thought], *Al-Muqtataf* 7 (1882): 33.
36. Amin Shumayyil, "Madhhab Darwin ʿand al-aqadimin" [Darwin's school of thought among the Ancients], "Munazarat wa-murasilat" [Views and letters], *Al-Muqtataf* 10 (1885): 145–46.
37. "Masaʾil wa-ajwibha" [Questions and answers], *Al-Muqtataf* 14 (1887): 359.
38. Rizqʾallah al-Birbari, "Fi asl al-insan" [On the origin of humanity], *Al-Muqtataf* 1 (1876): 231–32, 242–45, 279–81.
39. Ibid., 279–80; Charles Hodge, *What Is Darwinism?* (New York: Scribner, Armstrong, 1874).
40. Bishara Zilzal, "Al-insan" [Man], *Al-Muqtataf* 2 (1877): 202–5, 234–37, 254–57, 273–75.
41. Ibid., 203.
42. Ibid., 236.
43. Ibid., 257; Thomas Bendyshe makes references to Fabricius's 1721 refutation of Isaac de la Peyère's 1655 *Prae-Adamitae*; Thomas Bendyshe, "History of Anthropology," *Memoirs of the Anthropological Society* 1 (1863–64): 335–420. For more on the background to this, see David Livingstone, *Adam's Ancestors: Race, Religion, and the*

Politics of Human Origins (Baltimore, MD: Johns Hopkins University Press, 2008). See also Farag, "*Al-Muqtataf*," 249. Many nineteenth-century science writers similarly sought to square scripture with the geological findings of their day. See also Pietro Corsi, *Science and Religion: Baden Powell and the Anglican Debate, 1800–1860* (Cambridge: Cambridge University Press, 1988); Secord, *Victorian Sensation*, 57.

44. "Asl al-insan wa-atharuhu" [The origin and vestiges of man], *Al-Muqtataf* 4 (1879): 89.
45. Peter Bowler, *Evolution: The History of an Idea* (Berkeley and Los Angeles: University of California Press, 1983): 231; L. Leakey, Jack Prost, and Stephanie Prost, *Adam or Ape: A Sourcebook of Discoveries about Early Man* (Cambridge, MA: Schenkman, 1971).
46. "Asl al-insan wa-atharuhu," *Al-Muqtataf* 4 (1879): 92; *Al-Muqtataf* 23 (1898): 945.
47. Bowler, *Evolution*, 232–33; Peter Bowler, *Fossils and Progress: Paleontology and the Idea of Progressive Evolution in the Nineteenth Century* (New York: Science History Publications, 1976); Martin Rudwick, *The Meaning of Fossils: Episodes in the History of Palaeontology* (New York: Science History Publications, 1976); Secord, *Victorian Sensation*, 57.
48. For more on Sarruf's and Nimr's evolutionary notions of progress and their political views, see the next chapter.
49. Charles Darwin, Notebook M, in *On Evolution: The Development of the Theory of Natural Selection*, ed. Thomas Glick and David Kohn (Indianapolis: Hackett, 1996), 79.
50. Charles Darwin, *The Descent of Man, and Selection in Relation to Sex* (London: J. Murray, 1871), 34–69, 65; Charles Darwin, *The Expression of the Emotions of Man and the Animals* (London: John Murray, 1872).
51. Bishara Zilzal, "Al-insan" [Man], *Al-Muqtataf* 2 (1877): 235.
52. "Idrak al-hayawan ghayr al-natiq" [The intelligence of dumb animals], *Al-Muqtataf* 4 (1879): 92–96.
53. Ibid., 93.
54. "Ghariyzat al-hayawan" [Animal instincts], *Al-Muqtataf* 9 (1885): 513–18; G. J. Romanes, "The Darwinian Theory of Instinct," *Nineteenth Century*, September 1884. On Romanes, see Farag, "*Al-Muqtataf*," 263.
55. Darwin, *Descent of Man*, 71–72, 96, 100.
56. Zilzal, "Al-insan," *Al-Muqtataf* 2 (1877): 203, 236.
57. "Masa'il wa-ajwibha" [Questions and Answers], *Al-Muqtataf* 17 (1892): 138. Many of these articles were summaries of such late-nineteenth-century writings as John Lubbock, *Chapters in Popular Natural History* (London: National Society, 1883); John Lubbock, *On the Senses, Instincts and Intelligence of Animals* (London: Keegan, Paul, Trench, 1888); John Lubbock, "Intelligence in Animals," *Knowledge* 1 (1881): 28–30; G. J. Romanes, "Mental Evolution in Animals," *Nature* 24 (1884): 330–36, 404–28; J. M. Hayward, "Teaching Animals to Converse," *Nature* 24 (1884): 216, 547. For letters from readers, see *Al-Muqtataf* 9 (1885): 657; *Al-Muqtataf* 21 (1897): 184; Farag, "*Al-Muqtataf*," 266.
58. "Ghariyzat al-hayawan" [Animal instincts], *Al-Muqtataf* 9 (1885): 518.
59. "Sharils Darwin" [Charles Darwin], *Al-Muqtataf* 7 (1882): 2–6.
60. Ibid., 6.
61. Ibid.; Frederick Temple, *The Relations between Religion and Science: Eight Lectures Preached before the University of Oxford* (New York: Macmillan, 1884); James McCosh, *Christianity and Positivism: A Series of Lectures to the Times on Natural Theology*

and Apologetics (New York: R. Carter and Brothers, 1871); Richard Hofstadter, *Social Darwinism in American Thought, 1860–1915* (Boston: Beacon Press, 1955), 29.

62. "Al-madhhab al-Darwini," *Al-Muqtataf* 7 (1882): 33–40, 57–63 (quotation on 33); John Fiske, *Outlines of Cosmic Philosophy Based on the Doctrine of Evolution with Criticisms of the Positive Philosophy*, 2 vols. (New York: Houghton Mifflin, 1874). On Fiske, see also Mike Hawkins, *Social Darwinism in European and American Thought, 1860–1945* (Cambridge: Cambridge University Press, 1997), 106–8.
63. John Fiske, *Darwinism and Other Essays*, new ed. (Boston: Houghton Mifflin, 1885), 7–8.
64. "Madhhab Darwin," *Al-Muqtataf* 7 (1882): 62. Nevertheless, they themselves were not always consistent in their recognition of Darwin's theory of evolution as essentially non-progressive. This was particularly evident in their later writings on evolution, especially in their writings on social evolution (see next chapter).
65. Ibid., 39–40.
66. "Al-madhhab al-Darwini," *Al-Muqtataf* 7 (1882): 33.
67. Ibid., 35–36.
68. Ibid., 61–62.
69. Ibid., 62.
70. Ibid., 62–63.
71. Ibid., 63.
72. Ibid.
73. "Ta'liq 'ala Spinsir" [Comments on Spencer], *Al-Muqtataf* 22 (1898): 819–20.
74. On the background to this, see Theodore Dwight Bozeman, *Protestants in the Age of Science: The Baconian Ideal and Ante-bellum Religious Thought* (Chapel Hill: University of North Carolina Press, 1977).
75. James McCosh, *The Religious Aspects of Evolution* (New York: G. P. Putnam's Sons, 1888), 7; John Fiske, *Destiny of Man Viewed in Light of His Origin* (Boston: Houghton Mifflin, 1889). On McCosh's Calvinist views of "special selection" as "divine election" and Fiske's theistic interpretation of Spencer's positivism—and broader Presbyterian and Unitarian views on this—see Herbert Schneider, "The Influence of Darwin and Spencer on American Philosophical Theology," *Journal of the History of Ideas* 6 (1945): 3–18.
76. For the broader context to these debates, see James Moore, *The Post-Darwinian Controversies: A Study of the Protestant Struggle to Come to Terms with Darwin in Great Britain and America, 1870–1900* (Cambridge: Cambridge University Press, 1979); David Livingstone, *Darwin's Forgotten Defenders: The Encounter between Evangelical Theology and Evolutionary Thought* (Grand Rapids, MI: Eerdmans, 1987).
77. James Dennis, *Christian Missions and Social Progress: A Sociological Study of Foreign Missions* (New York: Fleming H. Revell, 1906) 3:3–4. See also T. O. Beidelman, *Colonial Evangelism* (Bloomington: Indiana University Press, 1982).
78. Antoine Rabbath, *Documents inédits pour servir à l'histoire du christianisme en orient*, 2 vols. (Paris: A. Picard et fils, 1905–14); Michel Jullien, *La nouvelle mission de la Compagnie de Jésus en Syrie, 1831–1895* (Tours: Mame, 1898); Université Saint-Josephe, *Les Jésuites en Syrie, 1831–1931* (Paris: Jesuit Press, 1931); Charles Frazee, *Catholics and Sultans: The Church and the Ottoman Empire, 1453–1923* (Cambridge: Cambridge University Press,, 1983).
79. Fisk announced this upon first landing in the Holy Land in 1819. Pliny Fisk, *The Holy Land an Interesting Field of Missionary Enterprise* (Boston: Samuel T. Armstrong,

1819), 24. See also John Davis, *The Landscape of Belief: Encountering the Holy Land in Nineteenth-Century American Art and Culture* (Princeton, NJ: Princeton University Press, 1996), 29.

80. O. W. Elsbree, *The Rise of the Missionary Spirit in America, 1790–1815* (Williamsport, PA: Williamsport Print and Binding Co., 1928), 132. On the First Great Awakening (c. 1720–40), see Frank Lambert, *Inventing the "Great Awakening"* (Princeton, NJ: Princeton University Press, 2001); on the second (c. 1795–1835), see Bary Hankins, *The Second Great Awakening and the Transcendentalists* (Westport, CT: Greenwald Press, 2004). See also Mark Noll, *The Rise of Evangelicalism: The Age of Edwards, Whitefeld and the Wesleys* (Leicester: InterVarsity Press, 2004).

81. See Clifton Jackson Phillips, *Protestant America and the Pagan World: The First Half Century of the American Board of Commissioners for Foreign Missions, 1810–1860* (Cambridge, MA: Harvard University Press, 1969): 20; and James Field, *America and the Mediterranean World* (Princeton, NJ: Princeton University Press, 1969), 89.

82. ABCFM, *First Ten Annual Reports of the American Board of Commissioners for Foreign Missions* (Boston: Crocker and Brewster, 1834), 28. The original churches represented on the ABCFM were largely Congregational, although there were also Dutch Reformed and Presbyterian churches. The latter church later played a more significant role after 1870, when the ABCFM was replaced by the Presbyterian Church Board of Foreign Missions (PCBFM). Rufus Anderson, *The History of the Missions of the American Board of Commissioners for Foreign Missions to the Oriental Churches* (Boston, 1872); Samuel Bartlett, *Historical Sketches of the Missions of the American Board* (New York, 1972); A. L. Tibawi, *American Interests in Syria, 1800–1901: A Study of Educational, Literary and Religious Work* (Oxford: Clarendon, 1966), 7, 10–11, 191.

83. *Missionary Herald* 5 (1810): 517.

84. By 1862 there were 1,925 students in American mission schools, and in 1876 there were 2,840; Lindsay Rao, "Nineteenth Century American Schools in the Levant: A Study of Progress" (PhD diss., University of Michigan, 1964), 95–101.

85. "Extracts from a Joint Letter of the Missionaries at Beyroot," *Missionary Herald* 33 (November 1837): 444.

86. Rao, "American Schools in the Levant," 40.

87. Eli Smith (1801–57) was a graduate of Yale and Andover Theological Seminary. He received his diploma in theology at the latter in 1826 and joined the ABCFM Syria mission the following year. He served as director of the mission press in Malta and, after 1834, in Beirut. He also founded one of the first Arabic newspapers in Beirut (*Majmuʿat Fawaʾid*) and authored and translated numerous works in Arabic, including the translation of the Bible, which he began with mission disciples Butrus al-Bustani and Nasif al-Yaziji under the auspices of the American Bible Society and the ABCFM in 1844. It was completed after his death in 1857—the Arabic New Testament was published in Beirut in 1860—by Cornelius Van Dyck and Al-Azhar graduate Shaykh Yusuf al-ʿAsir.

88. "Mission to Syria—Letter from Mr. Smith," *Missionary Herald* 31 (March 1835): 93.

89. "Extracts from a Joint Letter of the Missionaries at Beyroot," *Missionary Herald* 33 (November 1837): 444.

90. "American Board of Commissioners for Foreign Missions," *Journal of the American Oriental Society* 1 (1849): 61; Tibawi, *American Interests in Syria*, 27, 213.

91. Louis Cheikho, "Tarikh fann al-tibaʿa fi al-Sharq" [History of printing in the East], *Al-Mashriq* 3 (1900): 251–57, 500–503, 706–14, 800–807, 839–44; Louis Cheikho,

La littérature arabe au XIXe siècle (Beirut, 1908), 1:44–45; Tibawi, *American Interests in Syria*, 166; Ayalon, *Press*, 50.

92. Isaac Hall, "The Arabic Bible of Drs. Eli Smith and Cornelius Van Dyck," *Journal of the American Oriental Society* 11 (1885): 276–86.
93. See ABC 50: Eli Smith Arabic Papers (1821–57): Houghton Library Archives: box 1.
94. ABCFM, *Annual Report* (Boston: The Board, 1868), 48. On the broader cultural importance of calendars for the diffusion of ideas among "common readers," see James M. Brophy, "The Common Reader in the Rhineland: The Calendar as Political Primer in the Early Nineteenth Century," *Past and Present* 185 (2004): 119–59.
95. *Missionary Herald* 65 (1869): 410. On the events of 1860, see Leila Fawaz, *An Occasion for War: Civil Conflict, in Lebanon and Damascus in 1860* (Berkeley and Los Angeles: University of California Press); Ussama Makdisi, *The Culture of Sectarianism: Culture, Community, History and Violence in Ottoman Lebanon* (Berkeley and Los Angeles: University of California Press,, 2000); Ussama Makdisi, "After 1860: Debating Religion, Reform and Nationalism in the Ottoman Empire," *International Journal of Middle East Studies* 34 (2002): 601–17.
96. For more on this, see Ussama Makdisi, *Artillery of Heaven: American Missionaries and the Failed Conversion of the Middle East* (Ithaca, NY: Cornell University Press, 2008).
97. "Report of the Syria Mission for the Year 1841," *Missionary Herald* 38 (June 1842): 227.
98. Rufus Anderson (1796–1880) graduated from Bowdoin College in 1818 and from Andover Theological Seminary in 1822. He was ordained as a minister in 1826. He was assistant secretary of the ABCFM from 1824 to 1832, after which he became its secretary, in which office he remained until 1866. He authored numerous works on missions, and from 1867 to 1869 he also lectured on "foreign missions" at Andover. For more on Anderson and his mission strategy, see Paul William Harris, *Nothing but Christ: Rufus Anderson and the Ideology of Protestant Foreign Missions* (Oxford: Oxford University Press, 1999).
99. ABCFM, *The Divine Instrumentality for the World's Conversion* (Boston: n.p., 1856), 4, 10; Rao, "American Schools in the Levant," 154–56.
100. From the 1847 constitution in Edward Salisbury, "Syrian Society of Arts and Sciences," *Journal of the American Oriental Society* 3 (1853): 477–86. The majority of its members were nevertheless Christian. See Philippe Tarrazi, *Tarikh al-sihafa al-ʿArabiya* [History of the Arabic press] (Beirut: Dar Sadir, 1913–29), 4:112; and Fruma Zachs, *The Making of a Syrian Identity: Intellectuals and Merchants in Nineteenth Century Beirut* (Leiden: Brill, 2005), 137–45. The society lasted only about five years, from 1847 to 1852. It is possible that Eli Smith, one of the founding members, established the society on the model of the 1844 Syro-Egyptian Society of London. Smith was in regular correspondence with several of the council members of the latter society—including Asʿad Yaʿqub Khayyat and John Gordon Scott—and had copies of the society's circulars in his possession. ABC 60: Eli Smith Papers, folders 46, 72, and 144, Houghton Library Archives, Harvard University. The two societies also had many common goals: "Great efforts are making for the general diffusion of knowledge," read the *Prospectus of the Plan and Objects of the Syro-Egyptian Society*; "and though various benefits have been conferred, those who are acquainted with the modern condition of Egypt and Palestine and Syria are well aware how much remains to be done before they can be ranked among the civilized nations of the earth." Compare with: "The existence and prosperity of this society is an indication

most interesting to the philanthropist and the scholar that the culture of Western nations is exerting a great and happy influence upon minds in Syria, and even gives promise for the naturally fine intellect of the Arab race in the mould of modern civilization"; Salisbury, "Syrian Society of Arts and Sciences," *Journal of the American Oriental Society* 3 (1853): 478.

101. These lectures were delivered by Cornelius Van Dyck, Butrus al-Bustani, and Salim Nawfal respectively. Other papers included Nasif al-Yaziji's "On the Sciences of the Arabs"; Johannes Wortabet's "The Measure of the Progress of Knowledge in Syria at the Present Time and Its Causes"; and Mikha'il Mudawwar's "On the Origins of Commerce and Its Vicissitudes." See Salisbury, "Syrian Society of Arts and Sciences," *Journal of the American Oriental Society* 3 (1853): 477–86; and Butrus al-Bustani, *Aʿmal al-Jamʿiya al-Suriya* [Collected papers of the Syrian Society] (Beirut: n.p., 1852).

102. "ʿAbeih Station Report for 1849," *Missionary Herald*, August 1850, 262–63.

103. Al-Bustani worked with ABCFM missionaries as a schoolteacher, a private Arabic tutor, and translator and collaborator for mission station publications throughout the 1840s (he also worked briefly as dragoman for the US consulate in Beirut). He founded his own school, Al-Madrasa al-Wataniya (National School), in 1863. Missionaries at the Syrian Protestant College, which opened in 1866, had briefly considered using it as a "preparatory department" for their students, before deciding against it. Among other things, they felt that the school rivaled their own, and they suspected that al-Bustani had ambitions of his own. See the Board of Managers' discussion of a letter of complaint sent to them by al-Bustani: MSS AUB 33, page 55, AUB Library Archives. See also Jurji Zaydan, *Tarajim mashahir al-sharq* [Biographies of famous men of the East] (Cairo: Matbaʿat al-Hilal, 1911), 2:25–32; *al-Muqtataf* 8 (1883): 1–7; A. L. Tibawi, "The American Missionaries in Beirut and Butrus al-Bustani," *Middle Eastern Affairs* 3 (1963): 137–82; Abu-Manneh, "The Christians between Ottomanism and Syrian Nationalism"; Albert Hourani, "Bustani's Encyclopaedia," *Journal of Islamic Studies* 1 (1990): 111–19. For more on al-Bustani, see also Stephan Sheehi, "Inscribing the Arab Self: Butrus al-Bustani and Paradigms of Subjective Reform," *British Journal of Middle Eastern Studies* 27 (2000): 7–24.

104. Until 1853—when Protestantism was declared an official civil and ecclesiastical community by imperial decree—missionaries alleged that would-be converts had been deterred by the prospect of existing outside any clear ecclesiastical, legal, or political protection in the empire. As late as 1880, missionaries were expressing similar anxieties, complaining that the 1853 *firman* seemed to have been seen as "merely provisional" in the eyes of local authorities. See W. W. Eddy to Reverend Ellinwood, 3 May 1880, Correspondence, vol. 5, no. 42, Presbyterian Church in the U.S.A. Board of Foreign Missions, available on microfilm at Princeton Theological Seminary.

105. The combination of the 1838 Anglo-Turkish Convention and the Tanzimat reforms introduced by Sultan Abdülmecid in the 1840s triggered a period of rapid socioeconomic transformation and commercial growth. See Jens Hanssen, *Fin de Siècle Beirut: The Making of an Ottoman Provincial Capital* (Oxford: Oxford University Press, 2005), esp. 32–37.

106. According to an 1883 paper read in Arabic to the Syrian Scientific Society in Beirut (a descendant of the 1847 Syrian Society of Arts and Sciences), in the administrative district, or *sanjaq*, of Beirut alone, there about were 250 students to every 1,000

people (nearly six times the number of students in other districts throughout Lebanon). Henry Diab and Lars Wahlin, "The Geography of Education in Syria in 1882: With a Translation of 'Education in Syria' by Shahin Makarius," *Geografiska Annaler* 65 (1983): 105–28.

107. "Reasons for the Establishment of a Syrian Protestant College," ABCFM Papers, vol. 545: 1860–71, no. 110.

108. Albert Hourani's *Arabic Thought in the Liberal Age, 1798–1939* (Cambridge: Cambridge University Press, 1962) is a good example of this. See also Philip Hitti, *Lebanon in History from the Earliest Times to the Present* (London: Macmillan, 1957); Kamal Salibi, *The Modern History of Lebanon* (London: Weidenfeld and Nicolson, 1965), esp. chap. 7; George Antonius, *The Arab Awakening: The Story of the Arab National Movement* (New York: Capricorn Books, 1965), 35–43; Elie Kedourie, *The Chatham House Version and Other Middle Eastern Studies* (New York: Praeger, 1970), 289; Elie Kedourie, "The American University of Beirut," in his *Arab Political Memoirs and Other Studies* (London: Cass, 1974).

109. For a critical discussion of the uses—and abuses—of the term "modernity" outside the Anglo-American and continental European world, see Frederick Cooper, *Colonialism in Question* (Berkeley and Los Angeles: University of California Press, 2005), 113–49. For more on missions and modernity, see Peter Pels, "The Anthropology of Colonialism: Culture, History and the Emergence of Western Governmentality," *Annual Review of Anthropology* 26 (1997): esp. 171–74; Peter van der Veer, ed., *Conversion to Modernity: The Globalisation of Christianity* (New York: Routledge, 1996), esp. the editor's introduction, 1–22; Gauri Viswanathan, *Outside the Fold: Conversion, Modernity and Belief* (Princeton, NJ: Princeton University Press, 1998); Paul Sedra, *From Mission to Modernity: Evangelicals, Reform and Education in Nineteenth Century Egypt* (London: I. B. Tauris, 2011); Ussama Makdisi, "Reclaiming the Land of the Bible: Missionaries, Secularism and Evangelical Modernity," *American Historical Review* 102 (1997): esp. 683.

110. Bliss Family Papers, 1850–1981, box 5, addendum 1, folder 1 (1863), 110, Amherst College Library Archives and Special Collections.

111. See notices on public experiments in the annals in particular. On instrumentation, moral authority, and public experiments, see Simon Schaffer, "Self Evidence," *Critical Inquiry* 18 (1992): 327–62. On commerce and Christianity, see Andrew Porter, "'Commerce and Christianity': The Rise and Fall of a Nineteenth-Century Missionary Slogan," *Historical Journal* 28 (1985): 597–621.

112. On postmillennialism and the Social Gospel movement, see Walter Rauschenbusch, *A Theology for the Social Gospel* (New York: Abingdon Press, 1917); Charles Hopkins, *The Rise of the Social Gospel in American Protestantism, 1865–1915* (New Haven, CT: Yale University Press, 1940); Robert Handy, ed., *The Social Gospel in America, 1870–1920* (New York: Oxford University Press, 1966); and J. B. Quandt, "Religion and Social Thought: The Secularization of Postmillennialism," *American Quarterly* 25 (1973): 390–409.

113. Albeit weapons that, in the age of Darwin, had a propensity to backfire. See, e.g., Lynette Thistlethwayte, "The Role of Science in the Hindu-Christian Encounter," *Indo-British Review* 19 (1991): 73–82. For the literature on missionaries and science, see Sujit Sivasundaram, *Nature and the Godly Empire: Science and Evangelical Missions in the Pacific, 1795–1850* (Cambridge: Cambridge University Press, 2005), esp. 58–94; C. Peter Williams, "Healing and Evangelism: The Place of Medicine in Late Victo-

rian Protestant Missionary Thinking," in *The Church and Healing*, ed. W. J. Sheils (Oxford: Oxford University Press, 1982); Megan Vaughan, *Curing Their Ills: Colonial Power and African Illness* (Cambridge: Cambridge University Press, 1991): 55–76; Paul S. Landau, "Explaining Surgical Evangelism in Colonial Southern Africa: Teeth, Pain and Faith," *Journal of African History* 37 (1996): 261–81; John Comaroff and Jean Comaroff, *Of Revelation and Revolution* (Chicago: University of Chicago Press, 1997), 323–64; and Rosemary Fitzgerald, '"Clinical Christianity': The Emergence of Medical Work as a Missionary Strategy in Colonial India, 1800–1914," in *Health, Medicine and Empire*, ed. Bisamoy Pati and Mark Harrison (New Delhi: Orient Longman, 2001), 88–136.

114. "Reasons for the Establishment of a Syrian Protestant College," ABCFM Papers, vol. 545: 1860–71, no. 110.
115. For diverging nineteenth-century evangelical views on natural theology, see Aileen Fyfe, *Science and Salvation: Evangelical Popular Science Publishing in Victorian Britain* (Chicago: University of Chicago Press, 2004), 6–8. See also David Livingstone, D. G. Hart, and Mark Noll, eds., *Evangelicals and Science in Historical Perspective* (Oxford: Oxford University Press, 1999), esp. 120–74.
116. Rufus Anderson, *History of the Missions of the American Board of Commissioners for Foreign Missions to the Oriental Churches* (Boston: Congressional Publication Society, 1872), 387–88. For the full response and recommendations from the Prudential Committee, the ABCFM's executive committee, see Bliss Family Papers, 1850–1981, box 5, addendum 1, folder 1, Amherst College Library Archives and Special Collections.
117. Daniel Bliss was a graduate of Amherst College (1852) and Andover Theological Seminary (1855), where he was also ordained as a Congregational clergyman. He would later serve as president of the Syrian Protestant College from 1866 to 1902.
118. Syrian Protestant College Minutes of the Board of Trustees, book I (4 October 1866), 59, American University of Beirut (hereafter AUB) Library Archives.
119. Bliss Family Papers, 1850–1981, box 5, addendum 1, folder 1 (1863), 59–94, Amherst College Library Archives and Special Collections.
120. Both the Syrian Protestant College and Robert College in Istanbul were chartered in New York that year. Syrian Protestant College Minutes of the Board of Trustees, book I (18 April 1863), 3–4, AUB Library Archives.
121. "Annual Report of the Beirut Station, 1866," ABCFM Papers, vol. 545: 1860–1871, no. 36.
122. "Prospectus and Programme of the Syrian Protestant Collegiate Institute Beirut," 2, ABCFM Papers, vol. 545: 1860–1871, no. 110.
123. Syrian Protestant College, *Catalogue of the Syrian Protestant College*, 2 vols. (Beirut: Syrian Protestant College Press, 1887–88), 1:2–3.
124. Tuition for the college was initially set at 5 Turkish liras, and the cost of full board was 12 Turkish liras (or the equivalent of about $25 and $80 at the time); see "Education in Syria," *Hours at Home: A Popular Monthly of Instruction and Recreation* 11, no. 4 (1870): 328. Many students were aided moreover through college board funds, although the majority were encouraged to secure means toward their own support so as to foster "self-reliance." MSS AUB 33: Record of the Secretary of the Board of Managers of the Syrian Protestant College, 1864–1903, 23 January 1868, 89, AUB Library Archives.
125. Students in the preparatory department (admitted from the age of ten) engaged in regular Bible study plus ancient, modern, and scripture history and took lessons in

Arabic, English, Latin, and French or Greek, arithmetic, algebra, geometry, geography, geology, chemistry, physics, botany, zoology, physiology, and astronomy over a period of four years.

126. Students in the Medical Department took lessons in chemistry and anatomy (in year 1); materia medica, physiology, and zoology (year 2); and surgery, obstetrics, and pathology, alongside clinical work (years 3 and 4). In the Literary Department students studied Arabic, English, French, scripture criticism (*tafsir al-kitab al-muqadis*), and mathematics (in year 1); history and natural philosophy (regularly after year 2); Latin, chemistry, botany, and zoology (year 3); geology, physiology, logic, and moral philosophy (year 4).

127. *Missionary Herald* 58 (September 1862): 220; Rao, "American Schools in the Levant," 184. Bliss taught Bible studies, Latin, ethics (also called "moral science"), and philosophy.

128. Minutes of the Board of Trustees, book I (4 May 1864), 23, 27, AUB Library Archives.

129. Daniel Bliss to the Board, *Annual Reports, Board of Managers, Syrian Protestant College, 1866–67—1901–02* (Beirut: American University of Beirut Press, 1963), 27 June 1871: 18.

130. "Annual Report of the Beirut Station, 1866," ABCFM Papers, vol. 545: 1860–1871, no. 36.

131. *Fifty-Ninth Annual Report of the American Board of Commissioners for Foreign Missions* (Cambridge: The Board, 1869), 32–33.

132. Post wrote that he felt he had "entered a new era" in his life, one that "provided him with a broader or higher sphere of influence." George Post to the Board, June 1880, Correspondence and Reports, no. 47, PCBFM. Post (1838–1909) graduated from the Union Theological Seminary and the New York University Medical College in 1860. He was ordained by the Fourth Presbytery of New York the following year, and in 1863 the ABCFM stationed him in Tripoli, where he remained for five years, as a medical missionary. He authored and translated several natural history, botany, and medical texts—and collected and named countless botanical specimens in Syria—throughout his long career in Beirut and at the Syrian Protestant College. See Lutfi Saʿdi, *The Life and Works of George Edward Post (1838–1909)* (N.p.: Saint Catherine Press, 1938).

133. Minutes of the Board of Trustees, book I (1 November 1870), 98, AUB Library Archives.

134. Letter from F. A. Wood, 23 March 1877, Correspondences and Reports, letter no. 26, PCBFM. Instead, a theological department at the college was set up under the supervision of James Dennis to accommodate the needs of theology students, though they also enrolled in the regular course of study available through the Literary Department. Minutes of the Board of Trustees, book I (31 January 1873), 116, AUB Library Archives.

135. James Dennis, *A Sketch of the Syria Mission* (New York: Mission House, 1872), 21–22.

136. Daniel Bliss to the Board, in *Annual Reports*, 19 July 1877: 36. The observatory was supplied with a full set of instruments for a variety of meteorological observations, which were often discussed in the college's classrooms and publications.

137. Daniel Bliss, *Al-durus al-awliya fi al-falsafa al-ʿaqliya* [Primary lessons in rational philosophy] (Beirut: n.p., 1873), 1–2.

138. For a full listing of textbooks written by the faculty, see Suha Tamim, *A Bibliogra-*

phy of A.U.B. Faculty Publications, 1866–1966 (Beirut: American University of Beirut Press, 1967).

139. These textbooks were used, for instance, at the Madrasa al-Sultaniya; see chapter 4.
140. See Ayalon, *Press*, 31–39.
141. See Minutes of the Board of Trustees, book I (2 December 1874), 128, AUB Library Archives.
142. Among Van Dyck's works were *al-mirah al-wadiya fi al-kura al-ardiya, Kitab al-naqsh fi al-hajar, Kitab al-rawda al-zahriya fi al-usul al-jabriya, Usul al-kimiya, Usul ʿilm al-haya*; see Lutfi Saʿid, "Al-hakim Cornelius Van Allen Van Dyck," *Isis* 27 (1937): 20–45.
143. *Sirr al-najah* appeared in 1880 and was based upon Smiles's 1866 edition; excerpts of the translation were previously published in *Al-Muqtataf*; see, e.g., "Tarjamat Birnard Balsy" [The biography of Bernard Palissy], *Al-Muqtataf* 2 (1877): 73–75, 185–87; cf. Samuel Smiles, *Self-Help: With Illustrations of Character, Conduct and Perseverance* (Chicago: Belford, Clarke, 1881), 90–104. The Arabic translation proved wildly popular among Syrian and Egyptian students alike. "I had read parts of the book *The Secret of Success*," wrote Jurji Zaidan in his autobiography, "which Dr. Sarruf had translated into Arabic. Vigor and zeal sprang up in me" (*Autobiography*, 44). See also Tibawi, *American Interests in Syria*, 205; Farag, "*Al-Muqtataf*," 15; Mitchell, *Colonising Egypt*, 108–10.
144. Saʿid, "Al-hakim Cornelius Van Allen Van Dyck."
145. Van Dyck's contributions to *Al-Muqtataf* included "Tarikh al-tibaʿa fi al-Yunan wa al-Sharq" [History of publishing in Greece and Syria], *Al-Muqtataf* 1 (1876): 25–27, 49–51, 73–75, 97–98, 121–23, 145–46. On the suggestion of a "readers digest," see Farag, "*Al-Muqtataf*," 57; Zaydan, *Autobiography*, 57. On the director of publications, Khalil al-Khuri, see Cioeta, "Ottoman Censorship in Lebanon and Syria," 171.
146. See, e.g., MSS AUB 1: box 5, 1/3, folder 3: Letters from Dr. Stuart Dodge to Daniel Bliss, 1880–1883, 24 November 1881, AUB Library Archives.
147. From H. Jessup, 28 February 1877, Correspondence and Reports, 1877–1879, vol. 4 (90–91), letter no. 21, PCBFM. Henry Jessup (1832–1910) graduated from the Union Theological Seminary in 1855 and was ordained as a Presbyterian minister that same year. Between 1856 and 1860 he served in Tripoli as a missionary under the ABCFM and then moved to Beirut, where he remained till his death in 1910. He wrote numerous works on Syria and on mission work there, including *The Women of the Arabs* (New York: Dodd and Mead, 1873); *Syrian Home Life* (New York: Dodd and Mead, 1874); and *The Mohammedan Missionary Problem* (Philadelphia: Presbyterian Board of Publication, 1879).
148. In addition to his journalistic work, in 1877 Sarruf also translated a book on design in nature, said to have originally been composed by a Reverend Walker, and, as mentioned before, Samuel Smiles's *Self-Help* in 1880—both strong indications of his literary tastes and his fondness for high-minded literature. He later authored a textbook on astronomy, *Basaʾit ʿilm al-falak wa-suwar al-samaʾ*; a treatise on the soul, *Rasaʾil al-arwah*; a biography of businessmen, *Rajal al-mal*; and several novels: *Amina, Amir lubnan, Fatat fayyum*, and *Fatat misr*. "Al-duktur Yaʿqub Sarruf," *Al-Muqtataf* 71 (1927): 121–28; Muhammad Kurd ʿAli, "Al-ʿalama al-duktur Yaʿqub Sarruf" [Professor Yaʿqub Sarruf], *Majallat al-Majmaʿ al-ʿIlmi al-ʿArabi* 1 (1928): 57; Fuad Sarruf, *Yaʿqub Sarruf* (Beirut: Dar al-ʿIlm al-Malayin, 1960), 12; Henry Jessup, *Fifty Three Years in Syria* (New York: Fleming H. Revell, 1910), 430; Farag, "*Al-Muqtataf*," 17–20.

149. "Al-Duktur Faris Nimr" [Dr. Faris Nimr], *Al-Muqtataf* 120 (1952): 1–13; Farag, *"Al-Muqtataf,"* 20, 42–47. On these later enterprises, see the next chapter.

150. Farag, *"Al-Muqtataf,"* 48–53. On *Al-Lata'if*, see Louis Cheikho, "Al-sirr al-Masun fi shiʿat al-Frimasun" [Secrets of the Masons' and the Freemasons' lodge], *Al-Mashriq* 12 (1909): 721–31; Byron Cannon, "Nineteenth-Century Arabic Writings on Women and Society: The Interim Role of the Masonic Press in Cairo (*al-Lataif*, 1885–1895)," *International Journal of Middle East Studies* 17 (1985): 463–84.

151. Eli Smith, "Syrian Society of Arts and Sciences," *Journal of the American Oriental Society* 3 (1853): 477–86; Butrus al-Bustani, *Aʿmal al-Jamʿiya al-Suriya* [Proceedings of the Syrian Society] (Beirut: n.p., 1852). Among the lectures published in the proceedings were "The Measure of the Progress of Knowledge in Syria at the Present Time"; "On the Delights and Utilities of Science"; "On the Principles of the Laws of Nature"; "On the Sciences of the Arabs"; and "A Discourse on the Instruction of Women"; delivered by Johannes Wortabet, Cornelius Van Dyck, Salim al-Nafwal, Nasif al-Yaziji, and Butrus al-Bustani.

152. "Ghayyat al-Majmaʿ al-ʿIlmi al-Sharqi al-ʾAzmi" (Aims of the Eastern Scientific Society), *Al-Muqtataf* 8 (1883): 529–34; Farag, *"Al-Muqtataf,"* 45, 78–82. For a list of the proceedings see Jurji Zaydan, *Tarikh adab al-lugha al-ʿArabiya* [The history of Arabic literature] (Cairo: Matbaʿat al-Hilal, 1914), 4:85.

153. The papers read at the Shams al-Birr and at the Jamiʿat al-Sinaʿ (Industrial Society) were typically published in *Al-Muqtataf*; see, e.g., Faris Nimr, "Fasad falsafat al-maddiyin" [The corruption of materialist philosophy], *Al-Muqtataf* 7 (1882): 262–68; Khalil Effendi Shawwal, *"Asbab taʾkhir al-sinaʿa fi suriya"* [Reasons for Syria's technological lag], *Al-Muqtataf* 9 (1884): 85–87. On these societies, see Farag, *"Al-Muqtataf,"* 82–83.

154. "Al-Duktur Faris Nimr" [Dr. Faris Nimr], *Al-Muqtataf* 120 (1952): 8; Farag, *"Al-Muqtataf,"* 42–44, 77–78; George Antonius, *The Arab Awakening: The Story of the Arab National Movement* (London: H. Hamilton, 1938), 79; Zeine Zeine, *The Emergence of Arab Nationalism with a Background Study of Arab-Turkish Relations in the Near East* (Beirut: Khayats, 1966), 61; Butrus Abu-Manneh, "The Genesis of Midhat Pasha's Governorship in Syria, 1878–1880," in *The Syrian Land: Processes of Integration and Fragmentation: Bilād al-Shām from the 18th to the 20th Century*, ed. Thomas Philipp and Birgit Schaebler (Berlin: F. Steiner, 1998), 265–66.

155. William Curtis, *Today in Syria and Palestine* (Chicago: F. H. Revell, 1903), 39. Of the 120,000 inhabitants of Beirut, Curtis claimed there were 36,000 Muslims, 35,000 Greek Orthodox, 28,000 Maronites, 9,000 United Greeks, 2,500 Jews, 1,800 Roman Catholics, 2,100 Protestants, and, finally, representatives of five or six other religions.

156. Jullien, *La nouvelle mission de la Compagnie de Jésus en Syrie*; Joseph Burnichon, *La Compagnie de Jésus en France: Histoire d'un siècle, 1831–1914*, 3 vols. (Paris: G. Beauchesne, 1914–22); René Ristelhueber, *Traditions françaises au Liban* (Paris: F. Alcan: 1918), 74–92, 261–62; Antoine Rabbath, *Documents inédits pour servir à l'histoire du Christianisme en Orient* (Paris: Imprimerie Catholique, 1905), 1:30–70; John Spagnolo, "The Definition of a Style of Imperialism: The Internal Politics of the French Educational Investment in Ottoman Beirut," *French Historical Studies* 8 (1974): 563–84.

157. Virtually every issue of the journal, in its early years, contained some denunciation of magic and its "tricks"; see "Qiraʿ al-afkar" [Reading one's mind], *Al-Muqtataf*

1 (1876): 75–77; "ʿIlm qiraʿ al-afkar" [The science of reading one's mind], *Al-Muqtataf* 1 (1876): 98–100; "Khadaʿ al-ʿayn" [Optical illusions], *Al-Muqtataf* 1 (1876): 147–50, 177–80, "Khadaʿ al-ʿayn al-tabiʿiyin wa-baʿd darur bi-al-sihr" [Natural optical illusions and the harms of magic], *Al-Muqtataf* 1 (1987): 229–31; "Al-sihr" [Magic], *Al-Muqtataf* 2 (1877): 28; "Jawabna ʿala al-sihr" [Our answer to magic], *Al-Muqtataf* 2 (1877): 28–31; "Al-maghnatisiya al-hayawaniya" [Animal magnetism], *Al-Muqtataf* 2 (1877): 54–55, 146–48; "Al-fantrilikwist ay al-mutkalam min batinhu" [The ventriloquist], *Al-Muqtataf* 2 (1877): 220–22; "Al-haqq awla an yaqul" [The truth must be told], *Al-Muqtataf* 3 (1878): 29–32; "Al-zar al-Misri" [The Egyptian *zar*], *Al-Muqtataf* 3 (1878): 189; "Al-sihr ghash" [Magic is a scam], *Al-Muqtataf* 3 (1878): 205–10; "Al-sihr fi Afriqiya" [Magic in Africa], *Al-Muqtataf* 4 (1879): 113–18.

158. Louis Cheikho, "Al quwa al-ʿaqliya fi al-hayawan" [The power of reasoning in animals], *Al-Mashriq* 2 (1899): 753; Louis Cheikho, "ʿAql al-hayawan" [Animal reason], *Al-Mashriq* 2 (1899): 900; Louis Cheikho, "Al-insan wa-al-qurud" [Man and monkeys], *Al-Mashriq* 8 (1905): 626–27, referring to Yaʿqub Sarruf's "Al-insan wa-al-qurud" [Man and monkey], *Al-Muqtataf* 30 (1905): 498 and "Difaʿ al-kilab" [Canine defense], *Al-Muqtataf* 30 (1905): 460–63. See also "Iman *Al-Muqtataf* wa kufr *al-Bashir*" [*Al-Muqtataf*'s faith and *Al-Bashir*'s disbelief], *Al-Muqtataf* 8 (1883): 714–17, where they mention Husayn al-Jisr's support of the journal against *Al-Bashir*'s charges. Other minor controversies, such as over the nebular hypothesis, also erupted, instigated once again by *Al-Bashir*; see Farag, "*Al-Muqtataf*," 81. On Husayn al-Jisr, see chapter 4.

159. Louis Cheikho, "Nahnu wa-al-*Muqtataf* [We and *Al-Muqtataf*], *Al-Mashriq* 17 (1914): 696.

160. Edwin Lewis, "Al-ʿilm wa-al-maʿrifa wa-al-hikma," *Al-Muqtataf* 7 (1822): 158–67.

161. See Franz Rosenthal, "Muslim Definitions of Knowledge," in *The Conflict of Traditionalism and Modernism in the Muslim Middle East*, ed. Carl Leiden (Austin: University of Texas Press, 1966), 117–33; and Franz Rosenthal, *Knowledge Triumphant: The Concept of Knowledge in Medieval Islam* (Leiden: Brill, 1970). Dictionary entries for *ʿilm*, *maʿrifa*, and *hikma* reveal a complex genealogical and etymological trajectory for *ʿilm*, yet one in which it clearly emerges as the more extensive category. See, e.g., Muhammad Ibn Ahmad Khuwarizmi's tenth-century encyclopedic compendium of the sciences, *Mafatih al-ʿulum*; Muhammad Ibn Manzur's classic thirteenth-century dictionary, *Lisan al-ʿArab*; and Murtada al-Zabidi's eighteenth-century *Taj al-ʿarus min jawahir al-qamus*, one of the first modern dictionaries since Ibn Manzur's *Lisan al-ʿArab*. By the nineteenth century the idea that *ʿilm* could also refer more narrowly to science—as a more positive (and increasingly experimental) technical field of expertise and inquiry—emerged. See such dictionaries as Butrus al-Bustani's well-known *Muhit al-muhit* (Beirut: n.p., 1867); Ismail Mazhar's *Nahda Dictionary* (Cairo: Renaissance Bookshop, 1950); and the Majmaʿ al-Lugha al-ʿArabiya's (Cairo Academy of Language) *List of Scientific and Technical Terms* (Cairo: n.p., 1942).

162. For more on contemporary American neo-Baconian conceptions of science, see Theodore Dwight Bozeman, *Protestants in an Age of Science: The Baconian Ideal and Ante-bellum American Religious Thought* (Chapel Hill: University of North Carolina Press, 1977).

163. On the Lewis controversy, see Nadia Farag, "The Lewis Affair and the Fortunes of *al-Muqtataf*," *Middle Eastern Studies* 8 (1972): 73–83; Donald Leavitt, "Darwinism in

the Arab World and the Lewis Affair at the Syrian Protestant College," *Mulsim World* 71 (1981): 85–98; Shafiq Juha, *Darwin wa-azmat sanat 1882 fi al-kulliya al-tibbiya* [Darwin and the crisis of 1882 in the Medical Department at the Syrian Protestant College] (Beirut: American University of Beirut Press, 1991); Olivier Meïer, *"Al-Muqtataf" et le débat sur le darwinisme: Beyrouth, 1876–1885* (Cairo: CEDEJ, 1996).

164. Lewis, "Al-ʿilm wa-al-maʿrifa wa-al-hikma" [Science, knowledge, and wisdom], *Al-Muqtataf* 7 (1882): 158–67.
165. Minutes of the Board of Trustees, book II (1 December 1882), 51, AUB Library Archives.
166. Daniel Bliss to Howard and Willie Bliss, Beirut, 28 February 1883, Bliss Family Papers, 1850–1981, box 1, folder 7, Amherst College Library Archives and Special Collections.
167. MSS AUB 1: box 5, 1/3, folder 3: Letters from Dr. Stuart Dodge to Daniel Bliss, 1880–1883, 29 September 1882, AUB Library Archives. David Stuart Dodge (1836–1921) graduated from Yale and Union Theological Seminary in 1860 and was ordained in the Protestant Church in 1864. One of the founding sponsors of the college, he also served as the secretary, treasurer, and then president of the Board of Trustees. He also taught English and Latin (1869–73) and modern languages (1880–81) at the college.
168. James Dennis, "Al-madhhab al-Darwini" [The Darwinian school of thought], *Al-Muqtataf* 7 (1882): 233–36.
169. Edwin Lewis, "Al-madhhab al-Darwini" [The Darwinian school of thought], *Al-Muqtataf* 7 (1882): 287–90.
170. See HU 20.41, mfp, 1804–1888/9, reel 4, 1863/4–1873/4, Harvard University Archives.
171. MSS AUB 1: box 5, 1/1, folder 1: Letters from D. Stuart Dodge to Daniel Bliss, letter dated 16 November 1882, AUB Library Archives.
172. Yusuf al-Haʿik, "Al-madhhab al-Darwini" [The Darwinian school of thought], *Al-Muqtataf* 7 (1882): 290–92.
173. MSS AUB 1: boxes 5 and 14, AUB Library Archives. On Shibli Shumayyil and materialism, see chapter 3.
174. Reproduced in Stephen Penrose, *That They May Have Life: The Story of the American University of Beirut, 1866–1941* (New York: Trustees of the American University of Beirut, 1941), 309–10.
175. Lewis had an exchange with Darwin in January 1882, and then, shortly afterward, he sent him a copy of a paper he wrote, "On the Modification of a Race of Syrian Street Dogs of Beirut," which Darwin helped publish in the Zoological Society proceedings. See Darwin Correspondence Project, www.darwinproject.ac.uk: calendar no. 13645, C. R. Darwin to W. T. Van Dyck, 25 January 1882; 13710, W. T. Van Dyck to C. R. Darwin, 27 February 1882; and 13757, C. R. Darwin to W. T. Van Dyck, 3 April 1882 (Darwin died sixteen days later).
176. Sarruf and Nimr were also temporarily promoted to the newly created position of "adjunct professors," so as to allow for some instruction (in Arabic) to continue at the medical school. Yet they were also excluded from having any control over college affairs, which was left to other governing faculty and board members.
177. MSS AUB 1: box 5, 1/3, folder 3: Letters from Dr. Stuart Dodge to Daniel Bliss, 1880–1883, 1 August 1882, al-Kulliya," AUB Library Archives.
178. "Al-Madrasa al-Kulliya" [The Syrian Protestant College], *Al-Muqtataf* 7 (1882): 192;

"Al-Duktur Lewis" [Doctor Lewis], *Al-Muqtataf* 9 (1884): 183; Penrose, *That They May Have Life*, 47–48; A. L. Tibawi, "The Genesis and Early History of the Syrian Protestant College," *Middle East Journal* 21 (1967): 208.

179. Zaydan, *Autobiography*, 58–62.
180. Ibid., 64–71.
181. Missionaries at the college had worried earlier about Lewis's role in "stirring up these young men to ideas of some early prospect of advance, perhaps even to 'Professorship!'" Dodge thought this had been the strategy Lewis used to "gain control over them." "If the recent troubles indicate what may occur even with trusted men from America," he warned, "what might be our position with several natives in the Faculty." See MSS AUB 1: box 5, 1/1, folder 1: Letters from Dr. Stuart Dodge to Daniel Bliss, letter dated 29 September 1882, AUB Library Archives.
182. Samar Khalaf, "New England Puritanism and Liberal Education in the Middle East: The American University in Beirut as a Cultural Transplant," in *Cultural Transitions in the Middle East*, ed. Serif Mardin (Leiden: Brill, 1994), 64. Stephen Penrose (*That They May Have Life*, 65) later claimed that Sarruf's and Nimr's dismissal may have been due to their participation in the Secret Society and the fear that they played "too active a part in dangerous politics." After all, it had been just two years ago, between June and December 1880, that placards went up in Beirut, Tripoli, Sidon, and Damascus demanding constitutional reform. The actions of the Secret Society were immediately suspected, although officials in Istanbul and Syria had little to no information as to its membership. These placards caused quite a stir and eventually resulted in the dismissal of the governor-general of Syria, Midhat Pasha. Abu-Manneh, "Genesis of Midhat Pasha's Governorship," 265. A. L. Tibawi ("Genesis and Early History of the Syrian Protestant College," 211), however, says he found little to support the idea that Sarruf's and Nimr's participation in this society may have led to their dismissal from the college.
183. Jessup, *Mohammedan Missionary Problem*, 95; Elie Kedourie, *Arabic Political Memoirs and Other Studies* (London: Cass, 1974), 67. See also William Goodell, *The Old and the New: On the Changes of Thirty Years in the East with Some Allusions to Oriental Customs as Elucidating the Scriptures* (New York: M. W. Dodd, 1853).
184. Jessup, *Fifty Three Years in Syria*, 592.
185. Yaʿqub Sarruf, "Al-ʿilm wa-al-madaris al-jamiʿa" [Science and higher education], *Al-Muqtataf* 9 (1884): 468.
186. MSS AUB 1: box 6, 2/1, folder 1: Letters from Dr. Stuart Dodge to Daniel Bliss, 13 February 1885, AUB Library Archives.

CHAPTER TWO

1. James Dennis, *Christian Missions and Social Progress: A Sociological Study of Foreign Missions* (New York: Fleming H. Revell, 1906), 3:3–4.
2. Surprisingly, these connections are still hardly explored, and until recently, much of the literature on Arabic intellectual history of the nineteenth century has taken on board a largely uncritical attitude to the historical genealogy of the idea of an "Arab Renaissance." For notable exceptions to this, see Stephen Sheehi, "Toward a Critical Theory of *al-Nahdah*: Epistemology, Ideology and Capital," *Journal of Arabic Literature* 43 (2012): 269–98; and Shaden Tageldin, "Proxidistant Reading: Toward a Critical Pedagogy of the *Nahdah* in U.S. Comparative Literary Studies," *Journal of Arabic Literature* 43 (2012): 227–68.
3. See, e.g., their 1898 article on positivism, or what they called "al-madhhab al-

falsafi": *Al-Muqtataf* 22 (1898): 106–12. Here, they trace a broad genealogy from Comte to Kant, focusing on the latter's theory of metaphysics in particular (and largely following Spencer).

4. See Herbert Spencer, *Social Statics; or, The Conditions Essential to Human Happiness Specified, and the First of Them Developed* (London: John Chapman, 1851). See also Sydney Eisen, "Herbert Spencer and the Spectre of Comte," *Journal of British Studies* 7 (1967): 48–67.
5. Bernard Lightman, "Ideology, Evolution and Late-Victorian Agnostic Popularizers," in *History, Humanity and Evolution: Essays for John C. Greene*, ed. James Moore (Cambridge: Cambridge University Press, 1989), 294.
6. See chapter 7 for more on this.
7. Salama Musa, *The Education of Salama Musa*, trans. L. O. Schuman (Leiden: Brill, 1961), 33.
8. William E. Curtis, *Today in Syria and Palestine* (Chicago: F. H. Revell, 1903), 43.
9. Shibli Shumayyil, "Al-jara'id fi al-sharq" [Newspapers in the East], in *Majmuʿat al-Duktur Shibli Shumayyil* (Beirut: Dar Nazir ʿAbbud, 1983), 240; Susan Ziadeh, "A Radical in His Time: The Thought of Shibli Shumayyil and Arab Intellectual Discourse (1882–1917)" (PhD diss., University of Michigan, 1991), 14.
10. Earlier, of course, various missionary societies of the late Ottoman Empire also used the press for religious texts, and the press was selectively deployed throughout the empire for the circulation of pedagogical print since at least the eighteenth century. Joseph Szyliowicz, "The Printing Press in the Ottoman Empire," in *Transfer of Modern Science and Technology to the Muslim World*, ed. Ekmelleddin Ihsanoglu (Istanbul: Research Center for Islamic History, Art and Culture, 1992), 251–52.
11. J. Heyworth-Dunne, "Printing and Translation under Muhammad ʿAli of Egypt: The Foundation of Modern Arabic," *Journal of the Royal Asiatic Society* 3 (1940): 325–49; Jamal al-Shayyal, *Tarikh al-tarjama fi ʿasr Muhammad ʿAli* [History of the translation movement during the reign of Muhammad ʿAli] (Cairo: Dar al-Fikr al-ʿArabi, 1951).
12. Al-Shayyal, *Tarikh al-tarjama fi ʿasr Muhammad ʿAli*, 31–39. See also Stephen Sheehi, "Arabic Literary-Scientific Journals: Precedence for Globalization and the Creation of Modernity," *Comparative Studies of South Asia, Africa and the Middle East* 25 (2005): 438–48. The telegraph was not introduced in Syria and Egypt until the late 1860s; Roderic Davison, "The Advent of the Electric Telegraph in the Ottoman Empire," in *Essays in Ottoman and Turkish History, 1774–1923*, ed. Roderic Davison (Austin: University of Texas Press, 1990), 133–65; Eugene Rogan, "Instant Communication: The Impact of the Telegraph in Ottoman Syria," in *The Syrian Land: Processes of Integration and Fragmentation: Bilad al-Sham from the 18th to the 20th Century*, ed. Thomas Philipp and Birgit Scaebler (Stuttgart: F. Steiner, 1998), 113–28.
13. Ami Ayalon, *The Press in the Arab Middle East: A History* (New York: Oxford University Press, 1995), 39–49.
14. Ayalon, *The Press in the Arab Middle East*, 37–38, 46–49; Ahmed Bioud Abdelghani, *3200 Revues et journaux arabes de 1800 à 1965* (Paris: Bibliothèque Nationale, 1969). This was typical of many colonial centers of the new imperialism of the nineteenth century. See, e.g., C. A. Bayly, *Empire and Information: Intelligence Gathering and Social Communication in India, 1780–1870* (Cambridge: Cambridge University Press, 1996).
15. Martin Hartmann, *The Arabic Press of Egypt* (London: Luzac, 1899), 1.
16. Donald Cioeta, "Ottoman Censorship in Lebanon and Syria, 1876–1908," *International Journal of Middle East Studies* 10 (1979): 167–86.

17. Juan Cole, *Colonialism and Revolution in the Middle East: Social and Cultural Origins of Egypt's ʿUrabi Movement* (Princeton, NJ: Princeton University Press, 1993), 223–27.
18. E. Baring [Lord Cromer] to J. Pauncefort, 6 November 1887, FO 633/5, no. 220, National Archives, London.
19. *Al-Mu'ayyad* 3 and 4 (October 1904), cited from Abbas Kelidar, "Shaykh ʿAli Yusuf: Egyptian Journalist and Islamic Nationalist," in *Intellectual Life in the Arab East, 1830–1939*, ed. Marwan Buheiry (Beirut: American University of Beirut Press, 1981), 18.
20. Ayalon, *Press*, 221.
21. Shumayyil, "Al-jara'id fi al-sharq," in *Majmuʿat al-Duktur Shibli Shumayyil*, 240; Ziadeh, "Radical in His Time," 14; A. Albert Kudsi-Zadeh, "The Emergence of Political Journalism in Egypt," *Muslim World* 70 (1980): 47–55; Elisabeth Kendall, "The Marginal Voice: Journals and the *Avant-Garde* in Egypt," *Journal of Islamic Studies* 8 (1997): 216–38.
22. Cromer to Salisbury, 17 April 1899, FO 78/5023, no. 73, National Archives, London; Ministère des Affaires Étrangères, *Documents diplomatiques: Affaire du journal Le Bosphore Égyptien* (Paris: Imprimerie nationale, 1885), provides the relevant French diplomatic documents.
23. Hartmann, *Arabic Press*, 3.
24. Dodge to Bliss, 13 February 1885; see also "*Al-Muqtataf*, 1876–1900: A Study of the Influence of Victorian Thought on Modern Arabic Thought," (PhD diss., University of Oxford, 1969), 262.
25. For a full list of their publications, see ibid., appendix II, 396–97.
26. W. Fraser Rae, "The Egyptian Newspaper Press," *Nineteenth Century* 32 (1892): 20–21.
27. Abbas Kelidar, "The Political Press in Egypt, 1882–1914," in *Contemporary Egypt through Egyptian Eyes*, ed. Charles Tripp (London: Routledge, 1993), 1–21.
28. E. Baring to J. Pauncefote, 6 November 1887, FO 633/5, no. 220, National Archives, London.
29. E. Baring to J. Pauncefote, 24 October 1888, FO 633/5, no. 287, National Archives, London.
30. Ayalon, *Press*, 56. For more on *Al-Muqattam* and the British, see Taysir Abu ʿArja, *Al-Muqattam: Jaridat al-ihtilal al-Baritani fi Misr, 1889–1952* [*Al-Muqattam*: The newspaper of the British occupation in Egypt, 1889–1952] (Cairo: Al-Hay'a al-Misriya al-ʿAmma li-al-Kitab, 1997), esp. 17–34 (on its founding).
31. Farag, "*Al-Muqtataf*," 45, 140. On Boyle, see Robert Tignor, *Modernization and British Colonial Rule in Egypt, 1882–1914* (Princeton, NJ: Princeton University Press, 1966), 199–200 (for a letter from Nimr to Boyle, see 275).
32. Evelyn Baring to Colonel Trotter, 5 October 1891, FO 633/5, no. 496, National Archives, London; Cromer to Salisbury, 31 October 1896, FO 407/139, no. 27, National Archives, London. For more on their views of the occupation, see Abu ʿArja, *Al-Muqattam*, 123–80.
33. Annual Report for 1905, Cromer to Lansdowne, 15 March 1905, FO 407/164, no. 82, National Archives, London; "Khutbat Lurd Kromer fi Um Durman" [Lord Cromer's speech in Omdurman], *Al-Muqattam* 2975 (6 January 1899): 1.
34. Farag, "*Al-Muqtataf*," 47.
35. William Willcocks, *Sixty Years in the East* (London: W. Blackwood, 1935), 115–16.
36. Ayalon, *Press*, 56–57, 203–4; W. Frasier Rae, "The Egyptian Newspaper Press," *Nineteenth Century* 32 (1892): 220–21; and W. Frasier Rae, *Egypt Today* (London:

R. Bentley and Son, 1892), 250. A survey conducted in 1903 revealed that 51 percent of *Al-Muqattam* and *Al-Muqtataf* subscribers were property owners; 23 percent, civil servants; 10 percent, merchants; 4 percent, lawyers; and 2 percent, judges. See "Huquq al-suhuf wa-al-mushtarikun fiha" [The rights of journals and their subscribers], *Al-Muqtataf* 28 (1903): 666–68; Kelidar, "Political Press in Egypt," 13; Farag, "*Al-Muqtataf*," 103–5.

37. Farag, "*Al-Muqtataf*," 88–89, 106.
38. See forthcoming essays in *Global Spencerism*, ed. Bernard Lightman. Spencer's ideas were also used with a great deal of interpretive flexibility and, within any given vernacular tradition, toward sometimes seemingly contradictory ends. See, e.g., Michio Nagai, "Herbert Spencer in Early Meiji Japan," *Journal of Asian Studies* 14 (1954): 55–64; and Shruti Kapila, "Self, Spencer and *Swaraj*: Nationalist Thought and Critiques of Liberalism, 1890–1920," *Modern Intellectual History* 4 (2007): 109–27.
39. "Hayawan ha'il" [A prodigious], *Al-Muqtataf* 10 (1885): 118. See also Farag, "*Al-Muqtataf*," 125, 290.
40. "Al-tamaddun wa-al-tawahush" [Civilization and barbarism], *Al-Muqtataf* 9 (1885): 395.
41. Herbert Spencer, *The Principles of Sociology* (New York: D. Appleton, 1898), 3:321. See also Walter Simon, "Herbert Spencer and the 'Social Organism,'" *Journal of the History of Ideas* 21 (1960): 295.
42. As he later wrote: "Society is made up of individuals; all that is done in society is done by the combined actions of individuals; and therefore in individual actions only can be found the solutions of social phenomena. But the actions of individuals depend on the laws of nature; and their nature cannot be understood until these laws are understood." Herbert Spencer, *Education: Intellectual, Moral and Physical* (London: Williams and Norgate, 1906), 44.
43. "Hayawan ha'il" [prodigious], *Al-Muqtataf* 10 (1885): 118.
44. Farag, "*Al-Muqtataf*," 130–31.
45. "Al-masa'il" [Questions], *Al-Muqtataf* 15 (1890): 418.
46. Translations of *Principles of Sociology* cited in Farag, "*Al-Muqtataf*," 111–12; Musa, *Education of Salama Musa*, 37, 155.
47. "Al-ʿilm wa khayr al-balad" [Science and the good of the nation], *Al-Muqtataf* 11 (1886): 270.
48. "Afaid kitab fi al-tarbiya" [The most beneficial of books on education], *Al-Muqtataf* 28 (1903): 869.
49. This was one of Spencer's most popular books. For more on ʿAbduh and Spencer, see chapter 5.
50. "Al-faylasuf Hirbirt Sbinsir" [The philosopher Herbert Spencer], *Al-Muqtataf* 29 (1904): 1–9; "Sbinsir wa-falsafathu" [Spencer and his philosophy], *Al-Muqtataf* 29 (1904): 105–10; "Falsafat Sbinsir" [Spencer's philosophy], *Al-Muqtataf* 29 (1904): 281–85, 383–86; "Ra'y Sbinsir fi al-taʿlim" [Spencer's opinion on education], *Al-Muqtataf* 29 (1904): 289–95; "Al-masa'il" [Questions], *Al-Muqtataf* 29 (1904): 96.
51. "Al-faylasuf Hirbirt Sbinsir" [The philosopher Herbert Spencer], *Al-Muqtataf* 29 (1904): 1.
52. Ibid., 1; "Sbinsir wa-falsafathu" [Spencer and his philosophy], *Al-Muqtataf* 29 (1904): 105.
53. The editors reassured their readers that Najib Bey Shuqri was preparing to publish his translation of Spencer's work, as advertised in the journal earlier. "Al-masa'il" [Questions], *Al-Muqtataf* 29 (1904): 92–94.

54. "Al-masa'il" [Questions and answers], *Al-Muqtataf* 29 (1904): 451.
55. "Al-faylasuf Sbinsir" [The philosopher Spencer], *Al-Muqtataf* 29 (1904): 2. See also "Sbinsir wa-falsafathu" [Spencer and his philosophy], *Al-Muqtataf* 29 (1904): 105, 109; Hector Carsewell MacPherson, *Spencer and Spencerism* (New York: Doubleday, Page, 1900); "Falsafat Sbinsir" [Spencer's philosophy], *Al-Muqtataf* 29 (1904): 282–84, 383–86.
56. Ibid., 285.
57. "Ta'qib 'ala Sbinsir" [Commentary on Spencer], *Al-Muqtataf* 22 (1898): 818–21.
58. In this respect they followed the reconciliation between positivism and theism of American Protestant thinkers like John Fiske. See "Al-madhhab al-Darwini" [Darwinism], *Al-Muqtataf* 7 (1882): 33–57.
59. Shibli Shumayyil, "Tarikh al-ijtima' al-tabi'i" [The natural history of societies], *Al-Muqtataf* 9 (1884): 523–93.
60. "Hayawan ha'il" [prodigious], *Al-Muqtataf* 10 (1885): 182
61. Herbert Spencer, *The Study of Sociology* (London, 1873): 401. See also Farag, "*Al-Muqtataf*," 287.
62. "'Alaqat al-mashriq bi-al-gharb" [East and West], *Al-Muqtataf* 15 (1890): 572–76.
63. "Al-hayawan al-natiq wa-al-hayawan al-'ajam" [Rational and brute animals], *Al-Muqtataf* 13 (1888): 106.
64. Quoted in Sir John Marriott, *The Eastern Question: A Historical Study in European Diplomacy* (Oxford, 1917), iii.
65. "Al-amira al-Misriya" [The Egyptian princess], *Al-Muqtataf* 23 (1899): 66.
66. Qasim Amin, *The Liberation of Women*, trans. Samiha Sidhom Peterson (Cairo: American University in Cairo Press, 2000), 62–63.
67. Ibid., 63.
68. See George Stocking, *Race, Culture and Evolution: Essays in the History of Anthropology* (Chicago: University of Chicago Press, 1982), esp. 239–40. For a general review of some of the literature on scientific racial thought at this time, see also Marwa Elshakry and Sujit Sivasundaram, introduction to *Victorian Science and Literature: Science, Race and Imperialism*, ed. Marwa Elshakry and Sujit Sivasundaram (London: Pickering and Chatto, 2012), 6:ix–xli.
69. "Qaswa al-bashar" [The cruelty of humans], *Al-Muqtataf* 9 (1884): 186–87; "Sabagh al-baid sudan" [Dyeing white people black], *Al-Muqtataf* 11 (1886): 316; "Asl al-baid wa al-sud" [Origin of whites and blacks], *Al-Muqtataf* 12 (1887): 188; "Ikhfaq al-musa'i fi ard al-zunuj" [Hopes disappointed in the lands of the blacks], *Al-Muqtataf* 14 (1889): 758–60; "Abidad al-sud" [Whitening of blacks], *Al-Muqtataf* 18 (1893): 431–32; "Asl sukan Misr al-qadimun" [Origin of the ancient inhabitants of Egypt], *Al-Muqtataf* 22 (1898): 396; "Sukan Misr al-qadimun" [The ancient inhabitants of Egypt], *Al-Muqtataf* 23 (1899): 77; "Masra' al-zunuj" [The perdition of blacks], *Al-Muqtataf* 23 (1899): 601–6. On Egypt and the Sudan, see Eve Troutt Powell, *A Different Shade of Colonialism: Egypt, Great Britain, and the Mastery of the Sudan* (Berkeley and Los Angeles: University of California Press, 2003); and Heather Sharkey, *Living with Colonialism: Nationalism and Culture in the Anglo-Egyptian Sudan* (Berkeley and Los Angeles: University of California Press, 2003).
70. "Al-zunuj" [Blacks], *Al-Hilal* 4 (1895): 53; "Lun al-bashra" [Skin color], *Al-Hilal* 5 (1896): 142–43; "Sukan Amirika al-usliyun" [The natives of America], *Al-Hilal* 5 (1896): 624; Jurji Zaydan, *Tabaqat al-umam* (Cairo: Dar al-Hilal, 1912). See also Omnia El Shakry, *The Great Social Laboratory: Subjects of Knowledge in Colonial and Post-colonial Egypt* (Stanford, CA: Stanford University Press, 2007), 58–61.

71. Eve Troutt Powell, "Brothers along the Nile: Egyptian Concepts of Race and Ethnicity, 1895–1910," in *The Nile: Histories, Cultures, Myths*, ed. Haggai Erlich and Israel Gershoni (Boulder: Lynn Rienner, 2000), 171–81.
72. Farag, *"Al-Muqtataf,"* 372–73.
73. Powell, *Different Shade of Colonialism*, 160, 162–67. El Shakry, *Great Social Laboratory*, 58–61.
74. Farag, *"Al-Muqtataf,"* 372–73.
75. Demolins's book was also translated into Turkish in 1914 as *Anglosaksonlarin esbab-i faikiyeti nedir? Anglosaksonlar hakkinda tedkikat-i ictimaiye* and into Persian in 1923 as *Tafavvuq-i Anglosakson*.
76. Ahmad Fathi Zaghlul, *Sirr taqaddum al-Inkiliz al-Saksuniyin* (Cairo: Matbaʿat al-Jamaliya, 1899); Edmond Demolins, *A quoi tient la supériorité des Anglo-Saxons* (Paris: Libraire de Paris, 1897).
77. Demolins, *A quoi tient la supériorité des Anglo-Saxons*. See also "Kitab nafsiyan: Tahrir al-marʾa wa-sirr taqadum al-Inkiliz" [A psychological study: Women's liberation and the secret of England's advancement], *Al-Muqtataf* 23 (1899): 527–35; "Taqadum al-Inkiliz al-Saksuniyin ta'lif Admund Dimulan" [The superiority of the Anglo-Saxons by Edmonds Demolins], *Al-Muqtataf* 23 (1899): 551–53.
78. Zaghlul, *Sirr taqaddum al-Inkiliz al-Saksuniyin*, 20–30; Demolins, *Anglo-Saxon Superiority*, 111–34, 259–306; Timothy Mitchell, *Colonising Egypt* (Cambridge: Cambridge University Press, 1988), 110–11.
79. "Taqadum al-Inkiliz al-Saksuniyin taʿlif Admund Dimulan," *Al-Muqtataf* 23 (1899): 551–53; Edmond Demolins, *Anglo-Saxon Superiority: To What It Is Due* (New York: Scribner, 1898), xxi; Zaghlul, *Sirr taqaddum al-Inkiliz al-Saksuniyin*, 35–36.
80. "Asbab al-inhitat al-Sharq" [The causes of Eastern decline], *Al-Muqattam* 2991 (30 January 1899): 1; Mitchell, *Colonising Egypt*, 110–11 (citing Ahmad Lutfi al-Sayyid); M. Şükrü Hanioglu, *Preparation for a Revolution: The Young Turks, 1902–1908* (Oxford: Oxford University Press, 2001): 82–86 (citing Sabahaddin Bey).
81. Qasim Amin, *The Liberation of Women*, trans. Samiha Sidhom Peterson (Cairo, 2000), 71.
82. See the entry for "Al-Yaban" [Japan], in *Fihris Al-Muqtataf, 1876–1952* (Beirut: Al-Jamiʿa al-Amirikiya, 1967), 3:618–23.
83. Farag, *"Al-Muqtataf,"* 308–11. Musa, *Autobiography*, 34.
84. Renée Worringer, "Comparing Perceptions: Japan as Archetype for Ottoman Modernity, 1876–1918" (PhD diss., University of Chicago, 2001), chap. 7 passim.
85. "Al-ʿilm fi al-Yaban" [Science in Japan], *Al-Muqtataf* 7 (1882): 80–81 and 20 (1896): 70; Farag, *"Al-Muqtataf,"* 308–11.
86. "Masa'il wa-ajwibha" [Questions and answers], *Al-Muqtataf* 17 (1892): 136.
87. "Madrasat al-Yaban wa-sirr nijahha" [Schools in Japan and the secret of their success], *Al-Muqtataf* 11 (1886): 96; "Al-sinaʿa fi al-Yaban" [Industry in Japan], *Al-Muqtataf* 16 (1892): 717 and 22 (1898): 315 and 26 (1910): 573.
88. "Sabab taqadum al-Yaban" [Causes of the progress of Japan], *Al-Muqtataf* 21 (1897): 869. See also Farag, *"Al-Muqtataf,"* 315.
89. "Bilad al-Yaban wa-hukumatha" [Japan and its government], *Al-Muqtataf* 18 (1894): 361, 438; "Bilad al-Yaban wa-sabab irtiqa'ha" [Japan and the causes of its progress], *Al-Muqtataf* 22 (1899): 100; cited from Farag, *"Al-Muqtataf,"* 316.
90. "Masa'il wa-ajwibha" [Questions and answers], *Al-Muqtataf* 9 (1884): 182.
91. Cited by Farag, *"Al-Muqtataf,"* 317.
92. Cited in Worringer, "Comparing Perceptions," 383, 385.

93. See their review of his *Al-Islam wa-al-Nasraniya maʿa al-ʿilm wa-al-madaniya* (Science and civilization in Christianity and Islam): *"Al-Islam wa al-Nasraniya maʿ al-ʿilm wa al-madaniya* li-imam aʾimat al-Islam wa-hakim min hukama al-ʿalam" [*Science and Civilization in Christianity and Islam*, by the leader of all leaders and the wisest of the wise], *Al-Muqtataf* 28 (1903): 177–78. See also "Al-Shaykh Muhammad ʿAbduh" [Muhammad ʿAbduh's obituary], *Al-Muqtataf* 30 (1905): 593–96, 909–22, 985; "Al-islah fi al-Azhar" [The reform of Al-Azhar], *Al-Muqtataf* 30 (1905): 738–44; "Al-islah al-mahakim al-shariʿaya" [On Muhammad ʿAbduh's *shariʿa* court reforms project], *Al-Muqtataf* 24 (1900): 23–24. For a detailed analysis of ʿAbduh and his ideas, see chapter 5.
94. "Tatbiq al-diyana al-Islamiya ʿala nawamis al-madaniya" [The principles of civilization and Islam], *Al-Muqtataf* 23 (1899): 462–64. Despite their oblique criticisms of Wajdi, however, they were obviously not averse to including his ideas on science and religion in their own journal. A few years after their review of his book, they included an article written by Wajdi in a very similar spirit to his earlier work: Muhammad Farid Wajdi, "Al-Islam wa-al-ʿilm" [Islam and science], *Al-Muqtataf* 28 (1903): 175–76.
95. "An Egyptian View of the Egyptian Question: An Interview with an Egyptian Editor," *Pall Mall Gazette* 8289 (14 October 1891).
96. "Siyasat *Al-Muqattam*" [*Al-Muqattam*'s politics], *Al-Muqattam* 240 (1 August 1889): 1. See also Abu ʿArja, *Al-Muqattam*, 123–70.
97. "Jallaʿ al-Inkiliz ʿan Misr" [The British evacuation of Egypt], *Al-Muqattam* 1965 (9 September 1895): 1.
98. Paul Laity, *The British Peace Movement, 1870–1914* (Oxford: Oxford University Press, 2001), 94–96.
99. "Herbert Spencer's Advice to Japan," *Times*, 18 January 1904, 9; "Al-Yaban wa nasihat Sbinsir liha" [Japan and Spencer's advice], *Al-Muqtataf* 29 (1904): 113–16.
100. Cromer to Salisbury, 17 April 1899, FO 78/5023, no. 73, National Archives, London. Kelidar, "Shaykh ʿAli Yusuf." See also Abu ʿArja, *Al-Muqattam*, 53–92, for more on this.
101. Ayalon, *Press*, 54.
102. Ibid., 50–62; Kelidar, "Political Press in Egypt," 5–8.
103. Musa, *Autobiography*, 33.
104. Cromer to Salisbury, 27 August 1900, FO 407/155, no. 53, National Archives, London.
105. For a discussion of these competing ideologies in the late Ottoman Empire, see Kemal Karpat, *The Politicization of Islam* (Oxford: Oxford University Press, 2001); M. Şükrü Hanioglu, *The Young Turks in Opposition* (Oxford: Oxford University Press, 1995); and Hanioglu, *Preparation for a Revolution*.
106. "Al-ightirab wa-al-muhajara" [foreignization and emigration], *Al-Muqtataf* 16 (1891): 244; Farag, *"Al-Muqtataf,"* 354. Technically, Sarruf and Nimr were not considered Egyptians at this time: according to a legal decree issued in the 1890s, an "Egyptian" was anyone who was born in Egypt or any Ottoman subject who had spent fifteen years in residence there (Sarruf and Nimr, who moved to Egypt in 1885, were not naturalized therefore until 1900). Earl of Cromer, *Modern Egypt* (London: Macmillan, 1908), 2:217; Farag, *"Al-Muqtataf,"* 353.
107. "Baʿd sitin sanaʿ" [After sixty years], *Al-Muqtataf* 88 (1936): 570; Farag, *"Al-Muqtataf,"* 46.
108. Musa, *Autobiography*, 32.

109. "Siyasat *Al-Muqattam*" [*Al-Muqattam*'s politics], *Al-Muqattam* 240 (1 August 1889): 1.
110. Letter reproduced in Cromer to Salisbury, 31 October 1896, FO 407/139, no. 27, National Archives, London.
111. Letter dated 15 August 1896, reported in Cromer to Salisbury, 31 October 1896, FO 407, no. 27, National Archives, London.
112. "Tashiʿa *Al-Muqtataf* li-al-Inkiliz" [*Al-Muqtataf*'s predilection for the British], *Al-Muqtataf* 30 (1905): 563–65.

CHAPTER THREE

1. "Bidayat al-haya wa-al-muwt" [The beginnings of life and death], *Al-Muqtataf* 3 (1878): 15.
2. On earlier Arabic debates, see Remke Kruk, "A Frothy Bubble: Spontaneous Generation in the Medieval Islamic Tradition," *Journal of Semitic Studies* 35 (1990): 265–82.
3. "Fasad falsafat al-maddiyin" [The corrupt philosophy of the materialists], *Al-Muqtataf* 7 (1882): 262–68.
4. James Strick, *Sparks of Life: Darwinism and the Victorian Debates over Spontaneous Generation* (Cambridge: Cambridge University Press, 2000); John Farley, "The Spontaneous Generation Controversy (1859–1880)," *Journal of the History of Biology* 5 (1972): 285–319.
5. Henry Bastian, "Facts and Reasonings concerning the Heterogeneous Evolution of Living Things," *Nature*, 30 June 1870; Henry Bastian, *The Beginnings of Life, Being Some Account of the Nature, Modes of Origin and Transformation of Lower Organisms*, 2 vols. (London, 1872); Henry Bastian, *Evolution and the Origin of Life* (London, 1874); John Tyndall, "The Belfast Address," *Nature*, 20 August 1874, 172–89; Farley, "Spontaneous Generation Controversy," 293–303; Strick, *Sparks of Life*, 62–74. See also Bernard Lightman, "Scientists as Materialists in the Periodical Press: Tyndall's Belfast Address," in his *Evolutionary Naturalism in Victorian Britain: The "Darwinians" and Their Critics* (Farnham, UK: Ashgate / Variorum, 2009), 199–237; and John Farley and Gerald Geison, "Science, Politics and Spontaneous Generation in Nineteenth Century France: The Pasteur-Pouchet Debate," *Bulletin of the History of Medicine* 48 (1974): 161–98.
6. "Bidayat al-hayaʾ wa-al-muwt," *Al-Muqtataf* 3 (1878): 15; Strick, *Sparks of Life*, 129–78.
7. "Bidayat al-hayaʾ wa-al-muwt," *Al-Muqtataf* 3 (1878): 15; John Tyndall, "Spontaneous Generation," *Nineteenth Century* 3 (January 1878): 22–47.
8. Salama Musa, *The Education of Salama Musa*, trans. L. O. Schuman (Leiden: Brill, 1961), 164.
9. Shibli Shumayyil, "Al-Madrasa al-Kulliya" [The Syrian Protestant College], *Al-Muqattam*, 27 January 1909.
10. On the 1877 lecture, see Shibli Shumayyil, *Falsafat al-nushuʾ wa al-irtiqaʾ* [Philosophy of evolution and progress] (Cairo: Matbaʿat al-Muqtataf, 1910), 1:26.
11. Georges Haroun, *Šibli Šumayyil: Une pensée évolutionniste Arabe a l'époque d'an-Nahda* (Beirut, 1985), 55; Susan Ziadeh, "A Radical in His Time: The Thought of Shibli Shumayyil and Arab Intellectual Discourse" (PhD diss., University of Michigan, 1991): 16; M. Şükrü Hanioglu, "Blueprint for a Future Society: Late Ottoman Materialists on Science, Religion and Art," in *Late Ottoman Society: The Intellectual Legacy*, ed. Elisabeth Ozdalga (London: Routledge Curzon, 2005), 28–116.

12. Advertisement for Shibli Shumayyil, *Al-Hilal*, 4 (1895): 20; see Ziadeh, "A Radical in His Time," 22.
13. Shibli Shumayyil, "Hujatuna" [Our goals], *Al-Shifa'* 1 (1886): 1–5; "Inʿam Sultani," [Imperial award], *Al-Hilal* 3 (1894): 196; Ziadeh, "A Radical in His Time," 41.
14. See, e.g., Shumayyil's article on polluted water systems in Cairo, a health hazard that was indeed later rectified by the water company at the behest of the Health Department; discussed by Ziadeh, "A Radical in His Time," 41.
15. See *Al-Muqtataf* 10 (1886): 383; *Al-Muqtataf* 13 (1888): 360; and Shibli Shumayyil, "Khitam al-sana al-thaniya min al-Shifa'" [The end of the second year of *Al-Shifa'*], *Al-Shifa'* 2 (1888). See also Haroun, Šibli Šumayyil, 103.
16. See Vernon Egger, *A Fabian in Egypt: Salama Musa and the Rise of the Professional Classes in Egypt, 1909–1939* (Lanham, MD: University Press of America, 1986); and chapter 6 below for more on this journal.
17. Salama Musa, "ʿAda' *Al-Mustaqbal*" [The enemies of *Al-Mustaqbal*], *Al-Mustaqbal* 1 (1914): 116; Musa, *Education of Salama Musa*, 129–30; Adel Ziadat, *Western Science in the Arab World: The Impact of Darwinism, 1860–1930* (Basingstoke: Macmillan, 1986), 57.
18. "Al-Duktur Yaʿqub Sarruf," *Al-Muqtataf* 50 (1917): 108; Ziadat, *Western Science*, 30.
19. Rifʿat al-Saʿid, *Thalathat Lubnaniyin fi al-Qahira* [Three Lebanese in Cairo] (Cairo: Dar al-Thaqafa, 1973), 18.
20. See the list of contributors at the back of the 1910 edition of Shumayyil's *Falsafat al-nushu' wa-al-irtiqa'*.
21. His translation of Büchner's commentaries on Darwin appeared as *Taʿrib li-sharh Bukhnir ʿala madhhab Darwin fi intiqal al-anwaʿ wa-zuhur al-ʿalam al-ʿudwi wa-itlaq dhalika ʿala al-insan maʿa baʿdi tasarru fin* [A translation of Büchner's commentaries on Darwin on the transformation of species and the emergence of the organismic world and, from that, man] (Cairo: Matbaʿat Jaridat al-Mahrusa, 1884); other works included *Al-Rujhan* (Cairo, n.d.); *Risala fi al-hawa' al-asfar-wa al-wiqaya minhu wa-ʿilajuhu* (Cairo: n.p., 1890); a long "letter" addressed to Sultan Abdülhamid and entitled *Shakwa wa-amal* [Complaints and hopes] (Cairo: n.p., 1896); and *Suriya wa mustaqbaluha* [Syria and its future] (Cairo: n.p., 1915).
22. "Iʿtirad" [Objection], *Al-Muqtataf* 3 (1878): 174–75.
23. Ibid., 175.
24. Ibid.
25. Ibid., 175–76.
26. See, e.g., David Gooding, Trevor Pinch, and Simon Schaffer, eds., *The Uses of Experiment: Studies in the Natural Sciences* (Cambridge: Cambridge University Press, 1999). For more on the Young Guard's defense of Tyndall, see Strick, *Sparks of Life*.
27. "Al-haya' hiyrat al-ʿulama" [Life: The enigma of scientists], *Al-Muqtataf* 3 (1878): 178–79.
28. "Al-hiyra ʿala al-bahth" [A dilemma is cause for research], *Al-Muqtataf* 3 (1878): 243; "Fi mansha' al-haya" [On the genesis of life], *Al-Muqtataf* 3 (1878): 176.
29. "Al-Hiyra ʿala al-bahth," *Al-Muqtataf* 3 (1878): 243.
30. See Shumayyil's 1910 introduction to *Falsafat al-nushu' wa-al-irtiqa'*, 1:19.
31. "Al-hiyra ʿala al-bahth," *Al-Muqtataf* 3 (1878): 243; "Fi mansha' al-haya," *Al-Muqtataf* 3 (1878): 176.
32. See Alfred Russel Wallace, "Bastian on the Origin of Life," *Nature*, 6 July 1871, 178–79.
33. See esp. his articles: "Al-hass wa-anwaha al-mukhtalifa" [Sensibility and its diverse

forms], *Al-Muqtataf* 5 (1881): 294; "Al-hass fi al-madda" [Sensation in matter], *Al-Muqtataf* 6 (1881): 169; "Muqaddima fi al-tarikh al-tabiʿa" [An introduction to natural history], *Al-Muqtataf* 6 (1881): 221; "Kull al-sirr fi al-madda" [All secrets lie in matter], *Al-Muqtataf* 6 (1881): 282; "Al-haya wa al-ghazibiya" [Life and gravity], *Al-Muqtataf* 6 (1881): 423, 548; "Mulahizat fi al-haya" [Ideas on life], *Al-Muqtataf* 6 (1882): 677; "Al-haya fi ʿamaq al-miyaʾ" [Life at the bottom of the sea], *Al-Muqtataf* 6 (1882): 713; "Hayaʾa al-madda" [The life of matter], *Al-Muqtataf* 9 (1884): 477. He also repeats many of these arguments in his translation of Büchner.

34. Shumayyil, *Falsafat al-nushuʾ wa-al-irtiqaʾ*, 1:j; "Lesson Two," in ibid., 1:129.
35. Ibid., 1:j.
36. Farley, "Spontaneous Generation Controversy," 306; Ernst Haeckel, *The History of Creation; Or, The Development of the Earth and Its Inhabitants by the Action of Natural Causes*, trans. E. Lankester (New York: D. Appleton, 1876), 48.
37. Shumayyil, *Falsafat al-nushuʾ wa-al-irtiqaʾ*, 1:40–41.
38. *"Kull al-sirr fi al-madda"* [All mysteries lie in matter], *Al-Muqtataf* 6 (1881): 283–84.
39. Ibid., 284.
40. Shumayyil, *Falsafat al-nushuʾ wa-al-irtiqaʾ*, 1:50.
41. Ibid., 1:j.
42. Ibid., 1:39.
43. Ibid., 1:d.
44. Ludwig Büchner, *Sechs Vorleesungen über die Darwin'sche Theorie von der Verwandlung der Arten und die erste Entestehung der Organismenwelt* (Leipzig: T. Thomas, 1868); translated by Auguste Jacquot as *Conférences sur la théorie darwinienne de la transmutation des espèces et de l'apparition du monde organique* (Leipzig: T. Thomas, 1869).
45. See Ludwig Büchner, *Force and Matter: Empirico-philosophical Studies, Intelligibly Rendered*, trans. J. F. Collingwood (London: Trubner, 1864), viii; Farley, "Spontaneous Generation Controversy," 304.
46. Shumayyil, *Falsafat al-nushuʾ wa-al-irtiqaʾ*, 1:30.
47. Ibid.
48. Ibid., 1:31.
49. Hanioglu, "Blueprint for a Future Society," 35, 39.
50. John Lubbock, *The Origin of Civilisation and the Primitive Conditions of Man: Mental and Social Condition of Savages* (London: Longmans, Green, 1870); E. B. Tylor, *Primitive Culture: Researches into the Development of Mythology, Philosophy, Religion, Language, Art, and Custom* (New York: Holt, 1889); Herbert Spencer, *Principles of Sociology*, 3 vols. (New York: D. Appleton, 1898–99).
51. Shumayyil, *Falsafat al-nushuʾ wa-al-irtiqaʾ*, 1:44.
52. Ibid., 1:45–46.
53. Ibid.
54. Ibid., 1:48.
55. Ibid.
56. Ibid., 1:49–51.
57. Ibid., 1:49.
58. Ibid.
59. Ziadat, *Western Science*, 35; Musa, *Education of Salama Musa*, 164; cf. Haroun, *Šibli Šumayyil*, 81
60. Shumayyil, *Falsafat al-nushuʾ wa-al-irtiqaʾ*, 1:51–52.
61. Ibid., 1:51–52, 56.
62. Cited in Ziadeh, "A Radical in His Time," 305.

63. "Al-hass wa-anwaha al-mukhtalifal," *Al-Muqtataf* 5 (1881): 294; M. Oltramane, "Sensibility and Its Diverse Forms," *Science* 2 (21 May 1881): 230–32.
64. Shumayyil, *Falsafat al-nushu' wa-al-irtiqa'*, 1:25, 318; *Al-Muqtataf* 5 (1881): 236.
65. Shumayyil, *Falsafat al-nushu' wa-al-irtiqa'*, 1:23.
66. Ziadat, *Western Science*, 51.
67. "Ta'rib li-sharh Bukhnir 'ala madhhab Darwin" [Büchner's commentaries on Darwin], *Al-Muqtataf* 8 (1884): 638. See also Shafiq Juha, *Darwin and the Crisis of 1882 in the Medical Department* (Beirut: American University of Beirut Press, 2004), 143.
68. "Akhbar wa-iktashafat wa-ikhtara'at" [News and discoveries], *Al-Muqtataf* 9 (1884): 63.
69. Shumayyil, *Falsafat al-nushu' wa-al-irtiqa'*, 1:j.
70. For a full list of opponents and their critiques, see Haroun, *Šibli Šumayyil*, 318–99.
71. On *Al-Nashra al-Usbu'iya*, see Philippe Tarrazi, *Tarikh al-sihafa al-'Arabiya* (Beirut : Dar Sadir, 1914), 2:110–15. On Hurani, see Kamal al-Yaziji, *Al-Shaykh Ibrahim al-Hurani, 1844–1916* (Cairo: Jam'at al-Duwal al-'Arabiya, 1961).
72. Juha, *Darwin and the Crisis of 1882*, 42. Hurani had earlier written a testimony to God's work in nature that had been serialized in the early 1880s in *Al-Nashra al-Usbu'iya*: *Al-ayat al-bayinat fi ghara'ib al-ard wa-al-samawat* [Certain signs in the wonders of the heavens and earth] (Beirut: n.p., 1883). See the next chapter.
73. Juha, *Darwin and the Crisis of 1882*, 146.
74. Ibrahim Hurani, *Manahij al-hukama' fi nafyi al-nushu' wa-al-irtiqa'* (Beirut: n.p., 1884).
75. "*Al-haqiqa*: al-Duktur Shibli Shumayyil" [*The Truth* by Doctor Shibli Shumayyil], *Al-Muqtataf* 9 (1885): 704.
76. Ibrahim al-Hurani, *Al-haqq al-yaqin fi al-radd 'ala butl Darwin* [The certain truth in response to Darwin's errors] (Beirut: n.p., 1886); Shibli Shumayyil, *Al-haqiqa* (Cairo, 1885), republished in *Falsafat al-nushu' wa-al-irtiqa'*, vol. 1.
77. al-Hurani, *Al-haqq al-yaqin fi al-radd 'ala butl Darwin*, quoting Shumayyil, cited in Ziadeh, "A Radical in His Time,"131.
78. Ziadeh, "A Radical in His Time," 132.
79. Ibid., 134.
80. Najm Bezrigan, "The Islamic World," in *The Comparative Reception of Darwinism*, ed. Thomas Glick (Austin: University of Texas Press, 1972), 381.
81. Ziadeh, "A Radical in His Time," 134.
82. Shumayyil, *Falsafat al-nushu' wa-al-irtiqa'*, 1:263–64.
83. Ibid., 266–67; George Jackson Mivart, *On the Genesis of Species* (London: Macmillan, 1871); George Jackson Mivart, *Man and Apes* (New York: D. Appleton, 1874); Joe Burchfield, *Lord Kelvin and the Age of the Earth* (New York: Science History, 1975); Ziadeh, "A Radical in His Time," 137–38.
84. Bezrigan, "Islamic World," 380; Ziadeh, "A Radical in His Time," 132.
85. al-Hurani, *al-haqq al-yaqin fi al-radd 'ala butl Darwin*, 41. See also Ziadeh, "A Radical in His Time," 140.
86. Juha, *Darwin and the Crisis of 1882*, 147.
87. The full title in Arabic was rendered as *Risala fi ibtal madhhab al-dahriyin wa-bayan mafasidihim wa-ithbat anna al-din asas al-madaniya wa al-kufr fasad al-'umran*. 'Abduh was assisted in the translation by Afghani's manservant 'Abu Turab. For more on the Madrasa al-Sultaniya, see the next chapter.
88. On Sayyid Ahmad Khan, see Christian Troll, *Sayyid Ahmad Khan: A Reinterpretation*

of Muslim Theology (New Delhi: Vikas, 1978). Jean Lecerf, "Šibli Šumayyil: Métaphysicien et moraliste contemporain," *Bulletin d'Études Orientales* 1 (1931): 159.

89. Jamal al-Din al-Afghani, *Al-radd ʿala al-dahriyin*, trans. Muhammad ʿAbduh and ʿArif Effendi Abu Tarib (Cairo: Matbaʿat al-Mawsuʿat, 1320 A.H), 3–4. Cf. Jamal al-Din al-Afghani, *Haqiqat-i madhhab-i nichari va bayan hal nichariyan*, trans. Nikkie Keddie from the original Persian as *An Islamic Response to Imperialism: Political and Religious Writings of Sayyid Jamal al-Din "al-Afghani"* (Berkeley and Los Angeles: University of California Press, 1983), 130.
90. Afghani, *Al-radd*, 4. Cf. Afghani, *Haqiqat*, 131.
91. Qur'an, 76:1, 45:24; I. Goldziher and A. M. Grichon, "Dahriyya," in *Encyclopaedia of Islam*, 2nd ed., http://referenceworks. brillonline.com; Muhammad Ibn Manzur, *Lisan al-ʿArab* [The Arabic tongue] (Cairo: Bulaq, 1883–91).
92. Leon Zolondek, "The French Revolution in Arabic Literature of the Nineteenth Century," *Muslim World* 57 (1967): 203.
93. Afghani, *Haqiqat*, 160–61.
94. See chapter 7.
95. Afghani, *Al-radd*, 9–10. Cf. Afghani, *Haqiqat*, 135.
96. Afghani, *Al-radd*, 12. Cf. Afghani, *Haqiqat*, 137.
97. Jamal al-Din al-Afghani, "Ra'y fi madhhab al-nushu' wa-al-irtiqa' wa-in al-ʿArab sabaqhu" [An opinion on the school of evolution and how the Arabs foresaw it], in *Khatirat Jamal al-Din al-Afghani*, ed. Muhammad Makhzumi (Beirut: Al-Matbaʿa al-ʿIlmiya li-Yusuf Sadr, 1931), 181; Afghani, *Haqiqat*, 171.
98. Shibli Shumayyil, "Jamal al-Din al-Afghani fi nazariyat al-Duktur Shibli Shumayyil," in *Ara' Shibli Shumayyil* (Cairo: Matbaʿat al-Maʿarif, 1912), 79–80.
99. Ibid., 81.
100. See Margaret Kohn, "Afghani on Empire, Islam and Civilization," *Political Theory* 37 (2009): 398–422.
101. On the modernist invention of "religion," see Talal Asad, *Genealogies of Religion: Discipline and Reasons of Power in Christianity and Islam* (Baltimore, MD: John Hopkins University Press, 1993).
102. See Tomoko Masuzawa, *The Invention of World Religions; Or, How European Universalism Was Preserved in the Language of Pluralism* (Chicago: University of Chicago Press, 2005).
103. See, e.g., Jamal al-Din al-Afghani, "Answer of Jamal al-Din to Renan," reproduced in al-Afghani, *An Islamic Response to Imperialism*, trans. Keddie, and originally published in *Journal des Débats*, 18 May 1883, in response to Ernest Renan, "L'islamisme et la science," Sorbonne lecture originally delivered in 1882 and reprinted in *Œuvres completes* (Paris: Calmann-Levy, 1947).
104. Jamal al-Din al-Afghani, "The Materialists in India" (28 August 1884), in *An Islamic Response to: Imperialism*, trans. Nikkie Keddie, 178.
105. Ibid., 183.
106. Ibid., 184.
107. Ibid., 185. We should note too that ʿAbduh did not publish his translation of Afghani's treatise until 1902–3, when the charge of "materialism" was one the mufti was keen to support for reasons of his own (as we will see later). Yet its late publication date shows the resilience of the debate over materialism, just as it foreshadows the specter of this charge for generations of Muslim readers of Darwin to come. See chapter 5.
108. Haroun, *Šibli Šumayyil*, 318–99.

109. Jirjis Sfayr, *Kitab fi asl al-insan wa-al-ka'inat* [The origin of man and creation] (Beirut, 1890). Sfayr's account focuses on the debate about the origin of life itself (whereas al-Jisr, as shown in the next chapter, considered the descent of man to be the more important issue). Husayn al-Jisr will be the focus of chapter 4, while in chapter 5, which concentrates on Afghani's translator, ʿAbduh, I will show how the threat of a materialist view of nature was met by a reinvigoration of dialectical and natural theological methods, further underscoring the cross-confessional discourse constructed around the problem. There is perhaps no better example of this than the work of the Maronite Reverend Jirjis Sfayr, who published his *Kitab fi asl al-insan wa-al-ka'inat* in 1890. This took the form of a dialogue between a "materialist" and a "believer"; yet, in a curious twisting of inspirations, the very arrangement and argument of the book were also inspired by the Sufi theologian al-Jisr.
110. See the last two pages of Shumayyil, *Falsafat al-nushu' wa al-irtiqa'*.
111. Shumayyil, *Falsafat al-nushu' wa-al-irtiqa'*, 1:2, 3, 9.
112. Ibid., 1:2–3, 9. For more on Shumayyil's views on socialism, see chapter 6.
113. Shumayyil, *Falsafat al-nushu' wa-al-irtiqa'*, 1:9.
114. Ibid., 1:8–15.
115. Ziadeh, "A Radical in His Time," 331.
116. Ibid., 352.
117. Shibli Shumayyil, "Al-murasil wa-al-munazira" [Correspondence and views], *Al-Muqtataf* 48 (1916): 299–300. For Sarruf's reply, see ibid., 397. See also Shibli Shumayyil, *Hawadith wa khawatir* (Beirut: Dar al-Hamra, 1991); cited in Ziadat, *Western Science*, 55–57.
118. Shibli Shumayyil, "Al-falsafa al-maddiya: Haqiqatha wa-nata'ijiha" [The philosophy of materialism: Its truth and results], in *Majmuʿat al-Duktur Shibli Shumayyil* [Works of Dr. Shibli Shumayyil] (Cairo: Matbaʿat al-Muqtataf, 1910), 225–30.
119. Ziadeh, "A Radical in His Time," 393–99; cited from Shumayyil, *Hawadith wa-khawatir*, 227–28.

CHAPTER FOUR

1. As Bernard Lightman, Peter Bowler, and others have pointed out, designed evolution was a popular and recurrent theme, particularly in the 1860s. See Bernard Lightman, "Ideology, Evolution and Late-Victorian Agnostic Popularizers," in *History, Humanity and Evolution: Essays for John C. Greene*, edited by James Moore (Cambridge: Cambridge University Press, 1989), 285–310; Peter Bowler, "Darwinism and the Argument from Design: Suggestions for a Reevaluation," *Journal of the History of Biology* 10 (1977): 29–43; and Alvar Ellegard, *Darwin and the General Reader: The Reception of Darwin's Theory of Evolution in the British Periodical Press, 1859–1872* (Chicago: University of Chicago Press, 1990), 114–40.
2. Muhammad al-Jisr, in the introduction to his father's *Risala Hamidiya fi haqiqat al-diyana al-Islamiya wa-haqiqat al-shariʿa al-Muhammidiya* [A Hamidian treatise on the truth of Islam and the shariʿa] (Cairo: Idarat al-Tabaʿa al-Muniriya, 1933), b.
3. Muhammad al-Jisr, introduction to *Risala*, b; Johannes Ebert, *Religion und Reform in der arabischen Provinz: Husayn al-Gisr al-Tarablusi (1845–1909)—Ein islamischer Gelehrter zwischen Tradition und Reform* (Frankfurt: P. Lang, 1991), 74.
4. Ebert, *Religion und Reform*, 74–75.
5. Husayn al-Jisr, *Nuzhat al-fikr* (Beirut: Matbaʿat al-Adaniya, 1889).
6. Muhammad al-Jisr, introduction to *Risala*, d.
7. Husayn al-Marsafi, *Risalat al-kalam al-thaman* [Discourse on eight words] (Cairo:

n.p., 1881); Timothy Mitchell, *Colonising Egypt* (Cambridge: Cambridge University Press, 1988), 131–37.

8. See Hoda Yousef, "Reassessing Egypt's Dual System of Education under Isma'il: Growing *'Ilm* and Shifting Ground in Egypt's First Educational Journal, *Rawdat al-Madaris*, 1870–77," *International Journal of Middle East Studies* 40 (2008): 109.
9. Peter Gran also argues that the emphasis on *fiqh* and *kalam*, as opposed to *hadith* studies, follows the absolutist and "administrative logic of the state," particularly a strong, centralized state. See Peter Gran, *Islamic Roots of Capitalism: Egypt, 1760–1840* (Austin: University of Texas Press, 1979), 50. On the Khalwatiya order and the revival of logic, see Khaled El-Rouayheb, "Opening the Gate of Verification: The Forgotten Arab-Islamic Florescence of the 17th Century," *International Journal of Middle East Studies* 38 (2006): 263–81. On the mid-eighteenth-century migration of Syrian *Khalwatiya tariqa* among merchant-scholars in Egypt, see Gran, *Islamic Roots*, 42–49.
10. Ebert, *Religion und Reform*, 78.
11. See ibid., 80, on the possibility that Midhat Pasha's support was in response to the opening of a local Lazarist school. See also Mahmoud Haddad, "Syrian Muslim Attitudes toward Foreign Missionaries in the Late Nineteenth and Twentieth Centuries," in *Altruism and Imperialism: Western Religious and Cultural Missionary Enterprise in the Middle East*, ed. Eleanor Tejirian and Reeva S. Simon (New York: Middle East Institute, Columbia University, 2002), 253–74.
12. Ebert, *Religion und Reform*, 79.
13. Rashid Rida, *Al-Manar wa-al-Azhar* (Cairo: Matba'at al-Manar, 1934), 142; Albert Hourani, *Arabic Thought in the Liberal Age, 1798–1939* (Cambridge: Cambridge University Press, 1962), 223; Ebert, *Religion und Reform*, 80.
14. Ebert, *Religion und Reform*, 81.
15. Yasir Ibrahim, "Rashid Rida and the *Maqasid al-Shari'a*," *Studia Islamica* 102/103 (2006): 167.
16. Ebert, *Religion und Reform*, 80–82.
17. Butrus Abu-Manneh, "The Genesis of Midhat Pasha's Governorship in Syria, 1878–1880," in *The Syrian Land: Processes of Integration and Fragmentation: Bilad al-Sham from the 18th to the 20th Century*, ed. Thomas Philipp and Birgit Schaebler (Stuttgart: F. Steiner, 1998), 251–67; Najib Saliba, "The Achievements of Midhat Pasha as Governor of the Province of Syria, 1878–1880," *International Journal of Middle East Studies* 9 (1978): 307–23.
18. Hisham Nashabi, "Shaykh 'Abd al-Qadir al-Qabbani and *Thamarat al-Funun*," in *Intellectual Life in the Arab East, 1890–1939*, ed. Marwan Buheiry (Beirut: American University of Beirut Press, 1981), 84–91; Randi Deguilhem, "State Civil Education in Late Ottoman Damascus: A Unifying or a Separating Force?" in Philipp and Schaebler, *Syrian Land*, 245; Ami Ayalan, *The Press in the Arab Middle East* (New York: Oxford University Press, 1995), 32–33; 'Abd al-Qadir al-Qabbani's letter to Rashid Rida, and Rida's response in *Al-Manar* 12 (1920): 635; Adel Ziadat, *Western Science in the Arab World: The Impact of Darwinism, 1860–1930* (Basingstoke: Macmillan, 1986): 35.
19. Muhammad al-Jisr, introduction to *Risala*, h; Martin Strohmeier, "Muslim Education in the Vilayet of Beirut, 1880–1918," in *Decision Making and Change in the Ottoman Empire*, ed. Caesar E. Farah (Kirksville, MO: Thomas Jefferson University Press, 1993), 215–41.
20. Haddad, "Syrian Muslim Attitudes toward Foreign Missionaries," 255; *The True Dawn: Makassed Islamic and Philanthropic Association in Beirut: The First Annual*

Report—Translated from the Arabic Original Dated 1297 A.H. (1880) (Beirut: n.p., 1984).

21. Among the sciences they listed chemistry, medicine, surgery, and mathematics. Cited in Linda Schatkowski, "The Islamic Maqased of Beirut: A Case Study of Modernization in Lebanon" (thesis, American University of Beirut, 1981), 30.
22. Cited in Haddad, "Syrian Muslim Attitudes toward Foreign Missionaries"; see also Schatkowksi, "Islamic Maqased of Beirut," 31.
23. Haddad, "Syrian Muslim Attitudes toward Foreign Missionaries." Al-Nabhani wrote two treatises to Muslims advising them against the foreign missionaries: *Irshad al-hayara fi tahdhir al-Muslimin min madaris al-Nasara* [Advice to the perplexed: Warning the Muslims of Christian schools] in 1904 and *Mukhtasar kitab riyad al-salihin* [Selections of the Book of the Garden of the Righteous] in 1908.
24. See Rashid Rida, *Tarikh al-ustadh al-imam Muhammad ʿAbduh*; cited in Strohmeier, "Muslim Education," 216.
25. Benjamin Fortna, "Islamic Morality in Late Ottoman 'Secular' Schools," *International Journal of Middle East Studies* 32 (2000): 373.
26. Joseph Escovits, "'He Was the Muhammad Abduh of Syria': A Study of Tahir al-Jazairi and His Influence," *International Journal of Middle East Studies* 18 (1986): 293–310; David Dean Commins, "Religious Reformers and Arabists in Damascus, 1885–1914," *International Journal of Middle East Studies* 18 (1986): 412; and David Dean Commins, *Islamic Reform: Politics and Social Change in Late Ottoman Syria* (New York: Oxford University Press, 1990), 41–42.
27. Haddad, "Syrian Muslim Attitudes toward Foreign Missionaries"; Strohhmeier, "Muslim Education," 215.
28. Jens Hanssen, *Fin de Siècle Beirut: The Making of an Ottoman Provincial Capital* (Oxford: Oxford University Press, 2005), 172–73; Samir Sekaly, "Shaykh Yusuf al-Nabhani and the West," in *Les Européens vus par les Libanais à l'époque ottomane*, ed. Bernard Heyberger and Casten Walbiner (Beirut: Ergon Verlag, 2002), 175–81.
29. Jens Hanssen, *Fin de Siècle Beirut*, 174–76.
30. Ebert, *Religion und Reform*, 84.
31. Muhammad al-Jisr, introduction to *Risala*, u.
32. Yildiz Collection, Office of the Prime Minister, Y.PRK.AZJ 1/28: 1293 S 29.
33. Ebert, *Religion und Reform*, 87.
34. Midhat Pasha translated the *Risala* into Turkish and published it in his journal, *Tarjuman-i Haqiqat*, shortly after its Arabic publication; it was translated and published in Urdu in 1897 and in Chinese in 1951. See Muhammad Khalid Masud, "Iqbal's Approach to Islamic Theology of Modernity," *Al-Hikmat* 27 (2007): 1–36; and Haiyun Ma, "Patriotic and Pious Muslim Intellectuals in Modern China: The Case of Ma Jian," *American Journal of Islamic Social Science* 23 (2006): 61.
35. Husayn al-Jisr, *Jaridat al-Tarablus* 2 (1894): 6; cited in Ebert, *Religion und Reform*, 124.
36. Husayn al-Jisr, *Jaridat al-Tarablus* 9 (1900): 83; cited in Ebert, *Religion und Reform*, 117.
37. Husayn al-Jisr, *Al-risala al-Hamidiya fi haqiqat al-diyana al-Islamiya wa-haqiqat al-shariʿa al-Muhammadiya* [A Hamidian treatise on the truth of Islam and the shariʿa] (Beirut: Majlis al-Maʿarif, 1888 [AH 1305]). Note that the AH 1305 edition has sometimes been listed as appearing in 1887, but from internal references, it seems likely that it in fact appeared the following year. The edition used here is the Cairo edition of AH 1322 (Cairo: Al-Matbaʿa al-Hamidiya al-Misriya, 1904–5).

38. Verses 54:1–2. Al-Jisr, *Risala*, 18, 25–28. Al-Jisr's views on prophetic miracles was not, of course, shared by later, more thoroughgoing rationalists; see the discussion of miracles below.
39. Al-Jisr, *Risala*, 20–24.
40. Ibid., 30.
41. Ibid., 44.
42. Ibid., 45–95.
43. Ibid., 90, 310–20, 328–29.
44. Ibid., 96.
45. For more on this and the revival of this hermeneutic tool as a key interpretive stratagem for modern reformers of that generation and after, see Ibrahim, "Rashid Rida and the *Maqasid al-Shariʿa.*" Al-Jisr's former student Rashid Rida and Rida's mentor and literary associate Muhammad ʿAbduh, for example, would develop this point even further, as Ibrahim shows. See chapter 5 for more on this.
46. Al-Jisr, *Risala*, 46–74.
47. Ibid., 84–92.
48. Ibid., 93.
49. Ibid., 95–96.
50. See the next chapter for more on this.
51. Al-Jisr, *Risala*, 216.
52. Ibid., 211.
53. Ibid., 154–89.
54. See the epigraph at the beginning of this chapter (from ibid., 130–31).
55. Ibid., 133–34.
56. See the references in n. 1 of this chapter.
57. For more on these themes (with examples drawn from early-modern Europe), see Lorraine Daston and Katherine Park, *Wonders and the Order of Nature, 1150–1750* (New York: Zone Books, 1998).
58. Rizq'allah al-Birbari, "Fi asl al-insan" [On the origins of humanity], *Al-Muqtataf* 1 (1876): 242–44, 279–80.
59. Bishara Zalzal, "Al-insan" [Man], *Al-Muqtataf* 2 (1877): 203.
60. Yaʿqub Sarruf, "Al-Madhhab al-Darwiniya" [Darwin's theory], *Al-Muqtataf* 7 (1882): 33.
61. *Al-Muqtataf* 9 (1885): 704; see also Nadia Farag, "*Al-Muqtataf*, 1876–1900: A Study of the Influence of Victorian Thought on Modern Arabic Thought" (PhD diss., University of Oxford, 1969), 273.
62. Cited from Farag, "*Al-Muqtataf*," 160.
63. Ibrahim al-Hurani, *Al-ayat al-bayyinat fi gharaʾib al-ard wa-al-samawat* [Certain signs in the wonders of the heavens and earth] (Beirut: n.p., 1883), 1–5.
64. Jurji Effendi Baz, *Kitab al-rawda al-badiʿa fi tarikh al-tabiʿa* [The book of the magnificent garden of natural history] (Beirut: n.p., 1881), 5.
65. Fransis Fath Allah al-Marrash, *Shahadat al-tabiʿa fi wujud Allah wa-al-shariʿa* [Nature's testimonies to the existence of God and divine law] (Beirut Matbaʿat al-Amrikan, 1891); Sami al-Kayyali, *Muhadarat ʿan al-haraka al-adabiya fi halab* [Lectures on the literary movement of Aleppo] (Cairo: Jamiʿat al-Duwal al-ʿArabiya, 1956), 143; Jurji Zaydan, *Tarajim mashahir al-sharq* [Biographies of famous people of the East] (Cairo: Matbaʿat al-Hilal, 1910–11), 2:286.
66. Fransis Fath Allah al-Marrash, *Rihlat Baris* (Beirut: Al-Matbaʿa al-Sharifiya, 1867), 9–10; Haidar al-Haj Ismaʿil, *Fransis al-Marrash* (London: Riyad al-Rayyis, 1989), 21.

67. Marrash, *Rihlat Baris*, 28.
68. Ibid., 49–50.
69. Among his works were commentaries, poems, fictional accounts, and literary stories. He contributed—both in his lifetime and posthumously—to a variety of religious, literary, and scientific Syrian journals, such as Butrus al-Bustani's science journal *Al-Jinan* (The garden), the Jesuit university's weekly *Al-Bashir* (The herald), the Syrian Protestant College's weekly *Al-Nashra al-Usubuʿiya* (The weekly news), as well as lesser-known journals such as *Al-Zahra* (The splendor) and *Al-Nahla* (The bee). He published collections of poetry and several book-length tales and musings on political and social freedom, including the plays *Ghabat al-haqq* (Forest of truth) published in Aleppo in 1865 and *Mashhad al-ahwal* (A view on the state of things) published posthumously in Beirut in 1883. Most important, for our purposes here, were his treatises on natural science and theology, including *Al-miraʾat al-safiya fi al-mabadi al-tabiʿa* [The clear mirror of the principles of nature] (Aleppo: n.p., 1861), and his *Shahadat al-tabiʿa*. These themes also appear in Marrash, *Rihlat Baris*; Marrash, *Mashhad al-ahwal*. For more on his works, see Haidar al-Haj Ismail, *Fransis Marrash*, 21–22.
70. Marrash, *Shahadat al-tabiʿa*, 5.
71. Ibid., 13–41.
72. Alister McGrath, *Science and Religion* (Oxford: Blackwell, 1999), 102. See also Peter Bowler, *Evolution: The History of an Idea* (Berkeley and Los Angeles: University of California Press, 1982), 5; Dov Ospovat, *The Development of Darwin's Theory: Natural History, Natural Theology and Natural Selection, 1838–1859* (Cambridge: Cambridge University Press, 1981), 60–86; Peter Byrne, *Natural Religion and the Nature of Religion: The Legacy of Deism* (London: Routledge, 1989), 1–10.
73. Marrash, *Shahadat al-tabiʿa*, 15–21.
74. Ibid., 21–25.
75 Ibid., 26–41.
76. Ibid., 44–45.
77. Ibid., 41–42.
78. Ibid..
79. Ibid., 44–46.
80. Louis Foucher, *La philosophie catholique en France au XIXème siècle avant la renaissance thomiste et dans son rapport avec elle, 1800–1880* (Paris: J. Vrin, 1955), 11–50, 237–64. Aziz Al-Azmeh suggests that Marrash was greatly influenced by neo-Thomist apologist thought while studying at the medical school in Paris, particularly under the leadership of Dr. F. Frédault in the 1860s. See ibid., 256–64; F. Frédault, *Forme et matière* (Paris: E. Vaton, 1876); Aziz al-Azmeh, *Islams and Modernities* (London: Verso, 1993), 122.
81. Marrash, *Shahadat al-tabiʿa*, 6.
82. Ibid., 11.
83. Ibid., 58–61.
84. Ibid., 76–77.
85. Ibid., 54–55.
86. Al-Jisr, *Risala*, 128–30.
87. Ibid., 223–24.
88. Ibid., 115.
89. Ibid., 108–10.
90. Ibid., 119.

91. Ibid., 109.
92. Ibid., 128.
93. Ibid., 146–48.
94. Ibid., 138–45.
95. Ibid., 130–31, 133–34.
96. Ibid., 154–79.
97. Ibid., 179–89.
98. E.g., ibid., 112–13.
99. Ibid., 199.
100. Ibid., 190–92.
101. Ibid., 192.
102. Ibid., 224–25.
103. Ibid.
104. Ibid., 245–47.
105. Ibid., 228.
106. Majid Fakhry, "Classical Islamic Arguments for the Existence of God," *Muslim World* 47 (1957): 133–45.
107. Al-Jisr, *Risala*, 232, 238–42.
108. Ibid., 232–34.
109. Ibid., 235–37.
110. See Harry Austryn Wolfson, *The Philosophy of Kalam* (Cambridge, MA: Harvard University Press, 1976), 1.
111. On *ʿilm al-kalam* and its function as a defensive "apologia," see L. Gardet, "ʿIlm al-kalām," in *Encyclopaedia of Islam*, 2nd ed., http://referenceworks.brillonline.com/entries/encyclopaedia-of-islam-2/ilm-al-kalam-COM_0366.
112. See Richard M. Frank, "The Science of *Kalam*," *Arabic Sciences and Philosophy* 2 (1992): 9–11.
113. See Şerif Mardin, *The Genesis of Young Ottoman Thought: A Study in the Modernization of Turkish Political Ideas* (Princeton, NJ: Princeton University Press, 2000), 70.
114. For the background, see Thomas Prasch, "Which God for Africa? The Islamic-Christian Missionary Debate in Late-Victorian Africa," *Victorian Studies* 33 (1989): 51–73.
115. Cited in Christian Troll, *Sayyid Ahmad Khan: A Reinterpretation of Muslim Theology* (New Delhi: Vikas, 1978), 313; and M. K. Masud, "Iqbal's Approach to Islamic Theology of Modernity," *Al-Hikmat* 27 (2007): 4.
116. ʿAdullatif Kharputi, *Tarikh-i ilm-i kalam* (Istanbul, AH 1332); Izmirli Ismaʿil Haqqi, *Yeni ilm-i kalam*, 2 vols. (Istanbul, AH 1339–41 and 1340–43); and Izmirli Ismaʿil Hakki, *Muhassal al-kalam wa-al-hikma* (Istanbul, n.d); all works cited in M. Sait Özervarli, "Attempts to Revitalize *Kalam* in the Late 19th and Early 20th Centuries," *Muslim World* 89 (1999): 94–96, 100–104.
117. Commins, *Islamic Reform*, 48.
118. See Dimitry Zhantiev, "Islamic Factor in the Consolidation of the Ottoman Rule in the Arab Provinces during the Reign of Sultan Abdülhamid II (1876–1908)," in *Authority, Privacy and Public Order in Islam*, ed. B. Michalak-Pikulska and A. Pikulski (Leuven: Peeters, 2006). On al-Sayyidi, see Butrus Abu-Manneh, "Sultan Abdülhamid II and Shaikh Abulhuda al-Sayyadi," *Middle Eastern Studies* 15 (1979): 131–53; Thomas Eich, "The Forgotten Salafi: Abu'l-Huda as-Sayyadi," *Die Welt des Islams* 43 (2003): 61–87; Thomas Eich, "Abū l-Hudā l-Sayyādī—Still Such a Polarizing Figure (Response to Itzchak Weismann)," *Arabica* 55 (2008): 433–44; and Itzchak

Weismann, "Abu l-Huda l-sayyadi and the Rise of Islamic Fundamentalism," *Arabica* 54 (2007): 586–592.

119. Ebert, *Religion and Reform*, 91.
120. Council of the State, Department of Internal Affairs, Reports of the Ministry of Vakufs, Yildiz Collection, Y.A. RES 63/44: 1310 S 20, 9 February 1893.
121. Al-Jisr usually wrote the lead article on social and religious topics, while Muhammad Kamil al-Bukhayri, a local merchant, wrote the remaining two or three pages on political and world news.
122. Ebert, *Religion und Reform*, 158–61.
123. Al-Jisr, *Risala*, 26–28.

CHAPTER FIVE

1. Wilfrid Scawen Blunt Collection, box 55—Spencer, Herbert: Letters from 1898 to 1903: 27 July 1903 and 21 September 1903, West Sussex Record Office, Chichester, UK. For Blunt and Spencer's first meeting, see Wilfrid Scawen Blunt, *My Diaries* (New York: A. A. Knopf, 1921), 1:317 (27 March 1899).
2. See the essays in the forthcoming volume: Bernard Lightman, ed., *Global Spencerism* (London: Pickering and Chatto).
3. ʿAbduh translated Spencer's *On Education* sometime in 1889, while preparing to submit his proposal for the reform of the Egyptian educational system; Muhammad ʿAbduh, "Mashruʿa islah al-tarbiya fi Misr" [A project for Egyptian educational reform], in *Al-aʿmal al-kamila* [Complete works of Muhammad ʿAbduh], 5 vols. (Cairo: Dar al-Shuruq, 1993), 3:106–22; Rashid Rida, *Tarikh al-ustadh al-imam Muhammad ʿAbduh* [History of Imam Muhammad ʿAbduh], 3 vols. (Cairo: Matbaʿat al-Manar, 1908–31), 1:1031–34.
4. Blunt, *My Diaries*, 2:66–67 (August 1903).
5. Ibid., 2:67.
6. In other writings, ʿAbduh, like his later publicist, Rashid Rida, counts rationalism and freedom of thought among the advantages of European thought, and materialism one of its dangers. See "Manafiʿ al-urubiyin wa madarhum fi al-sharq" [The advantages and disadvantages of Europeans in the East], *Al-Manar* 10 (1907): 192–99 (originally published in *Al-Jarida*), which also makes this point.
7. Blunt, *My Diaries*, 2:67.
8. Ibid.
9. ʿAbduh, *Al-aʿmal al-kamila*, 3:510.
10. Ibid.
11. See Rida, *Tarikh al-ustadh al-imam Muhammad ʿAbduh*, 1:868; and ʿAbduh, *Al-aʿmal al-kamila*, 3:509.
12. See, e.g., Charles Adams, *Islam and Modernism in Egypt: A Study of the Modern Reform Movement Inaugurated by Muhammad ʿAbduh* (London: Oxford University Press, 1933); ʿUthman Amin, *Muhammad ʿAbduh: Essai sur ses ideés philosophiques et religieuses* (Cairo: Misr, 1944); Jamal Mohammed Ahmed, *The Intellectual Origins of Egyptian Nationalism* (London: Oxford University Press, 1960), 35–43; M. Zaki Badawi, *The Reformers of Egypt* (London: Croom Helm, 1976); Mahmudul Haq, *Muhammad ʿAbduh: A Study of a Modern Thinker of Egypt* (Aligarh: Institute of Islamic Studies, 1970); Zohair Husain, "Muhammad ʿAbduh (1849–1905): The Preeminent Muslim Modernist of Egypt," *Hamdard Islamicus* 9 (1986): 31–9; Albert Hourani, *Arabic Thought in the Liberal Age, 1798–1939* (Cambridge: Cambridge University Press, 1962); Malcolm Kerr, *The Political and Legal Theories of Muhammad ʿAbduh*

and Rashid Rida (Berkeley and Los Angeles: University of California Press, 1966); Nabeel Abdo al-Khoury and Abdo Baaklani, "Muhammad ʿAbduh: An Ideology of Development," *Muslim World* 69 (1979): 42–52; P. Vatikiotis, "Muhammad ʿAbduh and the Quest for a Muslim Humanism," *Arabica* 4 (1957): 54–72.

13. Or, more rarely, it was known as "Americanism" for its association with liberal American Catholic theologians.
14. See B. Valade, "Modernity," in *International Encyclopedia of the Social and Behavioral Sciences*, ed. Neil Smelser and Paul Bates (Amsterdam: Elsevier Science, 2001), 9939–44; Bernard Reardon, *Roman Catholic Modernism* (1970); Darrel Jodock, ed., *Catholicism Contending with Modernity* (Cambridge: Cambridge University Press, 2000). On ʿAbduh on science, see Adams, *Islam and Modernism*, 127–43; J. Jomier, *Le commentaire coranique du Manar: Tendences modernes de l'exégèse coranique en Égypte* (Paris, 1954), 150–59; John Livingston, "Muhammad ʿAbduh on Science," *Muslim World* 85 (1995): 215–34; Itzchak Weismann, "The Sociology of Islamic Modernism: Muhammad ʿAbduh, the National Public Sphere and the Colonial State," *Maghreb Review* 32 (2007): esp. 107.
15. For more on ʿAbduh in this light, see Samira Haj, *Reconfiguring Islamic Tradition: Reform, Rationality and Modernity* (Stanford, CA: Stanford University Press, 2009).
16. In fact, ʿAbduh could just as easily be counted as a "countermodernist," as the question of civilizational progress in modern times led to a number of parallel global counternarratives. See, e.g., Adam Webb, "The Countermodern Moment: A World-Historical Perspective on the Thought of Rabindranath Tagore, Muhammad Iqbal, and Liang Shuming," *Journal of World History* 19 (2008): 189–212.
17. ʿAbduh, *Al-aʿmal al-kamila*, 3:513.
18. Muhammad ʿAbduh, *The Theology of Unity*, trans. Ishaq Musaʿad and Kenneth Cragg (London: Allen and Unwin, 1966), 96.
19. On the idea of a Muslim discursive tradition, a methodological and conceptual approach that I follow in this chapter, see Talal Asad, *Genealogies of Religion: Discipline and Reasons of Power in Christianity and Islam* (Baltimore, MD: John Hopkins University Press, 1993).
20. ʿAbduh, *Theology of Unity*, 70 and 64 respectively.
21. Alexander Meyrick Broadley, *How We Defended Arabi and His Friends: A Story of Egypt and the Egyptians* (London: Chapman and Hall, 1884), 110–11.
22. From a large literature on the background to these events, see Juan Cole, *Colonialism and Revolution in the Middle East: Social and Cultural Origins of Egypt's ʿUrabi Movement* (Princeton, NJ: Princeton University Press, 1993); Alexander Schölch, *Egypt for the Egyptians: The Socio-political Crisis in Egypt, 1878–1882* (London: Ithaca Press, 1981).
23. See Adams, *Islam and Modernism*, 33–35, on how Afghani introduced him to science and the press.
24. Wilfrid Blunt, *The Secret History of the English Occupation of Egypt: Being a Personal Narrative of Events* (London: T. F. Unwin, 1907), 493.
25. For the letter between al-Jisr and ʿAbduh, see Khalil Ziyada, *Husayn al-Jisr, 1845–1909: Hayatahu wa fikrhu* (Tripoli, 1982), 82–84; Johannes Ebert, *Religion und Reform in der arabischen Provinz: Husayn al-Gisr al-Tarablusi (1845–1909)—Ein islamischer Gelehrter zwischen Tradition und Reform* (Frankfurt: P. Lang, 1991), 85. See the discussion of the broader debate in Ottoman Syria and in the empire at large in the previous chapter and Benjamin Fortna, *Imperial Classroom: Islam, the State, and Education in the Late Ottoman Empire* (Oxford: Oxford University Press, 2002), 81.

26. See Muhammad ʿAbduh, "Laʾihat islah al-taʿlim al-ʿUthmani" [Ottoman educational reform program], in *Al-aʿmal al-kamila* 3:73–91; and his related "Laʾihat islah al-qutr al-suri" [Reform program for the Syrian province], in ibid., 3:93–105.
27. For more on Cromer's views on ʿAbduh, whom he calls "my friend," likens to Sayyid Ahmad Khan, and suspects he "was in reality an agnostic," see Evelyn Baring Cromer, *Modern Egypt* (London: Macmillan, 1908), 2:179–80.
28. ʿAbduh also published a work on logic, *Sharh kitab al-basaʾir al-nasiriya tasnif al-qadi Zayn al-Din*, in 1898, which was a reprint of a twelfth-century work on Aristotelian logic by Zayn al-Din al-Sawi. ʿAbduh considered this to be the single best manual on formal logic in Arabic. See Ahmed Fouad El-Ehwany, "Present-Day Philosophy in Egypt," *Philosophy East and West* 5 (1956): 339.
29. Muhammad ʿAbduh and Rashid Rida, *Tafsir al-Qurʾan al-hakim al-mushtahir bi-ism Tafsir al-Manar* [Exegesis of the judicious Quʾran, widely known as "Tafsir al-Manar"], 12 vols. (Cairo: Dar al-Manar, 1906–35); the edition used here is *Tafsir al-Qurʾan al-hakim al-musamma Tafsir al-Manar* [Exegesis of the judicious Qurʾan, called "Tafsir al-Manar"], 12 vols. (Cairo: Al-Haya al-Misriya al-ʿAmm li-al-Kitab, 1972–75). Henceforth cited as *Tafsir al-Manar*. For more on the background to this, see Jomier, *Commentaire coranique du Manar*, 45–66.
30. These articles were based on lectures that ʿAbduh delivered at Al-Azhar between 1900 and 1901, which were then written up by Rida. After ʿAbduh's death, Rida continued the exegesis, from sura 4:125 to sura 12:107, until his own death in 1935. Though still incomplete, the exegesis was thousands of pages in length and filled 12 volumes. See Rashid Rida's 1927 introduction to the *Tafsir al-Manar*; Adams, *Islam and Modernism*, 190; and J. Jansen, *The Interpretation of the Koran in Modern Egypt* (Leiden: Brill, 1974), 24, 29.
31. ʿAbduh, *Theology of Unity*, 55.
32. Ibid., 55.
33. Christian Van Nispen Tot Sevenaer, *Activité humaine et agir de Dieu: Le concept de* "Sunan *de Dieu" dans le commentaire coranique du Manar* (Beirut: Dar el-Machreq, 1996), 63–104.
34. "Al-haqq, al-batil wa-al-qiwa" [Truth, falsehood and power], *Al-Manar* 9 (1906): 52–65.
35. See ʿAbduh, "Maʿani al-ʿilm" [The meanings of *ʿilm*], in *Al-aʿmal al-kamila*, 3: 145–48.
36. See Osman Amin, "Renaissance in Egypt: Muhammad ʿAbduh and his School," in *A History of Muslim Philosophy*, ed. M. M. Sharif (Karachi: Royal Book, 1983), 1497–99.
37. Cf. Serif Mardin's interpretation of the idea of natural law in Islam in his *The Genesis of Young Ottoman Thought: A Study in the Modernization of Turkish Political Ideas* (Princeton, NJ: Princeton University Press, 2000).
38. Van Nispen Tot Sevenaer, *Activité humaine et agir de Dieu*, 70–96.
39. ʿAbduh, *Theology of Unity*, 54.
40. Van Nispen Tot Sevenaer, in his study of Muhammad ʿAbduh and Rashid Rida's *Tafsir al-Manar*, claims that he has "never encountered a single passage [in the *Tafsir al-Manar*] in which evolution posed any sort of problem" (*Activité humaine et agir de Dieu*, 87).
41. "Tafsir al-Qurʾan al-hakim" [Exegesis of the judicious Qurʾan], *Al-Manar* 8 (1906): 929–30.

42. "Tafsir al-Qur'an al-hakim," *Al-Manar* 12 (1909): 483. Although several scholars have ascribed this passage to ʿAbduh (see Adams, *Islam and Modernism*, 139; and Livingston, "Muhammad ʿAbduh on Science," 227), it may have been Rida's.
43. For more, see Van Nispen Tot Sevenaer, *Activité humaine et agir de Dieu*, 87.
44. E.g., "Abuna Adam wa-madhhab Darwin: Min bab al-intiqad ʿala *Al-Manar*" (Our father Adam and Darwin's doctrine: Criticisms of *Al-Manar*), *Al-Manar* 8 (1906): 920.
45. Ibid.
46. Muhammad ʿAbduh, *Al-Islam wa al-Nasraniya maʿa al-ʿilm wa-al-madaniya* [Science and civilization in Islam and Christianity] (Cairo: n.p., 1905), 102–4.
47. "Al-haqq, al-batil wa-al-qiwa" [Truth, falsehood, and power], *Al-Manar* 9 (1906): 55.
48. Ibid. See also Adams, *Islam and Modernism*, 140–42, on ʿAbduh's conception of social laws.
49. ʿAbduh, *Al-Islam wa-al-Nasraniya*, 55–56.
50. Ibid., 57.
51. "Tafsir al-Qur'an," *Al-Manar* 7 (1904): 292.
52. "Tafsir al-Qur'an," *Al-Manar* 9 (1906): 334–35. This passage has been much discussed; see Adams, *Islam and Modernism*, 138; Livingston, "Muhammad ʿAbduh on Science," 226; Lutz Berger, "Esprits et microbes: L'interprétation des *ginn*-s dans quelques commentaires coraniques du XXe siècle," *Arabica* 47 (2000): 554–62. ʿAbduh, however, was not the first to liken *jinn* to microbes. Shaykh Taha al-Khalili, for instance, had earlier claimed that the Qur'an had "discovered microbes"; Taha al-Khalili, "Maziat al-din al-Islami fi al-ʿalam al-tamaddun," *Al-Islam* 4 (1897): 216–22, as cited by Indira Gesink, *Islamic Reform and Conservatism: Al-Azhar and the Evolution of Modern Sunni Islam* (London: I. B. Tauris, 2010), 162.
53. ʿAbduh, *Al-Islam wa-al-Nasraniya*, 171–72.
54. ʿAbduh, *Theology of Unity*, 107.
55. Osman Amin, "Renaissance in Egypt," 1496–97.
56. ʿAbduh, *Al-Islam wa-al-Nasraniya*, 62–63.
57. ʿAbduh, *Theology of Unity*, 31–32, 39.
58. ʿAbduh, *Al-aʿmal al-kamila*, 2:512–13.
59. Muhammad ʿAbduh, *Theology of Unity*, 103.
60. "Deaf dumb and blind," the unbelievers "return not; like a rainstorm from the sky wherein is darkness, thunder, and the flash of lightning. They thrust their fingers in their ears against the thunderclaps, for fear of death. . . . The lightning almost snatches away their sight. As often as it flashes, they walk, and when it darkens, they stand still" (sura 2:18–20). "Tafsir al-Qur'an," *Al-Manar* 9 (1906): 334. For more on this, see also Adams, *Islam and Modernism*, 137; and Livingston, "Muhammad ʿAbduh on Science," 226.
61. Muhammad ibn Ahmad al-Iskandarani al-Tabib, *Kashf al-asrar al-nuraniya al-Qur'aniya fi ma yata'allaqu bi-al-ajram al-samawiya wa-al-ardiya wa-al-hayawanat wa-al-nabatat wa-al-maʿdaniya* [Unveiling the luminous secrets of the Qur'an and their relation to the heavens, earth, animals, plants, and minerals] (Cairo: Al-Matbaʿa al-Wahbiya, 1880); and Muhammad ibn Ahmad al-Iskandarani al-Tabib, *Kitab al-asrar al-rabbaniya fi al-nabatat wa-al-maʿadin wa-al-khawass al-hayawaniya* [The book of the divine secrets of plants, minerals, and animal characteristics] (Damascus: Matbaʿat Majlis al-Maʿarif, 1882), 112. He also published a work on animals entitled *Kitab al-*

barahin al-bayyinat fi bayan haqa'iq al-hayawanat [The book of certain proofs revealing animal truths] (Damascus: Matbaʿat Majlis al-Maʿarif, 1886). For more on his training and career, see al-Iskandarani, *Kitab al-asrar*, 3.

62. They offered modern medical descriptions of man's constitution—from the physiology of digestion and the circulation of blood to details of respiration and the nervous system—as well as general discussions of the "three kingdoms of creation." They were therefore also intended to serve as religious treatises on ethical living and right conduct. *The Book of the Divine Secrets*, for instance, emphasizes the importance of proper nutrition, arguing that "good food gives men the power to worship." Similarly, it describes the need for adequate sleep and exercise and urges the restraint of certain passions. And alongside commentary on various verses of the Qur'an came sundry digressions, ranging from proper table manners and the nutritional value of okra, oxtail soup, and buffalo cream (*al-qishtah*) to the causes of impotence and corpulence. Al-Iskandarani, *Kitab al-asrar al-rabbaniya*, 33, 169 and 197, 34, 87, 241, 266–84 respectively.

63. Ibid., 112, citing sura 87:2. In fact, the idea for *Unveiling the Luminous Secrets of the Qur'an* emerged when al-Iskandarani was asked at a meeting of doctors in Damascus, most of whom were Christian, to find some reference to the formation of coal in the Qur'an. They themselves had determined, after extensive deliberation, that this could not be found in the Old and New Testaments. "Nothing have we omitted from the Book," was the claim of the Qur'an itself, yet, they challenged, was the claim true? Al-Iskandarani's response—and the result, he wrote, of years of extensive research—were the books that he published shortly before his death. Al-Iskandarani, *Kashf al-asrar al-nuraniya al-Qur'aniya*, 3–4, citing sura 6:38.

64. See Richard Netton, "Nature as Signs"; Denis Gril, "Miracles"; and Ahmad Dallal, "Science and the Qur'an"; all in *Encyclopaedia of the Qur'an*, Brill Online, 2007; Marwa Elshakry, "The Exegesis of Science in Twentieth Century Arabic Interpretations of the Qur'an," in *Nature and Scripture*, ed. Jitse van der Meer, 491–524 (Leiden: Brill, 2009).

65. See Muhammad Farid Wajdi's Azhari journal, *Nur al-Islam*, discussed in chapter 7. Wajdi also published the popular *Al-madaniya wa-al-Islam* [Civilization and Islam] (Cairo: Matbaʿat al-Waʿiz, 1901), which he wrote originally in French; a refutation of materialism, *ʿAla atlal al-madhhab al-maddi* (Cairo: Matbaʿat Dairat Maʿarif al-Qarn al-ʿIshrin, 1921); and responses to Hanotaux and Lord Cromer on Islam, "Maqallahu Hanutu wa-radd ahad a'immat al-din al-ʿalam ʿalayhi" and "Al-Lurd Krumar wa-al-Islam," both published in *Al-Dustur* in 1901 and 1908 respectively. See also Tantawi Jawhari, *Al-Jawahir fi tafsir al-Qur'an al-karim* [Gems of Qur'anic exegesis], 26 vols. (Cairo: Matbaʿit Mustafa al-Babi al-Halabi, 1925–35). Jawhari also published several works on Islam and science, including *Al-nizam wa-al-Islam* [Governance and Islam] (Cairo: Matbaʿat al-Jumhur, 1903), on the Qur'an and science; and *Bahjat al-ʿulum fi al-falsafa al-ʿarabiya wa-muwazanatih bi al-ʿulum al-ʿasriya* [Delights of the Arabic philosophical sciences and their relation to the contemporary sciences] (Cairo: n.p., 1936), on the Arabic sciences and philosophy and their relation to the contemporary sciences.

66. Wajdi, *Al-Islam wa-al-madaniya*, cited in Mansoor Moaddel and Karim Talatoff, eds., *Contemporary Debates in Islam* (New York: St. Martin's Press, 2000), 139.

67. Muhammad ʿAbduh, "Al-ʿulum al-kalamiya wa-al-daʿwa ila al-ʿulum al-ʿasriya" [The sciences of speculative theology and the call for modern sciences], *Al-Ahram* 36 (1877); reprinted in ʿAbduh, *Al-aʿmal al-kamila*, 3:15–22.

68. See Osman Amin, "Renaissance in Egypt," 1496.
69. ʿAbduh, *Al-aʿmal al-kamila*, 3:17–20.
70. Jamal al-Din al-Afghani, "Lecture on Teaching and Learning," delivered in Calcutta on 8 November 1882; cited in Nikki Keddie, *An Islamic Response to Imperialism: Political and Religious Writings of Sayyid Jamal ad-Din "al-Afghani"* (Berkeley and Los Angeles: University of California Press, 1983), 102–3; ʿAbduh, *Al-aʿmal al-kamila*, 3:19–20.
71. Citations from *Nineteenth Century* and *Fortnightly Review* in 1877 and 1884 respectively, cited from Paul Auchterlonie, "From the Eastern Question to the Death of General Gordon: Representations of the Middle East in the Victorian Periodical Press, 1876–1885," *British Journal of Middle Eastern Studies* 28 (2001): 11, 21.
72. Ibid.
73. Ernest Renan, "L'Islamisme et la science," Sorbonne lecture, 1882, in Renan, *Œuvres complètes* (Paris: Calmann-Levy, 1947), vol. 1. Jamal al-Din al-Afghani's response to Renan was published in *Journal des Débats*, 18 May 1883. See also Gabriel Hanotaux, "L'Islam," *Le Journal*, 21 March 1900; and Gabriel Hanotaux, "Encore l'Islam," *Le Journal*, 14 May 1900. Muhammad ʿAbduh's response to Hanotaux was published in *Al-Muʾayyad* 18 (6 May 1900) and 19 (13 May 1900). As mentioned, *Al-Dustur* also published a response by Muhammad Farid Wajdi, which he later published as *Ma qalahu Hanutu wa-radd ahad aʾimmat al-din al-aʿlam ʿalayhi* [What Hanotaux said and the response of a knowledgeable religious scholar] (Cairo: n.p., 1900). Meanwhile, Husayn al-Jisr also published a response to Hanotaux in his *Jaridat Tarablus*; see Ebert, *Religion und Reform*, 120; Cromer, *Modern Egypt*; response by Muhammad Farid Wajdi, published as *Lurd Krumar wa-al-Islam* (Cairo: Matbaʿat al-Waiz, 1908). Cromer's opinions on Islam were scorned by many Arab writers; even Shibli Shumayyil was moved to criticize him in 1909; see Shibli Shumayyil, *Majmuʿa* (Cairo: Matbaʿat al-Muqtataf, 1910), 2:293.
74. Renan, "L'Islamisme et la science."
75. Jamal al-Din al-Afghani, *Journal des Débats*, 18 May 1883; cited from Keddie, *Islamic Response*, 182.
76. Keddie, *Islamic Response*, 182.
77. See also Cemil Aydin, "Beyond Civilization: Pan-Islamism and Pan-Asianism and the Revolt against the West," *Journal of Modern European History* 4 (2006): 204–23.
78. Hanotaux, "L'Islam," *Le Journal*, 21 March 1900.
79. Gabriel Hanotaux and Muhammad ʿAbduh, *L'Europe et l'Islam*, reprint of articles from *Al-Muʾayyad* and *Le Journal* with a preface by Mohammad Talat Harb (Cairo, 1905), 29–30, 41, 44, 50–51; cf. Muhammad ʿAbduh, *Al-Islam: Din al-ʿilm wa al-madaniya* [Islam: A religion of science and civilization] (Cairo: n.p., 1905).
80. "Sirat al-ustadh al-imam," *Al-Manar* 8 (1905): 404; Hourani, *Arabic Thought in the Liberal Age*, 132; Adams, *Islam and Modernism*, 44.
81. Hanotaux and ʿAbduh, *L'Europe et l'Islam*, 58.
82. Farah Antun, *Tarikh al-Masih*, 4 vols. (Alexandria: Matbaʿat al-Jamiʿa, 1904) (translation of Ernest Renan's *Vie de Jésus*).
83. See Anke von Kugekgen, "A Call for Rationalism: 'Arab Averroists' in the Twentieth Century," *Alif* 16 (1996): 97–132.
84. See Donald Reid, *The Odyssey of Farah Antun* (Minneapolis: Bibliotheca Islamica, 1975).
85. Ibn Rushd was a prominent Maliki of the court at Andalusia who eventually lost favor with both Caliph Yaʿqub al-Mansur (1184–99) and the *fuqaha* (jurists), particu-

larly after his publication of *Tahafut al-tahfut*. For more on the philosopher's life and works, see Majid Fakhry, *Averröes (Ibn Rushd): His Life, Works and Influence* (Oxford: One World, 2001).

86. See Barry Kogan, *Averroes and the Metaphysics of Causation* (Albany: State University of New York Press, 1985); Kugekgen, "Call for Rationalism," 102–3.
87. Ernest Renan, *Avèroes et l'averoïsme* (Paris: Michel Lévy Frères, 1866); Farah Antun, *Ibn Rushd wa-falsafatuhu* (Alexandria: n.p., 1903), from the original serialized debate in *Al-Jamiʿa* 3 (1902) and 4 (1903); Reid, *Odyssey of Farah Antun*, 219–22; Muhammad al-Nasir Nafzawi, *Al-dawla wa al-mujtamaʾ: Min mihnat Ibn Rushd ila khusumihi Muhammad ʿAbduh, Farah Antun* [State and society: From the trial of Ibn Rushd to the controversies of Muhammad ʿAbduh and Farah Antun] (Tunis: Markaz al-Nashr al-Jamiʿi, 2000).
88. Reid, *Odyssey of Farah Antun*, 226. This was a course, as we will see later, that Farah Antun also recommended in his fictional *Al-din wa al-ʿilm wa al-mal* [Religion, science, and wealth] (Alexandria: n.p., c. 1903). See next chapter.
89. Antun, *Ibn Rushd*, dedication; Hourani, *Arabic Thought in the Liberal Age*, 254–55.
90. Antun, *Ibn Rushd*, 6–7, 119–20; Reid, *Odyssey of Farah Antun*, 222.
91. Muhammad ʿAbduh, *Falsafat Ibn Rushd* [Ibn Rushd's philosophy], in *Al-aʿmal al-kamila*, 3:515–29. See also Muhammad ʿAbduh, "Al-radd ʿala Farah Antun (al-Nasraniya wa-al-Islam) [Response to Farah Antun (Christianity and Islam)], in *Al-aʿmal al-kamila*, 3:258–368.
92. ʿAbduh, *Al-Islam wa-al-Nasraniya*, 7–8.
93. Ibid., 25–26.
94. Ibid., 23–30.
95. Ibid., 34–36.
96. Ibid., 48–49.
97. See references to Agent 294, for instance, in the letters and papers cited from the Khedive ʿAbbas Archives, Durham University, below.
98. ʿAbduh, *Al-Islam wa al-Nasraniya*, 103.
99. Ibid., 105–10.
100. Ibid. Draper's work was favorably received in Ottoman intellectual circles as well. See M. Şükrü Hanioglu, *The Young Turks in Opposition* (New York: Oxford University Press, 1995), 16. See also chapter 7 for more on this and Draper's influence among later Arab evolutionists.
101. ʿAbduh, *Al-Islam wa-al-Nasraniya*, 16.
102. Ibid., 100–101.
103. "Madhhab Darwin," *Al-Muqtataf* 7 (1882): 33.
104. ʿAbduh, *Al-Islam wa-al-Nasraniya*, 101.
105. Ibid., 112–45.
106. Ibid., 145–53.
107. Ibid., 160–61.
108. Ibid., 170–71.
109. Ibid., 169.
110. Ibid., 176.
111. Ibid., 170–76.
112. *Al-Jamiʿa* 3 (1902): 682; 5 (1905): 52–53, 145–57 (cited in Reid, *Odyssey of Farah Antun*, 224–25).
113. A. Chris Eccel, *Egypt, Islam and Social Change: Al-Azhar in Conflict and Accommodation* (Berlin: K. Schwartz, 1984), 145.

114. These articles appeared in *Al-Waqa'i*, *Al-Ahram*, *Misr*, and other journals in Egypt between 1876 and 1882.
115. Muhammad ʿAbduh, *Al-Tarbiya fi al-madaris wa-al-makatab al-amiriya*" [Education in national schools], *Al-Waqa'i al-Misriya* 957 (1880) (in ʿAbduh, *Al-aʿmal al-kamila*, 3:29–32); "Ta'thir al-taʿlim fi al-din wa-al-ʿaqida" [The effects of education on religion and faith], *Al-Waqa'i al-Misriya* (1881) (in ʿAbduh, *Al-aʿmal al-kamila*, 3:53–61).
116. Muhammad ʿAbduh, "Al-tarbiya fi al-madaris wa-al-makatib al-amiriya," in *Al-aʿmal al-kamila*, 3:29–32.
117. Adams, *Islam and Modernism*, 65–66. See also Badawi, *Reformers of Egypt*, 166–67, on Muhammad ʿAbduh and the importance of science in education as a means of warding off foreign missionaries.
118. ʿAbduh, *Theology of Unity*, 27.
119. Cited in Adams, *Islam and Modernism*, 22.
120. Muhammad ʿAbduh, "Sirat al-ustadh al-imam," *Al-Manar* 8 (1905): 381–82.
121. Rida, *Tarikh al-ustadh al-imam Muhammad ʿAbduh*, 1:11.
122. Herbert Spencer, *Education: Intellectual, Moral and Physical* (1861; New York: D. Appleton, 1864), 22–23.
123. Ibid., 163.
124. Qasim Amin, *The Liberation of Women*, trans. Samiha Sidhom Peterson (Cairo: American University in Cairo Press, 2000), 166, 171, 177. For more on this emerging discourse, see Omnia El Shakry, "Schooled Mothers, Structured Play: Childrearing in Turn of the Century Egypt," in *Remaking Women: Feminism and Modernity in the Middle East*, ed. Lila Abu Lughod (Princeton, NJ: Princeton University Press, 1998), 126–70.
125. Haj, *Reconfiguring Islamic Tradition*, 158–60.
126. See Talal Asad, *Formations of the Secular: Christianity, Islam, Modernity* (Stanford, CA: Stanford University Press, 2003); and Haj, *Reconfiguring Islamic Tradition*.
127. Wrote Spencer: "'True science and true religion,' says Professor Huxley at the close of a recent course of lectures, 'are twin-sisters, and the separation of either from the other is sure to prove the death of both. Science prospers exactly in proportion as it is religious; and religion flourishes in exact proportion to the scientific depth and firmness of its basis" (Spencer, *Education*, 90).
128. Rida, *Tarikh*, 1:11.
129. Muhammad ʿAbduh, "Al-islah al-haqiqi wa-al-wajib li-al-*Azhar*" [True reform and its necessity for *Al-Azhar*], *Al-Manar* 10 (1906): 758–65, cited in Moaddel and Talattoff, *Contemporary Debates in Islam*, 49.
130. See the entry on education in Egypt in the Oxford *Encyclopaedia of Islam* for more on this.
131. At some point, the shaykh would determine for himself whether a student had mastered the text under discussion. If satisfied with his pupil's progress, he would issue him a certificate: one kind granted the student the right to teach a specific text; another, more general certificate (*ijaza ʿamma*) granted the right to teach all the texts taught by the shaykh. When a student had acquired a number of these certificates and a group of students, then he could present himself as a teaching shaykh. His success depended upon his reputation, personal ties, or even charisma.
132. By 1865 enrollment had decreased to about 2,000, but by 1867 it had increased to 221 teachers and 4,712 students; by 1873, there were 314 teachers and 9,441 students, and in 1876, 325 teachers and 10,780 students. See Gesink, *Islamic Reform*,

16, 41; Eccel, *Egypt, Islam and Social Change*, 120, 126–27, 150–55, 232–36; Bayard Dodge, *Al-Azhar: A Millennium of Muslim Learning* (Washington, DC: Middle East Institute, 1961), 114; J. Heyworth-Dunne, *An Introduction to the History of Education in Modern Egypt* (London: Luzac, 1939), 28–29; Michael Reimer, "Views of Al-Azhar in the Nineteenth Century: Gabriel Charmes and ʿAli Pasha Muhabrak," in *Travelers in Egypt*, ed. Paul and Janet Starkey (London: I. B.Tauris, 1998).

133. In 1853, for instance, students of the *maghribi riwaq* (a dormitory for students from North Africa) broke out in violent protests over bread rations (bread was traditionally distributed to the students of Al-Azhar as a kind of stipend), and in 1860 *saʿidi* (Upper Egyptian) and Syrian students fought over sitting spaces for lectures, resulting in military intervention and the arrest of some thirty students. ʿAli Mubarak, *Al-khitat al-tawfiqiya al-jadida li-Misr al-Qahira wa-muduniha wa-biladiha al-qadima wa-al-shariʿa* [The new plans of (Khedive) Tawfiq for Cairo, Egypt], 20 vols. (Cairo: Al-Matbaʿa al-Kubr al-Amiriya, 1886–89; repr., Cairo: Al-Haya al-Misriya al-ʿAmma li-al-Kitab, 1980–88), 4:40–41; William Bromfield, *Letters from Egypt and Syria* (London: n.p., 1856), 150–51; Heyworth-Dunne, *Education in Modern Egypt*, 397; Gesink, *Islamic Reform*, 42–43; Eccel, *Egypt, Islam and Social Change*, 168.

134. Afaf Lutfi al-Sayyad Marsot, "The Beginnings of Modernization in Education among the Rectors of Al-Azhar, 1798–1879," in *The Beginnings of Modernization in the Middle East: The Nineteenth* Century, ed. William Polk and Richard Chambers (Chicago: University of Chicago Press, 1968), 277–78; Gesink, *Islamic Reform*, 47–50.

135. Mahmud Abu Ayun, *Al-Azhar: A Short Historical Survey*, trans. Mostafa Taha Habib (Cairo: Al-Azhar Press, 1949), 43.

136. Gesink, *Islamic Reform*, 53.

137. ʿAbd al-Mital al-Saʿidi, *Tarikh al-islah fi al-Azhar wa-safahat min al-jihad fi al-islah* [History of reform at Al-Azhar and chapters in the struggle for reform] (Cairo: Matbaʿat al-Iʿtimad, 1951), 49–50; Daniel Neil Crecelius, "The Ulama and the State in Modern Egypt" (PhD diss., Princeton University, 1967), 214; Dodge, *Al-Azhar*, 132; Pierre Arminjon, *L'enseignement, la doctrine et la vie dans les universités musulmanes d'Egypte* (Paris: F. Alcan, 1907), 273–91.

138. Mustafa Bayram, *Risala fi tarikh Al-Azhar* (Cairo: Matbaʿat al-Tamaddun, 1903), 27; Gesink, *Islamic Reform*, 116–17.

139. Bayram, *Al-Azhar*, 27–29; Dodge, *Al-Azhar*, 132–33.

140. See Louis Arminé Aroian, *The Nationalization of Arabic and Islamic Education in Egypt: Dar al-ʿUlum and Al-Azhar* (Cairo, 1983).

141. Muhammad ʿAbd al-Jawad, *Taqwim Dar al-ʿUlum, 1872–1947* (Cairo: Dar al-Maʿarif bi-Misr, 1952), 6; Eccel, *Egypt, Islam and Social Change*, 163.

142. Al-Jawad, *Taqwim Dar al-ʿUlum*, 6–11; Gesink, *Islamic Reform*, 115–16; Eccel, *Egypt, Islam and Social Change*, 162–65.

143. ʿAbd Allah Nadim, "Al-ʿulama wa-al-talim" [The *ʿulama* and learning], *Al-Muʾayyad* 18, January 1893; reprinted in *Al-Ustadh* 26 (1893): 603–19.

144. Blunt, *Secret History*, 164.

145. Nadim, "Al-ʿulama wa-al-talim," 609–10.

146. Ibid., 610–11.

147. Ibid., 612.

148. Nadim claimed that Muslim students had studied subjects from exegesis and philosophy to "arithmetic, algebra, geometry, history, agriculture, zoology and botany," from "Baghdad to Fez" (ibid., 606).

149. Ibid., 608–9.
150. Ibid., 619.
151. Abd al-Karim Salman, *A'mal majlis idarat Al-Azhar: Min ibtida' ta'sisihi sana 1312 ila ghayat sana 1322* [Reports of the Administrative Reform Committee at Al-Azhar from 1895 to 1905] (Cairo: n.p., 1905), 1–2; Gesink, *Islamic Reform*, 129–31; Eccel, *Egypt, Islam and Social Change*, 177–78.
152. Robert Tignor, *Modernization and British Colonial Rule in Egypt, 1882–1914* (Princeton, NJ: Princeton University Press, 1966): 158. For an interesting parallel concerning the state of traditional education in Istanbul, see Richard L. Chambers, "The Education of a Nineteenth-Century Ottoman Alim, Ahmed Cevdet Pasa," *International Journal of Middle East Studies* 4 (October 1973): 440–64.
153. Cited in Moaddel and Talatoff, *Contemporary Debates in Islam*, 49–50.
154. Salman, *A'mal majlis idarat Al-Azhar*, 1–5; Eccel, *Egypt, Islam and Social Change*, 170–71.
155. "Hadithat Al-Azhar" [Events at Al-Azhar], *Al-Muqattam* 2184 (1 June 1896), 2185 (2 June 1896), 2187 (4 June), 2188 (5 June), 2189 (6 June), 2194 (12 June), 2195 (13 June), 2198 (15 June), and 2202 (22 June); "Hadithat Al-Azhar" [Events at Al-Azhar], *Al-Muqtataf* 20 (1896): 558–59; Ahmad al-Shadhili, "Hadithat Al-Azhar al-sharif" [Events at the Noble Al-Azhar], *Al-Islam* 3 (1896): 101–21—cited in Gesink, *Islamic Reform*, 143–64.
156. Cromer to Salisbury, 1 June 1896, FO 407/137, no. 213, National Archives, London.
157. "Al-kulayra" [Cholera], *Al-Muqtataf* 20 (1896): 558. The cholera outbreaks of 1895–96 were generally seen to be the result of the pilgrimage to Mecca; see Cromer to Salisbury, 15 January 1897, FO 407/142, no. 14, National Archives, London. See also Daniel Panzac, "Endémies, épidémies et population en Égypte au XIXe siècle," in *Population et Santé dans L'Empire Ottoman (XVIIIe–XXe siècles)* (Istanbul, 1996), 57–77; Tignor, *Modernization and British Colonial Rule in Egypt*, 350–52.
158. Cromer to Salisbury, 6 June 1896, FO 407/137, no. 223, National Archives, London.
159. "Al-jami'a Al-Azhar" [The Azhar mosque], *Al-Muqtataf* 23 (1899): 758–59.
160. Donald Reid, *Cairo University and the Making of Modern Egypt* (Cairo: American University in Cairo Press, 1991), 17–22.
161. Hasan Husni al-Tuwayrani, "*Al-Azhar fi Misr*" [Al-Azhar in Egypt], *Al-Nil* 4 (1895): 4–5; cited in Indira Falk Gesink, "Beyond Modernisms: Opposition and Negotiation in the Azhar Reform Debate in Egypt, 1870–1911" (PhD diss., Washington University, 2000), 322.
162. Muhammad Sulaman al-Safti, "Al-Azhar," *Al-Islam* 1 (1895): 267–71, cited in Gesink, *Islamic Reform*, 133–39.
163. Eccel, *Egypt, Islam and Social Change*, 157, 183.
164. Ibid., 177.
165. This is especially true of the biographies and obituaries written shortly after 'Abduh's death, such as Rida's *Tarikh*; Jurji Zaydan's *Tarajim mashahir al-Sharq* [Famous men of the East], 2 vols. (Cairo: Matba'at al-Hilal, 1910), 300–309 (based on an *Al-Hilal* article written in 1903); and the articles and obituary of *Al-Muqtataf* 30 (1905): 593–96, 901–22, 985–991. This theme is also much repeated in the secondary literature on 'Abduh: Adams, *Islam and Modernism*; Mustafa 'Abd al-Raziq, *Muhammad 'Abduh* (Cairo, 1945); 'Uthman Amin, *Muhammad 'Abduh*; Hourani, *Arabic*

Thought in the Liberal Age; Kerr, *Political and Legal Theories*; Vatikiotis, "Muhammad ʿAbduh and the Quest for a Muslim Humanism"; see also Suhail ibn Salim Hanna, "Biographical Scholarship and Muhammad ʿAbduh," *Muslim World* 59 (1969): 300–307.

166. Blunt, *My Diaries*, 2:244.
167. For more on ideas of *islah* and *tajdid*, see John Voll, "Renewal and Reform in Islamic History: *Tajdid* and *Islah*," in *Voices of Resurgent Islam*, ed. John Esposito (New York: Oxford University Press, 1983), 32–45.
168. According to Lord Cromer, "[ʿAbduh's] associates, although they admitted his ability, were inclined to look askance at him as a 'filosouf.'" Note that for Lord Cromer, ʿAbduh's interest in philosophy also marked him off as a "modernist," against those "strictly orthodox" conservatives. "Now in the eyes of the strictly orthodox, one who studies philosophy, or, in other words, one who recognizes the difference between the seventh and the twentieth centuries, is on the high road to perdition" (Cromer, *Modern Egypt*, 2:180).
169. ʿAbduh, *Theology of Unity*, 36. After the first publication of the *Risalat*, *Al-Manar* quietly excised any mention of the Qurʾan's "createdness" from subsequent editions until 1925; Gesink, *Islamic Reform*, 170.
170. The Muʿtazila were a branch of theologians who rose to and fell from power during the Abbasid era (750–1258) and were later renowned among mainstream Sunni Muslims for advancing a heretical adherence to rationalism and a dangerous speculative theology; D. Gimaret, "Muʿtazila," *Encyclopaedia of Islam*; Baber Johansen, *Contingency in a Sacred Law: Legal and Ethical Norms in the Muslim* Fiqh (Leiden, 1999), introduction.
171. Blunt, *My Diaries*, 244.
172. "Sirat al-Ustadh al-Imam," *Al-Manar* 8 (1905): 391; Kerr, *Islamic Reform*, 58–60.
173. See his compilation of fatwas in Muhammad Ibn Ahmad ʿIllish, *Fath al-ʿali al-malik fi al-fatawa ʿala madhhab al-Imam Malik* [The princely key to the fatwa of the Maliki School] (Cairo: Al-Maktaba al-Tijariya al-Kubra, 1937), 18, 51.
174. Gesink, *Islamic Reform*, 94–95.
175. Ibid., 99–106.
176. Nevertheless, the place of *ijtihad* in legal reasoning was quite complex; it was a subject of debate and an object of practice throughout the nineteenth-century Muslim world, as it had been for centuries before: see Wael Hallaq, "Was the Gate of Ijtihad Closed?," *International Journal of Middle Eastern Studies* 16 (1984): 3–41; Rudolph Peters, "*Idjtihad* and *Taqlid* in Eighteenth and Nineteenth Century Islam," *Die Welt des Islams* 20 (1980): 131–45.
177. ʿIllish cited from Gesink, "Beyond Modernisms," 250. This subject has given rise to extensive historiographical debate; see esp. Frank Vogel, "The Closing of the Door of *Ijtihad* and the Application of the Law," *American Journal of Islamic Social Science* 10 (1993): 396–401; Lutz Wiederhold, "Legal Doctrines in Conflict: The Relevance of *Madhhab* Boundaries to Legal Reasoning in the Light of an Unpublished Treatise on *Taqlid* and *Ijtihad*," *Islamic Law and Society* 3 (1996): 234–89.
178. Blunt, *Secret History*, 326–27.
179. Eccel, *Egypt, Islam and Social Change*, 84–85.
180. Adams, *Islam and Modernism*, 30.
181. ʿIllish, *Fatawa*, 2–5. The fatwas also contain *hamash* (marginal notations) by Ibrahim al-Din al-Maliki al-Madani (d. AH 799); ʿIllish's collection of fatwas was quite

popular as a legal manual for Maliki ʿulama and *fuqaha*, and it was reprinted several times throughout the late nineteenth and early twentieth centuries.

182. ʿIllish, *Fatawa*, 21, 27, 30.
183. Cited in Gesink, *Islamic Reform*, 97.
184. File 52/202, Mahmoud Sadiq to the khedival palace, 17 August 1905, Khedive ʿAbbas Archives, Durham University.
185. File 62/75, Khedive ʿAbbas Archives, Durham University.
186. Eccel, *Al-Azhar*, 172.
187. Wiederhold, "Legal Doctrines in Conflict."
188. ʿIllish, *Fatawa*, 71.
189. File 52/193, Khedive ʿAbbas Archives, Durham University.
190. Ahmad Pasha Shafiq, *Mudhakkirati fi nisf qarn* [Memoirs of half a century], 2 vols. (Cairo: 1934–36), 2:279.
191. Tawfik al-Bakry file, letter dated 7/3/1899, Khedive ʿAbbas Archives, Durham University.
192. Cromer to Salisbury, 6 June 1899, FO 78/5023, no. 105, National Archives, London. That summer, ʿAbduh began a state-commissioned tour to investigate the state of shariʿa courts throughout Lower and Upper Egypt. Dismayed by the general disarray of the courts, he immediately recommended, in his "Takrir al-mahakim al-shariʿa" (A report on the shariʿa courts), greater government supervision and regulation of court deputies, judges, and scribes, as well as the codification of shariʿa law and practice. Muhammad ʿAbduh, "Takrir al-mahakim al-shariʿa," in *Al-aʿmal al-kamila*, 2:211–304; submitted to the khedival government in November 1899, a mere five months after he became grand mufti. The codification of the shariʿa for all *madhahib* was first attempted under Khedive Ismaʿil by Muhammad Qadri Pasha in 1875 but never legislated; Filib Ibn Yusuf Jallad, *Al-Qamus al-ʿamm li-al-idara wa-al-qada*, 3:2800, 5:3621; Eccel, *Egypt, Islam and Social Change*, 91.
193. See, e.g., Blunt, *My Diaries*, 492.
194. Cromer to Grey, 5 March 1906, p. 85, FO 407/165, no. 143, National Archives, London.
195. Cromer to Lansdowne, 26 February 1903, FO 407/161, no. 7, National Archives, London; David Steele, "Britain and Egypt, 1882–1914: The Containment of Islamic Nationalism," in *Imperialism and Nationalism in the Middle East: The Anglo-Egyptian Experience, 1882–1982*, ed. Keith Wilson (London, 1983), 18. On science, see Cromer, *Modern Egypt*, 2:530.
196. In 1899 the nationalist revolutionary Mustafa Kamil gave a speech in which he referred to the 1897 war as evidence that the Ottoman Empire was "not dead and remains faithful to Islam" ("Turkey, the Khedive, and Egyptians are but one," he announced); Cromer to Salisbury, 23 December 1899, FO 407/155, no. 1, National Archives, London.
197. Cromer to Salisbury, 27 April 1900, FO 407/155, no. 53, National Archives, London.
198. 28/65, 4/11/1901, signed "294"; 28/3, no date, signed "294"; and 28/19, no date, signed "294"; Khedive ʿAbbas Archives, Durham University.
199. Blunt, *My Diaries*, 76–77, 492–93.
200. Ibid., 496.
201. On the khedive's espionage, see Amira El Azhary Sonbol, *The Last Khedive of Egypt: Memoirs of Abbas Hilmi II* (Reading, UK: Ithaca Press, 1998).
202. Ahmad Shafiq, *Mudhakkirat fi nisf qarn*, 3 vols. (Cairo, 1934–36), 2:35–40.

203. Blunt, *My Diaries*, 2:88.
204. See next chapter and Haj, *Reconfiguring Islamic Tradition*, 144, for more on this.
205. See for instance "Manafiʿ al-urubiyin wa-madarhum fi al-sharq," *Al-Manar* 10 (1907): 192–99.
206. Cited in Gesink, *Islamic Reform*, 181.
207. Ibrahim al-Muwilhi, "Al-Imam fi al-tafarnuj," *Misbah al-Sharq* 4 (1901): 3, cited in Gesink, *Islamic Reform*, 183.
208. Cromer to Lansdowne, 23 May 1902, FO 407/159, no. 58, National Archives, London.
209. 28/5, from "294" Khedive ʿAbbas Archives, Durham University.
210. 28/99, Khedive ʿAbbas Archives, Durham University.
211. See Amal Ghazal, "Sufism, Ijtihad and Modernity: Yusuf al-Nabhani in the Age of ʿAbd al-Hamid II," *Archivum Ottomanicum* 19 (2001): 269.
212. His own early experiences with Sufi thought possessed the qualities of a quest for moral earnestness and an interiorization of faith that was very much in line with the Sufi views of al-Ghazali, a figure ʿAbduh himself much admired and cited. His interest in mysticism was further enhanced by his relationship to Afghani, whom ʿAbduh often referred to in Sufi fashion as a "sage."
213. Ibid., 261.
214. See the articles on *taqlid* and *ijtihad* cited earlier.
215. See Ghazal, "Sufism, Ijtihad and Modernity."
216. Ibid., 265.
217. Ibid., 263.
218. Chibli Mallat, "The Debate on Riba and Interest in Twentieth Century Jurisprudence," in *Islamic Law and Finance*, ed. Chibli Mallat (London, 1988), 74.
219. Cited in Gesink, *Islamic Reform*, 175–83. In 1903 another fatwa that ʿAbduh issued, this time for a Muslim petitioner in the Transvaal in South Africa, further fanned the flames of controversy. See Charles Adams, "Muhammad ʿAbduh and the Transvaal Fatwa," in *The MacDonald Presentation Volume: A Tribute to Duncan Black Macdonald* (Princeton, NJ: Princeton University Press, 1933), 11–30. The petitioner asked a series of questions about the problems of leading a Muslim way of life in a predominantly non-Muslim society: he asked about the permissibility of wearing Western-style hats, eating meat slaughtered by Christians and Jews, and following communal prayers led by a shaykh of a *madhhab* not one's own. ʿAbduh ruled that, given the circumstances, all three practices could be said to be permissible. See also Blunt, *My Diaries*, 498.
220. Quoted from Gesink, *Islamic Reform*, 190; and Gesink, "Beyond Modernisms," 447.
221. Muhammad al-Ahmadi al-Zawahri, *Al-ʿIlm wa-al-ʿulama wa-nizam al-tadris* (Cairo: n.p., 1904), 89; cited in Gesink, *Islamic Reform*, 208. See also Eccel, *Al-Azhar, Egypt, Islam and Social Change*, 148.
222. Gesink, *Islamic Reform*, 209; "Al-faylasuf Herbert Sbinsir" (The philosopher Herbert Spencer), *Al-Muqtataf* 29 (1904): 1–8; "Ra'y Sbinsir fi al-ta'lim" (Spencer's opinions on education), *Al-Muqtataf* 29 (1904): 289–95.
223. This was originally published as an anonymous letter, signed "one of the Azhar ʿulama"; *Al-Muqattam* 4856 (18 March 1905): 1; Gesink, *Islamic Reform*, 209.
224. A. B. De Guerville, *New Egypt* (London: E. P. Dutton, 1906), 158–62.
225. Eccel, *Egypt, Islam and Social Change*, 184.
226. Shortly before his death, ʿAbduh had discussed the idea with Rashid Rida, Ahmad Fathi Zaghlul (Saʿad's brother), and Ahmad Pasha al-Minshawi (a wealthy notable

who later agreed to finance the project, offering the land and monies required to build it). See Reid, *Cairo University*, 25; Adams, *Islam and Modernism*, 195–98; Eccel, *Egypt, Islam and Social Change*, 184.

227. In Egypt today, ʿAbduh is widely considered something of a nationalist pioneer, a daring reformer, and wronged hero. Currently, the assembly hall of Al-Azhar bears his name. Eccel, *Egypt, Islam and Social Change*, 156. Indeed, immediately after his death, even staunch nationalist critics like Mustafa Kamil commemorated ʿAbduh's achievements. See, e.g., Mustafa Kamil, "Khatb jalal" [A tragic event], *Al-Liwa* 1869 (12 July 1905).

228. Yaʿqub Artin, *Considérations sur l'instruction publique en Égypte* (Cairo: n.p., 1894), 166–67; Jurji Zaydan, "Al-madrasa al-kulliya al-Misriya" [The Egyptian College], and "Al-Jamiʿa al-Misriya" [The Egyptian University], *Al-Hilal* 15 (1906): 67–88; "Al-Jamiʿa al-Misriya" [The Egyptian University], *Al-Muqtataf* 31 (1906): 873–80; Mustafa Kamil, "Hayat al-shaʿb fi al-shaʿb," *Al-Liwa*, 25 January 1900; Reid, *Cairo University*, 22–26. Zaydan was later commissioned to teach at the Egyptian University; see ʿAbdin, archive box 231: *Taʿlim al-jamiʿa al-misriya*, 1906–41, Egyptian National Archives, for relevant documents.

229. Reid, *Cairo University*, 31–32, 36, 54–55.

230. E.g., Shaykh Al-Maraghi. See Adams, *Islam and Modernism*.

231. On al-Maraghi, see Rainer Brunner, "Education, Politics and the Struggle for Intellectual Leadership: Al-Azhar between 1927 and 1945," in *Guardians of Faith in Modern Times*, ed. Meir Hatim (Leiden: Brill, 2009), 109–40. On al-Shaltut, see Kate Zeibiri, *Mahmud Shaltut and Islamic Modernism* (Oxford: Oxford University Press, 1993).

232. Rida's *Tarikh*, based largely on the series of articles entitled "Sirat al-ustadh al-imam" in *Al-Manar*, was begun shortly after ʿAbduh's death and based on ʿAbduh's unpublished and incomplete autobiography. See also Jurji Zaydan's biography of ʿAbduh in *Tarajim mashahir al-sharq* [Biographies of famous men of the East] (Cairo: Matbaʿat al-Hilal, 1910–11).

233. Examples are the works of Muhammad Farid Wajdi and Tantawi Jawhari, both younger than ʿAbduh. See n. 65 for references to their works. For more on this, see the afterword.

234. Rashid Rida argued this in his reply to various letters he had received from readers who had objected to the article "Al-din fi nazar al-ʿaql" [A rationalist view of religion]. "Abuna Adam wa-madhhab Darwin: Min bab al-intiqad ʿala *Al-Manar*," *Al-Manar* 8 (1906): 920.

CHAPTER SIX

1. It was originally entitled *Al-Jamiʿa al-ʿUthmaniya* (the Ottoman collective).

2. Farah Antun, *Al-din wa-al-ʿilm wa-al-mal aw-al-mudun al-thalath*, published as a supplement to *Al-Jamiʿa* 4 (1903): 1. Articles on socialism had appeared in the Arabic press more than a decade earlier, but this was one of the first literary treatments of the topic.

3. See ibid., 1–50 (found after p. 258). See also Matti Moosa, *The Origins of Modern Arabic Fiction* (Boulder, CO: Lynne Rienner, 1997); Donald Reid, *The Odyssey of Farah Antun: A Syrian Christian's Quest for Secularism* (Minneapolis: Bibliotheca Islamica, 1975), 114–16; and Donald Reid, "The Syrian Christians and Early Socialism in the Arab World," *International Journal of Middle Eastern Studies* 5 (1974): 181–82.

4. Antun, *Al-din wa-al-ʿilm wa-al-mal*, 40.

5. Richard Walzer, trans., *Al-Farabi on the Perfect State: Abu Nasr al-Farabi's "Mabadi ara' ahl al-madina al-fadila"; A Revised Text with Introduction, Translation, and Commentary* (Oxford: Clarendon Press, 1998), 255.
6. Luc Racine, "Paradise, the Golden Age, the Millennium and Utopia: A Note on the Differentiation of Forms of the Ideal Society," *Diogenes* 31 (1983): 131. Despite the technocratic, organicist, and scientistic vision of the Saint-Simonians, they were seldom discussed in Arabic utopian socialist works until after the Second World War. They are also seldom treated in secondary (Arabic or other) sources, despite their importance in technical schools and public engineering works in Egypt. For a study of them in the Maghreb, see Osama W. Abi-Mershed, *Apostles of Modernity: Saint-Simonians and the Civilizing Mission in Algeria* (Stanford, CA: Stanford University Press, 2010). One reason for their dismissal by early Arab socialists may be that, particularly after 1848, they too came to be associated with radical, revolutionary change.
7. Antun, *Al-din wa-al-ʿilm wa-al-mal*, 39.
8. For more on this and on other expressions of the relation between capitalism, socialism, religion, and discussions of the "social problem" at the time, see Charles Tripp, *Islam and the Moral Economy: The Challenge of Capitalism* (Cambridge: Cambridge University Press, 2006), 13–45, esp. 35.
9. Jamal al-Din al-Afghani, translated in Sami Hanna and George Gardner, eds., *Arab Socialism: A Documentary Survey* (Leiden: Brill, 1969), 269–70.
10. Cited in Zachary Lockman, "British Policy toward Egyptian Labor Activism, 1882–1936," *International Journal of Middle East Studies* 20 (1988): 265. See also Joel Beinin, "Formation of the Egyptian Working Class," *MERIP Reports* 94 (1981): 15–16; and John Chalcraft, "Coalheavers of Port Said: State Making and Worker Protest, 1869–1914," *International Labour and Working Class History* 60 (2001): 110–24.
11. Aristotle, *Nicomachean Ethics*, bk. 4. This work was discussed extensively in Arabic publications at the turn of the century and translated into Arabic in 1924: *ʿIlm al-akhlaq ila Niqumakhus* (Cairo: Matbaʿat Dar al-Kutub al-Misriya, 1924).
12. Samira Haj, *Reconfiguring Islamic Tradition: Reform, Rationality and Modernity* (Stanford, CA: Stanford University Press, 2009), 143–46.
13. Alfred Russel Wallace, "Man's Place in the Universe," *Independent* 55 (1903): 483. "Al-samaʿ wa-al-ard wa-al-insan," *Al-Jamiʿa* 4 (1903): 109–11.
14. Mustafa al-Mansuri, *Tarikh al-madhahab al-ishtirakiya* [History of socialisms] (Cairo: n.p., 1915).
15. E.g., Farabi's discussion of the virtuous republic, which "resembles the perfect and healthy body, all of whose limbs cooperate to make the life of the animal perfect and to preserve it in this state" (Farabi, *On the Perfect State*, 231.
16. See Stefan Wild, "'Das Kapital' in arabischen Übersetzungen," in *Festgabe für Hans Wehr* (Wiesbaden, 1969), 97–111.
17. See Ilham Khuri-Makdisi, *The Eastern Mediterranean and the Making of Global Radicalism* (Berkeley and Los Angeles: University of California Press, 2010); and Eric Hobsbawm, *How to Change the World: Tales of Marx and Marxism* (London: Little, Brown, 2011), 211–60, quotation from 215.
18. Hobsbawm, *How to Change the World*, 245.
19. See Khuri-Makdisi, *Eastern Mediterranean and the Making of Global Radicalism*.
20. The chapter focuses on Egypt, though many of the intellectuals discussed in this chapter traveled—both in person and in print—across the Arab world at this time. Socialist thought was equally important among intellectuals writing from Beirut, Damascus, Mosul, and Baghdad. Yet for much of the period discussed in this chap-

ter—roughly the late 1890s to the start of the Second World War—Cairo and Alexandria were major hubs of evolutionary socialist ideas.

21. Shibli Shumayyil, *Al-ishtirakiya* [Socialism], in *Arab Socialists: A Documentary Survey*, ed. Sami Hanna and George Gardner (Leiden: Brill, 1969), 289–96.
22. For more on the subject see Daniel Stoltz, "From Shellfish to Apes: Socialism and the Changing Politics of Evolution in *al-Muqtataf*, 1899–1914" (B.A. thesis, Harvard University, 2007).
23. Charles Darwin, *The Descent of Man, and Selection in Relation to Sex* (London: J. Murray, 1871), chaps. 4 and 5. For more on this subject, see Robert Richards, *Darwin and the Emergence of Evolutionary Theories of Mind and Behavior* (Chicago: University of Chicago Press, 1989); and Thomas Dixon, *The Invention of Altruism: Making Moral Meanings in Victorian Britain* (Oxford: Oxford University Press, 2008).
24. This journal was published in Cairo between 1930 and 1939. See the next chapter for more on Abu Shadi.
25. For instance, Salama Musa, a key figure in this story, had a close relationship with the editors of *Al-Muqtataf*, and he published some of his earliest works in their journal (despite their disagreement with many of his ideas). He was also the editor of *Al-Hilal* from 1923 to 1929.
26. This new class of *effendiya* (middle-class bureaucrats, professionals, and self-labeled "modern" urbanites) formed a large part of the Arabic-reading public of the early twentieth century, and many of them scripted their own version of an Egyptian modernity around such figures as Darwin, Marx, and Freud. See Lucie Ryzova, "Egyptian Modernity through the 'New Effendiya': Social and Cultural Constructions of the Middle Class in Egypt under the Monarchy," in *Re-envisioning Egypt: 1919–1952*, ed. Arthur Goldschmidt and Amy Johnson, 124–63 (Cairo: American University in Cairo Press, 2005).
27. Mourad Magdi Wahba, "The Meaning of *Ishtirakiyah*: Arab Perceptions of Socialism in the Nineteenth Century," *Alif* 10 (1990): 44–52.
28. See Farah Antun, "Urushlim al-jadida wa ara' al-rusafa" [New Jerusalem and opinions of the Brethren], *Al-Jami*ʿ*a* 4 (1904): 371–74; and Reid, "Syrian Christians and Early Socialism in the Arab World," 182.
29. Ali Dessouki, "The Origins of Socialist Thought in Egypt, 1882–1922" (PhD diss., McGill University, 1972), 83–88.
30. The term *al-shuyu*ʿ*iya* was used to describe the Communist Party and movement in 1922 in *Al-Muqtataf* and features regularly in *Al-Hilal* after 1923. See Reid, "Syrian Christians and Early Socialism in the Arab World," 191.
31. *Al-Hilal* 5 (1897): 290–94, cited in Khuri-Makdisi, *Eastern Mediterranean and the Making of Global Radicalism*, 19–20; Tarek Ismael and Rifa'at El-Sa'id, *Communist Movement in Egypt* (Syracuse, NY: Syracuse University Press, 1990), 1–2, 10; and Malak Badrawi, *Political Violence in Egypt, 1910–1924: Secret Societies, Plots and Assassinations* (London: Curzon, 2000), 28.
32. See Khuri-Makdisi, *Eastern Mediterranean and the Making of Global Radicalism*, for more on this.
33. Cited in Zachary Lockman, "Imagining the Working Class: Culture, Nationalism and Class Formation in Egypt, 1899–1914," *Poetics Today* 15 (1994): 172–73. See also Anthony Gorman, "Anarchists in Education: The Free Popular University in Egypt (1901)," *Middle Eastern Studies* 41 (2005): 311.
34. Cited in Wahba, "Meaning of *Ishtirakiyah*," 52. See "Al-ishtirakiyun wa fasadhum" [Socialists and their corruption], *Al-Muqtataf* 14 (1890): 361–64; and "Al-

ishtirakiyun wa-al-fawdiyun" [Socialists and anarchists], *Al-Muqtataf* 18 (1894): 721–29, 801–7.

35. Dessouki, "Socialist Thought in Egypt," 102.
36. See *Al-Muqtataf* 13 (1889): 590.
37. "Al-faqr wa-al-ghani," *Al-Muqtataf* 13 (1889): 577; and *Al-Muqtataf* 15 (1890): 53; "Al-fuqara' wa-al-aghna," *Al-Muqtataf* 20 (1896): 794; and 32 (1907): 125. See also Nadia Farag, "*Al-Muqtataf*, 1876–1900: A Study of the Influence of Victorian Thought on Modern Arabic Thought" (PhD diss., University of Oxford, 1969), 171.
38. Dessouki, "Socialist Thought in Egypt," 103–4.
39. See *Al-Hilal* 6 (1897): 293; Dessouki, "Socialist Thought in Egypt," 105; Khuri-Makdisi, *Eastern Mediterranean and the Making of Global Radicalism*, 38, 51.
40. Zaglul translated Demolins's *À quoi tient le superiorité des Anglo-saxons?*, Le Bon's *Les lois psychologiques de l'évolution des peuples*, and Bentham's *An Introduction to the Principles of Morals and Legislation* and produced an unpublished translation of Spencer's *Man versus the State*. See also chapter 2.
41. Samah Selim, *The Novel and the Rural Imaginary in Egypt: 1880–1965* (New York: Routledge, 2004), 8–9.
42. Gustave Le Bon, *Sirr tatawwur al-ummam*, trans. Ahmad Fathi Zaghlul (Cairo, 1913), 1–8.
43. Dessouki, "Socialist Thought in Egypt," 108–13.
44. Thabit was sent by the editors to the Sudan in 1903 to run their latest journalistic venture there, *Al-Sudan*.
45. "Al-ishtirakiyun al-dimaqratiyun" [Democratic socialists], *Al-Muqtataf* 25 (1900): 146. See Khuri-Makdisi, *Eastern Mediterranean and Making of Global Radicalism*, 38–39, for more on this. As *Al-Muqtataf*'s article title also suggests, many presented socialism as a higher form of democracy. As late as 1915, Mustafa al-Mansuri made the same point in his *Tarikh al-madhahab al-ishtirakiya*: socialist governments succeeded in those nations that already had well-established democratic governments, as the examples from Western Europe show.
46. *Al-Muqtataf*'s response to a reader's question in "Tanazuʿ al-baqaʾ" [The struggle for survival], *Al-Muqtataf* 25 (1900): 465–66. See "Al-tanazuʿ wa al-taʿawun," [Struggle and cooperation], *Al-Muqtataf* 25 (1900): 219; "Al-ghairiya" [Altruism], *Al-Muqtataf* 25 (1900): 505; and "Kitab fi al-ijtimaʿ" [Books on society], *Al-Muqtataf* 40 (1912): 97.
47. For more on contemporary critiques (and meanings) of individualism, see Steven Lukes, "The Meanings of 'Individualism,'" *Journal of the History of Ideas* 32 (1971): 63.
48. Stefan Collini, *Liberalism and Sociology: L. T. Hobhouse and the Political Argument in England, 1880–1914* (Cambridge: Cambridge University Press, 1979), 150.
49. P. Kropotkin, *Mutual Aid: A Factor of Evolution* (New York: McClure Phillips, 1904), 6. See also Shibli Shumayyil, "Al-ishtirakiya al-sahihaʾ," *Al-Muqtataf* 42 (1913): 9–16, translated as "True Socialism" in Hanna and Gardner, *Arab Socialism*, 289–96. See also Salama Musa, "Al-ishtirakiya," in Hanna and Gardner, *Arab Socialism*, 273–88; Ismaʿil Mazhar, "Darwin wa-tanazuʿ al-baqaʾ" [Darwin and the struggle for existence], *Al-Muqtataf* 63 (1923): 60–62.
50. On Ibrahim al-Wardani's anarchist connections, see Badrawi, *Political Violence in Egypt*, 26.
51. Jurji Zaydan, "Al-insaniya al-qadima" (The future of humanity), *Al-Hilal* 23 (1915): 464.

52. *Al-Muqtataf* 47 (1915): 3; cited in Thomas Philipp, "Perceptions of the First World War," in *Ottoman Reform and Muslim Regeneration: Studies in Honour of Butrus Abu-Manneb* (London: I. B. Tauris, 2005), 271.
53. Ya'qub Sarruf, *Al-Muqtataf* 48 (1916): 397, reply to Shibli Shumayyil, *Al-Muqtataf* 48 (1916): 299–300; cited in Adel Ziadet, *Western Science in the Arab World: The Impact of Darwinism, 1860–1930* (Basingstoke: Macmillan, 1986), 55–57.
54. Rashid Rida cited in Philipp, "Perceptions of the First World War," 221; and Musa in Ziadet, *Western Science*, 58.
55. Shibli Shumayyil, *Hawadith wa-khawatir* [Conversations and speeches] (Beirut: Dar al-Hamra, 1991), 227–28.
56. Salama Musa, "Naza'a jadida fi al-'ilm: Min al-madiya ila al-ruhiya" (New movements in science: From materialism to spiritualism), *Al-Hilal* 28 (1919): 132–35.
57. Rashid Rida cited in Philipp, "Perceptions of the First World War," 220.
58. See Ziadet's discussion of this in *Western Science*, 60.
59. Jurji Zaydan, "Al-insaniya al-qadima," *Al-Hilal* 23 (1915): 463–68 (published posthumously); Ziadet, *Western Science*, 57.
60. See Erez Manela, *The Wilsonian Moment* (New York: Oxford University Press, 2007).
61. Rashid Rida cited in Philipp, "Perceptions of the First World War," 221.
62. Niqula Haddad, *Al-ishtirakiya* [Socialism] (Cairo: Matba'at al-Hilal, 1920), 60–61.
63. Ibid., 87.
64. Dessouki, "Socialist Thought in Egypt," 223.
65. See esp. Shumayyil, in Hanna and Gardner, *Arab Socialism*, 290.
66. Georges Haroun, *Sibli Sumayyil: Une pensée évolutionniste arabe à l'époque d'an-Nahda* (Beirut: Université Libanaise, 1985), 242
67. Shumayyil, in Hanna and Gardner, *Arab Socialism*, 290.
68. Ibid., 296.
69. Shibli Shumayyil, *Majmu'at al-Duktur Shibli Shumayyil* [Works of Dr. Shibli Shumayyil] (Beirut: Dar Nazir 'Abbud, 1983), 187–89; Dessouki, "Socialist Thought in Egypt," 217–18.
70. Shumayyil, in Hanna and Gardner, *Arab Socialism*, 296.
71. See Kenneth Cunno, "The Origins of Private Ownership of Land in Egypt: A Reappraisal," *International Journal of Middle East Studies* 12 (1980): 245–75.
72. See Samir Saul, "European Capital and Its Impact on Land Redistribution in Egypt: A Quantitative Analysis (1900–1914)," in *Colonialism and the Modern World*, ed. Gregory Blue et al. (Armonk, NY: N. E. Sharpe, 2002), 126.
73. See Bent Hansen, "Interest Rates and Foreign Capital in Egypt under the British," *Journal of Economic History* 43 (1983): 867–84.
74. See Adam Sabra, *Poverty and Charity in Medieval Islam* (Cambridge: Cambridge University Press, 2001), for more on these early networks.
75. 'Abduh, in Hanna and Gardner, *Arab Socialism*, 206.
76. Ibid., 207.
77. See Tripp, *Islam and the Moral Economy*, 28.
78. See Tripp, *Islam and the Moral Economy*. 'Abduh quoted in Hanna and Gardner, *Arab Socialism*, 216. 'Abduh also recommended the "dissemination of knowledge on the sciences, on the expansion of industries and arts, and the strengthening of the spirit of true education" among its people (Hanna and Gardner, *Arab Socialism*, 216).
79. Tarik Yousef, "Egypt's Growth Performance under Economic Liberalism: A Reassessment with New GDP Estimates, 1886–1945," *Review of Income and Wealth* 48 (2002): 561–79.

80. Mustafa al-Mansuri, *Al-taqaddum wa al-faqr* [Translation of Henry George's *Progress and Poverty*] (Cairo: Matbaʿat al-Saʿada, 1919), frontispiece.
81. Al-Mansuri, *Tarikh al-madhahab al-ishtirakiya*; Amin ʿIzz al-Din, *Al-Mansuri: Sirat muthaqaf thawri* [Al-Mansuri: Biography of a revolutionary intellectual] (Cairo: Dar al-Ghad al-ʿArabi, 1984), 31.
82. Al-Mansuri, *Tarikh al-madhahab al-ishtirakiya*, 10.
83. Ibid., 110–17; and Dessouki, "Socialist Thought in Egypt," 215.
84. Al-Mansuri, *Tarikh al-madhahab al-ishtirakiya*, 9–12.
85. Ibid., 82–84.
86. Ibid.
87. ʿIzz al-Din, *Al-Mansuri*, 17–18. Badrawi, *Political Violence in Egypt*, 120.
88. Donald Reid, "Turn-of-the-Century Egyptian School Days," *Comparative Education Review* 27 (1983): 374–93.
89. Giuseppe Contu, *Gli aspetti positivi e i limiti del laicismo in Salama Musa (1887–1958)* (Naples: Istituto Orientale di Napoli, 1980), 6.
90. Salama Musa, *The Education of Salama Musa*, trans. L. O. Schuman (Leiden: Brill, 1961), 60, 135.
91. Membership and Speakers' Panel Cards: "Moussa, S," LSE: Fabian Society Archives, London.
92. On Shaw, the Webbs, and Wallas, see George Bernard Shaw, ed., *Fabian Essays* (London: W. Scott, 1889). On MacDonald, see Ramsay MacDonald, *Socialism and Society* (London: Independent Labour Party, 1905).
93. See George Bernard Shaw, *Man and Superman* (London: Constable, 1903), for more on these themes.
94. Salama Musa, "Taʿlim al-umma," *Al-Muqtataf* 37 (1910): 906.
95. FO 371/1113, Egypt 1911, no. 12729, cited in Contu, *Gli aspetti positivi*, 8.
96. Salama Musa, "Taʿlim al-umma," *Al-Muqtataf* 37 (1910): 906.
97. "Memorandum by Sir E. Gorst respecting the Press in Egypt," 16 September 1908, FO 78/3987, National Archives, London.
98. Cited in Contu, *Gli aspetti positive*, 74.
99. The memorandum was signed by Blunt, Henry Cotton, and L. T. Hobhouse: "Petition to His Highness Abbas II, Khedive of Egypt," 1914, Wilfred Scawen Blunt collection, box 67, West Sussex Record Office, Chichester, UK; Moussa-Blunt, 6 July 1910, Wilfred Scawen Blunt collection, box 44, West Sussex Record Office, Chichester, UK. See also Contu, *Gli aspetti positivi*, 8–9; Vernon Egger, *A Fabian in Egypt: Salama Musa and the Rise of the Professional Classes in Egypt, 1909–1939* (Lanham, MD: University Press of America, 1986), 58–59; Musa, *Education of Salama Musa*, 129–30.
100. Musa, "Socialism," in Hanna and Gardner, *Arab Socialism*, 283.
101. Ibid., 277.
102. Egger, *Fabian in Egypt*, 50.
103. Musa, "Socialism," 288.
104. "Islah nasl al-insan" [Improving the human stock], *Al-Muqtataf* 26 (1901): 1088–92; "Islah al-nasl" [Improvement of the stock], *Al-Muqtataf* 29 (1904): 697–99; Daniel Stoltz, "From Shellfish to Apes," 86–89. For more on the institutional, and international, background to eugenics, see the collected essays in Alison Bashford and Philippa Levine, eds., *The Oxford Handbook of the History of Eugenics* (Oxford: Oxford University Press, 2010).
105. Donald MacKenzie, "Eugenics in Britain," *Social Studies of Science* 6 (1976): 503,

512. See also Véronique Mottier, "Eugenics and the State: Policy Making in Comparative Perspective," in Bashford and Levine, *Oxford Handbook of the History of Eugenics*, 134–53.

106. See Salama Musa, "Birnard Shu wa-ruwayyatuhu" [Bernard Shaw and his novels], *Al-Muqtataf* 35 (1909): 1178; "Nitsha' wa-ibn al-insan" [Nietzsche and the son of man], *Al-Muqtataf* 34 (1909): 570; "Kutub Wills wa-riwayatuhu" [Wells's works and novels], *Al-Muqtataf* 36 (1910): 119, and 289; "Nazariyat al-nushu' al-hadira" [Current theories of evolution], *Al-Muqtataf* 36 (1910): 437.

107. Salama Musa, *Muqadimmat al-Suberman*, trans. in Abdullah al-Omar, "The Reception of Darwinism in the Arab World" (PhD diss., Harvard University, 1982), 316–18.

108. Salama Musa, "Al-Abyad wa-al-zinj" [White and black], *Al-Muqtataf* 36 (1910): 561–63; quotation from Musa, *Muqadimmat al-Suberman*, 321. See also Christian Geulen, "The Common Grounds of Conflict: Racial Visions of World Order 1880–1940" in Sebastian Conrad and Dominic Sachsenmaier (eds.), *Competing Visions of World Order: Global Moments and Movements, 1880s–1930s* (New York, 2007): 69–96.

109. Musa, *Muqadimmat al-Suberman*, 319.

110. Ibid., 321–25. For Musa, eugenic marriage was to take place between contracting parties and under a socialist system, paralleling the history of eugenic politics and thought elsewhere. See Michael Freeden, "Eugenics and Progressive Thought: A Study in Ideological Affinity," *Historical Journal* 22 (1979): 664.

111. For parallels elsewhere, see, e.g., Suchetana Chattopadhyay, "The Bolshevik Menace: Colonial Surveillance and the Origins of Socialist Politics in Calcutta," *South Asia Research* 26 (2006): 165–79.

112. FO 141/779/9065, p. 45, National Archives, London; Ismael and El-Sa'id, *Communist Movement in Egypt*, 14.

113. FO 141/779/9065: Cairo, 26 August 1919, 70; "Translation of a Report by Sherif Mahmud on the Political Situation in Egypt and the Sudan," 9 September 1919.

114. "Tanaquʿ al-Bulshifiya iyha al-Misriyun," FO 141/779/9065, pp. 98–110; FO 141/779/9065, p. 46; "Extract from 'Al Ahali' dated 21 August 1919," FO 141/779/9065, p. 50; National Archives, London.

115. Dessouki, "Socialist Thought in Egypt," 231; al-Saʿid, *Tarikh al-haraka al-ishtiraki fi Misr*, 260–63.

116. See *al-Ahram*, 18 August 1921, cited in al-Saʿid, *Tarikh al-haraka al-ishtiraki fi Misr*, 273.

117. Ismael and El-Sa'id, *Communist Movement in Egypt*, 16; Salama Musa, *Al-Ahram*, 17 August 1921, cited in al-Saʿid, *Tarikh al-haraka al-ishtiraki fi Misr*, 280, 287–90.

118 Egger, *Fabian in Egypt*, 76–80.

119. Dessouki, "Socialist Thought in Egypt," 232–33.

120. Egger, *Fabian in Egypt*, 84–85.

121. "Egyptian Socialist Party," 11 November 1922, FO 141/779/9065, pp. 311–14, National Archives, London.

122. ʿAbdin 576 (1925), Egyptian National Archives.

123. Ami Ayalon, *The Press in the Arab Middle East* (New York, 1995), 81.

124. Musa, *Nazariyat al-tatawwur wa-asl al-insan* [The theory of evolution and the origin of man] (Cairo: Matbaʿa al-ʿAsriya, 1925), 10.

125. Musa, *Nazariyat al-tatawwur wa-asl al-insan* (Cairo, 1925): 24–25; Musa, *Education of Salama Musa*, 86–87.

126. Musa, *Nazariyat al-tatawwur*, 23.

127. Ibid., 205.
128. Salama Musa, *Ahlam al-falasifa* (Cairo, 1926), 107–24. For a discussion, see Egger, *Fabian in Egypt*, 94–96.
129. Egger, *Fabian in Egypt*, 90.
130. Israel Gershoni and James Jankowski, *Egypt, Islam and the Arabs: The Search for Egyptian Nationhood, 1900–1930* (New York, 1986), 108–16.
131. Omnia El Shakry, *The Great Social Laboratory: Subjects of Knowledge in Colonial and Post-colonial Egypt* (Stanford, CA: Stanford University Press, 2007), 55–66.
132. The letter in question revealed that in 1929, in exchange for being granted a license to set up his own paper, Musa had been willing to support the government position on treaty negotiations with the British. Egger, *Fabian in Egypt*, 170, 179.
133. Salama Musa, *Ma hiya al-Nahda?* [What is the Nahda?] (1935; Beirut: Maktabat al-Ma'arif, 1962).
134. See his *Kitab al-thawrat* [The book of revolutions] (Cairo: n.p., 1954).
135. Musa, *Education of Salama Musa*, 213.
136. Ibid., 215; M. Baraka, *The Egyptian Upper Class between Revolutions: 1919–1952* (Reading, UK: Ithaca Press, 1998), 283.

CHAPTER SEVEN

1. Isma'il Mazhar, *Asl al-anwa' wa-nushu'iha bi-al-intikhab al-tabi'i; wa-hifz al-sufuf al-ghaliba fi al-tanahur 'ala al-baqa'* [The origin of species and their evolution by means of natural selection; or, The preservation of the victorious ranks in the struggle for survival] (Cairo: Wizarat al-Thaqafa wa al-Irshad al-Qawmi, 1928), 2:197, 203, 219–31. Unless otherwise indicated, all references will be to the 1928 edition. This translation was based on the sixth edition: Charles Darwin, *The Origin of Species by Means of Natural Selection; or, The Preservation of Favoured Races in the Struggle for Life* (London: John Murray, 1872).
2. The word "evolved" does appear in the last sentence, however.
3. The subtitle similarly substitutes the closely allied "struggle for survival" for "struggle for life," which Darwin also used interchangeably.
4. Mazhar continued to work on translating *Origin* for much of his life, but the resulting work only appeared after his death in 1962. The first five chapters comprised "Variation under Domestication"; "Variation under Nature"; "The Struggle for Existence"; "Natural Selection or 'The Survival of the Fittest'"; and "The Laws of Variation." The next four included "Difficulties on Theory"; "Instinct"; "Hybridism"; and "On the Imperfection of the Geological Record." The two final chapters of the full edition were actually translated by Muhammad Yusuf Hasan and not Mazhar.
5. See chapter 5 of *Origin of Species*.
6. Entry 32 in the 1928 glossary (see quotation above): Mazhar, *Asl al-anwa'*, 2:224.
7. Darwin, *Origin of Species*, 42.
8. Mazhar, *Asl al-anwa'*, 2:203.
9. Another entry, "Sexual Reproduction," included the English term alongside a novel Arabic translation (*al-tanasul al-jinsi*) and mediated between a familiar and a new lexicon. The glossary defined it, much as Darwin did, as the source of all variation in nature (contra asexual reproduction), yet in its very linguistic construction, the Arabic term also brought together the traditional word for reproduction, *al-tanasul* (which implies the adjective "sexual" already) with the more modernist *jinsi*. For more on this, see the literature on the emergence of "sexuality"—*al-jinsiya*—in nineteenth-century Arabic discourse. See Khaled El-Rouayheb, *Before Homosexuality*

in the Arab-Islamic World, 1500–1800 (Chicago: University of Chicago Press, 2005), 158–61; and Joseph Massad, *Desiring Arabs* (Chicago: University of Chicago Press, 2005), 32n106, 171–72.

10. See Mazhar, *Asl al-anwaʿ*, introduction.
11. Ayyub ʿIsa Abu Diyah, *Ismaʿil Mazhar, 1891–1962: Min al-ishtirakiya ila al-Islam* [Ismaʿil Mazhar, 1891–1962: From socialism to Islam] (Amman: Dar Ward li-al-Nashr wa al-Tawziʿ, 2005), 45.
12. Mazhar, *Asl al-anwaʿ*, 12.
13. For an exception to this, see Abu Diyah, *Ismaʿil Mazhar*.
14. Both Mazhar's father and paternal grandfather were engineers. His father was also head of Majlis al-shura, or Consultative Council, while his paternal grandfather had been among the first students sent to Paris in the 1820s, as part of L'école egyptienne. Yusuf Daghir, *Masadir al-dirasat al-adabiya* [Sources in literary studies] (Sayda, Lebanon: Matbaʿat Dayr al-Mukhlis, 1950), 1094. He was also said to be quite proud of the fact that he had a French and a Circassian grandmother. Abu Diyah, *Ismaʿil Mazhar*, 115.
15. Mazhar, *Asl al-anwaʿ*, 5.
16. These included Ibn Miskawayh's treatise on the soul and his *Tahdhib al-akhlaq* and Ibn Sina's *On the Soul*.
17. Mazhar, *Asl al-anwaʿ*, 6–7.
18. Ibid., 11. Mazhar would later return to Socrates's trial in 1922, translating sections of the *Apology* for readers of *Al-Muqtataf*. See Ismaʿil Mazhar, "Al-Abulujiya" [The *Apology*], *Al-Muqtataf* 61 (1922): 165–70, 253–56. See also his definition of philosophy in *Malqa al-sabil fi madhhab al-nushuʾ wa-al-irtiqa wa-atharuhu fi al-fikr al-hadith* [Evolution at the crossroads: The theory of evolution and its impact on modern thought] (Cairo: Al-Matbaʿa al-ʿAsriya, 1924), 28–34.
19. It read as follows: "It is impossible to answer your question briefly; and I am not sure that I could do so, even if I wrote at some length. But I may say that the impossibility of conceiving that this grand and wondrous universe, with our conscious selves, arose through chance, seems to me the chief argument for the existence of God; but whether this is an argument of real value, I have never been able to decide. I am aware that if we admit a first cause, the mind still craves to know whence it came and how it arose. Nor can I overlook the difficulty from the immense amount of suffering through the world. I am, also, induced to defer to a certain extent to the judgment [*sic*] of the many able men who have fully believed in God; but here again I see how poor an argument this is. The safest conclusion seems to me that the whole subject is beyond the scope of man's intellect; but man can do his duty." Darwin Correspondence Project, www.darwinproject.ac.uk: letter 8837, C. R. Darwin to N. D. Doedes, 2 April 1873.
20. Mazhar, *Asl al-anwaʿ* (1918 ed.), 12.
21. "Kitab Asl al-anwaʿ" [The book *Origin of Species*], *Al-Muqtataf* 55 (1919): 73.
22. Ibid.
23. European (particularly German) scholarship on al-Jahiz began as early as the 1830s, while he had something of a revival in the Arabic press in the first years of the twentieth century: see the 1906 Cairo edition of the *Kitab*, for instance, and the late-nineteenth-century Egyptian publication of *Kitab al-bayan wa-al-tabyin* [The book of eloquence and demonstration].This interest in al-Jahiz was also more widespread; see, e.g., Muhammad Iqbal's 1906 discussion in *Development of Metaphysics in Persia* (the thesis he published at the Faculty of Philosophy of the Ludwig-Maximilians-Universität, Munich, under the supervision of Friedrich Hommel).

24. See M. McDonald, "Animal Books as a Genre of Arabic Literature," *Bulletin of the British Society for Middle East Studies* 15 (1989): 3–10.
25. Mazhar, *Asl al-anwaʿ*, 41.
26. There were a Latin reedition of Ibn Miskawayh's works in 1869 edited by F. De Jong and M. J. Goeje, a French translation of his work on the *Tablet of Cebes*, and Arabic and Persian editions of these and other works beginning in the 1870s. His writings were also published, alongside al-Ghazali's commentaries, in the early 1900s: *Tahdhib al-akhlaq wa-ta'thir al-ara'q: Wa bihamishi kitab al-adab fi al-din* (Cairo: Al-Matbaʿa al-Khayriya, 1904). *Tahdhib al-akhlaq* and *Tarajib al-umam* were his most widely published works (and discussed in numerous Arabic, English, and other writings throughout the late nineteenth and early twentieth centuries).
27. Ibn Miskawayh, *The Refinement of Character* (Beirut: American University of Beirut, 1968), 157; translation of *Tahdhib al-akhlaq*.
28. They also had a revival in the nineteenth-century global press: there are editions of the *Ikhwan al-Safa'* in Calcutta as early as 1812, German editions in the 1860s, and a full publication in Arabic in 1888.
29. "The unity and complexity of his soul and body make him the 'antipode of God.'" Good and evil qualities equally distributed among creatures created further distinctions in the hierarchy. As with Ibn Miskawayh, intermediate forms were also listed: in this case, minerals were said to be connected below to water and earth, and their lowest types were alum and vitriol; while red gold was classed among the highest of the minerals, approaching the world of plants. Among plants, moss was considered the lowest order, while the palm tree once again stood between the plant and animal worlds. Among animals, the snail was mentioned as being closest to the plant world, and the elephant—with its great intelligence—the highest and nearest to man. See Sayyed Hossein Nasr, *An Introduction to Islamic Cosmological Doctrines* (Albany: State University of New York Press, 1993), 69.
30. Epistle 22 of the *Rasa'il Ikhwan al-Safa'* (Treatise of the Brethren of Purity) takes up this theme and elaborates more specifically upon man's unique place in creation. In a dialogue between animals and humans, where the former complain to the king of the *jinn* about their unjust treatment at the hands of humans, the king makes the case that what marked humans off from the rest of creation, and what therefore ultimately secured their supremacy over other created beings, was their unique immortality of the soul and, more crucially, their ability to become like angels—among men, there are "saints of God, the choice flower of his creation." Ikhwan al-Safa', *Epistles of the Brethren of Purity*, ed. Nader El-Birzi (Oxford: Oxford University Press, 2008), 273.
31. Mazhar quotes Ibn Khaldun's complaint that people had previously imagined that black-skinned people were children of the son of the accursed Ham and hence consigned to a lifetime of slavery (this was a common Christian reading, too). The Pentateuch describes the curse of Ham and his enslavement to his brother but not the darkened skin, corrected Ibn Khaldun. Moreover, he was a slave only to his brother and not to all men. Temperature and air, he argued, were the true causes of skin coloration. Mazhar cites the passages by Ibn Khaldun on the effect of these conditions on the disposition of peoples as another example of the "Lamarckian" view on the inheritance of acquired characteristics.
32. See Omnia El Shakry, *The Great Social Laboratory: Subjects of Knowledge in Colonial and Post-colonial Egypt* (Stanford, CA: Stanford University Press, 2007).

33. See Evans De Lacy O'Leary, *Arabic Thought and Its Place in History* (London: Kegan Paul, Trench, Trubner, 1922). The *Muqtataf* articles were later published as a book: Isma'il Mazhar, *Tarikh al-fikr al-'Arabi* [History of Arabic thought] (Cairo: Dar al-'Usur li-al-Taba' wa-al-Nashr, 1928).
34. See Isma'il Mazhar, *Tajdid al-'Arabiya* [The revival of Arabic] (Cairo: Maktabat al-Nahda al-Misriya, 1950).
35. Jurji Zaydan, *Tarikh al-lugha al-'Arabiya: Bi-i'itibar anaha ka'in hay* (Cairo: Matba'at al-Hilal, 1904).
36. Mazhar, *Asl al-anwa'*, 15.
37. Ibid., 12–13.
38. For more on these strategies, see Jaroslav Stetkevych, *The Modern Literary Arabic Language* (Chicago: University of Chicago Press, 1970). See also Adrian Gully, "Arabic Linguistic Issues and Controversies of the Late Nineteenth and Early Twentieth Centuries," *Journal of Semitic Studies* 42 (1997): 99.
39. "Muqaddimat al-sana al-khamisa 'ashar" [Introduction to volume 15], *Al-Muqtataf* 15 (1890): 1–2; "Al-amira al-Misriya" [The Egyptian princess], *Al-Muqtataf* 23 (1899): 66.
40. "Al-kalimat al-a'jamiya" [Foreign words], *Al-Muqtataf* 15 (1891): 52–53.
41. Mazhar, *Tajdid al-'Arabiya*, 10–13.
42. Mazhar, *Asl al-anwa'*, 4.
43. For other transcendentalist reactions to positivism, see references below in n. 50. For Mazhar's views on pragmatism, see his "Al-brajmatism" [Pragmatism], *Al-Muqtataf* 89 (1936): 359.
44. Mazhar pointed out that Richter, like William Thompson (later Lord Kelvin), speculated that life first descended to earth from outer space.
45. He thought the second camp, in particular, simply moved one from "the darkness of simple ignorance to the pitch-black darkness of a complicated ignorance." As for the third—though it might in fact "perhaps be a correct view concealed from us in our times"—such a view was, at the moment, merely a conjecture, unsupported by scientific evidence or experimentation. Mazhar, *Asl al-anwa'*, 56–57.
46. Ibid., 56.
47. Ibid., 60, 62.
48. Isma'il Mazhar, "Uslub al-fikr al-'ilmi" [The method of scientific thought], *Al-Muqtataf* 68 (1926): 140. Mazhar's 1923 translation of John Theodore Merz's introduction to *A History of European Thought in the Nineteenth Century* (1907)—published as *Naz'at al-fikr al-Urubbi fi al-qarn al-tasi' 'ashr* (Cairo: Matba'at al-Muqtataf wa al-Muqattam, 1923)—and his *Malqa al-sabil* defined positivism in much the same way.
49. "Al-madhhab al-falsafi" [The philosophical school, or "positivism"], *Al-Muqtataf* 22 (1898): 106–12.
50. For more on this, see Antonio Aliotta, *The Idealist Reaction against Science* (London: Macmillan, 1914); and J. W. Burrow, *The Crisis of Reason: European Thought, 1848–1914* (New Haven, CT: Yale University Press, 2000), 56–67. Cf. Vladimir Solovyov, *Crisis in Western Philosophy: Against the Positivists* (Hudson, NY: Lindisfarne, 1996).
51. Isma'il Adham, *La madha ana mulhid* [Why I am an apostate] (Cairo, [publisher] 1937). Adham was a Turk living in Egypt who published a number of outspoken articles on his faith in scientific rationalism and his rejection of a personal God. The term *mulhid* (apostate) was a label that still held considerable charge, as tech-

nically it was punishable by death; indeed, its legal status in Egypt has remained uncertain long after the abolishment of religious courts by Gamal ʿAbd al-Nasser in 1952.

52. Ibrahim Muhammad Sayhi, *Falsafat al-takwin* [Philosophy of creation] (Alexandria: Al-Matbaʿa al-Misriya, 1920), introduction.

53. Sayhi offered a potted history of theories of matter, energy, motion, and the origin of life while citing English textbooks on physics, chemistry, and natural philosophy, classical Arabic works by Ibn Rushd and al-Farabi, and modern Arabic works like Husayn al-Jisr's *Al-risala al-Hamidaya fi haqiqat al-diyana al-Islamiya wa haqiqat al-shariʿa al-Muhamadiya* [A Hamidian treatise on the truth of Islam and the shariʿa] (Beirut: Majlis al-Maʿarif, 1888).

54. Roughly the first half of *Falsafat al-takwin* is a fairly straightforward account of contemporary theories of matter and motion, touching in particular on the subjects of energy, entropy, and astronomy. He then turns to such subjects as "the truth about matter" and presents a long discussion of cosmogony, which Sayhi reckoned to be one of the most important of the "contemporary Western sciences."

55. Muhammad Farid Wajdi published his first book at the age of twenty: *Kitab al-falsafa al-haqqa fi badaʾi al-akwan* [The true philosophy of the marvels of the cosmos] (Cairo: Al-Matbaʿa al-Amira al-ʿUthmaniya, 1895). Later works include a critique of Lord Cromer's *Modern Egypt* and a *tafsir* of the Qurʾan. His four-volume work *ʿAla atlal al-madhhab al-maddi* [On the ruins of materialism] (Cairo: Matbaʿat Dairat al-Maʿarif, 1921) was his most popular.

56. Anwar Moghith, "Sur les ruines du matérialisme," *Égypte: Monde Arabe* 4–5 (2001): 311–15.

57. The genre of *tafsir ʿilmi* (scientific exegesis), which Wajdi and others were helping to promote in the 1920s and 1930s, provides the most obvious example of this.

58. See Mazhar, *Malqa al-sabil*, 4. Arabic readers were also given a highly detailed account of modern geological debates—from catastrophism to uniformitarianism and the latest debates on fossil remains of early man. It was here too that they were first introduced to the latest research in evolutionary biology and the emerging field of genetics, from the work of Hugo De Vries, William Bateson, Herbert Spencer Jennings, and D'Arcy Thompson.

59. Adel Ziadat, *Western Science in the Arab World: The Impact of Darwinism, 1860–1930* (Basingstoke: Macmillan, 1986), 119.

60. Mazhar, *Malqa al-sabil*, 78.

61. George Smith, "Herbert Spencer's Theory of Causation," *Journal of Libertarian Studies* 5 (1981): 116.

62. Bernard Lightman, "Ideology, Evolution and Late-Victorian Agnostic Popularizers," in *History, Humanity and Evolution: Essays for John C. Greene*, ed. James Moore (Cambridge: Cambridge University Press, 1989), 294.

63. John Beattie Crozier, *Civilization and Progress* (London: Longmans, 1888), 186.

64. Mazhar, *Malqa al-sabil*, 63–66.

65. Ibid., 286–96; and Abdullah Al-Omar, "The Reception of Darwinism in the Arab World" (PhD diss., Harvard University, 1982), 268–71.

66. Mazhar, *Malqa al-sabil*, 88.

67. Ibid., 41, 49–53.

68. Ismaʿil Mazhar, "Darwin wa-tanazuʿ al-baqaʾ" [Darwin and the struggle for survival], *Al-Muqtataf* 63 (1923): 60–62.

69. Leonard T. Hobhouse, *Social Evolution and Political Theory* (New York: Columbia University Press, 1911), 23.
70. Mazhar, *Malqa al-sabil*, 95–97, 101–2; and Al-Omar, "Reception of Darwinism in the Arab World," 250–52.
71. Mazhar, *Malqa al-sabil*, 99–100. Discussing the work of De Vries and others, which he compared with Darwin's theory of pangenesis, Mazhar argued (taking the view opposite to that of his contemporary Salama Musa) that the idea of acquired characteristics was still an open question.
72. On his criticisms of eugenics, see Mazhar, *Malqa al-sabil*, 94–95.
73. J. B. S. Haldane, "Science and Ethics," *Harper's Magazine* 157 (1928): 1–10; Isma'il Mazhar, "Al-'ilm wa-al-ikhlaq" [Science and ethics], *Al-Muqtataf* 73 (1928): 128–33.
74. "Did we Rise?," *Al-'Usur* 3 (1928): 161–68.
75. Isma'il Mazhar, "Hurriyat al-fikr" [Rationalism or freedom of thought], *Al-'Usur* 2 (1928): 1172–75.
76. Mazhar, *Bayn al-'ilm wa al-din* [Between religion and science] (Cairo: n.p., n.d.), 12–13.
77. Ibid., 28.
78. Ibid., 13.
79. Ibid., 19–20.
80. Ibid., 27–29. See also Mazhar's "Hurriyat al-fikr," 1179; Ziadet, *Western Science in the Arab World*, 115; and Isma'il Mazhar, *Fakk al-ighlal* [Release from oppression] (Cairo: n.p., 1946), also discussed by Abu Diyah in *Isma'il Mazhar*, 10.
81. Isma'il Mazhar, "Falsafat al-inqilab al-Turkiya al-haditha" [The philosophy of the recent Turkish revolution], *Al-Majalla al-Jadida* 2 (1931): 1213.
82. See Isma'il Mazhar, "Falsafat al-inqilab al-Turkiya al-haditha" [The philosophy of the recent Turkish revolution], *Al-'Usur* 1 (1927): 113–32; later published in *Al-Majalla al-Jadida* 2 (1931): 1212–28.
83. These were by Qabil Adam, Jamal Nuri Bey, and Rafiq Sadiq Bey. Only the first of these was covered in the earlier article. Isma'il Mazhar, *Wathbat al-sharq* [Advance of the East] (Cairo: Dar al-'Usur, 1929).
84. On reactions in the Egyptian press, see Richard Hattemer, "Atatürk and the Reforms in Turkey as Reflected in the Egyptian Press," *Journal of Islamic Studies* 11 (1999): 21–42.
85. Mazhar, *Wathbat al-sharq*, 8–17.
86. "Ta'amulat fi al-adab wa-al-haya" [Reflections on literature and life], *Al-'Usur* 1 (1927): 25. The original terms were provided in both Arabic and English, and the article was signed by Filawpuns, a pseudonym, perhaps for philo-*penser*, or the "love of thinking."
87. See Samah Selim, "Languages of Civilization: Nation, Translation and the Politics of Race," *Translator* 15 (2009): 139–56.
88. Ahmad Fathi Zaghlul, *Sirr tatawwur al-umam* [Secret of evolution of nations] (Cairo: Matba'a al-Ma'arifa, 1913); Muhammad Rustum, *Al-hadara al-Misriya* [Egyptian civilization] (Cairo: Matba'at al-'Asriya, 1924). Le Bon's *Psychologie du socialisme* was also translated in 1925: Muhammad 'Adil Zu'aytar, *Ruh al-ishtirakiya* [The spirit of socialism] (Cairo: Matba'at al-'Asriya, 1925). For more on this, see Israel Gershoni and James Jankowski, *Egypt, Islam and the Arabs: The Search for Egyptian Nationhood, 1900–1930* (Oxford: Oxford University Press, 1987), 131–36.
89. On the anthropological construction of racial identities as well as conceptions

of race more generally in Egypt at this time, see El Shakry, *Great Social Laboratory*, 55–86.

90. Gershoni and Jankowski, *Egypt, Islam and the Arabs*, 264–66.
91. Isma'il Mazhar, "Uslub al-fikr al-'ilmi" [The method of scientific thought], *Al-Muqtataf* 68 (1926): 137–45. See also the critique of Afghani in *Malqa al-sabil*.
92. Amin al-Khuli, "Uslub al-fikr al-'ilmi" [The method of scientific thought], *Al-Muqtataf* 68 (1926): 440–44; Mustafa al-Shihabi, "Al-'Arab wa-al-bahth al-'ilmi" [Arabs and scientific research], *Al-Muqtataf* 69 (1926): 319–21; Isma'il Mazhar, "Hawal uslub al-fikr al-'ilmi" [On the method of scientific thought], *Al-Muqtataf* 69 (1926): 431–33; Mustafa al-Shihabi, "Al-'Arab wa-al-uslub al-'ilmi" [The Arabs and the scientific method], *Al-Muqtataf* 70 (1927): 87–88; Isma'il Mazhar, "Uslub al-'Arab al-'ilmi" [The Arabs' scientific method], *Al-Muqtataf* 70 (1927): 328.
93. See Stephen Wild, "Ottomanism versus Arabism: The Case of Farid Kassab (1884–1970)," *Die Welt des Islams* 28 (1988): 626.
94. Salama Musa, *The Education of Salama Musa*, trans. L. O. Schuman (Leiden: Brill, 1961), 87.
95. *Al-'Usur* 1 (1927): 15–16.
96. *"Al-'Usur," Al-Muqtataf* 71 (1927): 343; "'Ala uslub al-'Arabi" [On the Arabic style], *Al-'Usur* 3 (1928): 134–37; "Al-hubb wa-al-jamal" [Love and beauty], *Al-'Usur* 3 (1928): 137–41.
97. *Al-Muqtataf* 71 (1927): 343.
98. "Al-Jami'a al-Azhar," *Al-'Usur* 2 (1928): 1237–38.
99. Ibid., 1242–44.
100. For more on this, see the afterword below and my "The Exegesis of Science in Twentieth Century Arabic Interpretations of the Qur'an," in *Nature and Scripture*, ed. Jitse van der Meer (Leiden: Brill, 2009).
101. Isma'il Mazhar, "I'jaz al-Qur'an," *Al-'Usur* 2 (1928): 966.
102. See Tomoko Masuzawa, *The Invention of World Religions; Or, How European Universalism Was Preserved in the Language of Pluralism* (Chicago: University of Chicago Press, 2005).
103. Isma'il Mazhar, "I'jaz al-Qur'an," *Al-'Usur* 2 (1928): 962–63.
104. Ibid., 964.
105. For the background to this journal, see Wilfred Cantwell Smith, *Islam in Modern History* (Princeton, NJ: Princeton University Press, 1957), 122.
106. Cited in Isma'il Mazhar, "Al-iman bi-Allah" [Faith in God], *Al-'Usur* 6 (1929): 162.
107. Ibid., 163–64.
108. Ibid., 166–79.
109. Muhammad Farid Wajdi, "Satwa al-adyan 'ala hurriyat al-fikr" [The assault of religions on rationalism] *Al-'Usur* 7 (1930): 206–12.
110. Daghir, *Masadir al-dirasat*, 1094. According to Isma'il Mazhar Jr., Mazhar's grandson, Mazhar had also suffered serious financial losses by then, particularly after the stock market crisis of 1929, which affected Egypt as elsewhere. Personal correspondence, 11 February 2013.
111. Isma'il Mazhar, "Sa'ad Zaghlul," *Al-'Usur* 1 (1927): 71–74.
112. Isma'il Mazhar, *Al-ishtirakiya ta'uqu irtiqa' al-naw' al-insani* [Socialism hinders human evolution] (Cairo: Matba'at al-Taraqqi, 1927).
113. Isma'il Mazhar, *Mashru' li-ta'sis hazb al-fallah al-Masri* [Proposal for the formation of an Egyptian peasants' party] (Cairo: Dar al-'Usur, 1929), cover.
114. Ibid., 21.

115. Isma'il Mazhar, *Risalat al-fikr al-hurr* [Essays on the freedom of thought] (Beirut: Dar al-Thaqafa, 1963), 11 (introduction by Jallal Mazhar).
116. See Rached Hamzaoui, *L'Académie de langue arabe du Caire: Histoire et œuvre* (Tunis: University of Tunis, 1975), 150.
117. Louise Robbins, "Publishing American Values: The Franklin Book Programs as Cold War Cultural Diplomacy," *Library Trends* 55 (2007): 638–50.
118. See Muhammad Rashad Hamzawi, *Majma' al-lugha al-'arabiya bi-damashq* (Tunis, [publisher]1988); Muhammad Rashad Hamzawi, *L'Académie Arabe de Damas* (Leiden: Brill, 1965); and Anwar Chejne, *Arabic Language: Its Role in History* (Minneapolis: University of Minnesota Press, 1989), 112. In 1938 the Royal Academy of Cairo became known as the Fu'ad I Arabic Academy of Language and after the revolution of 1952 simply as the Arabic Academy of Language, or Al-Majma' al-Lugha al-'Arabiya. See *Al-Muqtataf* 77 (1930): 249–55; and Chejne, *Arabic Language*, 104–5.
119. Fearing the academy would serve as an effective second parliament—using the platform of language reform to overthrow the monarchy—the government brought together a carefully coordinated mix of "conservatives" and "loyalists," Azhari shaykhs and university educators, and Arab nationalists and Orientalists. These included the pro-court and pro-British civil servants Muhammad Tawfiq Rif'at Pasha and Mansur Fahmi and the traditionally trained Azhari shaykhs Ahmad al-Iskandari and Muhammad Husayn (who had, incidentally, also been responsible for issuing a fatwa for Fuad I declaring the monarchy legitimate in Islam); and Muhammad Kurd 'Ali (the original director of the Damascus Academy), Father Anistas al-Karmili (the founder of the Baghdad language journal *Lughat al-'Arab*), Faris Nimr, 'Abd al-Qadir al-Maghribi, H. A. R. Gibb, and Louis Massignon.
120. See Mazhar, *Tajdid al-Arabiya*, 7, 17 (on Iskandarani's attitude to *naht* in particular).
121. Giora Eliraz, "Tradition and Change: Egyptian Intellectuals and Linguistic Reform, 1919–1930," *Asian and African Studies* 20 (1986): 254. Musa had also earlier championed the idea that Arabic scientific terminology was best rendered in Latin script. See his "Al-khatt al-Latini li-al-lugha al-'Arabiya" [Latin script for the Arabic language], *Al-Majalla al-Jadida* 4 (1935): 79–80. On the subject of language reform and modernization, Musa also published *Al-Balagha al-'asriya wa-al-lugha al-'Arabiya* [Modern rhetoric and the Arabic language] (Cairo: Matba'a al-'Asriya, 1940) and *Al-adab li-al-sha'b* [Literature for the people] (Cairo: n.p., 1956). He had, moreover, been a longtime champion of the use of the colloquial in written Arabic.
122. Mazhar, *Tajdid al-'Arabiya*, 10–13.
123. Chejne, *Arabic Language*, 106.
124. Mazhar, *Tajdid al-'Arabiya*, 7–8.
125. Oles Smolansky, "Moscow-Cairo Crisis, 1959," *Slavic Review* 22 (1963): 713–26.
126. Isma'il Mazhar, *Al-din fi zill al-shuyu'iya* (Cairo: n.p., 1958).
127. Isma'il Mazhar, "Mutual Responsibility, Not Communism," in *Arab Socialism*, ed. Sami Hanna and George Gardner (Leiden: Brill, 1969), 172–76.
128. Ibid., 196; on Shaltut, see his "Socialism and Islam," in *Islam in Transition: Muslim Perspectives*, ed. John Donohue and John Esposito (Oxford: Oxford University Press, 1982), 99–102. More generally, see Fruma Zebiri, *Mahmud Shaltut and Islamic Modernism* (Oxford: Oxford University Press, 1993), 55–57.

AFTERWORD

1. These were *ahadith*, or Judeo-Christian prophetic tales, many of which can be traced to Ibn Hisham, the ninth-century biographer of Muhammad. Shahin also examines

Qisat al-anbiya [Tales of the Prophet]—a collections of stories of the lives of the prophets, from Abraham to Jesus, from various sources—as part of this Israʿiliyat tradition. He denounces them all as mere "folklore." He gives the example of the difference between their dating of human history since the creation of Adam and that of modern scientific estimates. These stories, he concludes, are at odds with both a rational and a scientific account of creation.

2. According to Shahin, he created two types of being: the tangible and the intangible (*al-shihada wa-al-ghayb*); only the former had attributes of time and place. And while the latter remain among the divine secrets, knowledge of their existence and attributes can sometimes be gleaned from the tangible world. God might also reveal hidden knowledge to a few select people who could be said to have a clear vision or "discovery" of latent knowledge; Shahin expresses the hope that his book might count as an example of this gnostic revelation of the origins of creation. Yet his own account is very much rooted in the traditional methods of *tafsir*, or Qurʾanic exegesis. Much of the hermeneutics he deploys, for instance, is bound up with linguistic analysis of the Qurʾan. Shahin makes only passing mention of *ahadith*, even those considered "sound," or of previous works of *tafsir*. Caught between a traditional approach and a more modernist construction, his exegesis examines all the verses on creation in the Qurʾan, stressing throughout that his interpretation never strays from the usual meaning of the *sunan*, or traditions.
3. ʿAbd al-Sabur Shahin, *Abi Adam* (Cairo: Maktabat al-Shabab, 1998).
4. Charles Hirschkind, "Heresy or Hermeneutics: The Case of Nasr Hamid Abu Zayd," *Stanford Humanities Review* 5 (1995): 35–49; Baber Johansen, "Apostasy as Objective and Depersonalized Fact: Two Recent Egyptian Court Judgments," *Social Research: An International Quarterly* 70 (2003): 687–710; Maurits S. Berger, "Apostasy and Public Policy in Contemporary Egypt: An Evaluation of Recent Cases from Egypt's Highest Courts," *Human Rights Quarterly* 25 (2003): 720–40; Talal Asad, Judith Butler, Saba Mahmoud, and Wendy Brown, *Is Critique Secular? Blasphemy, Injury, and Free Speech* (Berkeley and Los Angeles: University of California Press, 2009).
5. This case was discussed extensively in the Egyptian press the year it took place and for years after. For a translation into English of some of these discussions, see *Akhir Saʿah* (3 February 1999); *Al-Ahrar* (12 and 19 February and 4 May 1999); *Al-Liwa al-Islami* (11 February and 29 April 1999); *Ruz al-Yusuf* (4 January and 8 February 1999); and *Al-Wafd* (13 September 1999), available at www.arabwestreport.com.
6. ʿAbbud Khalidi, *Al-ʿaql wa-al-khalq* (Baghdad: Dar al-Hawra, 2006).
7. Of particular interest here are those journals published by new institutes and academic departments of science, like *Tatawwur al-Insan* (The evolution of man), which brought together evolutionary science and psychology.
8. See, e.g., Gamal Nkrumah, "Tree of Life," *Ahram Weekly* 973 (19–25 November 2009), www.weekly.ahram.org.eg/2009/973/feature.htm.
9. From an abundant literature, see Muhmmad ʿAli Yusuf, *Masraʿ al-Darwiniya* (Jiddah: Dar al-Shuruq, 1983); Mahfuz ʿAli ʿAzzam, *Nazariyat al-tatawwur ʿinda mufakkiri al-Islam: Dirasa muqarana* (Cairo: Dar al-Hidaya, 1986); Mahfuz ʿAli ʿAzzam, *Fi al-falsafa al-tabiʿi ʿinda al-Jahiz* (Cairo: Dar al-Hidaya li-al-Tibaʿa wa-al-Nashr wa-al-Tawziʿ, 1986); and Mahfuz ʿAli ʿAzzam, *Mabdaʾ al-tatawwur al-hayawi lada falsafat al-Islam* (Beirut: Al-Muʾassasa al-Jamiʿa li-al-Dirasat wa-al-Nashr wa-al-Tawziʿ, 1996).
10. Best known of these perhaps is the Turkish author Adnan Oktar, better known as Harun Yahya, who originally trained as an interior designer but who has now developed a cottage industry around questions of creationism and Islam. His *Atlas of Cre-*

ation was sent to academics around the world, and his website contains discussions of such things as the connection between evolution and Zionism and between militarism and capitalism. One of his followers testified at a hearing in Kansas against the teaching of evolution in schools, highlighting the new global links among these groups. See Ron Numbers, "Global Creationism," in his *The Creationists* (Cambridge, MA: Harvard University Press, 2006).

11. On the influence of Christian biblical criticism on Rashid Rida in particular, see Umar Ryad, *Islamic Reformism and Christianity: A Critical Reading of the Works of Muhammad Rashid Rida and His Associates (1898–1935)* (Leiden: Brill, 2009).
12. Cited in Wilfred Cantwell Smith, *Islam in Modern History* (Princeton, NJ: Princeton University Press, 1957), 139.
13. See Muhammad Farid Wajdi, *Al-Islam wa al-madaniya* (Civilization and Islam) (Cairo: Matbaʿat al-Waʿiz, 1901), for more on this.
14. Tantawi Jawhari had made a similarly ambivalent argument in his "Madhhab Darwin ʿand al-ʿArab" [Darwin among the Arabs], published in *Jaridat Misr al-Fata* in 1909. See also Samera Esmeir, *Juridical Humanity: A Colonial History* (Stanford, CA: Stanford University Press, 2012).
15. I have written more extensively on this subject elsewhere: see my "The Exegesis of Science in Twentieth Century Arabic Interpretations of the Qurʾan," in *Nature and Scripture*, ed. Jitse van der Meer (Leiden: Brill, 2009), 491–524.
16. J. Jomier, *Le commentaire coranique du Manar* (Paris: G. P. Maisonneuve, 1954).
17. Tantawi Jawhari, *Kitab al-insan* [Book of humanity] (Cairo: Maktabat wa Matbaʿat al-Maʿarif, 1911). See also the discussion of this in Esmeir, *Juridical Humanity*.
18. Tantawi Jawhari, *Kitab al-arwah* (Cairo: Matbaʿat al-Saʿada, 1920).
19. See Tantawi Jawhari, *Risalat al-tawhid wa-al-taʿdid* [Treatise on unity] (Cairo: n.p., n.d.), picture on p. 9.
20. Maurice Barbanell, *Philosophy of Silver Birch* (London: Spiritualist Press, 1969).
21. See Jane Smith, "The Concourse between the Living and the Dead in Islamic Eschatological Literature," *History of Religions* 19 (1980): 224–36.
22. Jawhari, *Al-Jawahir fi tafsir al-Qurʾan al-karim: Al-mushtamil ʿala ʿajaib badaiʿ al-mukawwanat wa-gharaʾib al-ayat al-bahirat* [Gems of Qurʾanic exegesis], 26 vols. (Cairo: Matbaʿat Mustafa al-Babi al-Halabi, 1922–35), 1:10–11.
23. Ahmad Abu Hajar, *Al-tafsir al-ʿilmi li-al-Qurʾan fi al-mizan* (Beirut: Dar Qutayba, 2000), 297–99.
24. He goes on: "There were flaws in the original theory, which was conceived at a time when nothing was known of the genes which carry hereditary properties and distinguish one species from another. Several aspects of Darwin's theory have since been disproved and many others are still a matter of debate." Sayyid Qutb, *In the Shade of the Qurʾan* (Leicester: Islamic Foundation, 1999), 1:203.
25. Ami Ayalon, *The Press in the Arab Middle East: A History* (New York: Oxford University Press, 1995), 141–42, 151.
26. See James Whidden, "The Generation of 1919," in *Re-envisioning Egypt: 1919–1952*, ed. Arthur Goldschmidt and Amy Johnson (Cairo: American University of Cairo Press, 2005).

BIBLIOGRAPHY

PRIMARY SOURCES

Archives and Special Collections

American Board of Commissioners for Foreign Missions Archives, Houghton Library, Harvard University, Cambridge, MA

American University of Beirut (AUB) Library Archives, Beirut, Lebanon

Amherst College Library Archives and Special Collections, Amherst, MA

Egyptian National Archives (Dar al-Watha'iq al-Qawmiya), Cairo, Egypt

Fabian Society Archives (LSE), London, UK

Khedive ʿAbbas Archives, Durham University, Durham, UK

National Archives, London, UK

Yildiz Collection, Ottoman Archives, Istanbul, Turkey

Presbyterian Church in the U.S.A. Board of Foreign Missions, 1833–1911, Princeton Theological Seminary, Princeton, NJ

West Sussex Record Office, Chichester, UK

Journals

Al-Hilal

Al-Jamiʿa

Al-Jinan

Journal of the American Oriental Society

Al-Liwa

Al-Majalla al-Jadida

Majallat al-Azhar

Majallat al-Majmaʿ al-ʿIlmi al-ʿArabi

Al-Manar

Al-Mashriq

Missionary Herald

Al-Mu'ayyad

Al-Muqattam

Al-Muqtataf

Al-Nashra al-Usbuʿiya

Nineteenth Century

Nur al-Islam

Popular Science Monthly
Rawdat al-Madaris
Al-Ustadh
Al-ʿUsur

Published Sources

ʿAbduh, Muhammad. *Al-aʿmal al-kamila* [Complete works of Muhammad ʿAbduh]. 5 vols. Cairo: Dar al-Shuruq, 1993.

———. *Al-Islam: Din al-ʿilm wa-al-madaniya* [Islam: A religion of science and civilization]. Cairo: n.p., 1905.

———. *Al-Islam wa-al-Nasraniya maʿa al-ʿilm wa-al-madaniya* [Science and civilization in Islam and Christianity]. Cairo: n.p., 1905.

———. *The Theology of Unity*. Translated by Ishaq Musaʿad and Kenneth Cragg. London: Allen and Unwin, 1966. Translation of *Risalat al-tawhid*.

ʿAbduh, Muhammad, and Rashid Rida. *Tafsir al-Qurʾan al-hakim al-mushtahir bi-ism Tafsir al-Manar* [Exegesis of the judicious Qurʾan, widely known as "Tafsir al-Manar"]. 12 vols. Cairo: Dar al-Manar, 1906–35. Reprinted as *Tafsir al-Qurʾan al-hakim al-musamma Tafsir al-Manar*, 12 vols. (Cairo: Al-Haya al-Misriya al-ʿAmma li-al-Kitab, 1972–75).

Adham, Ismaʿil. *La madha ana mulhid* [Why I am an atheist]. Cairo: n.p., 1937.

Afghani, Jamal al-Din al-. *An Islamic Response to Imperialism: Political and Religious Writings of Sayyid Jamal al-Din "al-Afghani."* Translated by Nikkie Keddie. Berkeley and Los Angeles: University of California Press, 1983. Translation of *Haqiqat-i madhhab-i nichari va bayan hal nichariyan* [in Persian].

———. *Al-radd ʿala al-dahriyin* [A refutation of the materialists]. Translated by Muhammad ʿAbduh and ʿArif Effendi Abu Tarib. Cairo: Matbaʿat al-Mawsuʿat, c. 1902.

American Board of Commissioners for Foreign Missions. *Annual Reports*. Boston: The Board, 1834–1940.

———. *The Divine Instrumentality for the World's Conversion*. Boston: n.p., 1856.

———. *First Ten Annual Reports of the American Board of Commissioners for Foreign Missions*. Boston: Crocker and Brewster, 1834.

Amin, Qasim. *The Liberation of Women; and, The New Woman*. Translated by Samiha Sidhom Peterson. Cairo: American University in Cairo Press, 2000. Translation of *Tahrir al-marʾa* (Cairo, 1899).

Anderson, Rufus. *History of the Missions of the American Board of Commissioners for Foreign Missions to the Oriental Churches*. Boston: Congressional Publication Society, 1872.

Antun, Farah. *Al-din wa-al-ʿilm wa-al-mal* [Religion, science, and wealth]. Alexandria: n.p., c. 1903.

———. *Ibn Rushd wa-falsafatuhu* [Ibn Rushd and his philosophy]. Alexandria: n.p., 1903.

———. *Tarikh al-Masih*. 4 vols. Alexandria: Matbaʿat al-Jamiʿa, 1904. Translation of Ernest Renan's *Vie de Jésus*.

Arminjon, Pierre. *L'enseignement, la doctrine et la vie dans les universités musilmanes d'Egypte*. Paris: F. Alcan, 1907.

Artin, Ya'qub. *Considérations sur l'instruction publique en Egypte*. Cairo: Al-Matbaʿa al-Kubra al-Amiriya, 1894.

Bastian, Henry. *The Beginnings of Life, Being Some Account of the Nature, Modes of Origin and Transformation of Lower Organisms*. 2 vols. London: n.p., 1872.

———. *Evolution and the Origin of Life*. London: Macmillan, 1874.

———. "Facts and Reasonings concerning the Heterogeneous Evolution of Living Things." *Nature*, June 30, 1870.

Bayram, Mustafa. *Al-Azhar*. Cairo: Matbaʿat al-Tamaddun, 1903.

Baz, Jurji Effendi. *Kitab al-rawda al-badiʿa fi tarikh al-tabiʿa*. [The book of the magnificent garden of natural history]. Beirut: n.p., 1881.

Bendyshe, Thomas. "History of Anthropology." *Memoirs of the Anthropological Society* London 1 (1863–64): 335–420.

Bliss, Daniel. *Al-durus al-awliya fi al-falsafa al-ʿaqliya* [Primary lessons in rational philosophy]. Beirut: n.p., 1873.

Blunt, Wilfrid Scawen. *My Diaries*. 2 vols. New York: A. A. Knopf, 1921.

———. *Secret History of the English Occupation of Egypt: Being a Personal Narrative of Events*. London: T. F. Unwin, 1907.

Bromfield, William. *Letters from Egypt and Syria*. London: n.p., 1856.

Büchner, Ludwig. *Force and Matter: Empirico-philosophical Studies, Intelligibly Rendered*. Translated by J. F. Collingwood. London: Trubner, 1864. Translation of *Kraft und Stoff*.

———. *Sechs Vorlesungen über die Darwinische Theorie von der Verwandlung der Arten und die erste Entstehung der Organismenwelt*. Leipzig: T. Thomas, 1868. Translated by Auguste Jacquot as *Conférences sur la théorie darwinienne de la transmutation des espèces et de l'apparition du monde organique* (Leipzig: T. Thomas, 1869).

Burnichon, Joseph. *La Compagnie de Jésus en France: Histoire d'un siècle, 1831–1914*. 3 vols. Paris: G. Beauchesne, 1914–22.

Bustani, Butrus al-. *Aʿmal al-Jamʿiya al-Suriya* [Papers of the Syrian Society]. Beirut: n.p., 1852.

———. *Muhit al-muhit* [Sea of seas]. Beirut: n.p., 1867.

Bustani, Butrus, Eli Smith, and Cornelius Van Dyck. Translation of *Kitab al-ʿahd al-jadid* [New Testament]. Beirut: n.p., 1860.

———. Translation of *Kitab al-ʿahd al-qadim* [Old Testament]. Beirut: American Bible Society, 1864–65.

Chambers, Robert. *Vestiges of the Natural History of Creation*. London: J. Churchill, 1844.

Cheikho, Louis. *La littérature arabe au XIXe siècle*. 2 vols. Beirut: Imprimerie Catholique, 1908–10.

Cromer, (Evelyn Baring) Earl of. *Modern Egypt*. 2 vols. London: Macmillan, 1908.

Crozier, John Beattie. *Civilization and Progress*. London: Longmans, 1888.

Curtis, William E. *Today in Syria and Palestine*. Chicago: F. H. Revell, 1903.

Darwin, Charles. *The Descent of Man, and Selection in Relation to Sex*. London: John Murray, 1871.

———. *The Expression of the Emotions of Man and the Animals*. London: John Murray, 1872.

———. *On Evolution: The Development of the Theory of Natural Selection*. Edited by Thomas Glick and David Kohn. Indianapolis: Hackett, 1996.

———. *The Origin of Species by Means of Natural Selection; or, The Preservation of Favoured Races in the Struggle for Life*. 1859. London: John Murray, 1872.

Darwin Correspondence Project. Online database at www.darwinproject.ac.uk.

De Guerville, A. B. *New Egypt*. London: E. P. Dutton, 1906.

Demolins, Edmond. *À quoi tient la supériorité des Anglo-Saxons*. Paris: Libraire de Paris, 1897. Translated by Louis B. Lavigne as *Anglo-Saxon Superiority: To What It Is Due* (New York: Scribner, 1898).

Dennis, James. *Christian Missions and Social Progress: A Sociological Study of Foreign Missions*. 3 vols. New York: Fleming H. Revell, 1899–1906.

———. *A Sketch of the Syria Mission*. New York: Mission House, 1872.

Dozy, Reinhart. *Supplément aux dictionnaires arabes*. Leiden: Brill, 1881.

"Education in Syria." *Hours at Home: A Popular Monthly of Instruction and Recreation* 11 (1870): 131–34, 244–51, 327–31.

"An Egyptian View of the Egyptian Question: An Interview with an Egyptian Editor." *Pall Mall Gazette* 8289 (14 October 1891).

Fisk, Pliny. *The Holy Land, an Interesting Field of Missionary Enterprise*. Boston: Samuel T. Armstrong, 1819.

Fiske, John. *Darwinism and Other Essays*. 1879. New ed., Boston: Houghton Mifflin, 1885.

———. *Destiny of Man Viewed in the Light of His Origin*. Boston: Houghton Mifflin, 1889.

———. *Outlines of Cosmic Philosophy Based on the Doctrine of Evolution with Criticisms of the Positive Philosophy*. 2 vols. New York: Houghton Mifflin, 1874.

Frédault, Felix. *Forme et matière*. Paris: E. Vaton, 1876.

Ghalib, ʿUthman. *ʿIlm al-hayawan* [Zoology]. Cairo: n.p., 1886.

Ghazzi, Kamil al-. *Nahr al-dhahab fi tarikh halab* [Golden river of Aleppine history]. Aleppo: Al-Matbaʿa al-Maruniya, 1924.

Goodell, William. *The Old and the New: On the Changes of Thirty Years in the East with Some Allusions to Oriental Customs as Elucidating the Scriptures*. New York: M. W. Dodd, 1853.

Guizot, François. *History of Civilisation in Europe*. 2 vols. London: David Bogue, 1846. Translation of *Histoire de la civilisation en Europe*.

Haddad, Niqula. *Al-ishtirakiya* [Socialism]. Cairo: Matbaʿat al-Hilal, 1920.

Haeckel, Ernst. *The History of Creation; or, The Development of the Earth and Its Inhabitants by the Action of Natural Causes*. Translated by E. Lankester. New York: D. Appleton, 1876.

———. *Last Words on Evolution: A Popular Retrospect and Summary*. Translated by Joseph McCabe. London: A. Owen, 1906. Translation of *Der Kampf um den Entwicklungsgedanken* (1905).

Haldane, J. B. S. "Science and Ethics." *Harper's Magazine* 157 (1928): 1–10.

Hanna, Sami, and George Gardner, eds. *Arab Socialism: A Documentary Survey*. Leiden: Brill, 1969.

Hanotaux, Gabriel, and Muhammad ʿAbduh. *L'Europe et l'Islam*. Cairo: Impr. J. Politis, 1905.

Hartmann, Martin. *The Arabic Press of Egypt*. London: Luzac, 1899.

Hobhouse, Leonard T. *Social Evolution and Political Theory*. New York: Columbia University Press, 1911.

Hodge, Charles. *What Is Darwinism?* New York: Scribner, Armstrong, 1874.

Hurani, Ibrahim. *Al-ayat al-bayyinat fi ghara'ib al-ard wa al-samawat* [Certain signs in the wonders of the heavens and earth]. Beirut: n.p., 1883.

———. *Manahij al-hukamaʾ fi nafy al-nushuʾ wa-al-irtiqaʾ* [A philosophical program for the refutation of evolution]. Beirut: n.p., 1884.

Hurani, Ibrahim al-. *Al-haqq al-yaqin fi al-radd ʿala butl Darwin* [The certain truth in response to Darwin's errors]. Beirut: n.p., 1886.

Husayn, Hasan. *Fasl al-maqal fi falsafat al-nushuʾ wa-al-irtiqaʾ* [The definitive treatise on the philosophy of evolution and progress]. Cairo: n.p., 1924. Translation of Ernst Haeckel, *Last Words on Evolution: A Popular Retrospect and Summary*, translated by Joseph McCabe (London: A. Owen, 1906).

Huxley, Thomas Henry. *Collected Essays*. 5 vols. Cambridge: Cambridge University Press, 2011.

Ibn Khaldun. *Al-muqaddima*. Translated by Franz Rosenthal. Princeton, NJ: Princeton University Press, 1969.

Ikhwan al-Safa'. *Epistles of the Brethren of Purity*. Edited by Nader El-Bizri. Oxford: Oxford University Press, 2008.

'Illish, Muhammad Ibn Ahmad. *Fath al-ʿali al-malik fi al-fatawa ʿala madhhab al-Imam Malik* [The princely key to the fatwa of the Maliki School]. Cairo: Al-Maktaba al-Tijariya al-Kubra, 1937.

Iskandarani, Muhammad Ibn Ahmad al-. *Kashf al-asrar al-nuraniya al-Qur'aniya fima yataʿallaqu bi-al-ajram al-samawiya wa-al-ardiya wa-al-hayawanat wa-al-nabatat wa-al-maʿdaniya* [Unveiling the luminous secrets of the Qur'an and their relation to the heavens, earth, animals, plants, and minerals]. Cairo: Al-Matbaʿa al-Wahbiya, 1880.

———. *Kitab al-asrar al-rabbaniya fi al-nabatat wa-al-maʿadin wa-al khawass al-hayawaniya* [The book of the divine secrets of plants, minerals, and animal characteristics]. Damascus: Matbaʿat Majlis al-Maʿarif, 1882.

———. *Kitab al-barahin al-bayyinat fi bayan haqa'iq al-hayawanat* [The book of certain proofs revealing animal truths]. Damascus: Matbaʿat Majlis al-Maʿarif, 1886.

Jallad, Filib Ibn Yusuf. *Al-qamus al-ʿamm li-al-idara wa-al-qada'* [General dictionary of administration and legislation]. 6 vols. Alexandria: Al-Matbaʿa al-Buhariya, 1890–1902.

Jawhari, Tantawi. *Bahjat al-ʿulum fi al-falsafa al-ʿarabiya wa-muwazanatih bi-al-ʿulum al-ʿasriya* [Delights of the Arabic philosophical sciences and their relation to the contemporary sciences]. Cairo: n.p., 1936.

———. *Baraʿat al-ʿAbbasa ukht Rashid* [The innocence of ʿAbbasa, sister of Rashid]. Cairo: Mustafa al-Babi al-Halabi, 1936.

———. *Al-jawahir fi tafsir al-Qur'an al-karim: Al-mushtamil ʿala ʿajaib badaiʿ al-mukawwanat wa-ghara'ib al-ayat al-bahirat* [Gems of Qur'anic exegesis]. 26 vols. Cairo: Matbaʿat Mustafa al-Babi al-Halabi, 1922–35.

———. *Kitab al-arwah* [Book of spirits]. Cairo: Matbaʿat al-Saʿada, 1920.

———. *Kitab ayna al-insan* [The book of man]. Cairo: Maktabat wa Matbaʿat al-Maʿarif, 1911.

———. *Al-nizam wa-al-Islam* [Governance and Islam]. Cairo: Matbaʿat al-Jumhur, 1903.

Jessup, Henry. *Fifty Three Years in Syria*. New York: Fleming H. Revell, 1910.

———. *The Mohammedan Missionary Problem*. Philadelphia: Presbyterian Board of Publication, 1879.

———. *Syrian Home Life*. New York: Dodd and Mead, 1874.

———. *The Women of the Arabs*. New York: Dodd and Mead, 1873.

Jisr, Husayn al-. *Kitab nuzhat al-fikr* [Intellectual diversions]. Beirut: Matbaʿa al-Adaniya, 1889.

———. *Al-risala al-Hamidiya fi haqiqat al-diyana al-Islamiya wa-haqiqat al-shariʿa al-Muhammadiya* [A Hamidian treatise on the truth of Islam and the shariʿa]. With an introduction by Muhammad al-Jisr. Cairo: Idarat al-Tibaʿa al-Muniriya, 1933.

———. *Al-risala al-Hamidiya fi haqiqat al-diyana al-Islamiya wa haqiqat al-shariʿa al-Muhammadiya* [A Hamidian treatise on the truth of Islam and the shariʿa]. Beirut: Majlis al-Maʿarif, 1888; Cairo: Al-Matbaʿa al-Hamidiya al-Misriya, AH 1322.

Jowett, William. *Christian Researches in Syria and the Holy Land, in 1823 and 1824*. Boston: Crocker and Brewster, 1826.

Jullien, Michel. *La nouvelle mission de la Compagnie de Jésus en Syrie, 1831–1895*. Tours: Mame, 1898.

Khalidi, ʿAbbud. *Al-ʿaql wa-al-khalq*. Baghdad: Dar al-Hawra, 2006.

Khuwarizmi, Muhammad Ibn Ahmad. *Mafatih al-ʿulum* [Keys of knowledge]. Cairo: n.p., n.d.

Kropotkin, Pyotr. *Mutual Aid: A Factor of Evolution*. New York: McClure Phillips, 1904.

Le Bon, Gustave. *Ruh al-ishtirakiya* [The spirit of socialism]. Translated by Muhammad ʿAdil Zuʿaytar. 1925. Cairo: Matbaʿat al-ʿAsriyah, 1970. Translation of *Psychologie du socialisme*.

———. *Sirr tatawwur al-umam* [Secret of the evolution of nations]. Translated by Ahmad Fathi Zaghlul. Cairo: Matbaʿat al-Maʿarifa, 1913. Translation of *Les lois psychologiques de l'évolution des peuples*.

Lubbock, John. *Chapters in Popular Natural History*. London: National Society, 1883.

———. *On the Senses, Instincts and Intelligence of Animals*. London: Keegan, Paul, Trench, 1888.

———. *The Origin of Civilisation and the Primitive Conditions of Man: Mental and Social Condition of Savages*. London: Longmans, Green, 1870.

MacDonald, Ramsay. *Socialism and Society*. London: Independent Labour Party, 1905.

MacPherson, Hector Carsewell. *Spencer and Spencerism*. New York: Doubleday, Page, 1900.

Mansuri, Mustafa al-. *Al-taqaddum wa-al-faqr* [Translation of Henry George's *Progress and Poverty*]. Cairo: Matbaʿat al-Saʿada, 1919.

———. *Tarikh al-madhahab al-ishtirakiya* [History of socialisms]. Cairo: n.p., 1915.

Manzur, Muhammad Ibn. *Lisan al-ʿArab* [The Arabic tongue]. Cairo: n.p., n.d.

Marrash, Fransis Fath Allah al-. *Ghabat al-haqq* [The forest of truth]. Aleppo: n.p., 1865.

———. *Mashhad al-ahwal* [A view on the state of things]. Beirut: Al-Matbaʿa al-Kulliya, 1883.

———. *Al-miraʾat al-safiya fi al-mabadi al-tabiʿa* [The clear mirror of the principles of nature]. Aleppo: n.p., 1861.

———. *Rihlat Baris* [A journey to Paris]. Beirut: Al-Matbaʿa al-Sharifiya, 1867.

———. *Shahadat al-tabiʿa fi wujud Allah wa-al-shariʿa* [Nature's testimonies to the existence of God and divine law]. Beirut: Maktabat al-Amrikan, 1891.

Marriott, John. *The Eastern Question: A Historical Study in European Diplomacy*. Oxford: Clarendon Press, 1917.

Marsafi, Husayn al-. *Risalat al-kalim al-thaman* [Discourse on eight words]. Cairo: n.p., 1881.

Mazhar, Ismaʿil. *Asl al-anwaʿ wa nushuʾiha bi al-intikhab al-tabiʿi wa hifz al-sufuf al-ghaliba fi al-tanahur ʿala al-baqaʾ* [The origin of species and their evolution by natural selection and the preservation of the victorious orders in the struggle for life]. 2 vols. Cairo: Al-Matbaʿa al-Misriya, 1918. Partial translation of Charles Darwin's *Origin of Species*.

———. *Asl al-anwaʿ wa nushuʾiha bi al-intikhab al-tabiʿi wa hifz al-sufuf al-ghalibah fi al-tanahur ʿala al-baqaʿ* [The origin of species and their evolution by natural selection and the preservation of the victorious orders in the struggle for life]. 2 vols. Cairo: Dar al-ʿUsur, 1928. Partial translation of Charles Darwin's *Origin of Species*.

———. *Bayn al-ʿilm wa-al-din* [Between religion and science]. Cairo: Dar al-ʿUsur, 1930. Partial translation of Andrew Dickson White's *History of the Warfare of Science with Theology in Christendom*.

———. *Malqa al-sabil fi madhhab al-nushuʾ wa-al-irtiqaʾ wa-atharuhu fi al-fikr al-hadith* [Evolution at the crossroads: The theory of evolution and its impact on modern thought]. Cairo: Al-Matba ʿa al-ʿAsriya, 1924.

———. *Mashruʿ li-taʾsis hizb al-fallah al-Misri* [Proposal for the formation of an Egyptian peasants' party]. Cairo: Dar al-ʿUsur, 1929.

———. *Nahda Dictionary*. Cairo: Renaissance Bookshop, 1950.

———. *Risalat al-fikr al-hurr* [Essays on the freedom of thought]. Beirut: Dar al-Thaqafa, 1963.

———. *Tajdid al-ʿArabiya* [The revival of Arabic]. Cairo: Maktabat al-Nahda al-Misriya, 1950.

———. *Tarikh al-fikr al-ʿArabi* [History of Arabic thought]. Cairo: Dar al-ʿUsur li-al-Tabaʿ wa-al-Nashr, 1928.

———. *Wathbat al-sharq*. [Advance of the East]. Cairo: Dar al-ʿUsur, 1929.

McCosh, James. *Christianity and Positivism: A Series of Lectures to the Times on Natural Theology and Apologetics*. New York: R. Carter and Brothers, 1871.

———. *The Religious Aspects of Evolution*. New York: G. P. Putnam's Sons, 1888.

Merz, John Theodore. *A History of European Thought in the Nineteenth Century*. Edinburgh: Blackwood and Sons, 1896.

Ministère des Affaires Étrangères. *Documents diplomatiques: Affaire du journal Le Bosphore Égyptien*. Paris: Imprimerie nationale, 1885.

Mivart, George Jackson. *Man and Apes*. New York: D. Appleton, 1874.

———. *On the Genesis of Species*. London: Macmillan, 1871.

Mubarak, ʿAli. *Al-khitat al-tawfiqiya al-jadida li-Misr al-Qahira* [The new plans of (Khedive) Tawfiq for Cairo, Egypt]. 20 vols. Cairo: Al-Matbaʿa al-Kubr al-Amiriya, 1886–89. Reprint, Cairo: Al-Haya al-Misriya al-ʿAmma li-al-Kitab, 1980–88.

Musa, Salama. *Al-adab li-al-shaʿb* [Literature for the people]. Cairo: n.p., 1956.

———. *Ahlam al-falasifa* [Dreams of the philosophers]. Cairo: Matbaʿat al-Hilal, 1926.

———. *Al-balagha al-ʿasriya wa-al-lugha al-ʿArabiya* [Contemporary rhetoric and the Arabic language]. Cairo: Matbaʿat al-ʿAsriya, 1940.

———. *The Education of Salama Musa*. Translated by L. O. Schuman. Leiden: Brill, 1961. Translation of *Tarbiyat Salama Musa*.

———. *Al-ishtirakiya* [Socialism]. Translated in *Arab Socialism: A Documentary Survey*, edited by Sami Hanna and George Gardner (Leiden: Brill, 1969), 273–88.

———. *Kitab al-thawrat* [The book of revolutions]. Cairo: n.p., 1954.

———. *Ma hiya al-Nahda?* [What is the Nahda?]. 1935. Beirut: Maktabat al-Maʿarif, 1962.

———. *Nazariyat al-tatawwur wa-asl al-insan* [The theory of evolution and the origin of man]. Cairo: Matbaʿat al-ʿAsriya, 1925.

Nabhani, Yusuf al-. *Mukhtasar irshad al-hayara fi tahdhir al-Muslimin min madaris al-Nasara* [Concise advice to the perplexed: Warning Muslims of the Christian schools]. Beirut: n.p., 1901.

O'Leary, Evans DeLacy. *Islam at the Crossroads*. London: K. Paul, Trench, Trubner, 1923.

———. *Arabic Thought and Its Place in History*. London: Kegan, Paul, Trench, Trubner, 1922.

Oltramane, M. "Sensibility and Its Diverse Forms." *Science* 2 (21 May 1881): 230–32.

Post, George. *Mabadi ʿilm al-nabat* [Principles of botany]. Beirut: n.p., 1871.

———. *Nizam al-halaqat fi silsilat dhawat al-faqarat* [Classification of vertebrates]. Beirut: n.p., 1869.

Qutb, Sayyid. *In the Shade of the Qurʾan*. Translated by M. A. Salahi and A. A. Shamsi. Leicester: Islamic Foundation, 1999.

Rabbath, Antoine. *Documents inédits pour servir à l'histoire du christianisme en Orient*. 2 vols. Paris: A. Picard et fils, 1905–14.

Rauschenbusch, Walter. *A Theology for the Social Gospel*. New York: Abingdon Press, 1917.

Renan, Ernest. *Averroès et l'averroïsme: Essai historique*. Paris: Michel Lévy Frères, 1866.

———. "L'islamisme et la science." Sorbonne lecture originally delivered in 1882 and reprinted in *Œures complètes* (Paris: Calmann-Levy, 1947).

Rida, Rashid. *Al-Manar wa al-Azhar*. Cairo: Matbaʿat al-Manar, 1934.

———. *Tarikh al-ustadh al-imam Muhammad ʿAbduh* [History of Imam Muhammad ʿAbduh]. 3 vols. Cairo: Matbaʿat al-Manar, 1908–31.

Rustum, Muhammad. *Al-hadara al-Misriya* [Egyptian civilization]. Cairo: Al-Matbaʿa al-ʿAsriya, 1924. Partial translation of Gustave Le Bon's *Les premières civilisations*.

Salman, Abd al-Karim. *Aʿmal majlis idarat Al-Azhar: Min ibtida' ta-sisihi sana 1312 ila ghayat sana 1322* [Reports of the Administrative Reform Committee at Al-Azhar from 1895 to 1905]. Cairo: n.p., 1905.

Sayhi, Ibrahim Muhammad. *Falsafat al-takwin* [Philosophy of creation]. Alexandria: Al-Matbaʿa al-Misriya, 1920.

Sfayr, Jirjis. *Kitab fi asl al-insan wa-al-ka'inat*. Beirut: Matbaʿat al-Aba al-Yausyin, 1890.

Shadudi, As'ad Ibrahim al-. *Al-ʿarus al-badiʿa fi ʿilm al-tabiʿa* [The marvelous bride of natural science]. Beirut: n.p., 1873.

Shafiq, Ahmad Pasha. *Mudhakkirati fi nisf qarn* [Memoirs of half a century]. 2 vols. Cairo: Matbaʿat Misr, 1934–37.

Shahin, ʿAbd al-Sabur. *Abi Adam* [My father Adam]. Cairo: Maktabat al-Shabab, 1998.

Shaw, George Bernard, ed. *Fabian Essays in Socialism*. London: W. Scott, 1889.

———. *Man and Superman*. London: Constable, 1903.

Shumayyil, Shibli. *Falsafat al-nushu' wa-al-irtiqa'* [Philosophy of evolution and progress]. Cairo: Matbaʿat al-Muqtataf, 1910.

———. *Al-haqiqa* [The truth]. Cairo: n.p., 1885. Reprinted in *Falsafat al-nuhu wa-al-irtiqa'*, vol. 1 (Cairo: Matbaʿat al-Muqtataf, 1910).

———. *Hawadith wa-khawatir* [Conversations and speeches]. Beirut: Dar al-Hamra, 1991.

———. *Majmuʿat al-Duktur Shibli Shumayyil* [Works of Dr. Shibli Shumayyil]. Cairo: Matbaʿat al-Muqtataf, 1910.

———. *Majmuʿat al-Duktur Shibli Shumayyil* [Works of Dr. Shibli Shumayyil]. Beirut: Dar Nazir ʿAbbud, 1983.

———. *Taʿrib li-sharh Bukhnir ʿala madhhab Darwin fi intiqal al-anwaʿ wa-zuhur al-ʿalam al-ʿudwi wa-itlaq dhalika ʿala al-insan maʿa baʿdi tasarrufin fih* [A translation of Büchner's commentaries on Darwin on the transformation of species and the emergence of the organismic world and, from that, man]. Cairo: Matbaʿat Jaridat al-Mahrusa, 1884.

Smiles, Samuel. *Self-Help: With Illustrations of Character, Conduct and Perseverance*. 1859. Chicago: Belford, Clarke, 1881.

Spencer, Herbert. *Education: Intellectual, Moral and Physical*. 1861. New York: D. Appleton, 1864.

———. *Principles of Psychology*. London: Longman, Brown, Green and Longmans, 1855.

———. *The Principles of Sociology*. 3 vols. 1876. New York: D. Appleton, 1898–99.

———. *Social Statics; or, The Conditions Essential to Human Happiness Specified, and the First of Them Developed*. London: John Chapman, 1851.

———. *The Study of Sociology*. London: Kegan, Paul, Trench, Trubner, 1873.

Syrian Protestant College. *Annual Reports, Board of Managers, Syrian Protestant College, 1866–67—1901–02*. Beirut: American University of Beirut Press, 1963.

———. *Catalogue of the Syrian Protestant College*. 2 vols. Beirut: Syrian Protestant College Press, 1887–88.

Tahtawi, Rifaʿa Rafiʿ al-. *Takhlis al-ibriz fi talkhis Bariz* [Purification of ore in Parisian lore]. Cairo: Bulaq, 1834. Translated by Daniel Newman as *An Imam in Paris* (London: Saqi, 2004).

Tarrazi, Philippe. *Tarikh al-sihafa al-ʿArabiya* [History of the Arabic press]. 4 vols. Beirut: Dar Sadir, 1913–29.

Temple, Frederick. *The Relations between Religion and Science: Eight Lectures Preached before the University of Oxford*. New York: Macmillan, 1884.

Tylor, E. B. *Primitive Culture: Researches into the Development of Mythology, Philosophy, Religion, Language, Art, and Custom*. 1871. New York: Holt, 1889.

Tyndall, John. "The Belfast Address." *Nature*, 20 August 1874, 172–89.

———. "Spontaneous Generation." *Nineteenth Century* 3 (January 1878): 22–47.

Université Saint-Joseph. *Faculté de medecine, 1883–1908*. Beirut: Jesuit Press, 1908.

———. *Les Jésuites en Syrie, 1831–1931*. Paris: Jesuit Press, 1931.

Van Dyck, Cornelius. *Kitab al-miraʾ al-wadiya fi al-kura al-ardiya* [Riparian reflections of the terrestrial globe]. Beirut: n.p., 1852.

———. *Kitab al-naqsh fi al-hajar* [Discourse on geology]. 8 vols. Beirut: Al-Matbaʿa al-Adabiya, 1886–87.

———. *Kitab fi usul al-handasiya* [Euclid's geometry]. Beirut: Matbaʿat al-Amrikan, 1857.

———. *Usul al-kimiya* [Principles of chemistry]. Beirut: n.p., 1869.

———. *Usul ʿilm al-haya* [Principles of astronomy]. Beirut: n.p., 1874.

Wajdi, Muhammad Farid. *ʿAla atlal al-madhhab al-maddi* [On the refutation of materialism]. 4 vols. Cairo: Matbaʿat Dairat Maʿarif al-Qarn al-ʿIshrin, 1921.

———. *Al-madaniya wa-al-Islam* [Civilization and Islam]. Cairo: Matbaʿat al-Waiz, 1901.

Wallace, Alfred Russel. "Man's Place in the Universe." *Independent* 55 (1903): 473–83.

White, Andrew Dickson. *Autobiography*. 2 vols. New York: Dent, 1907.

———. *Autobiography*. 2 vols. New York: Century, 1922.

———. *History of the Warfare of Science with Theology in Christendom*. New York: D. Appleton, 1896.

Zaghlul, Ahmed Fathi. *Sirr taqaddum al-Inkiliz al-Saksuniyin* [The secret of the progress of the Anglo-Saxons]. Translation of Edmond Demolins's *À quoi tient la superiorité des Anglo-Saxons*. Cairo: Matbaʿat al-Jamaliya, 1899.

Zawahiri, Muhammad al-Ahmadi al-, and Fakhr al-Din al-Ahmadi. *Al-ʿilm wa-al-ʿulama wa-nizam al-tadris* [Science and the *ʿulama* and pedagogy]. Cairo: n.p., 1904.

Zaydan, Jurji. *Autobiography of Jurji Zaydan: Including Four Letters to His Son*. Translated by Thomas Philipp. Washington, DC: Three Continents Press, 1990.

———. *Tabaqat al-umam* [Categories of peoples]. Cairo: Matbaʿat al-Hilal, 1912.

———. *Tarajim mashahir al-sharq* [Biographies of famous men of the East]. 2 vols. Cairo: Matbaʿat al-Hilal, 1910–11.

———. *Tarikh adab al-lugha al-ʿArabiya* [The history of Arabic literature]. Cairo: Matbaʿat al-Hilal, 1914.

———. *Tarikh al-lugha al-ʿArabiya: Bi-iʿitibar anaha kaʿin hay* [History of the Arabic language as a living entity]. Cairo: Matbaʿat al-Hilal, 1904.

———. *Tarikh al-tamaddun al-Islami* [History of Islamic civilization]. Cairo: Matbaʿat al-Hilal, 1902–6.

Zilzal, Bishara. *Tanwir al-adhhan* [The evolution of mind]. Alexandria: n.p., 1879.

SECONDARY SOURCES

Abattouy, Jürgen Renn, and Paul Weinig. "Transmission as Transformation: The Translation Movements in the Medieval East and West in a Comparative Perspective." *Science in Context* 14 (2001): 1–12.

ʿAbd al-Raziq, Mustafa. *Muhammad ʿAbduh*, Cairo: Dar al-Maʿarif li-al-Tibaʿa, 1946.

Abdelghani, Ahmed Bioud. *3200 Revues et journaux arabes de 1800 à 1965*. Paris: Bibliothèque Nationale, 1969.

Abdo al-Khoury, Nabeel, and Abdo Baaklani. "Muhammad ʿAbduh: An Ideology of Development." *Muslim World* 69 (1979): 42–52.

Abi-Mershed, Osama. *Apostles of Modernity: Saint-Simonians and the Civilizing Mission in Algeria*. Stanford, CA: Stanford University Press, 2010.

Abu ʿArja, Taysir. *Al-Muqattam: Jaridat al-ihtilal al-Baritani fi Misr, 1889–1952* [*Al-Muqattam*: The newspaper of the British occupation in Egypt, 1889–1952]. Cairo: Al-Hayʾa al-Misriya al-ʿAmma li-al-Kitab, 1997.

Abu Al Eyun, Mahmud. *Al-Azhar: A Short Historical Survey*. Translated by Mostafa Taha Habib. Cairo: Al-Azhar Press, 1949.

Abu Diyah, Ayyub ʿIsa. *Ismaʿil Mazhar, 1891–1962: Min al-ishtirakiya ila al-Islam* [Ismaʿil Mazhar, 1891–1962: From Socialism to Islam]. Amman: Dar Ward li-al-Nashr wa-al-Tawziʿ, 2005.

Abu-Manneh, Butrus. "The Christians between Ottomanism and Syrian Nationalism: The Ideas of Butrus al-Bustani." *International Journal of Middle East Studies* 11 (1980): 287–304.

———. "The Genesis of Midhat Pasha's Governorship in Syria, 1878–1880." In *The Syrian Land: Processes of Integration and Fragmentation: Bilād al-Shām from the 18th to the 20th Century*, edited by Thomas Philipp and Birgit Schaebler, 251–67. Stuttgart: F. Steiner, 1998.

———. "Sultan Abdulhamid II and Shaikh Abulhuda al-Sayyadi." *Middle Eastern Studies* 15 (1979): 131–53.

Adams, Charles. *Islam and Modernism in Egypt: A Study of the Modern Reform Movement Inaugurated by Muhammad ʿAbduh*. London: Oxford University Press, 1933.

———. "Muhammad ʿAbduh and the Transvaal Fatwa." In *The Macdonald Presentation Volume: A Tribute to Duncan Black Macdonald*, 11–30. Princeton, NJ: Princeton University Press, 1933.

Ahmed, Jamal Mohammed. *The Intellectual Origins of Egyptian Nationalism*. London: Oxford University Press, 1960.

Aliotta, Antonio. *The Idealist Reaction against Science*. London: Macmillan, 1914.

Allan, Michael. "The Limits of Secular Criticism: Reflections on Literary Reading in a Colonial Frame." *Townsend Center for the Humanities Newsletter*, February 2007.

Amin, Osman. "Renaissance in Egypt: Muhammad ʿAbduh and His School." In *A History of Muslim Philosophy*, edited by M. M. Sharif, 1497–99. Karachi: Royal Book, 1983.

Amin, ʿUthman. *Muhammad ʿAbduh: Essai sur ses ideés philosophiques et religieuses*. Cairo: Imprimerie Misr, 1944.

Anderson, Lisa. "Nineteenth Century Reform in Ottoman Libya." *International Journal of Middle East Studies* 16 (1984): 325–48.

Antonius, George. *The Arab Awakening: The Story of the Arab National Movement*. London: H. Hamilton, 1938. Reprint, New York: Capricorn Books, 1965.

Aroian, Lois Arminé. *The Nationalization of Arabic and Islamic Education in Egypt: Dar al-ʿUlum and Al-Azhar*. Cairo: American University in Cairo Press, 1983.

Asad, Talal. *Formations of the Secular: Christianity, Islam, Modernity*. Stanford, CA: Stanford University Press, 2003.

———. *Genealogies of Religion: Discipline and Reasons of Power in Christianity and Islam*. Baltimore, MD: John Hopkins University Press, 1993.

Asad, Talal, Judith Butler, Saba Mahmoud, and Wendy Brown. *Is Critique Secular? Blas-*

phemy, Injury and Free Speech. Berkeley and Los Angeles: University of California Press, 2009.

Auchterlonie, Paul. "From the Eastern Question to the Death of General
Gordon: Representations of the Middle East in the Victorian Periodical Press, 1876–1885." *British Journal of Middle Eastern Studies* 28 (2001): 5–24.

Ayalon, Ami. "Modern Texts and Their Readers in Late Ottoman Palestine." *Middle Eastern Studies* 38 (2002): 17–40.

———. *The Press in the Arab Middle East: A History*. New York: Oxford University Press, 1995.

Aydin, Cemil. "Beyond Civilization: Pan-Islamism and Pan-Asianism and the Revolt against the West." *Journal of Modern European History* 4 (2006): 204–23.

Azmeh, Aziz al-. *Islams and Modernities*. London: Verso, 1993.

Badawi, M. Zaki. *The Reformers of Egypt*. London: Croom Helm, 1976.

Badrawi, Malak. *Political Violence in Egypt, 1910–1924: Secret Societies, Plots and Assassinations*. Richmond, Surrey: Curzon, 2000.

Baraka, Magda. *The Egyptian Upper Class between Revolutions: 1919–1952*. Reading, UK: Ithaca Press, 1998.

Barbanell, Maurice. *Philosophy of Silver Birch*. London: Spiritualist Press, 1969.

Bartlett, Samuel. *Historical Sketches of the Missions of the American Board*. Boston: The Board, 1876.

Basalla, George. "The Spread of Western Science: A Three-Stage Model Describes the Introduction of Modern Science into Any Non-European Nation." *Science* 156 (1967): 611–22.

Bashford, Alison, and Philippa Levine, eds. *The Oxford Handbook of the History of Eugenics*. Oxford: Oxford University Press, 2010.

Bayly, Chris. *Empire and Information: Intelligence Gathering and Social Communication in India, 1780–1870*. Cambridge: Cambridge University Press, 1996.

Beidelman, T. O. *Colonial Evangelism*. Bloomington: Indiana University Press, 1982.

Beinin, Joel. "Formation of the Egyptian Working Class." *MERIP Reports* 94 (1981): 14–23.

Bell, Duncan. *The Idea of Greater Britain: Empire and the Future of World Order, 1860–1900*. Princeton, NJ: Princeton University Press, 2007.

Bellomy, Donald C. "'Social Darwinism' Revisited." *Perspectives in American History* 1 (1984): 1–129.

Bennett, Scott. "Revolutions in Thought: Serial Publications and the Mass Market for Readings." In *The Victorian Periodical Press: Samplings and Soundings*, edited by Joanne Shatock and Michael Wolff, 225–60. Leicester: Leicester University Press, 1982.

Berger, Lutz. "Esprits et microbes: L'interprétation des *ginns* dans quelques commentaires coraniques du XXe siècle." *Arabica* 47 (2000): 554–62.

Berger, Maurits S. "Apostasy and Public Policy in Contemporary Egypt: An Evaluation of Recent Cases from Egypt's Highest Courts." *Human Rights Quarterly* 25 (2003): 720–40.

Bezirgan, Najm. "The Islamic World." In *The Comparative Reception of Darwinism*, edited by Thomas Glick, 375–87. Austin: University of Texas Press, 1972.

Bowden, Brett. *Empire of Civilization: The Evolution of an Imperial Idea*. Chicago: University of Chicago Press, 2009.

Bowler, Peter. "Darwinism and the Argument from Design: Suggestions for a Reevaluation." *Journal of the History of Biology* 10 (1977): 29–43.

———. *Evolution: The History of an Idea*. Berkeley and Los Angeles: University of California Press, 1984.

———. *Fossils and Progress: Paleontology and the Idea of Progressive Evolution in the Nineteenth Century*. New York: Science History, 1976.

Bozeman, Theodore Dwight. *Protestants in an Age of Science: The Baconian Ideal and Antebellum American Religious Thought*. Chapel Hill: University of North Carolina Press, 1977.

Broadley, Alexander Meyrick. *How We Defended Arabi and His Friends: A Story of Egypt and the Egyptians*. London: Chapman and Hall, 1884.

Brophy, James M. "The Common Reader in the Rhineland: The Calendar as Political Primer in the Early Nineteenth Century." *Past and Present* 185 (2004): 119–59.

Brown, C. Mackenzie. "Western Roots of Avataric Evolutionism in Colonial India." *Zygon* 42 (2007); 423–48.

Browne, Janet. "Biogeography and Empire." In *Cultures of Natural History: From Curiosity to Crisis*, edited by Nicholas Jardine, James Secord, and Emma Spary, 305–21. Cambridge: Cambridge University Press, 1996.

Brunner, Rainer. "Education, Politics and the Struggle for Intellectual Leadership: Al-Azhar between 1927 and 1945." In *Guardians of Faith in Modern Times*, edited by Meir Hatim, 109–40. Leiden: Brill, 2009.

Burchfield, Joe. *Lord Kelvin and the Age of the Earth*. New York: Science History, 1975.

Burrow, J. W. *The Crisis of Reason: European Thought, 1848–1914*. New Haven, CT: Yale University Press, 2000.

Byrne, Peter. *Natural Religion and the Nature of Religion: The Legacy of Deism*. London: Routledge, 1989.

Campbell, Mary Baine, Lorraine Daston, Arnold Davidson, John Forrester, and Simon Goldhill. "Enlightenment Now: Concluding Reflections on Knowledge and Belief." *Common Knowledge* 13 (2007): 429–50.

Cannon, Byron. "Nineteenth-Century Arabic Writings on Women and Society: The Interim Role of the Masonic Press in Cairo (*al-Lataif*, 1885–1895)." *International Journal of Middle East Studies* 17 (1985): 463–84.

Chakrabarti, Pratik. *Western Science in Modern India: Metropolitan Methods, Colonial Practices*. Delhi: Permanent Black, 2004.

Chalcraft, John. "Coalheavers of Port Said: State Making and Worker Protest, 1869–1914." *International Labour and Working Class History* 60 (2001): 110–24.

Chambers, Richard L. "The Education of a Nineteenth-Century Ottoman Alim: Ahmed Cevdet Pasa Alim." *International Journal of Middle East Studies* 4 (1973): 440–64.

Chattopadhyay, Suchetana. "The Bolshevik Menace: Colonial Surveillance and the Origins of Socialist Politics in Calcutta." *South Asia Research* 26 (2006): 165–79.

Chejne, Anwar. *The Arabic Language: Its Role in History*. Minneapolis: University of Minnesota Press, 1989.

Cioeta, Donald. "Ottoman Censorship in Lebanon and Syria, 1876–1908." *International Journal of Middle East Studies* 10 (1979): 167–86.

Cole, Juan. *Colonialism and Revolution in the Middle East: Social and Cultural Origins of Egypt's ʿUrabi Movement*. Princeton, NJ: Princeton University Press, 1993.

Collini, Stefan. *Liberalism and Sociology: L. T. Hobhouse and the Political Argument in England, 1880–1914*. Cambridge: Cambridge University Press, 1979.

Comaroff, John, and Jean Comaroff. *Of Revelation and Revolution*. Chicago: University of Chicago Press, 1997.

Commins, David Dean. *Islamic Reform: Politics and Social Change in Late Ottoman Syria*. New York: Oxford University Press, 1990.

———. "Religious Reformers and Arabists in Damascus, 1885–1914." *International Journal of Middle East Studies* 18 (1986): 405–25.

Contu, Giuseppe. *Gli aspetti positivi e i limiti del laicismo in Salama Musa (1887–1958)*. Naples: Istituto Orientale di Napoli, 1980.

Cooper, Frederick. *Colonialism in Question*. Berkeley, and Los Angeles: University of California Press, 2005.

Corsi, Pietro. *Science and Religion: Baden Powell and the Anglican Debate, 1800–1860*. Cambridge: Cambridge University Press, 1988

Crecelius, Daniel Neil. "The Ulama and the State in Modern Egypt." PhD diss., Princeton University, 1967.

Crook, Paul. "Social Darwinism: The Concept." *History of European Ideas* 22 (1996): 261–74.

Crozet, Pascal. *Les sciences modernes en Égypte: Transfert et appropriation, 1805–1902*. Paris: Geuthner, 2008.

Cummingham, Andrew, and Perry Williams. "De-centering the 'Big Picture': The Origins of Modern Science and the Modern Origins of Science." *British Journal for the History of Science* 26 (1993): 407–32.

Cunno, Kenneth. "The Origins of Private Ownership of Land in Egypt: A Reappraisal." *International Journal of Middle East Studies* 12 (1980): 245–75.

Daghir, Yusuf. *Masadir al-dirasat al-adabiya* [Sources in literary studies]. Sayda, Lebanon: Matbaʿat Dayr al-Mukhlis, 1950.

Daly, Nicholas. *Literature, Technology and Modernity, 1860–2000*. Cambridge: Cambridge University Press, 2004.

Daston, Lorraine, and Katherine Park. *Wonders and the Order of Nature, 1150–1750*. New York: Zone Books, 1998.

Davis, John. *The Landscape of Belief: Encountering the Holy Land in Nineteenth-Century American Art and Culture*. Princeton, NJ: Princeton University Press, 1996.

Davison, Roderic. "The Advent of the Electric Telegraph in the Ottoman Empire." In *Essays in Ottoman and Turkish History, 1774–1923*, edited by Roderic Davison, 133–65. Austin: University of Texas Press, 1990.

Deguilhem, Randi. "State Civil Education in Late Ottoman Damascus: A Unifying or a Separating Force?" In *The Syrian Land: Processes of Integration and Fragmentation: Bilad al-Sham from the 18th to the 20th Century*, edited by Thomas Philipp and Birgit Schaebler, 221–50. Stuttgart: F. Steiner, 1998.

De Jong, F. "The Works of Tantawi Jawhari (1862–1940): Some Bibliographical and Biographical Notes." *Bibliotheca Orientalis* 34 (1977): 153–61.

Dessouki, Ali. "The Origins of Socialist Thought in Egypt, 1882–1922." PhD diss., McGill University, 1972.

Diab, Henry, and Lars Wahlin. "The Geography of Education in Syria in 1882: With a Translation of 'Education in Syria' by Shahin Makarius." *Geografiska Annaler* 65 (1983): 105–28.

Dixon, Thomas. *The Invention of Altruism: Making Moral Meanings in Victorian Britain*. Oxford: Oxford University Press, 2008.

Dodge, Bayard. *Al-Azhar: A Millennium of Muslim Learning*. Washington, DC: Middle East Institute, 1961.

Dodson, Michael. "Translating Science, Translating Empire: The Power of Language in Colonial North India." *Comparative Studies in Society and History* 47 (2005): 809–35.

Donohue, John, and John Esposito, eds. *Islam in Transition: Muslim Perspectives*. Oxford: Oxford University Press, 1982.

Dupont, Anne-Laure. *Gurgī Zaydan, 1861–1914: Écrivain réformiste et témoin de la renaissance arabe*. Damascus: IFPO, 2006.

———. "L'histoire de l'Islam au regard des autres histoires." *Arabica* 43 (1996): 486–93.

Ebert, Johannes. *Religion und Reform in der arabischen Provinz: Husayn al-Gisr al-Tarablusi (1845–1909)—Ein islamischer Gelehrter zwischen Tradition und Reform*. Frankfurt: P. Lang, 1991.

Eccel, A. Chris. *Egypt, Islam and Social Change: Al-Azhar in Conflict and Accommodation*. Berlin: K. Schwartz, 1984.

Egger, Vernon. *A Fabian in Egypt: Salama Musa and the Rise of the Professional Classes in Egypt, 1909–1939*. Lanham, MD: University Press of America, 1986.

Eich, Thomas. "Abu l-Huda l-Sayydi—Still Such a Polarizing Figure (Response to Itzchak Weismann)." *Arabica* 55 (2008): 433–44.

———. "The Forgotten Salafi: Abu'l-Huda as-Sayyadi." *Die Welt des Islams* 43 (2003): 61–87.

Ellegard, Alvar. *Darwin and the General Reader: The Reception of Darwin's Theory of Evolution in the British Periodical Press, 1859–1872*. Chicago: University of Chicago Press, 1990.

El-Ehwany, Ahmed Fouad. "Present Day Philosophy in Egypt." *Philosophy East and West* 5 (1956): 339–47.

El-Rouayheb, Khaled. *Before Homosexuality in the Arab-Islamic World, 1500–1800*. Chicago: University of Chicago Press, 2005.

———. "Opening the Gate of Verification: The Forgotten Arab-Islamic Florescence of the 17th Century." *International Journal of Middle East Studies* 38 (2006): 263–81.

Elsbree, Oliver Wendell. *The Rise of the Missionary Spirit in America, 1790–1815*. Williamsport, PA: Williamsport Print and Binding, 1928.

Elshakry, Marwa. "The Exegesis of Science in Twentieth Century Arabic Interpretations of the Qur'an." In *Nature and Scripture*, edited by Jitse van der Meer, 491–524. Leiden: Brill, 2009.

———. "Global Darwin: Eastern Enchantment." *Nature* 461 (2009): 1200–1201.

———. "The Gospel of Science and American Evangelism in Late Ottoman Beirut." *Past and Present* 196 (August 2007): 173–214.

———. "Knowledge in Motion: The Cultural Politics of Modern Science Translations in Arabic." *Isis* 99 (2008): 701–30.

———. "When Science Became Western: Historiographical Reflections." *Isis* 101 (2010): 98–109.

El Shakry, Omnia. *The Great Social Laboratory: Subjects of Knowledge in Colonial and Postcolonial Egypt*. Stanford, CA: Stanford University Press, 2007.

———. "Schooled Mothers, Structured Play: Childrearing in Turn of the Century Egypt." In *Remaking Women: Feminism and Modernity in the Middle East*, edited by Lila Abu-Lughod, 126–70. Princeton, NJ: Princeton University Press, 1998.

Escovits, Joseph. "'He Was the Muhammad Abduh of Syria': A Study of Tahir al-Jazairi and His Influence." *International Journal of Middle East Studies* 18 (1986): 293–310.

Esmeir, Samera. *Juridical Humanity: A Colonial History*. Stanford, CA: Stanford University Press, 2012.

Fahmy, Ziad. *Ordinary Egyptians: Creating the Modern Nation through Popular Culture*. Stanford, CA: Stanford University Press, 2011.

Fakhry, Majid. "Classical Islamic Arguments for the Existence of God." *Muslim World* 47 (1957): 133–45.

Fan, Fa-ti. "The Global Turn in the History of Science." *East Asian Science, Technology, and Society* 6 (2012): 249–58.

Farag, Nadia. "The Lewis Affair and the Fortunes of *al-Muqtataf*." *Middle Eastern Studies* 8 (1972): 73–83.

———. "*Al-Muqtataf*, 1876–1900: A Study of the Influence of Victorian Thought on Modern Arabic Thought." PhD diss., University of Oxford, 1969.

Farley, John. "The Spontaneous Generation Controversy (1859–1880)." *Journal of the History of Biology* 5 (1972): 285–319.

Farley, John, and Gerald Geison. "Science, Politics and Spontaneous Generation in Nineteenth Century France: The Pasteur-Pouchet Debate." *Bulletin of the History of Medicine* 48 (1974): 161–98.

Fawaz, Leila. *An Occasion for War: Civil Conflict, in Lebanon and Damascus in 1860*. Berkeley and Los Angeles: University of California Press, 1994.

Field, James. *America and the Mediterranean World*. Princeton, NJ: Princeton University Press, 1969.

Fitzgerald, Rosemary. "'Clinical Christianity': The Emergence of Medical Work as a Missionary Strategy in Colonial India, 1800–1914." In *Health, Medicine and Empire*, edited by Bisamoy Pati and Mark Harrison, 88–136. New Delhi: Orient Longman, 2001.

Forbes, Geraldine Hancock. *Positivism in Bengal: A Case Study in the Transmission and Assimilation of an Ideology*. Calcutta: Minerva Associates, 1975.

Fortna, Benjamin. *Imperial Classroom: Islam, the State, and Education in the Late Ottoman Empire*. Oxford: Oxford University Press, 2002.

———. "Islamic Morality in Late Ottoman 'Secular' Schools." *International Journal of Middle East Studies* 32 (2000): 369–93.

Foucher, Louis. *La philosophie catholique en France au XIXème siècle avant la renaissance thomiste et dans son rapport avec elle, 1800–1880*. Paris: J. Vrin, 1955.

Frank, Richard M. "The Science of *Kalam*." *Arabic Sciences and Philosophy* 2 (1992): 9–11.

Fraser, Nancy. "Rethinking the Public Sphere: A Contribution to the Critique of Actually Existing Democracy." In *Habermas and the Public Sphere*, edited by Craig Calhoun, 109–42. Cambridge: MIT Press, 1992.

Frazee, Charles. *Catholics and Sultans: The Church and the Ottoman Empire, 1453–1923*. Cambridge: Cambridge University Press, 1983.

Freeden, Michael. "Eugenics and Progressive Thought: A Study in Ideological Affinity." *Historical Journal* 22 (1979): 645–71.

Frolova, Yevegeniya. "The Problem of Knowledge and Belief, Science and Religion in Arab Thought." In *Muslim Philosophy in Soviet Studies*, edited by Marietta Stepanyants, 117–32. New Delhi: Sterling, 1988.

Fyfe, Aileen. *Science and Salvation: Evangelical Popular Science Publishing in Victorian Britain*. Chicago: University of Chicago Press, 2004.

Gershoni, Israel, and James Jankowski. *Egypt, Islam and the Arabs: The Search for Egyptian Nationhood, 1900–1930*. New York: Oxford University Press, 1986.

Gesink, Indira Falk. "Beyond Modernisms: Opposition and Negotiation in the Azhar Reform Debate in Egypt, 1870–1911." PhD diss., Washington University, 2000.

———. *Islamic Reform and Conservatism: Al-Azhar and the Evolution of Modern Sunni Islam*. London: I. B. Tauris, 2010.

Geulen, Christian. "The Common Grounds of Conflict: Racial Visions of World Order, 1880–1940." In *Competing Visions of World Order: Global Moments and Movements,*

1880s–1930s, ed. Sebastian Conrad and Dominic Sachsenmaier, 69–96. New York: Palgrave Macmillan, 2007.

Ghazal, Amal. "Sufism, Ijtihad and Modernity: Yusuf al-Nabhani in the Age of ʿAbd al-Hamid II." *Archivum Ottomanicum* 19 (2001): 239–72.

Glick, Thomas. "Reception Studies since 1974." In *The Comparative Reception of Darwinism*, edited by Thomas Glick, xi–xxviii. Chicago: University of Chicago Press, 1988.

Godart, Gerard. "Darwin in Japan: Evolutionary Theory and Japan's Modernity (1820–1970)." PhD diss., University of Chicago, 2009.

Goldschmidt, Arthur. *Biographical Dictionary of Modern Egypt*. Cairo: American University in Cairo Press, 2000.

Gorman, Anthony. "Anarchists in Education: The Free Popular University in Egypt (1901)." *Middle Eastern Studies* 41 (2005): 305–20.

Gosling, David. "Darwin and the Hindu Tradition: 'Does What Goes Around Come Around?'" *Zygon* 46 (2011): 345–69.

Graham, Mark. "'The Enchanter's Wand': Charles Darwin, Foreign Missions, and the Voyage of H.M.S. *Beagle*." *Journal of Religious History* 31 (2007): 131–50.

Gran, Peter. *Islamic Roots of Capitalism: Egypt, 1760–1840*. Austin: University of Texas Press, 1979.

Habib, S. Irfan, and Dhruv Raina. Introduction to *Social History of Science in Colonial India*, edited by S. Irfan Habib, Dhruv Raina, and Zaheer Baber. Oxford: Oxford University Press, 2007.

Haddad, Mahmoud. "Syrian Muslim Attitudes toward Foreign Missionaries in the Late Nineteenth and Twentieth Centuries." In *Altruism and Imperialism: Western Religious and Cultural Missionary Enterprise in the Middle East*, 253–74. New York: Middle East Institute, Columbia University, 2002.

Haj, Samira. *Reconfiguring Islamic Tradition: Reform, Rationality, and Modernity*. Stanford, CA: Stanford University Press, 2009.

Hallaq, Wael. "Was the Gate of Ijtihad Closed?" *International Journal of Middle Eastern Studies* 16 (1984): 3–41.

Hamza, Dyala, ed. *The Making of the Arab Intellectual: Empire, Public Sphere and the Colonial Coordinates of Selfhood*. New York: Routledge, 2013.

Hamzaoui, Rached. *L'Academie de langue arabe du Caire: Histoire et œuvre*. Tunis: University of Tunis, 1975.

Hamzaoui, Muhammad Rashad. *L'Académie arabe de Damas*. Leiden: Brill, 1965.

———. *Majmaʿ al-lugha al-ʿArabiya bi-damashq*. Tunis: Dar al-Turki, 1988.

Handy, Robert, ed. *The Social Gospel in America, 1870–1920*. New York: Oxford University Press, 1966.

Hanioglu, M. Şükrü. "Blueprint for a Future Society: Late Ottoman Materialists on Science, Religion and Art." In *Late Ottoman Society: The Intellectual Legacy*, edited by Elisabeth Ozdalga, 28–116. New York: Routledge Curzon, 2005.

———. *Preparation for a Revolution: The Young Turks, 1902–1908*. Oxford: Oxford University Press, 2001.

———. *The Young Turks in Opposition*. New York: Oxford University Press, 1995.

Hankins, Barry. *The Second Great Awakening and the Transcendentalists*. Westport, CT: Greenwald Press, 2004.

Hanna, Suhail ibn Salim. "Biographical Scholarship and Muhammad ʿAbduh." *Muslim World* 59 (1969): 300–307.

Hansen, Bent. "Interest Rates and Foreign Capital in Egypt under the British." *Journal of Economic History* 43 (1983): 867–84.

Hanssen, Jens. *Fin de Siècle Beirut: The Making of an Ottoman Provincial Capital*. Oxford: Oxford University Press, 2005.

Haq, Mahmud. *Muhammad ʿAbduh: A Study of a Modern Thinker of Egypt*. Aligarh: Institute of Islamic Studies, 1970.

Haroun, Georges. *Šibli Šumayyil: Une pensée évolutionniste Arabe a l'époque d'an-Nahda*. Beirut: Université Libanais, 1985.

Harris, Paul William. *Nothing but Christ: Rufus Anderson and the Ideology of Protestant Foreign Missions*. Oxford: Oxford University Press, 1999.

Harrison, Mark. "Decentering 'Colonial Science.'" *Metascience* 16 (2007): 543–47.

Hart, Roger. "Translating the Untranslatable: From Copula to Incommensurable Worlds." In *Tokens of Exchange: The Problem of Translation in Global Circulations*, edited by Lydia Liu, 45–73. Durham, NC: Duke University Press, 2000.

Hattemer, Richard. "Atatürk and the Reforms in Turkey as Reflected in the Egyptian Press." *Journal of Islamic Studies* 11 (1999): 21–42.

Hawkins, Mike. *Social Darwinism in European and American Thought, 1860–1945*. Cambridge: Cambridge University Press, 1997.

Hayes, Carlton J. *A Generation of Materialism, 1871–1900*. New York: Harper, 1941.

Helmy, M. A. "Notes on the Reception of Darwinism in Some Islamic Countries." In *Science in Islamic Civilisation*, edited by Ekmeleddin Ihsanoglu and Geza Gunergun, 245–55. Istanbul: IRCICA, 2000.

Heyworth-Dunne, J. *An Introduction to the History of Education in Modern Egypt*. London: Luzac, 1939.

———. "Printing and Translation under Muhammad ʿAli of Egypt: The Foundation of Modern Arabic." *Journal of the Royal Asiatic Society* 3 (1940): 325–49.

Hirschkind, Charles. *The Ethical Soundscape: Cassette Sermons and Islamic Counterpublics*. New York: Columbia University Press, 2006.

———. "Heresy or Hermeneutics: The Case of Nasr Hamid Abu Zayd." *Stanford Humanities Review* 5 (1995): 35–49.

Hitti, Philip. *Lebanon in History from Earliest Times to the Present*. London: Macmillan, 1957.

Hobsbawm, Eric. *How to Change the World: Tales of Marx and Marxism*. London: Little, Brown, 2011.

Hofstadter, Richard. *Social Darwinism in American Thought, 1860–1915*. Boston: Beacon Press, 1955.

Hopkins, Charles. *The Rise of the Social Gospel in American Protestantism, 1865–1915*. New Haven: Yale University Press, 1940.

Hourani, Albert. *Arabic Thought in the Liberal Age, 1798–1939*. Cambridge: Cambridge University Press, 1962.

———. "Bustani's Encyclopaedia." *Journal of Islamic Studies* 1 (1990): 111–19.

Howard, Damian. *Being Human in Islam: The Impact of the Evolutionary Worldview*. New York: Routledge, 2011.

Husain, Zohair. "Muhammad ʿAbduh (1849–1905): The Preeminent Muslim Modernist of Egypt." *Hamdard Islamicus* 9 (1986): 31–39.

Ibrahim, Yasir. "Rashid Rida and the *Maqasid al-Shariʿa*." *Studia Islamica* 102/103 (2006): 157–98.

Ismael, Tarek, and Rifa'aat El-Sa'id. *Communist Movement in Egypt*. Syracuse, NY: Syracuse University Press, 1990.

Ismail, Haidar al-Haj. *Fransis al-Marrash*. London: Riyad al-Rayyis, 1989.

Issawi, Charles. *The Economic History of Turkey, 1800–1914*. Chicago: University of Chicago Press, 1980.

———. *The Fertile Crescent, 1800–1914*. New York: Oxford University Press, 1988.

ʿIzz al-Din, Amin. *Al-Mansuri: Sirat muthaqaf thawri* [Al-Mansuri: Biography of a revolutionary intellectual]. Cairo: Dar al-Ghad al-ʿArabi, 1984.

Jansen, J. *The Interpretation of the Koran in Modern Egypt*. Leiden: Brill, 1974.

Jawad, Muhammad ʿAbd al-. *Taqwim Dar al-ʿUlum, 1872–1947* [The organization of Dar al-ʿUlum, 1872–1947]. Cairo: Dar al-Maʿarif bi Misr, 1952.

Jodock, Darrel, ed. *Catholicism Contending with Modernity*. Cambridge: Cambridge University Press, 2000.

Johansen, Baber. "Apostasy as Objective and Depersonalized Fact: Two Recent Egyptian Court Judgments." *Social Research: An International Quarterly* 70 (2003): 687–710.

———. *Contingency in a Sacred Law: Legal and Ethical Norms in the Muslim* Fiqh. Leiden: Brill, 1999.

Jolivet, Jean. "Classifications of the Sciences." In *Encyclopedia of the History of Arabic Science*, edited by Roshdi Rashed, 1008–25. London: Routledge, 1996.

Jomier, J. *Le commentaire coranique du Manar: Tendences modernes de l'exégèse coranique en Égypte*. Paris: G. P. Maisonneuve, 1954.

Jones, Aled. *Powers of the Press: Newspapers, Power and the Public in Nineteenth-Century England*. Aldershot: Ashgate, 1996.

Juha, Shafiq. *Darwin wa azmat sanat 1882 fi al-kulliya al-tibbiya* [Darwin and the crisis of 1882 in the Medical Department at the Syrian Protestant College]. Beirut: American University of Beirut Press, 1991. Translated as *Darwin and the Crisis of 1882 in the Medical Department* (Beirut: American University of Beirut Press, 2004).

Kapila, Shruti, "Self, Spencer and *Swaraj*: Nationalist Thought and Critiques of Liberalism, 1890–1920." *Modern Intellectual History* 4 (2007): 109–27.

Karpat, Kemal. *The Politicization of Islam*. Oxford: Oxford University Press, 2001.

Kayyali, Sami al-. *Muhadarat ʿan al-haraka al-adabiya fi halab* [Lectures on the literary movement of Aleppo]. Cairo: Jamiʿat al-Duwal al-ʿArabiya, 1956.

Kedourie, Elie. *Arabic Political Memoirs and Other Studies*. London: Cass, 1974.

———. *The Chatham House Version and Other Middle Eastern Studies*. New York: Praeger, 1970.

Kelidar, Abbas. "The Political Press in Egypt, 1882–1914." In *Contemporary Egypt through Egyptian Eyes*, edited by Charles Tripp, 1–21. London: Routledge, 1993.

———. "Shaykh ʿAli Yusuf: Egyptian Journalist and Islamic Nationalist." In *Intellectual Life in the Arab East, 1890–1939*, edited by Marwan Buheiry, 10–20. Beirut: American University of Beirut Press, 1981.

Kendall, Elisabeth. "The Marginal Voice: Journals and the *Avant-Garde* in Egypt." *Journal of Islamic Studies* 8 (1997): 216–38.

Kerr, Malcolm. *Islamic Reform: The Political and Legal Theories of Muhammad ʿAbduh and Rashid Rida*. Berkeley and Los Angeles: University of California Press, 1966.

Khalaf, Samir. "New England Puritanism and Liberal Education in the Middle East: The American University in Beirut as a Cultural Transplant." In *Cultural Transitions in the Middle East*, edited by Serif Mardin, 50–85. Leiden: Brill, 1994.

Khuri-Makdisi, Ilham. *The Eastern Mediterranean and the Making of Global Radicalism*. Berkeley and Los Angeles: University of California Press, 2010.

Kogan, Barry. *Averroes and the Metaphysics of Causation*. Albany: State University of New York Press, 1985.

Kohn, Margaret. "Afghani on Empire, Islam and Civilization." *Political Theory* 37 (2009): 398–422.

Kruk, Remke. "A Frothy Bubble: Spontaneous Generation in the Medieval Islamic Tradition." *Journal of Semitic Studies* 35 (1990): 265–82.

Kudsi-Zadeh, A. Albert. "The Emergence of Political Journalism in Egypt." *Muslim World* 70 (1980): 47–55.

Kugekgen, Anke von. "A Call for Rationalism: 'Arab Averroists' in the Twentieth Century." *Alif* 16 (1996): 97–132.

Laity, Paul. *The British Peace Movement, 1870–1914*. Oxford: Oxford University Press, 2001.

Lambert, Frank. *Inventing the "Great Awakening."* Princeton, NJ: Princeton University Press, 2001.

Lamoureaux, David. "Between 'The Origin of Species' and 'The Fundamentals': Toward a Historiographical Model of the Evolutionary Reaction to Darwinism in the First Fifty Years." PhD diss., Toronto School of Theology, 1991.

Landau, Paul S. "Explaining Surgical Evangelism in Colonial Southern Africa: Teeth, Pain and Faith." *Journal of African History* 37 (1996): 261–81.

Leakey, L., Jack Prost, and Stephanie Prost. *Adam or Ape: A Sourcebook of Discoveries about Early Man*. Cambridge, MA: Schenkman, 1971.

Leavitt, Donald. "Darwinism in the Arab World and the Lewis Affair at the Syrian Protestant College." *Muslim World* 71 (1981): 85–98.

Lecerf, Jean. "Šibli Šumayyil: Métaphysicien et moraliste contemporain." *Bulletin d'Études Orientales* 1 (1931): 153–86.

Lightman, Bernard. *Evolutionary Naturalism in Victorian Britain: The "Darwinians" and Their Critics*. Farnham, UK: Ashgate / Variorum, 2009.

———. "Ideology, Evolution and Late-Victorian Agnostic Popularizers." In *History, Humanity and Evolution: Essays for John C. Greene*, edited by James Moore, 285–310. Cambridge: Cambridge University Press, 1989.

———. *Victorian Popularizers of Science: Designing Nature for New Audiences*. Chicago: University of Chicago Press, 2007.

Livingston, John. "Muhammad 'Abduh on Science." *Muslim World* 85 (1995): 215–34.

Livingstone, David. *Adam's Ancestors: Race, Religion, and the Politics of Human Origins*. Baltimore, MD: Johns Hopkins University Press, 2008.

———. *Darwin's Forgotten Defenders: The Encounter between Evangelical Theology and Evolutionary Thought*. Grand Rapids, MI: Eerdmans, 1987.

———. *Putting Science in Its Place: Geographies of Scientific Knowledge*. Chicago: University of Chicago Press, 2003.

Livingstone, David, D. G. Hart, and Mark Noll, eds. *Evangelicals and Science in Historical Perspective*. Oxford: Oxford University Press, 1999.

Lockman, Zachary. "British Policy toward Egyptian Labor Activism, 1882–1936." *International Journal of Middle East Studies* 20 (1988): 265–85.

———. "Imagining the Working Class: Culture, Nationalism and Class Formation in Egypt, 1899–1914." *Poetics Today* 15 (1994): 157–90.

Lukes, Steven. "The Meanings of 'Individualism.'" *Journal of the History of Ideas* 32 (1971): 45–66.

Ma, Haiyun. "Patriotic and Pious Muslim Intellectuals in Modern China: The Case of Ma Jian." *American Journal of Islamic Social Science* 23 (2006): 54–70.

MacKenzie, Donald. "Eugenics in Britain." *Social Studies of Science* 6 (1976): 499–532.

MacLeod, Roy. "On Visiting the Moving Metropolis: Reflections on the Architecture of

Imperial Science." In *Scientific Colonialism: A Cross-Cultural Comparison*, edited by Nathan Reingold and Marc Rothenberg, 217–50. Washington, DC: Smithsonian Institution Press, 1987).

MacLeod, Roy, and Philip F. Rehbock, eds. *Nature in Its Greatest Extent: Western Science in the Pacific*. Honolulu: University of Hawaii Press, 1988.

Makdisi, Ussama. "After 1860: Debating Religion, Reform and Nationalism in the Ottoman Empire." *International Journal of Middle East Studies* 34 (2002): 601–17.

———. *American Missionaries and the Failed Conversion of the Middle East*. Ithaca, NY: Cornell University Press, 2008.

———. *The Culture of Sectarianism: Culture, Community, History and Violence in Ottoman Lebanon*. Berkeley and Los Angeles: University of California Press, 2000.

———. "Reclaiming the Land of the Bible: Missionaries, Secularism and Evangelical Modernity." *American Historical Review* 102 (1997): 680–713.

Makhzumi, Muhammad, ed. *Khatirat Jamal al-Din al-Afghani*. Beirut: Al-Matbaʿa al-ʿIlmiya li-Yusuf Sadr, 1931.

Mallat, Chibli. "The Debate on Riba and Interest in Twentieth Century Jurisprudence." In *Islamic Law and Finance*, edited by Chibli Mallat, 69–88. London: Graham and Trotman, 1988.

Manela, Erez. *The Wilsonian Moment*. New York: Oxford University Press, 2007.

Manguel, Alberto. *A History of Reading*. London: Flamingo, 1997.

Mardin, Şerif. *The Genesis of Young Ottoman Thought: A Study in the Modernization of Turkish Political Ideas*. Princeton, NJ: Princeton University Press, 2000.

Margoliouth, D. S. "Recent Arabic Publications." *Journal of the Royal Asiatic Society of Great Britain and Ireland*, 1904, 571–86.

Marsot, Afaf Lutfi al-Sayyad. "The Beginnings of Modernization in Education among the Rectors of Al-Azhar, 1798–1879." In *The Beginnings of Modernization in the Middle East: The Nineteenth Century*, edited by William Polk and Richard Chambers, 267–80. Chicago: University of Chicago Press, 1968.

Massad, Joseph. *Colonial Effects: The Making of National Identity in Jordan*. New York: Columbia University Press, 2001.

———. *Desiring Arabs*. Chicago: University of Chicago Press, 2005.

Masters, Bruce. *Christians and Jews in the Ottoman Arab World: The Roots of Sectarianism*. Cambridge: Cambridge University Press, 2001.

Masud, Muhammad Khalid. "Iqbal's Approach to Islamic Theology of Modernity." *Al-Hikmat* 27 (2007): 1–3.

Masuzawa, Tomoko. *The Invention of World Religions; or, How European Universalism Was Preserved in the Language of Pluralism*. Chicago: University of Chicago Press, 2005.

McClellan, James, III. "Science and Empire Studies and Postcolonial Studies: A Report from the Contact Zone." In *Entangled Knowledge: Scientific Discourses and Cultural Difference*, edited by Klaus Hock and Gesa Mackenthun, 51–75. Münster: Deutsche Nationalbibliothek, 2012.

McDonald, M. "Animal Books as a Genre of Arabic Literature." *Bulletin of the British Society for Middle East Studies* 15 (1989): 3–10.

McGrath, Alister. *Science and Religion*. Oxford: Blackwell, 1999.

Meïer, Olivier. *"Al-Muqtataf" et le débat sur le darwinisme: Beyrouth, 1876–1885*. Cairo: CEDEJ, 1996.

Messick, Brinkley. *Calligraphic State: Textual Domination and History in a Muslim Society*. Berkeley and Los Angeles: University of California Press, 1993.

Mitchell, Timothy. *Colonising Egypt*. Cambridge: Cambridge University Press, 1988.

Moaddel, Mansoor, and Karim Talatoff, eds. *Contemporary Debates in Islam*. New York: St. Martin's Press, 2000.

Moghith, Anwar. "Sur les ruines du matérialisme." *Égypte: Monde arabe* 4–5 (2001): 311–15.

Montgomery, Scott. *Science in Translation: Movements of Knowledge through Cultures and Time*. Chicago: University of Chicago Press, 2000.

Moore, James. *The Post-Darwinian Controversies: A Study of the Protestant Struggle to Come to Terms with Darwin in Great Britain and America, 1870–1900*. Cambridge: Cambridge University Press, 1979.

———. "Socializing Darwinism: Historiography and the Fortunes of a Phrase." In *Science as Politics*, edited by Les Levidow, 38–80. London: Free Association Books, 1986.

Moosa, Matti. *The Origins of Modern Arabic Fiction*. Boulder, CO: Lynne Rienner, 1997.

Musa, Ra'uf Salama. *Salama Musa—abi* [Salama Musa—my father]. Cairo: Dar wa Matabi' al-Mustaqbal, 1992.

Nafzawi, Muhammad al-Nasir. *Al-dawla wa-al-mujtama': Min mihnat Ibn Rushd ila khusum-ihi Muhammad 'Abduh, Farah Antun* [State and society: From the trial of Ibn Rushd to the controversies of Muhammad 'Abduh and Farah Antun]. Tunis: Markaz al-Nashr al-Jami'i, 2000.

Nagai, Michio. "Herbert Spencer in Early Meiji Japan." *Far Eastern Quarterly* 14 (November 1954): 55–64.

Nashabi, Hisham. "Shaykh 'Abd al-Qadir al-Qabbani and *Thamarat al-Funun*." In *Intellectual Life in the Arab East, 1890–1939*, edited by Marwan Buheiry, 84–91. Beirut: American University of Beirut Press, 1981.

Nasr, Seyyed Hossein. *An Introduction to Islamic Cosmological Doctrines*. Albany: State University of New York Press, 1993.

Noll, Mark. *The Rise of Evangelicalism: The Age of Edwards, Whitefield and the Wesleys*. Leicester: InterVarsity Press, 2004.

Numbers, Ronald. *The Creationists*. Cambridge, MA: Harvard University Press, 2006.

Omar, Abdullah al-. "The Reception of Darwinism in the Arab World." PhD diss., Harvard University, 1982.

O'Neil, Patrick. *Polyglot Joyce: Fictions of Translation*. Toronto: University of Toronto Press, 2005.

Ospovat, Dov. *The Development of Darwin's Theory: Natural History, Natural Theology and Natural Selection, 1838–1859*. Cambridge: Cambridge University Press, 1981.

Özervarli, M. Sait. "Attempts to Revitalize Kalam in the Late 19th and Early 20th Centuries." *Muslim World* 89 (1999): 89–105.

Panzac, Daniel. *Population et santé dans l'empire Ottoman (XVIIIe–XXe siècles)*. Istanbul: Editions Isis, 1996.

Pels, Peter. "The Anthropology of Colonialism: Culture, History, and the Emergence of Western Governmentality." *Annual Review of Anthropology* 26 (1997): 163–83.

Penrose, Stephen. *That They May Have Life: The Story of the American University of Beirut, 1866–1941*. New York: Trustees of the American University of Beirut, 1941.

Peters, Rudolph. "*Idjtihad* and *Taqlid* in Eighteenth and Nineteenth Century Islam." *Die Welt des Islams* 20 (1980): 131–45.

Petrou, Georgia. "Translation Studies and the History of Science: The Greek Textbooks of the 18th Century." *Science and Education* 15 (2006): 823–40.

Philipp, Thomas. "Perceptions of the First World War." In *Ottoman Reform and Muslim Regeneration: Studies in Honour of Butrus Abu-Manneb*. London: I. B. Tauris, 2005.

Phillips, Clifton Jackson. *Protestant America and the Pagan World: The First Half Century of the American Board of Commissioners for Foreign Missions, 1810–1860*. Cambridge, MA: Harvard University Press, 1969.

Pollard, Lisa. *Nurturing the Nation: The Family Politics of Modernizing, Colonizing, and Liberating Egypt (1805–1923)*. Berkeley and Los Angeles: University of California Press, 2005.

Pollock, Sheldon. *The Language of the Gods in the World of Men: Sanskrit, Culture, and Power in Premodern India*. Berkeley and Los Angeles: University of California Press, 2006.

Porter, Andrew. "'Commerce and Christianity': The Rise and Fall of a Nineteenth-Century Missionary Slogan." *Historical Journal* 28 (1985): 597–621.

Powell, Eve Troutt. "Brothers along the Nile: Egyptian Concepts of Race and Ethnicity, 1895–1910." In *The Nile: Histories, Cultures, Myths*, edited by Haggai Erlich and Israel Gershoni, 171–81. Boulder, CO: Lynn Rienner, 2000.

———. *A Different Shade of Colonialism: Egypt, Great Britain, and the Mastery of the Sudan*. Berkeley and Los Angeles: University of California Press, 2003.

Prasch, Thomas. "Which God for Africa? The Islamic-Christian Missionary Debate in Late-Victorian Africa." *Victorian Studies* 33 (1989): 51–73.

Price, Leah. "Reading: The State of the Discipline." *Book History* 7 (2004): 303–20.

Pusey, James. *China and Charles Darwin*. Cambridge, MA: Harvard University Press, 1983.

Quandt, J. B. "Religion and Social Thought: The Secularization of Postmillennialism." *American Quarterly* 25 (1973): 390–409.

Quataert, Donald. "Population." In *An Economic and Social History of the Ottoman Empire, 1300–1914*, edited by Halil Inalick and Donald Quataert, 777–97. Cambridge: Cambridge University Press, 1994.

Racine, Luc. "Paradise, The Golden Age, the Millennium and Utopia: A Note on the Differentiation of Forms of the Ideal Society." *Diogenes* 31 [1983]: 119–36.

Rae, W. Fraser. "The Egyptian Newspaper Press." *Nineteenth Century* 32 (1892): 213–23.

———. *Egypt Today*. London: R. Bentley and Son, 1892.

Ragep, Jamil and Sally Ragep, eds. *Tradition, Transmission, Transformation*. Leiden: Brill, 1996.

Raj, Kapil. *Relocating Modern Science: Circulation and the Construction of Knowledge in South Asia and Europe, 1650–1900*. New York: Palgrave Macmillan, 2007.

Rao, Lindsay. "Nineteenth Century American Schools in the Levant: A Study of Progress." PhD diss., University of Michigan, 1964.

Rashed, Roshdi, ed. *Encyclopedia of the History of Arabic Science*. London: Routledge, 1996.

Reardon, Bernard. *Roman Catholic Modernism*. Stanford, CA: Stanford University Press, 1970.

Reid, Donald. *The Odyssey of Farah Antun: A Syrian Christian's Quest for Secularism*. Minneapolis: Bibliotheca Islamica, 1975.

———. "The Syrian Christians and Early Socialism in the Arab World." *International Journal of Middle Eastern Studies* 5 (1974): 177–93.

———. "Turn-of-the-Century Egyptian School Days." *Comparative Education Review* 27 (1983): 374–93.

Reimer, Michael. "Views of Al-Azhar in the Nineteenth Century: Gabriel Charmes and 'Ali Pasha Muhabrak." In *Travelers in Egypt*, edited by Paul Starkey and Janet Starkey. London: I. B. Tauris, 1998.

Renn, Jürgen, ed. *The Globalization of Knowledge in History*. Berlin: Edition Open Access, 2012.

Reynolds, David. "Redrawing China's Intellectual Map: Images of Science in Nineteenth Century China." *Late Imperial China* 12 (1991): 27–61.

Richards, Robert. *Darwin and the Emergence of Evolutionary Theories of Mind and Behavior*. Chicago: University of Chicago Press, 1989.

Ristelhueber, René. *Traditions françaises au Liban*. Paris: F. Alcan, 1918.

Rizk, Karim, and Dominique Avon. *La Nahda: Réveils de la pensée en langue arabe, approches, perspectives; Collloque organisé à l'USEK les 28–29 octobre 2004*. Jounieh, Lebanon: Université Saint Esprit de Kaslik, 2009.

Robbins, Louise. "Publishing American Values: The Franklin Book Programs as Cold War Cultural Diplomacy." *Library Trends*, Winter 2007, 638–50.

Rogan, Eugene. "Instant Communication: The Impact of the Telegraph in Ottoman Syria." In *The Syrian Land: Processes of Integration and Fragmentation; Bilad al-Sham from the 18th to the 20th Century*, edited by Thomas Philipp and Birgit Schaebler, 113–28 Stuttgart: F. Steiner, 1998.

Rose, Jonathan. *The Intellectual Life of the British Working Classes*. New Haven, CT: Yale University Press, 2001.

Rosenthal, Franz. *Knowledge Triumphant: The Concept of Knowledge in Medieval Islam*. Leiden: Brill, 1970.

———. "Muslim Definitions of Knowledge." In *The Conflict of Traditionalism and Modernism in the Muslim Middle East*, edited by Carl Leiden, 117–33. Austin: University of Texas Press, 1966.

Rudwick, Martin. *The Meaning of Fossils: Episodes in the History of Palaeontology*. New York: Science History, 1976.

Rupke, Nicolas. "Translation Studies in the History of Science: The Example of Vestiges." *British Journal for the History of Science* 33 (2000): 209–22.

Russell, Mona. *Creating the New Egyptian Woman: Consumerism, Education, and National Identity, 1863–1922*. New York: Palgrave, 2004.

Ryad, Umar. *Islamic Reformism and Christianity: A Critical Reading of the Works of Muhammad Rashid Rida and His Associates (1898–1935)*. Leiden: Brill, 2009.

Ryzova, Luci. "Egyptianizing Modernity through the 'New Effendiya': Social and Cultural Constructions of the Middle Class in Egypt under the Monarchy." In *Re-envisioning Egypt: 1919–1952*, edited by Arthur Goldschmidt and Amy Johnson, 124–63. Cairo: American University of Cairo Press, 2005.

Sabra, A. I. "The Appropriation and Subsequent Naturalization of Greek Science in Medieval Islam." *History of Science* 25 (1987): 223–43.

Sabra, Adam. *Poverty and Charity in Medieval Islam*. Cambridge: Cambridge University Press, 2001.

Saʿid, ʿAbd Allah Ibrahim. *Al al-Jisr fi Tarabulus, 1757–1980: Min irshad al-dini ila al-ʿamal al-siyasi* [The al-Jisrs in Tripoli, 1757–1980: From religious guidance to political works]. Tokyo: Maʿhad al-Abhath fi Lughat wa-Thaqafat Asiya wa-Afriqiya, 2007.

Saʿid, Lutfi. "Al-Hakim Cornelius Van Allen Van Dyck." *Isis* 27 (1937): 20–45.

———. *The Life and Works of George Edward Post (1839–1909)*. N.p.: Saint Catherine Press, 1938.

Saʿid, Rifʿat al-. *Thalathat Lubnaniyin fi al-Qahira* [Three Lebanese in Cairo]. Cairo: Dar al-Thaqafa, 1973.

Saʿidi, ʿAbd al-Mital al-. *Tarikh al-islah fi Al-Azhar wa-safahat min al-jihad fi al-islah* [History of reform at Al-Azhar and chapters in the struggle for reform]. Cairo: Matbaʿat al-Iʿtimad, 1951.

Saliba, Najib. "The Achievements of Midhat Pasha as Governor of the Province of Syria, 1878–1880." *International Journal of Middle East Studies* 9 (1978): 307–23.

Salibi, Kamal. *The Modern History of Lebanon*. London: Weidenfeld and Nicolson, 1965.

Salvatore, Armando. *The Public Sphere: Liberal Modernity, Catholicism, Islam*. New York: Palgrave Macmillan, 2007.

Sarkis, Yusuf. *Muʿjam al-matbuʿat al-Arabiya wa-al-Muʿarraba* [Arabic publications and translations]. 2 vols. Beirut: Dar Sadir, 1990.

Sarruf, Fuad. *Yaʿqub Sarruf*. Beirut: Dar al-ʿIlm al-Malayin, 1960.

Sarukkai, Sundar. *Translating the World: Science and Language*. Lanham, MD: University Press of America, 2002.

Saul, Samir. "European Capital and Its Impact on Land Redistribution in Egypt: A Quantitative Analysis (1900–1914)." In *Colonialism and the Modern World: Selected Studies*, edited by Gregory Blue, Martin Bunton, and Ralph Croizier, 120–44. Armonk, NY: M. E. Sharpe, 2002.

Schaffer, Simon. "Self Evidence." *Critical Inquiry* 18 (1992): 327–62.

Schatkowksi, Linda. "The Islamic Maqased of Beirut: A Case Study of Modernization in Lebanon." Thesis, American University of Beirut, 1981.

Schivelbusch, Wolfgang. *The Railway Journey: The Industrialization of Time and Space in the Nineteenth Century*. Berkeley and Los Angeles: University of California Press, 1986.

Schneider, Herbert. "The Influence of Darwin and Spencer on American Philosophical Theology." *Journal of the History of Ideas* 6 (1945): 3–18.

Schölch, Alexander. *Egypt for the Egyptians: The Socio-political Crisis in Egypt, 1878–1882*. London: Ithaca Press, 1981.

Schulte, Reinhart, and John Biguenet, eds. *Theories of Translation: An Anthology of Essays from Dryden to Derrida*. Chicago: University of Chicago Press, 1992.

Secord, James. "Global Darwin." In *Darwin*, edited by William Brown and Andrew Fabian, 31–57. Cambridge: Cambridge University Press, 2010.

———. *Victorian Sensation: The Extraordinary Publication, Reception, and Secret Authorship of "Vestiges of the Natural History of Creation."* Chicago: University of Chicago Press, 2000.

Sedra, Paul. *From Mission to Modernity: Evangelicals, Reformers and Education in Nineteenth Century Egypt*. London: I. B. Tauris, 2011.

Sekaly, Samir. "Shaykh Yusuf al-Nabahani and the West." In *Les Européens vus par les Libanais à l'époque ottomane*, edited by Bernard Heyberger and Casten Walbiner, 175–81. Beirut: Ergon Verlag, 2002.

Selim, Samah. "Languages of Civilization: Nation, Translation and the Politics of Race in Colonial Egypt." *Translator* 15 (2009): 139–56.

———. *The Novel and the Rural Imaginary in Egypt: 1880–1965*. New York: Routledge, 2004.

———. "The People's Entertainments: Translation, Popular Fiction, and the Nahdah in Egypt." In *Other Renaissances: A New Approach to World Literature*, edited by Brenda Deen Schildgen, Gang Zhou, and Sander Gilman, 35–58. New York: Palgrave Macmillan, 2007.

Shapin, Steve. "Placing the View from Nowhere: Historical and Sociological Problems in the Location of Science." *Transactions of the Institute of British Geographers* 23 (1998): 5–12.

Sharkey, Heather. *Living with Colonialism: Nationalism and Culture in the Anglo-Egyptian Sudan*. Berkeley and Los Angeles: University of California Press, 2003.

Shayyal, Jamal al-. *Tarikh al-tarjama fi ʿasr Muhammad ʿAli* [History of the translation movement during the reign of Muhammad ʿAli]. Cairo: Dar al-Fikr al-ʿArabi, 1951.

Sheehi, Stephen. "Arabic Literary-Scientific Journals: Precedence for Globalization and the Creation of Modernity." *Comparative Studies of South Asia, Africa and the Middle East* 25 (2005): 438–48.

———. *Foundations of Modern Identity*. Gainesville: University Press of Florida, 2004.

———. "Inscribing the Arab Self: Butrus al-Bustani and Paradigms of Subjective Reform." *British Journal of Middle Eastern Studies* 27 (2000): 7–24.

———. "Toward a Critical Theory of *al-Nahda*: Epistemology, Ideology and Capital." *Journal of Arabic Literature* 43 (2012): 269–98.

Simon, Walter. "Herbert Spencer and the 'Social Organism.'" *Journal of the History of Ideas* 21 (1960): 295.

Sivasundaram, Sujit. *Nature and the Godly Empire: Science and Evangelical Missions in the Pacific, 1795–1850*. Cambridge: Cambridge University Press, 2005.

Smith, George. "Herbert Spencer's Theory of Causation." *Journal of Libertarian Studies* 5 (1981): 113–52.

Smith, Jane. "The Concourse between the Living and the Dead in Islamic Eschatological Literature." *History of Religions* 19 (1980): 224–36.

Smith, Wilfred Cantwell. *Islam in Modern History*. Princeton, NJ: Princeton University Press, 1957.

Smolansky, Oles. "Moscow-Cairo Crisis, 1959." *Slavic Review* 22 (1963): 713–26.

Sonbol, Amira El Azhary. *The Last Khedive of Egypt: Memoirs of Abbas Hilmi II*. Reading, UK: Ithaca Press, 1998.

Spagnolo, John. "The Definition of a Style of Imperialism: The Internal Politics of the French Educational Investment in Ottoman Beirut." *French Historical Studies* 8 (1974): 563–84.

Standaert, Nicolas. "The Classification of Sciences and the Jesuit Mission in Late Ming China." In *Linked Faiths: Essays on Chinese Religions and Traditional Culture*, edited by Jan De Meyer and Peter Engelfriet, 287–317. Leiden: Brill, 2000.

Steele, David. "Britain and Egypt, 1882–1914: The Containment of Islamic Nationalism." In *Imperialism and Nationalism in the Middle East: The Anglo-Egyptian Experience, 1882–1982*, edited by Keith Wilson, 1–25. London: Mansell, 1983.

Stetkevych, Jaroslav. *The Modern Arabic Literary Language: Lexical and Stylistic Developments*. Chicago: University of Chicago Press, 1970.

Stocking, George. *Race, Culture and Evolution: Essays in the History of Anthropology*. Chicago: University of Chicago Press, 1982.

Stoltz, Daniel. "From Shellfish to Apes: Socialism and the Changing Politics of Evolution in *al-Muqtataf*, 1899–1914." B.A. thesis, Harvard University, 2007.

Strick, James. *Sparks of Life: Darwinism and the Victorian Debates over Spontaneous Generation*. Cambridge: Cambridge University Press, 2000.

Strohmeier, Martin. "Muslim Education in the Vilayet of Beirut, 1880–1918." In *Decision Making and Change in the Ottoman Empire*, edited by Caesar Farah, 215–41. Kirksville, MO: Thomas Jefferson University Press at Northeast Missouri State University, 1993.

Szyliowicz, Joseph. "The Printing Press in the Ottoman Empire." In *Transfer of Modern Science and Technology to the Muslim World*, edited by Ekmelleddin Ihsanogl, 251–52. Istanbul: Research Center for Islamic History, Art, and Culture, 1992.

Tageldin, Shaden. *Disarming Words: Empire and the Seductions of Translation in Egypt*. Berkeley and Los Angeles: University of California Press, 2011.

———. "Proxidistant Reading: Toward a Critical Pedagogy of the *Nahdah* in U.S. Comparative Literary Studies," *Journal of Arabic Literature* 43 (2012): 227–68.

Tamim, Suha. *A Bibliography of A.U.B. Faculty Publications, 1866–1966*. Beirut: American University of Beirut Press, 1967.

Thistlethwayte, Lynette. "The Role of Science in the Hindu-Christian Encounter." *Indo-British Review* 19 (1991): 73–82.

Tibawi, A. L. *American Interests in Syria, 1800–1901: A Study of Educational, Literary and Religious Work*. Oxford: Clarendon, 1966.

———. "The American Missionaries in Beirut and Butrus al-Bustani." *Middle Eastern Affairs* 3 (1963): 137–82.

———. "The Genesis and Early History of the Syrian Protestant College." *Middle East Journal* 21 (1967): 1–15.

Tignor, Robert. *Modernization and British Colonial Rule in Egypt*. Princeton, NJ: Princeton University Press, 1966.

Tripp, Charles. *Islam and the Moral Economy: The Challenge of Capitalism*. Cambridge: Cambridge University Press, 2006.

Troll, Christian. *Sayyid Ahmad Khan: A Reinterpretation of Muslim Theology*. New Delhi: Vikas, 1978.

The True Dawn: Makassed Islamic and Philanthropic Association in Beirut: The First Annual Report—Translated from the Arabic Original Dated 1297 A.H. (1880). Beirut: n.p., 1984.

Übersetzung: Ein internationales Handbuch zur Übersetzungsforschung. Berlin: W. de Gruyter, 2004.

Valade, B. "Modernity." In *International Encyclopedia of the Social and Behavioral Sciences*, edited by Neil Smelser and Paul Bates, 9939–44. Amsterdam: Elsevier Science, 2001.

Van Nispen Tot Sevenaer, Christian. *Activité humaine et agir de Dieu: Le concept de* "Sunan *de Dieu" dans le commentaire coranique du Manar*. Beirut: Dar el-Machreq, 1996.

Vatikiotis, P. J. "Muhammad ʿAbduh and the Quest for a Muslim Humanism." *Arabica* 4 (1957): 54–72.

Vaughan, Megan. *Curing Their Ills: Colonial Power and African Illness*. Cambridge: Cambridge University Press, 1991.

Veer, Peter van der, ed. *Conversion to Modernities: The Globalization of Christianity*. New York: Routledge, 1996.

Venuti, Lawrence. *Scandals of Translation: Towards an Ethics of Difference*. London: Routledge, 1998.

———. *The Translation Studies Reader*. New York: Routledge, 2000.

———. *The Translator's Invisibility: A History of Translation*. London: Routledge, 1995.

Vikør, Knut. *Sufi and Scholar on the Desert Edge: Muhammad b. ʿAlī al-Sanūsī and His Brotherhood*. Evanston, IL: Northwestern University Press, 1995.

Viswanathan, Gauri. *Outside the Fold: Conversion, Modernity, and Belief*. Princeton, NJ: Princeton University Press, 1998.

Vogel, Frank. "The Closing of the Door of *Ijtihad* and the Application of the Law." *American Journal of Islamic Social Science* 10 (1993): 396–401.

Voll, John. "Renewal and Reform in Islamic History: *Tajdid* and *Islah*." In *Voices of Resurgent Islam*, edited by John Esposito, 32–45. New York: Oxford University Press, 1983.

Wahba, Mourad Magdi. "The Meaning of *Ishtirakiyah*: Arab Perceptions of Socialism in the Nineteenth Century." *Alif* 10 (1990): 44–52.

Walzer, Richard, trans. *Al-Farabi on the Perfect State: Abu Nasr al-Farabi's "Mabadi ara ahl al-madina al-fadila"; A Revised Text with Introduction, Translation, and Commentary*. Oxford: Clarendon Press, 1998.

Warner, Michael. *Publics and Counterpublics*. New York: Zone, 2002.

Webb, Michael. "The Countermodern Moment: A World-Historical Perspective on the

Thought of Rabindranath Tagore, Muhammad Iqbal, and Liang Shuming." *Journal of World History* 19 (2008): 189–212.

Weismann, Itzchak. "Abu l-Huda l-Ṣayyadi and the Rise of Islamic Fundamentalism." *Arabica* 54 (2007): 586–92.

———. "The Sociology of Islamic Modernism: Muhammad ʿAbduh, the National Public Sphere and the Colonial State." *Maghreb Review* 32 (2007): 104–21.

Whidden, James. "The Generation of 1919." In *Re-envisioning Egypt: 1919–1952*, edited by Arthur Goldschmidt and Amy Johnson, 19–45. Cairo: American University of Cairo Press, 2005.

Wiederhold, Lutz. "Legal Doctrines in Conflict: The Relevance of *Madhhab* Boundaries to Legal Reasoning in the Light of an Unpublished Treatise on *Taqlid* and *Ijtihad*." *Islamic Law and Society* 3 (1996): 234–89.

Wild, Stefan. "'Das Kapital' in arabischen Übersetzungen." In *Festgabe für Hans Wehr*, edited by W. Fischer, 97–111. Wiesbaden: n.p., 1969.

Wild, Stephen. "Ottomanism versus Arabism: The Case of Farid Kassab (1884–1970)." *Die Welt des Islams* 28 (1988): 626.

Willcocks, William. *Sixty Years in the East*. London: W. Blackwood, 1935.

Williams, C. Peter. "Healing and Evangelism: The Place of Medicine in Late Victorian Protestant Missionary Thinking." In *The Church and Healing*, edited by W. J. Sheils, 271–85. Oxford: Oxford University Press, 1982.

Wisnovsky, Robert. *Avicenna's Metaphysics in Context*. Ithaca, NY: Cornell University Press, 2003.

Wolfson, Robert. *The Philosophy of the Kalam*. Cambridge: Cambridge University Press, 1976.

Wright, David. "The Translation of Modern Western Science in Nineteenth-Century China, 1840–1895." *Isis* 89 (1998): 653–73.

Xiaosui Xiao. "China Encounters Darwinism: A Case of Intercultural Rhetoric." *Quarterly Journal of Speech* 81 (1995): 83–99.

Yaziji, Kamal, al-. *Al-Shaykh Ibrahim al-Hurani, 1844–1916*. Cairo: Jamʿat al-Duwal al-ʿArabiya, 1961.

Yousef, Hoda. "Reassessing Egypt's Dual System of Education under Ismaʾil: Growing ʿ*Ilm* and Shifting Ground in Egypt's First Educational Journal, *Rawdat al-Madaris*, 1870–77." *International Journal of Middle East Studies* 40 (2008): 109–30.

Yousef, Tarik. "Egypt's Growth Performance under Economic Liberalism: A Reassessment with New GDP Estimates, 1886–1945." *Review of Income and Wealth* 48 (2002): 561–79.

Zachs, Fruma. *The Making of Syrian Identity: Intellectuals and Merchants in Nineteenth Century Beirut*. Leiden: Brill, 2005.

Zebiri, Fruma. *Mahmud Shaltut and Islamic Modernism*. Oxford: Oxford University Press, 1993.

Zeine, Zeine. *The Emergence of Arab Nationalism with a Background Study of Arab-Turkish Relations in the Near East*. Beirut: Khayats, 1966.

Zhantiev, Dimitry. "Islamic Factor in the Consolidation of the Ottoman Rule in the Arab Provinces during the Reign of Sultan Abdulhamid II (1876–1908)." In *Authority, Privacy and Public Order in Islam*, edited by B. Michalak-Pikulska and A. Pikulski, 453–59. Leuven: Peeters, 2006.

Ziadat, Adel. *Western Science in the Arab World: The Impact of Darwinism, 1860–1930*. Basingstoke: Macmillan, 1986.

Ziadeh, Susan. "A Radical in His Time: The Thought of Shibli Shumayyil and Arab Intellectual Discourse (1882–1917)." PhD diss., University of Michigan, 1991.
Zirikli, Khayr al-Din al-. *Al-Aʿlam*. Beirut: Dar al-ʿIlm al-Malayin, 1927.
Ziyada, Khalid. *Husayn al-Jisr, 1845–1909: Hayatahu wa-fikru* [Husayn al-Jisr, 1845–1909: His life and thought]. Tripoli: Dar al-Insha' li-al-Sihafa wa-al-Tibaʿa wa-al-Nashr, 1982.
Zolondek, Leon. "The French Revolution in Arabic Literature of the Nineteenth Century." *Muslim World* 57 (1967): 202–11.

INDEX

ʿAbbas, Khedive, 207, 210–11, 213
ʿAbbas II, 171, 238
ʿAbbas Hilmi II, 95
ʿAbbasid era, 273
ʿAbduh, Muhammad, 8, 9, 14, 19, 21–22, 32, 77, 83, 93, 100, 113, 119–20, 124–25, 135–36, 158–59, 180, 200, 206, 215, 228–29, 231, 243, 264, 282, 288, 295, 304, 312–13, 353n107, 362n30, 371n192; and Adam and Eve, 175; Antun, response to, 188; background of, 168–69, 171, 194–95; British, friendly relations with, 207–9, 213–14, 217; on Christianity, as irrational, 189; Christianity, and Protestantism, distinguishing between by, 189; criticism of, 203–4, 206, 209–10, 213; and Darwin, 166–67; on education, 193–96; and educational reform, 208; on Egyptian poor, 237–38; evolution, attitude toward, 175–76, 191, 218; exile of, 169; on family, 196–97; fatwas of, 213, 223; heresy, reputation for, 214; Ibn Rushd, debate over, 185–86; and ʿIllish, differences between, 204–6; impiety, accusation of toward, 209; and *jinn* (spirits), 177; *kalam*, revitalization of, 181; and laws of nature (*sunan*), 173; and laws of societies, 176; and Le Bon, 190; legacy of, as far-ranging, 216–18; man's descent, interpretation of, 175; and miracles (*ayat*), 174; missionary schools, as threat, 194; as "modernist," 370n168; as *mujtahid* (reformer), 204; on Muslim civilization, decline of, 192; as Muʿtazilite, accusations of, 204–5; as nationalist pioneer, 373n227; natural knowledge, and faith, 172; and natural laws, 176; nature, causation in, debate over, 173–74; on Noah and the Flood, 178; nominalism, appeal to, 174–75; pardoning of, 170; pedagogy, views on, 193, 196; as pragmatist, 174; and predestination (*qada'*), 174; as rationalist, 167, 178, 360n6; reason, conception of, 178; religious instruction, rationalization of, 198; religious reform, call for, 170–71; and science, 164–66, 170, 173, 179; scriptural references, interpretation of, 177; and Spencer, 161–64, 166, 175, 195–97; and Sufism, 212; *taqlid*, attitude toward, 205, 212; and *tawhid* (unity), 211; as traditionalist, 204; *ʿulama*, conflicts with, 192–93, 195, 198, 201, 205–7, 214; *ʿulama*, rote learning of, as disagreeing with, 194; *umma*, concern with, 165; and ʿUrabi revolt, 169; Wahhabi, condemned as, 165, 190
Abdülaziz, 76
Abdülhamid, Sultan, 95, 113, 124, 126, 132, 137, 157, 169, 208, 210, 213, 255
Abdül Mecid, 334n107
Abi Adam (My father Adam) (Shahin), 307
abiogenesis, 102
adab (proper manners, morals, taste), 19, 39, 297

Addresses (Owen), 228
Adham, Ismaiʿil, 280, 282, 284, 383n51
ʿadl (justice), 133
Afghani, Jamal al-Din al, 22, 97, 100, 119–20, 137, 156, 166, 169, 182, 184–85, 189, 213, 217–18, 221, 223, 280, 282, 292, 310, 353n107; on Darwin, 121–22, 124–25, 167; Islam, and Muslims, distinguishing between by, 183; on materialism, 121–22, 124; on science, and Islam, 183; on socialism, 222
Africa, 35, 45, 87, 89, 255; Islam and Christianity, civilizing influence of in, debate over, 155
Afro-Asian Peoples' Solidarity Organization, 259
Agassiz, Louis, 28, 68, 278
agnosticism (*al-ladriya*), 118, 162, 164
Agrarian Party, 299
ahadith tales, 387n1, 387n2
ahl al-ʿilm (scientists), 220–21
Ahlam al-falasifa (Dreams of the philosophers) (Musa), 254
ʿAla atlal al-madhhab al-maddi (On the refutation of materialism) (Wajdi), 281–82
Al-Ahali (journal), 249
Al-Ahram (journal), 79, 95, 105, 250, 293
Al-ʿarus al-badiʿa fi ʿilm al-tabiʿa (The Marvelous Bride of Natural Science) (Shadudi), 57
Al-Azhar University, 19, 170, 171, 181–82, 193, 195, 206, 240, 283, 295–96, 313; Administrative Council at, 201; cholera outbreak, 202; curricular reform, as source of conflict, 202–3; deteriorating infrastructure of, 198–99; as madrasa, 203; new subjects (*al-ʿulum al-haditha*), introduction of, controversy over, 199; overcrowding at, 199; reforms, battles over at, 201, 204, 214–15, 217–18; rioting at, 202, 210; sciences, decline of, 201; spy system at, 209; students, and lack of discipline, 200–201; *ʿulama* in and reformers, battle between, 199–200
Al-Balagh (newspaper), 293
Al-Bashir (journal): *Al-Muqtataf*, attacks on, 63, 65
Al-din al-ʿilm wa al-mal aw al-mudun al-thalath (Religion, science, and wealth: The three cities) (Antun), 219–20, 223–24; "three cities" and city-states, connection to, 221
Al-din fi zill shiuʿaya (Religion in the shadow of communism) (Mazhar), 303
Al-Duhur (The ages) (journal), 298
Aleppo (Syria), 31
Alexandria (Egypt), 22, 25, 30–32, 76, 78, 230, 250, 310
Al-Fayyum (Egypt), 31
Al-haqiqa (The truth) (Shumayyil)
Al-Hilal (journal), 78–79, 88, 104–5, 167, 226, 228–29, 252–53, 255–56; socialism, discussion of in, 230–31
ʿAli, Muhammad, 20, 75–76, 83, 264
ʿAli, Muhammad Kurd, 387n119
Aligarh Muslim University, 120
Al-ʿilm wa-al-ʿulama (Zawahri), 214
ʿalim, 139, 140, 148–50, 153
Al-ishtirakiya taʿwuqa al-nuʿa al-insani (Socialism retards human progress) (Mazhar), 298
Al-Islam (journal), 203
Al-Islam la al-shiuʿaya (Islam, not communism) (Mazhar), 304
Al-Islam wa al-Nasraniya maʿa al-ʿilm wa al-madaniya (Science and civilization in Islam and Christianity) (ʿAbduh), 171, 186
Al-Islam wa usul al-hukm (Islam and the foundations of governance) (Raziq), 290
al-Jahiz, 140, 268–69
Al-Jamiʿa (The collective) (journal), 186, 188, 193, 238; social programs, approach to, 219
Al-Jarida (journal), 243
Al-jawahir fi tafsir al-Qurʾan (Gems of the exegesis of the munificent Qurʾan) (Jawhari), 312
Al-Jinan (journal), 30, 49, 58, 76, 103, 228
al-Jisr, Husayn, 5, 8–9, 12–14, 63, 131–32, 143, 145, 169–73, 181, 280–81, 310, 311; antimaterialism of, 149–50; background of, 132–34, 136–37; and First Cause, 149, 153; and four laws of evolution, 153–54; Islam, rationality of, 153; and *kalam* tradition, 148–49; on

man from animals, distinguishing from, theory of, 151–52; Marrash, differences with, 147; Paley, watchmaker analogy, 324n44; reform, and Ottoman state, approach to, as cautious, 157–58; on religious belief, as "most perfect science," 150; Rida, exchange between, 158; and *Risala*, 137–41; scholarship on, 381n23; and *shariʿa*, 156; and *ta'wil* (analogic reasoning), 150; on theory of evolution, as insufficiently proven, 151; Western learning, and Islam, 137–38

al-Jisr, Muhammad, 132–33

Al-Jumhuriya (newspaper), 303

Allah, Shah Wali, 212

Al-Lata'if (magazine), 4, 61

Al-Liwa (journal), 91, 210, 243

al-Maʿari, 266

Al-Mahrusa (journal), 32

Al-Majalla al-Jadida (The new magazine), 252

Al-Manar (The lighthouse) (journal), 78, 113, 158, 171, 175, 176, 180, 185, 211, 217, 292, 312; *tarbiya* (child-rearing and general education), articles on in, 197; and Wilson's speeches, reprinting of in, 234

Al-Mashriq (journal), 65

Al-Minya (Egypt), 31

Al-Mu'ayyad (newspaper), 77, 95, 105, 200

Al-Muqattam (newspaper), 4, 61, 79, 81, 105, 214, 215; circulation of, 80; and Demolins, praise of in, 90; pro-British stance of, 80, 94–96, 222

Al-Muqtataf (*The digest*) (magazine), 4–5, 8, 13, 20, 22–23, 25, 58, 61, 63, 68, 75–76, 78, 81, 90, 94, 105–6, 113–14, 117, 119, 129, 140, 153, 167, 172, 186, 190–91, 202, 215, 224, 226, 238, 242–43, 253, 262, 268, 271, 275–76, 282, 288, 292, 294, 301, 310, 315, 328n18; antipathy toward, 97; appeal of, reasons for, 32; Arabic reading, and popular science writings, rise of in, 30; Beirut phase of, 141; Bergson speech, publishing of, 128; British racial scientists, discoveries of, reporting on in, 87–88; circulation, increase in, 79; closing of, 303, 317; Darwin, as indebted to, 44; Darwin, first mention of in, 33, 142; Darwin, ideas of, and popularization, role of in, 26–27; Darwin, obituary of in, 40; Darwin, praise of in, 40; Darwin, theory of, implications of, and Christian defenses of scripture, tension between in, 34; (*The digest*) (publishing house), 77, 104, 293; distribution of, 31; early issues of, 142; on Eastern Question, 86; evolution (*tatawwur*), translation of, as first to popularize, 32; format of, 30; and Francis Galton, 245–46; geology, discussions of in, 35; glossaries (*mu-ʿaribat*) in, as staple, 32; Hanotaux, rebuttal of by, 184–85; human descent, writings of in, 36; Japan, on success of in, 91–93; Lewis speech in, 67; missionary culture, as shaped by, 44; older forms, connection to, stressing of in, 32; on origin of life, 101; on *Origin of Species* translation, 267; popularization of, commitment to in, 28; popular readership of, 31; popular science writings, 30; positivism, use of in, 279; praise of, 32; price of, 30; pro-British stance of, 317; prose style, in Arabic, use of in, 32; race, issue of in, 88, 246; science, popularizing of in, 32, 74; socialism, discussion of in, 230, 232; on "The Social Organism" (article), 82; sources of, 28; Spencer, death of, 83–84; Spencer, defense of in, 82–83, 85, 214; Sudan, war in, reporting on, 87; translation, of scientific terminology, 32; on Tyndall-Bastian controversy, 106–7; as unique, 30–31; Van Dyck, as ally of, 60; West, borrowing from in, 274

Al-Mustaqbal (The future) (paper), 104; censoring of, 243; *taqlid* (imitation), relinquishing, urging of in, 243

al-Nashra al-Usbuʿiya (College weekly news), 58, 63, 144

Al-Nil (journal), 203

Al-Shaʿab (The people) (journal), 265

Al-Shifa' (Curatives) (journal), 30, 78, 104

Al-Sihha (journal), 30

Al-Sudan (Sudan Times) (newspaper), 61

al-taʿawan al-ijtimaʿi (social cooperation), 228

Al-Tabib (The Physician) (journal), 58
al-Taftazani, 205
Al-Urwa al-Wuthqa (newspaper), 32
Al-Ustadh (newspaper), 95
Al-ʿUsur (The times) (magazine), 293, 299
Al-Waqaʿi al-Misriya (journal), 32, 169, 193
Al-Watan (journal), 32, 79, 105
al-wujudiya (existentialism), 314
American Artisan (magazine), 28
American Baconianism, 15
American Board of Commissioners for Foreign Missions (ABCFM), 45, 47–48, 51–52, 56
American Journal of Science (magazine), 28
American Revolution, 259
American University of Beirut, 4, 53. *See also* Syrian Protestant College
Americas, 34–35, 45, 76
Amin, Qasim, 197, 216, 282; liberation of women, call for, 87, 90, 196
anarchism, 240
Ancient Stone Implements (Evans), 36
Anderson, Rufus, 48–49, 52, 56, 333n100; reforms of, as failure, 50
Andover Theological Seminary, 45. *See also* Calvinist Divinity College
Anglo-Egyptian Condominium, 87
Anglo-Turkish Convention, 334n107
animal instincts: and animal intelligence, 38–39; and heritable experience (*al-ikhtibar al-mawruth*), 39; and materialism, 40
"Animal Instincts" (article), 40
Anthropology (Tylor), 82
anticolonialism, 208, 241
Antiquity of Man (Lyell), 35
Antun, Farah, 9, 14, 78, 173, 186–89, 219, 228–29, 231, 239–40; ʿAbduh, rebuttals to, 193; reform, preference for, 220; science, as intermediary, between religion and capital, 221; on socialism, 222; Wallace, support of, 223–24
Aqqad, Mahmud ʿAbbas al-, 293
"Arab Awakening" (Nahda), 73. *See also* Arab Renaissance; Golden Age; Nahda
Arab Civilization and Its Impact upon Europe (Sayhi), 280
Arab East, 15, 86, 192
Arabic Academy of Language, 301
Arabic Golden Age. *See* Golden Age
Arabic press, censorship of, 77; expansion of, 22, 75–76; financial backing, withdrawal of from, 77; public affairs, as powerful medium for debating, 77–78; radicalism, coverage of, 229
Arabic print culture, 76; Darwin, reading of, and classical Arabic texts, revival of interest in, 270; middle class, rise of, 22; popular science writings, rise of interest in, 30; rapid spread of, 75; reading communities in, as distinct, 23
Arabic Thought and Its Place in History (O'Leary), 271
Arabism, 255
Arab Renaissance (Nahda), 185, 342n2. *See also* "Arab Awakening"; Golden Age; Nahda
Argentina, 31
Aristotle, 5, 18, 84, 114, 121, 142, 176, 181, 188
Artin, Yaʿqub, 216
Ascent of Man (Drummond), 234
Ashʿari school (*madhhab*), 205, 281
Asia, 34–35, 45, 217, 255, 291
Asian-African Conference, 259
ʿAsir, Shaykh Yusuf al-, 63, 332n89
Asyut (Egypt), 31
Atatürk, Mustafa Kemal, 89, 91, 210, 216, 241, 254, 289–90, 313, 371n196
atheism (*al-muʿattila*), 118, 239, 265, 280, 282, 284
Atlas of Creation (Yahya), 388n10
atomism, 99, 109, 121–22, 140
Australasia, 6
Averroès el l'averroïsme (Renan), 186
Ayalon, Ami, 328n18
Azhari, 189, 240, 296

Bacon, Francis, 190, 240, 281
Badri, Shaykh Yusuf al-, 308–9
Baghdad (Iraq), 31
Bakhit, Muhammad, 248
Balkans, 87
Balkan Wars, 126
Banna, Muhammad al-, 199
Barakat, Dawud, 293
Barbanell, Maurice, 314
Basaʾir al-Nasiriya (Saqi), 181

Bashrun, Abu Bakr ibn, 124
Bastian, Henry, 28, 102, 105–7, 278
Bayram, Muhammad, 199
Becker, Carl, 290
The Beginnings of Life (Bastian), 102
"The Beginnings of Life and Death" (article), 101–2
Beirut (Lebanon), 22, 25, 28, 31, 49, 62, 72, 75–77, 80, 83, 100, 114, 119, 132, 135–36, 144, 169–70, 194, 339n157; American missionaries in, 45–46; American Protestant missionaries in, 47; Catholic missionaries in, 52; Catholic-Protestant rivalry in, 63; foreign missionaries in, and "new sciences," 27, 50–51; printing presses in, 47–48, 58; reputation of, as center of liberal thought, 157
Beirut High School for Boys, 46, 49–50; closing of, 48
Beirut Masonic Lodge, 62
Bendyshe, Thomas, 34
Bentham, Jeremy, 82, 231
Bergson, Henri, 128
Bey, Ahmed Nashid, 97
Bey, Sabahaddin, 90
Bichat, Marie François Xavier, 108
The Biology of the Spirit (*Hayat al-ruh fi diwa' al-ʿilm*) (Sinnott), 302, 305
Birbari, Rizq'allah al-, 33–34, 38, 142
Bird, William, 48
Bliss, Daniel, 51–52, 56–57, 67, 70, 72, 336n119
Blumenbach, Johann Friedrich, 34
Blunt, Wilfrid Scawen, 161–64, 168, 200, 209, 244
Board of Foreign Missions, 56–57
Bolshevik Revolution, 241, 247, 251, 298
Bolshevism, 226, 238, 245, 287, 298; in Egypt, 248–49, 300
Book of Animals (al-Jisr), 131–32, 268
The Book of the Divine Secrets, 364n62
The Book of the Magnificent Garden of Natural History (Jesuit mission press), 144
Booth, William, 68
Bose, Jagadish Chandra, 312
Le Bosphore Égyptien (newspaper), 78
Boyle, Harry, 80
Bozzano, Ernesto, 283
Brazil, 31, 229
British Agency, 80
British and Foreign Bible Society, 47
Büchner, Ludwig, 18, 100, 104–5, 107–8, 110–11, 116–18, 125, 128, 138, 169–70, 264–65, 267, 280
Buckle, Thomas, 257
Buenos Aires (Argentina), 22
Buffon, George-Louis Leclerq, 34
Bukhayri, Shaykh al-, 203
Bulaq (Egypt), 76
Bustani, Butrus al-, 30, 47, 59, 62, 76, 103, 228, 334n105; new sciences (*al-ʿulum al-haditha*), as key figure in, 49

Cairo (Egypt), 22, 25, 30–32, 61–62, 72, 75–79, 81, 120, 126, 170, 230, 250, 311, 313; newspapers in, rise of, 95; strike in, 209
Cairo Law School: strike at, 242
Cairo Scientific Society, 28
Cairo University, 216, 308. *See also* Egyptian University
Calcutta (India), 120
Calvinist Divinity College: Brethren fraternity at, 45. *See also* Andover Theological Seminary
Canada, 31, 88
Capuchins, 45
Carmelites, 45
Carnot Sadi, 230
Catholic modernism, 164
Catholics: Muqtataf group, attacks on, 63
causality, 187
Certain Signs in the Wonders of the Heavens and Earth (Hurani), 143; marvels of creation, emphasis on in, 144
Chain of Being, 142; man's place in, as unquestioned, 146
Character (Smiles), 92
Cheikho, Father Louis, 63, 69, 125; Darwin's theory of evolution, as absurd, 65
China, 6, 51, 93, 263
Christianity, 8–9, 128, 155–56, 171, 178, 218; and Christian West v. Muslim East, 182; church and state, separation of, 188–89; and free will, 184; Islam, chasm between, 182; and Protestantism, 189; and science, 165, 186, 188–89

Church Missionary Society, 95
Circassia, 88
civilization: and civilizational progress, 291; "clash of civilizations," 183; and law of nature, 90; as social progression, 12
Civilization and Progress (Crozier), 284
collectivism, 129, 223–24; and Darwin, 226; early Muslims, as model for, 304; and mutual aid, 227; social cooperation, focus on, 226; and World War I, 232
colonialism, 87, 317
"The Coming of the Superman" (Musa), 246
"Commentary on Spencer" (article), 85
communism, 229; Arab solidarity, as threat to, 304
Communist Party, 251
Comte, Auguste, 73, 176, 279, 291, 297
Condillac, Étienne Bonnot de, 38
Contemporary Review (magazine), 231
Copernican astronomy, 46
"The Corrupt Philosophy of the Materialists" (speech), 114
creation, 8, 13; of man, 218
Creation (*al-khalq al-mustaqil*), 42
creationism, 309–11
Cresson, Andre, 283
Crimean War, 25, 198
Cromer, Lord, 18, 77–80, 95, 128, 171, 183, 202, 207–9, 211, 215, 222, 241, 370n168
Crozier, John Beattie, 284–85, 288, 297
Cuba, 31
Czechoslovakia, 256

dahriya, 121
dahriyin (materialism and impiety), 121
Damanhur (Egypt), 31
Damascus (Syria), 31, 80
Danawi, al-Hajj al-, 134
Dar al-ʿUlum (Teachers' Training College), 193, 200, 313
Darwin, Charles, 1, 4–5, 9–10, 13–16, 18, 28, 43–44, 58, 65–70, 81–83, 90, 97, 100, 101–3, 105, 108, 110, 118–19, 121–22, 124–25, 128, 140–42, 144, 149, 153, 163–64, 166–67, 170, 191, 214, 232, 235, 245–46, 253, 264, 270, 275–76, 280–83, 286, 288, 305, 320n13, 320n14, 353n107, 389n24; "as accomplice of the devil," 310; as ambiguous, 7; Arabic readers of, 7–8, 12, 20–23, 132; and cooperation, 233–34; and court case, 307–8; criticism of, 117; and First Cause, 279; genealogical ancestry of, by Mazhar, 267–68; global appeal of, 6–7; intelligence, and adaptive change, 36; intelligence, in animals, 36; interwar reading of, 129; and "law of mutual aid," 226; liberalism, association with, 256; M and N, and materialism, 131; and morality, 39; and natural laws, 240; and natural selection, 87, 270; nature, as model for collective advancement, 226–27; notebooks of, 36; *Origin of Species*, translation of, 261–63; popularization of, 26–27; as rationalist, 266–67; religious sentiments, as highest form of man's mental abilities, 38; skepticism toward, 309; and socialism, 224, 239; theories of, as hypothetical, 176; and theory of descent, 33–34; and theory of evolution, 33, 42; waning appeal of, 258–59. *See also* evolution; theory of evolution
Darwinian Revolution, 256
Darwinism (*al-madhhab al-Darwini*), 41, 67, 81, 166–67
Darwinism and Other Essays (Fiske), 41
Davis, Natalie Zemon, 22
Debs, Eugene, 229, 235, 241
Degeneration (Nordau), 234
deism (*al-ilahiya*), 118, 296
democratic socialism, 231, 240–41
Democritus, 122
Demolins, Edmond, 89–90, 97, 231
Dennis, Reverend James, 44, 57, 67–69, 337n136
Descartes, René, 38
The Descent of Man (Darwin), 12, 33, 36, 38, 69, 226, 285
De Vries, Hugo, 247
Dewey, John, 195
d'Holbach, Baron, 104

Dijwi, Shaykh Yusuf al-, 296–97
Dinshaway case, 231; and anticolonialism, 241; "problem of the peasantry" (*mushkillat al-fallah*), 241
Discourse on Eight Words (Marsafi), 19
Divine Comedy (Dante), 296
Divine Instrumentality for the World's Conversion (circular), 48
divine laws, 165
Dodge, David Stuart, 67, 72, 79, 341n169
Draper, John William, 1, 5, 190–92
dustur (constitutional regulation), 303

East: "Awakening" of, 11
East Africa, 155
Easternism, 292
Eastern League, 255
Eastern Question (*al-mas'ala al-sharqiya*), 10, 86–87
Eastern Scientific Society, 28
Eastern Scientific Society (*Al-Majma' al-'Ilmi al-Sharqi*), 62
Economist (magazine), 28
education, 18–20, 23, 44, 46, 50–53, 56–57, 61, 71, 83, 89–92, 97, 103, 114, 126, 133–37, 144, 146, 150, 157, 164–68, 170, 171, 182, 190, 193–98, 199, 203, 206–8, 215–18, 220, 230, 232, 235, 240–41, 243, 246–47, 250, 252, 263–66, 287, 289, 291, 293, 298, 300, 303, 310, 313, 316, 345n46, 345n48, 345n40, 360n3, 367n115
Education: Intellectual, Moral and Physical (Spencer), 83
effendiya class, 227, 256, 264, 317, 375n26
Egypt, 1, 4, 11, 18, 25, 30–31, 56, 72–73, 80, 83, 90–91, 104–5, 124, 128, 133, 166, 170–71, 194, 196, 213, 215–17, 231, 252–53, 255, 258, 287–88, 291, 298–300, 304, 307, 309–10, 386n110; Arabic, books translated into in, 75; and Bolshevism, 248–49, 300; British occupation, and imprisonment, of journalists, 79; British occupation, as welcome, 113; British occupation of, 10, 19–21, 87, 94–95, 127, 155, 164, 169, 182, 209, 317; British occupation of, and Darwin's ideas, impact of on, 75; caliphate in, question of, 290, 292; and debt crisis, 77; Easternism in, 292; Europe, social order in, and parallels between, 220; and Fabians, 241; *kalam*, revitalization of in, 181; labor unions in, 229–30; labor unrest and strikes in, 222, 230, 234, 249; land ownership in, 236; land reform, promotion of in, 239; literacy rates in, 22, 316, 329n23; Nahda in, 256; nationalist movement in, 95–96, 244, 257, 268, 271; newspaper readership in, expansion of, 253, 316; political violence, rise of in, 317; poor in, 237; press censorship in, 77–78; as print capital, of Arab world, 76; race, issue of in, 88, 246–47; radicalism, British monitoring of in, 248; religion, public role of in, 289; *savants* in, 50; socialism in, 225, 227, 229, 245, 247, 250–51, 254; and Sudan, 88–89; Syrian journalists in, 78; and 'Urabi crisis, 172
Egyptian Academy for Scientific Culture, 293
Egyptian Gazette (newspaper), 80
Egyptian National Bank (Bank Misr), 293
Egyptian Revolution, 231, 234, 274
Egyptian Socialist Party, 241, 249–52, 257
Egyptian University, 32, 216, 287, 313. *See also* Cairo University
Egyptology, 255
empiricism, 311
Encyclopaedia (Bustani), 228
England, 86, 113, 141, 155, 239–40, 242, 250, 282, 300; Islam, interest of in, 156; literacy rates in, 22; and materialism, 163; superiority of, explanation for, 89–90. *See also* Great Britain
Epicureanism, 121, 279
epiphenomenalism, 108
epistemological anarchism, 283
ethics (*'ilm al-ikhlaq*), 39
eugenics (*'ilm al-ujaniya*), 245, 254, 256–57, 286; and Fabians, 246
Eugenics Education Society, 246
Eugenics Laboratory, 246

Europe, 6, 8, 11, 22, 28, 34–35, 41, 52, 76, 84, 87, 89, 91, 113, 124, 127, 131, 156, 161, 183–84, 190, 193, 206, 218, 225, 231, 236, 239, 242, 249, 254, 256, 281–82, 296; civilizational progress in, 188; individualism, attack on in, 232; materialism, and moral decline of, 163, 185; as model, 243; rise of, and science, 10, 182, 192; social disorder in, and Egypt, parallels between, 220; spontaneous generation, debate over in, 101–2; superiority of, 12
Evangelical Alliance, 69
evolution, 8, 90, 101, 108, 112, 132, 144, 165, 195, 256, 309; discourse over, and fate of "civilizations," 10–11; and evangelism, 44; and fossil forms, 153–54; four laws of, 153; and materialism, 13, 69; and origin of life, 278; and Qur'an, 175; as "scientific Sufism," 253; as social development, 176; teaching of, 310. *See also* theory of evolution
Evolution and the Origin of Life (Bastian), 102
evolutionary collectivism, 235
evolutionary materialism, 131–32
evolutionary socialism, 225, 228, 231, 235, 238, 242, 257; collectivism, and mutual aid, 227; "soft" radicalism of, 226
evolutionism, 15, 259

Fabian Essays, 243
Fabianism, 257
Fabian Party, 241–42
Fabians, 227, 229, 241, 245; and eugenics, 246
Fabian socialism, 243, 252
Fabian Society, 241, 250
Fahmi, Mansur, 387n119
faith (*al-iman*), 150
Falaki, Mahmud al-, 31
falsafa, 14
Falsafat al-nushu' wa al-irtiqa' (Philosophy of evolution and progress) (Shumayyil), 265, 276
Falsafat al-takwin (Philosophy of creation), 280, 384n54
Farabi, Abu Nasr al-, 154, 221
Faraday, Michael, 28
Farag, Nadia, 328n18
Fatah, Shaykh Safwan Abu, 250
Fénelon, François, 197
Ferrer, Francisco, 229
fetishism, 111, 123
Fichte, Johann Gottlieb, 118
Filawpuns, 294
fiqh (jurisprudence), 133, 140
First Cause, 13, 149, 153, 266–67, 279, 281, 283, 297
First International, 239
Firuzabadi, 275
Fiske, John, 41, 43
Fiske, Willard, 2, 319n6
Fortnightly Review (magazine), 28, 223
fossil discoveries: and prehistoric men, 35
France, 16, 103, 128, 146, 184, 239, 255, 282
Franco-Prussian War, 228
Franklin Book Program, 301
Frazer, James, 111
Freemasonry, 61–62
French Enlightenment, 256
French Revolution, 192, 259, 298
Freud, Sigmund, 254, 257
Fuad I, 302, 387n119
Fuat, Beşir, 111

Galapagos Islands, 262
Galileo, 9, 189, 287–88
Galton, Francis, 245–46, 257
Gandhi, Mahatma, 241, 255
"Geber," 271
Geley, Gustave, 283
Generation of Animals (Aristotle), 268
geography, 46
George, Henry, 238–40
Germany, 233, 239, 249, 256
germ theory, 102
Ghali, Butrus Pacha, 232
Ghalib, 'Uthman, 27
Ghazali, Abu Hamid al-, 8, 14–15, 99, 100, 121–22, 134, 139–40, 148–49, 165, 167, 179, 186–87, 314
Ghazir (Lebanon), 63
Glisson, Francis, 108
gnosticism, 307
Goethe, Johann Wolfgang von, 262
Golden Age, 11, 186, 191, 221, 271–72,

322n31. *See also* "Arab Awakening"; Arab Renaissance; Nahda
Gordon, Charles George, 155
Gorst, Eldon, 243
Gospels, 138
gradualism, 21
Graves, Philip, 80
Gray, Asa, 68
Great Awakenings: and Second Coming of Christ, 45
Great Britain, 52, 91, 102. *See also* England
Greater Syria, 28, 31, 44, 52; Christian missionaries in, 45, 47
Greece, 1, 121, 191
group selection, 233. *See also* social selection
Guizot, François, 11–12, 82, 123

hadara (civilization), 323n38
Haddad, Ibrahim al-, 298
Haddad, Niqula, 228–29, 235, 249; on individualism, 234
Hadiqat al-Akhbar (journal), 76
hadith, 138–39, 228
Haeckel, Ernst, 5, 8, 17, 27–28, 108, 118, 128–29, 278, 280
Haʿik, Yusuf al-, 68
Hakki, Izmirli Ismaʿil, 156
Haldane, J. B. S., 286
Hama (Syria), 31
Hanafi, Abu Saud al-, 140
Hanafi school (*madhhab*), 207, 212
Hanotaux, Gabriel, 171, 183, 282; Islam and Muslims, distinction between, 184
Hardie, Keir, 242
Hartmann, Martin, 78
Haykal, Muhammad Husayn, 241, 293, 303
"The Heavens and Earth (*al-samaʿ wa al-ard*) and Man" (Antun), 223
Hegel, Georg Wilhelm Friedrich, 84
heterogenesis, 101–2
High Council on Islamic Affairs, 315
hisba law, 308
History of Animals (Aristotle), 18
History of Civilization in Europe (Guizot), 184
The History of Science and the New Humanism (Tarikh al-ʿilm wa al-insaniya al-jadida) (Sarton), 302
History of Socialism (Kirkup), 239
History of the Conflict between Religion and Science (Draper), 1, 190
History of the Warfare of Science with Theology in Christendom (White), 4, 9, 288–89, 311
Hobhouse, L. T., 232, 244, 286
Hodge, Charles, 34
Homs (Syria), 31
huduth (temporality), 152
hukuma (government), 133
Humarat Monyati (journal), 210
Hume, David, 8, 14, 187
Hungary, 248–49
Hurani, Ibrahim al-, 57–58, 125, 141, 143, 147; Shumayyil, criticism of, 117–18
huriya (freedom), 133
Husayn, Hasan, 17, 19
Husayn, Muhammad, 387n119
Husayn, Taha, 282, 287
Husayn affair, 287, 289
Huxley, Thomas Henry, 27, 35–36, 102, 108, 118, 190, 197, 280, 320n13, 320n14, 367n127
hylozoism, 108, 279

Ibn ʿAbbas, 152
Ibn Arabi, 99, 133
Ibn Haytham, 200
Ibn Hayyin, Jabir, 271
Ibn Hisham, 387n1
Ibn Khaldun, 8–9, 11–12, 82, 123, 167, 176, 322n31, 323n38, 382n31; modern sociological inquiry, as forefather of, 270
Ibn Miskawayh, 196, 269–70, 382n26, 382n29
Ibn Qayyim, 314
Ibn Rushd, 5, 8, 14–15, 173, 176, 181, 185–86; and causality, 187; defense of, 189; and Prime Mover, 188
Ibn Rushd wa falsafatuhu (Ibn Rushd and his philosophy), 186
Ibn Saʿud, 211
Ibn Sina, 181, 274
Ibn Taymiyya, 139, 165, 167, 213
Ibrahim, Hafiz, 293

Ibrahim, Jindi, 79
idealism, 279
Ihya ulum al-din (Ghazali), 134
iʿjaz al-Qurʾan, 295–96
ijmaʿ (consensual law), 139
ijtihad (independent juridical and theological opinions), 205, 212–13; dangers of, 137
Ikhwan al-Safaʾ (Brethren of Purity), 5, 8, 270
ʿIllish, Shaykh Muhammad, 204–6
ʿilm (knowledge), 16, 73, 173, 189, 220, 297; science, distinguishing between, 65–66
ʿilm al-kalam (speculative theology), 154, 181–82
ʿilm al-mantiq (logic), 134
ʿilm usul al-din, 154
imkan (possibility), 152
imperialism, 210, 243, 245
impiety (*kufr*), 35
ʿInan, ʿAbd Allah, 250
Inbabi, Shams al-Din al-, 199–200
India, 6, 51, 81, 93, 95, 120, 155–56, 183–84, 238, 242, 248, 255, 257, 263
individualism: attack on, 232; socialism, as opposite of, 234; utilitarian ethics, morality of, 226
Inge, Dean, 286
Inquisition, 189
intangible soul (*al-nafs ghair mahsusa*), 39
intelligent design, 309
International Congress of Orientalists, 28
International Organization of Islamic Brotherhood, 313
intuitive knowledge (*ʿilm badihiya*): and moral faculties, 38
Iphigenie (Racine), 105
Iqbal, Muhammad, 156
Iraq, 238
Ishtirakiya (*see also* socialism), 228–29
Iskandarani, Muhammad Ibn Ahmad al-, 179–80, 364n63, 387n119
islah (reform), 165, 185
Islam, 8–9, 89, 128, 132, 133, 137–38, 164, 167, 171, 180, 192, 204, 218, 248, 303; Christianity, chasm between, 182; civilization, as kind of, 168, 182–84; Golden Age of, 191; and *ijtihad* (reasoning against conventional interpretation), 139; as impediment, to science, 183; missionaries, attack on, 155–56; rationality of, 153, 159; and reason, 185; reason, and nature, reconciling with, 178, 181; and science, 93, 156, 158, 165, 177, 181–83, 186, 188; socialism, as compatible with, 239; and Third Way, 304–5
Islam and Civilization (Wajid), 93
Islamic Charitable Society, 231
Islamicism, 255
Ismaʿil, Khedive, 20, 30, 76–77, 79, 199
Israʿiliyat tales, 307
Istanbul (Turkey), 22, 75, 80, 126, 135, 156–57, 170, 194
istifrad (individualism), 223
Italy, 249
itifaqiyin (coincidentalists), 145
Ittihad Party, 311

Jabarti, ʿAbd al-Rahman al-, 200
Jabarti, Hasan al-, 200
Jamiʿat al-Funun (Society of Arts), 134
Jamiʿat al-Maqasid al-Khayriya (Charitable Aims Society), 134–35
Japan, 6, 81, 95, 216; success of, 91–93
Jaridat al-Tarablus (Tripoli news) (journal), 157
Jawhari, Tantawi al-, 180, 312–15
Jazaʾiri, Tahir al-, 135
Jerusalem (Israel), 31; American missionaries in, 45
Jessup, Henry, 51–52, 60, 68, 71, 319n4, 338n149
Jesuits, 45, 47–48, 50, 63, 116
jihad, 137
jinn (spirits), 177
Journal des Débats (journal), 183
Judaism, 178; Jews, mass conversion of, 45
jurthum (germ plasm), 122

kalam (scholastic theology), 14, 133, 148–49, 154, 181–82; scholastic disputation (*gadal*), connotations of, 155
Kamal, Husayn, 238
Kampf um den Entwickelungs-Gedanken (Haeckel), 17
Kaneko, Baron, 95

Kant, Immanuel, 84, 297
Das Kapital (Marx), 224
Karmili, Anistas al-, 387n119
Kelvin, Lord, 118. *See also* William Thompson
Kemalism, 251, 254
Khalwatiya movement, 133
Khan, Genghis, 121
Khan, Sayyid Ahmad, 120, 156
Khartoum (Sudan), 61, 80
Khayyat, As'ad Ya'qub, 333n102
Khedival School, 313
Khuli, Amin al-, 292
Khuri, Khalil al-, 76–77
Kidd, Benjamin, 232, 286, 298–99
Kirkup, Thomas, 239
Kitab al-tawhid (Wahhab), 211
Kitab al-thawrat (The book of revolutions) (Musa), 259
Kitab nizam al-halaqat fi silsilat dhawat al-fiqarat (Classification of Vertebrates) (Post), 57
Kitchener, Herbert, 87
knowledge (*ma'rifa*): belief, distinction between, 14; and rationalism, 14–15; science (*'ilm*), distinguishing between, 65–66; and trust, 14
Knowledge (magazine), 28
Kropotkin, Piotr, 28, 227, 232, 245, 286, 299, 320n14; mutual aid idea of, 304
Kun, Bela, 248

Lamarck, Jean-Baptiste, 8, 41, 253, 262, 270, 282; and theory of transformism, 41
L'âme est immortelle (Vinnault), 282
The Land and the Book (Thomson), 52
language: organicist conception of, 273
Lausanne (Switzerland), 229
Law of Association (1923), 251
Law on Public Education (1869), 136
League of Nations, 255, 300; international cooperation, as new world order, 234
Le Bon, Gustave, 89, 183, 190, 231, 291, 297; and "evolution of races," 12
"Lecture on Teaching and Learning" (Afghani), 182
Leipzig (Germany), 47
Lenin, Vladimir, 249
Lewis, Edwin, 68, 342n183; commencement speech of, 65–67; Darwin, praise of, 66–67; dismissal of, 69–70; *'ilm*, reorientation of by, 66
Lewis affair, 65, 71, 102, 117–18
L'Humanité (journal), 242
Liberal Constitutional Party, 293, 300
liberalism, 226, 231; Darwin, association with, 256; decline in, 257–58
Libya, 126
The Life of Charles Darwin (Bettany), 266
Life of Matter (Synott), 279
Linnaeus, Carl, 34
Locke, John, 36, 38
Lois psychologiques de l'évolution des peuples (Le Bon), 291. See also *Sirr tatawwur al-umam* (Secret of evolution of nations)
London (England), 22, 47, 229, 241
Lubbock, John, 28, 35–36, 111, 227, 312
Luther, Martin, 123
Lyell, Charles, 35–36

Ma'arri, Abu al-'Ala al-, 99, 112, 124
Mabadi' ara' ahl al-madina al-fadila (Principles of the people of the virtuous cities) (Farabi), 221
Mabadi 'ilm al-nabat (Principles of Botany) (Post), 57
MacDonald, Ramsay, 242, 250, 252, 257
MacPherson, Hector Carsewell, 84
madaniya (civilization), 323n38
La madha ana mulhid (Why I am an apostate) (Adham), 280
Madhhab al-nushu' wa al-irtiqa' (The theory of evolution and progress) (Mazhar), 284
"Madhhab Darwin" (article), 191
Madrasa al-Sultaniya (Sultan's School), 119, 136, 143, 169, 172; shut down of, 170–71
Madrasa al-Wataniya (National School), 334n105; and American missionaries, 134–35; curriculum, complaints of, 134; science-oriented curriculum of, 135
Maghribi, 'Abd al-Qadir al-, 387n119
Mahdiya, 87, 89
Mahfouz, Naguib, 252
Mahometanism, 182–83

Majallat al-Azhar (journal), 283. See also *Nur al-Islam* (journal)

Makarius, Shahin, 61, 79, 327n10; British, ties to, as strong, 80; Egypt, move to, 72; political radicalism of, 62; professional and political clubs, founding of by, 62

Malqa al-sabil fi madhab al-nushu' wa al-irtiqa' (Mazhar), 284, 286

Malta, 47

Mamlakat al-Nahl (The bee kingdom) (Shadi), 227, 294

man: reason, and animals, 38; speech, and divine intelligence, as proof of, 38; uniqueness of, 42–43

Manifesto of German Intellectuals, 128–29

Man's Place in Nature (Huxley), 35

"Man's Place in the Universe" (Wallace), 223

Mansur, Caliph, 186

Mansuri, Mustafa al-, 227–28; and evolutionary socialism, sympathies for, 238–39; nature, theology of, 240

maqasid (practical and ethical aims), 138–39

Maqasid society, 136

Maraghi, Mustafa al-, 217

Maraghi, Shaykh Mustafa al-, 295

Marrash, Fransis Fath Allah al-: anxiety of, over materialism, 145–46; background of, 144–45; and chain of descent (*silsilat al-tanazul*), 146; and miracles, 147; religion, and human nature, as fundamental part of, 147; and revelation, 146–47

Marsafi, Husayn al-, 133

Marsafi, Shaykh al-Azhar al, 19

Marsafi, Shaykh ʿAli al-, 264

Marun, Antun, 250

Marx, Karl, 10, 21, 221, 225, 232, 239, 241, 243, 245, 254, 257–59, 304, 320n14; and radicalism, 224; and socialism, 224

Marxism, 221, 225

maskh, 269

maslaha (interest), 139

materialism (*al-maddiya*), 14, 85, 103, 108, 110, 119–21, 123–25, 127, 131, 210, 284, 353n107; and animal instincts, 40; and animal intelligence, 39; criticism of, 114, 116, 128, 138, 140, 148–50, 152, 280, 282; and Darwin, 131; defense of, 99–100, 109, 129; and German aggression, 128; and moral decline, 163, 184; popularity of, 146; and positivism, 279; and theory of evolution, 13, 69, 310; World War I, effect on, 233

Mawardi-al, 152

Mazhar, Ismaʿil, 8, 15, 17, 20, 128, 232, 253, 261, 263, 283, 305, 309–11, 313, 317, 382n31, 386n110; Afghani, criticism of, 284, 292; *Al-ʿUsur* (The times), magazine of, 293–95; Anglo-Saxon political model, preference for, 295, 300; and Arabicization, 302; Arab solidarity, embracing of, 301; arrest of, 300–301; Asiatic mentality, and human mind, universal development of, 291; background of, 264; Bolshevism, arguments against, 298; communism, criticism of, 304; cooperation, as greater natural law, 286; Darwin, and Ikhwan, comparison of by, 270; Darwin, as causalist, 279; Darwin, as rationalist, 266–67; Darwin, genealogical ancestry of by, 267–68; *dustur* (constitutional regulation), of language, call for by, 303; East and West, differences between, reasons for, 289–90; as *effendiya*, 264; on evolution, 284–86; evolution, and origin of life, 278; freedom of thought, defense of, 287; glossaries, reliance on, 275; Islam, and Arab identity, and mutual responsibility, 304; and mutual aid, 286; natural selection, terms for, 273; *Origin of Species*, as "manual for belief," 272; positivism tendencies of, 278–79, 291; on progress and *jummud* (stagnation), as coexisting, 288; and rationalism, 287; reform measures, call for, in Egypt, 300; religion, and mutual aid, 286; religion, social function of, 285–86, 289; science, popularization of, 293, 296, 301; on science and religion, as natural instincts, 288; Shumayyil, criticism of, 284; socialism, argument against, 298–99; Socrates, influence on, 266; soul, belief in, 285; and spontaneous generation, 276; stagnation of thought, as cardinal sin, 292; theory of evolution, and religion, 276, 278; tran-

scendentalism of, 279–80; translation, philosophy of, 272–76; translation, second edition of, 271–72; *ʿulama*, as critical of, 295; Unknowable, belief in, 284; Wajdi, attack on, 297
McCabe, Joseph, 17
McCosh, James, 40, 43
mechanism, 108
Mediterranean, 45
Memoirs of the Anthropological Society, 34
Mesopotamia, 248
Mexico, 31, 229
Middle East, 163, 255
Milan (Italy), 75
Mill, John Stuart, 11–12, 73, 82, 320n14
Milne-Edwards, Henri, 313
Minshaway, Ahmad Pasha al-, 97
miracles (*aqat*), 174
Misbah al-Sharq (newspaper), 95, 210
Misr al-Fatat (newspaper), 105
missionaries, 334n106; attacks on, 135; and Beirut, Catholic missionaries in, 53; competition among, 50; conversion, difficulty of, and science, use of, 57; and Copernican astronomy, 46; in Greater Syria, 44, 47; and Islam, attack on, 155–56; in Jerusalem, 45; and journals, 58; and Madrasa al-Wataniya (National School), 134–35; and natural science experiments, 47; and "new men," 44; and pan-Islamic press, 95–96; printing press, and Old and New Testaments, 47; proselytizing of, 27, 44–46; Protestants and Catholics, rivalry between, 63; schools, a threat, in Muslim lands, 194; and Social Gospel movement, 51
Mivart, George Jackson, 68, 118
mizan (Aristotelian golden mean), 223
Modern Egypt (Cromer), 183
monism, 108
monotheism, 112, 178
Montesquieu, 12
moral education, 19
morality: and instinct, 39
moral thought (*adab*), 39
La mort et son mystère (Flammarion), 282
Mount Lebanon, 48, 50
Mubarak, Ali, 31
Muhammad, 138–39, 150
Muhammadiya, 205
Müller, Max, 28
Muqtataf group, 231, 311
Musa, Salama, 20, 31, 83, 91, 95–96, 104, 113, 129, 224, 232–33, 264, 276, 287, 291, 293–94, 302, 313, 317, 387n121; *Al-Hilal*, as editor of, 252; *Al-Majalla al-Jadida* (The new magazine), founding of, 252; arrest of, 258; autobiography of, 258; background of, 241; British racial prejudice, hatred of, 242; Darwin, as popularizer of, 256; Egyptian Socialist Party, denouncing of, 252; and eugenics, 246, 254, 257; evolutionism, commitment to, 259; expulsion of, 251; heresy, accusation of, 256; and natural selection, 245; Pharaonism, proponent of, 247; race, views on, 246–47; reputation, decline of, 258; science, popularization of, as mission of, 253; sexual equality, importance of to, 247; sexual selection, promotion of, 247; social engineering, interest in, 256; on socialism, 243–45, 258; Society of the Egyptians for the Egyptians, boycotting of, 255–56; state intervention, support of, 247, 257; and theory of evolution, 253–54; as utopian, 252; and West, as model, 256
Muslim: stagnation (*jumud*) of, 192
Muslim Brotherhood, 313, 315, 317
Muslim code of ethics: fasting: *hijab* (veiling), 138; *khuttub* (sermons), 138; laws of inheritance and punishment, 138; prayer, 138
Mussolini, Benito, 298
mutakallimun, 8, 14
muʿtazila, 8, 167, 204–5, 370n170
Müteferrika, Ibrahim, 75
Mutran, Khalil, 293
mutual aid, 226; and collectivism, 227; and mutual responsibility (*al-takaful al-ishtirakiya*), 304; and religion, 286
Mutual Aid (Kropotkin), 227, 232
Muwaylihi, Ibrahim al-, 95, 210

Nabhani, Yusuf al-, 135–36, 212–13
Nadim, ʿAbd Allah, 200
Nadim, Abdullah, 95

Nägeli, Carl, 107
Nahas, Mustafa al-, 299–300
Nahda (Renaissance), 1, 21, 31, 145, 256, 272, 287; first one, 11; second one, 11; as term, 133. *See also* "Arab Awakening"; Arab Renaissance: Golden Age
Najd, 157, 213
Napoleon, 50, 298
Nasif, ʿAssam al-Din Hifni, 294
Nasser, Gamal Abdel, 218, 258–59, 303–4
nationalism, 21, 95, 244, 255; as anti-British sentiment, 96
Nationalist and Subject Races Committee, 244
Nationalist Party, 91, 243, 313
National Socialist Party (Nazis), 256
natural history (*qism bi tarikh al-tabiʿiya*), 39, 143
"The Natural History of Societies" (*Tarikh al-ijtimaʿ al-tabiʿi*) (article), 85
naturalism, 108, 120–21, 140
natural law, 108, 142, 176; and causation, 174; civilizations, rise and fall of, 86
natural science, 126
natural selection (*al-intikhab al-tabiʿi*), 17, 39, 41, 82, 87, 117–18, 153, 176, 267, 276; dismissal of, 142; political implications of, 245
natural theology, 141–44; and *shahada* (nature's testimony), 145
nature: and knowledge, 173
Nature's Testimony (Marrash), 144, 146
nayshiriya (naturalists), 120–21
Nazariyat al-tatawwur wa asl al-insan (The theory of evolution and the origin of man) (Musa), 253
Near Eastern Crisis, 155, 182
Neoplatonism, 279
New Testament, 189; Arabic translation of, 47
Newton, Isaac, 40, 281, 287
New York (New York), 22, 229
Nietzsche, Friedrich, 241, 246
Nimr, Faris, 1, 2, 4, 27, 36, 38, 44, 58-61, 75, 78, 81, 83, 86, 88, 93, 97, 108, 142, 215–16, 327n10, 341n178, 387n119; and animal instincts, 39–40; as anti-nationalist, 96; British, ties to, as strong, 80; on British occupation, as opportunity, 74; on Büchner translation, by Shumayyil, 116; dismissal of, 342n184; Egypt, move to, 72, 74, 79; as foreign intruder (*dukhala*), 96; Hamidian regime, critical of, 74; materialism, opposition to, 114, 116; political radicalism of, 62; process of evolution, and design in nature, as evidence of, 43; professional and political clubs, founding of by, 62; promotion, rescinding of, 71; religion, public role of, anxiety about, 93; science and scripture, tension between, 43; Spencer, appeal of to, 73–74; on spontaneous generation, 101–3; suspicion toward, by Egyptian elite, 94; theory of evolution, 42; theory of natural selection, 41
Nineteenth Century (magazine), 28, 87, 102
North Africa, 217, 291
North America, 28
Nuʿmani, Shibli, 156
Nur al-Islam (Light of Islam) (journal), 283, 296. See also *Majallat al-Azhar* (journal)

Oktar, Adnan, 388n10. *See also* Yahya, Harun
O'Leary, De Lacy, 271
Omdurman (Sudan), 87
On Education: Intellectual, Moral, and Physical (Spencer), 19, 161, 195
On the Origin of Species (Darwin), 6–7, 26, 69, 101, 153, 253, 267, 268, 272, 276; glossary of, 262–63; and species, 16–17, 263; translation of, 261–62, 265
Origin of Civilisation and the Primitive Conditions of Man (Lubbock), 36
origin of life, 106, 108
Ottoman Empire, 25, 30, 81, 86–87, 89, 113, 124, 135, 211, 241, 371n196; new reading community in, 44; pan-Islamic ideas, spread of in, 93; press censorship in, 77; printing presses, introduced in, 75
Ottomanism, 21
Ottoman Party of Administrative Decentralization, 127
Owen, Richard, 68
Owen, Robert, 221

paideia, 19
Palestine, 234, 252
Paley, William, 142; and watchmaker analogy of, 132, 141, 145, 149, 324n44
Pall Mall Gazette (newspaper), 87, 94
pan-Islamism, 243
pan-psychism, 279
Paris (France), 22; materialism, popularity of in, 146
Paris Commune, 228
Paris Peace Conference (1919), 234
Pasha, Artin, 319n4
Pasha, Fu'ad, 76
Pasha, Ibrahim, 76
Pasha, Isma'il Sidqi, 295
Pasha, Midhat, 62–63, 69, 134, 136
Pasha, Muhammad Tawfiq Rif'at, 387n119
Pasha, Riaz, 31, 319n4
Pasteur, Louis, 28, 68, 102, 276, 278
Penrose, Stephen, 342n184
People's (*Umma*) Party, 91
Permanent Council for the Development of National Production, 259
Pharaonism, 247, 255, 291
phenomenalism, 279
A Philosophical Account of the Refutation of Evolution (Hurani), 117
Pius X, 164
Plato, 121, 265–66
Popular Science Monthly (magazine), 4, 28, 81–84
positivism, 13, 15, 102, 311, 313; as term, 279; as transcendental, 279
Post, George, 56, 57, 67, 103, 337n134
Pouchet, Félix Archimède, 68
poverty, 238; as social problem, 237
Prehistoric Times (Lubbock), 35
prehistory, 36
Les premières civilisations (Le Bon), 190, 291
Presbyterian Church, 56–57
Presbyterian Church in the U.S.A. Board of Foreign Missions (PCBFM), 56–57, 70
Primary Lessons in Rational Philosophy (Bliss), 57
Prime Mover, 188, 281
Principles of Sociology (Spencer), 82–83
Progress and Poverty (George), 238
Prolegomena (Ibn Khaldun), 184
proliberalism, 21
Protestantism: and Christianity, 189
Proudhon, Pierre, Joseph, 221
psalms (*mazamir*), 138
Psychic News (newspaper), 314
public sphere, 23

Qabbani, 'Abd al-Qadir al-, 134
Qisat al-anbiya (Tales of the Prophet), 388n1
qiyas (analogic cases), 139, 148
A quoi tient la supériorité des Anglo-Saxons (Demolins), 89
Qur'an, 75, 120–21, 137–39, 147, 150–52, 154, 161, 173, 192, 222, 234, 278, 281, 297–98, 304, 314; evolution, compatibility of, 175; evolution, references in, as pointless, 316; *hamd* (praise), use of in, 313; human creation, interpretation of in, 307–8, 388n2; *i'jaz al-Qur'an*, popularity of, 295; and miracles, 174, 296, 315; "scientific exegesis" of (*tafsir 'ilmi*), 218; scientific principles in, 165, 218, 315–16; and social animals, 227; and Surat al-Baqara, interpretation of, 175, 179; "two books," theme of, 177, 180
Qutb, Sayyid, 295, 315–16

race, 87; and Egyptian national identity, 246–47; global discourse on, 246–47; racial differences, politics of, as ambiguous, 88
Radcliffe de, Stratford, 87
radicalism, 224, 229
radical materialism, 131
Rafi'i, Mustafa Sadiq al-, 296
Rasa'il Ikhwan al-Safa' (Treatise of the Brethren of Purity), 382n30
Rashid, Shaykh, 210, 211
rationalism, 287, 311, 313, 360n6; and knowledge, 14–15
Razi, Fakhr al-Din al-, 140, 148, 152–53, 274, 314
Raziq, 'Ali 'Abd al-, 290, 292; and Raziq affair, 287
reason, 14; intuition, 38; and reasoning, 155; and revelation, 178; sensation and perception, 38; understanding, faculty of, 38

Reformation, 113, 123, 189, 218; as failure, 192
reformist socialism (*al-ijtima'iya*), 230
A Refutation of the Materialists (Al-radd 'ala al-dahriyin) (Afghani), 119, 124–25, 169, 200, 229
religion, 16; and belief, 15; civilizations, progress of, 164; evolution of, 178; and reason, 178; and science, 187
religious modernism, 165–66
Renaissance, 256, 273
Renan, Ernest, 14, 183–84, 186–87, 189, 193
The Republic (Plato), 221, 228
Richter, Hermann, 278, 383n44
Rida, Rashid, 63–64, 78, 104, 113, 139, 157–59, 163–64, 166, 170–71, 175, 185–86, 197, 204–5, 211, 213, 215, 217, 233–34, 248, 292, 295, 311–13, 315, 360n6, 362n30; evolution, attitude toward, 218
Rifa'iya Sufi order, 157
Risala (al-Jisr), 119; Beirut, as product of, 141; classical and contemporary themes, as fusion of, 140–41; creation, examples of in, 149–50; God's singularity (*tawhid*), case for in, 149; *ijtihad* (independent juridical and theological opinions), dangers of, 137; *'ilm al-kalam* (speculative theology), as larger exercise in, 154; Islam, misconceptions about abroad, as defense of against, 155–56; on *jihad*, 137; as logical, 147–48; materialism, refutation of, 138, 140, 148–49; on miracles, 158; Muslim code of ethics in, 138; "new theology," as part of, 156; and polygamy, 138; popularity of, 137; on *shari'a*, 138–39
Risalat al-tawhid (Theology of unity) ('Abduh), 171–72, 181, 204, 211
The Rising Sun (Kamil), 91
Robinet, Jean-Baptiste, 282
Romanes, G. J., 39
Rome (Italy), 47, 75, 191
Rosenthal, Joseph, 250–51
Rousseau, Jean-Jacques, 12, 197
Royal Arabic Language Academy, 302
Royal Asiatic Society, 28
Royal Geographic Society of Egypt, 28
Royal Society of London, 28
Russell, Bertrand, 287
Russia, 91, 234, 238–39, 249, 304. *See also* USSR
Russian Revolution, 250–51, 259, 298
Russo-Japanese War, 91
Rustum, Sadiq, 291

Saba'i, Muhammad al-, 83
Safa, Ikhwan al-, 269
Saint-Hilaire, Isidore Geoffroy, 262
Saint-Simon, 221, 374n6
salaf (forefathers), 211
Sanusi, Ibn Abd al-Wahhab Ahmad al-, 212
Sanusi, Maghribi Muhammad al-, 205
Saqi, al-, 181
Sarruf, Fu'ad, 293, 301
Sarruf, Ya'qub, 27, 36, 38, 44, 49, 58–61, 75, 78, 82–83, 86, 97, 108, 113, 128, 142–43, 215–16, 233, 327n10, 338n150, 341n178; and animal instincts, 39–40; as anti-nationalist, 96; British, ties to, as strong, 80; and British occupation, as opportunity, 74; on Büchner translation, by Shumayyil, 116; dismissal of, 342n184; Egypt, move to, 72, 74, 79; as foreign intruder (*dukhala*), 96; Hamidian regime, critical of, 74; materialism, opposition to, 114; process of evolution, and design in nature, as evidence of, 43; professional and political clubs, founding of by, 62; promotion, rescinding of, 71; religion, public role of, anxiety about, 93; science and scripture, tension between, 43; Spencer, appeal of to, 73–74; on spontaneous generation, 101–3; suspicion toward, by Egyptian elite, 94; and theory of evolution, 42; theory of natural selection, 41
Sarton, George, 263, 302
Sayhi, Ibrahim Muhammad, 280; and transcendental positivism, 281
Sayyid, Abu al-Huda al-, 157, 212
Sayyid, Ahmad Lutfi al-, 90–91, 231, 241
Sayyid, Mikha'il al-, 79
Schiller, Friedrich, 197
Schopenhauer, Arthur, 246
science (*'ilm*), 6, 23, 123, 131, 190, 206, 227; and empire, 9–12; and faith, 190,

192; and gravity, 191; and Islam, 165, 177; and knowledge, 168; and politics, 235; popularization of, 25, 27–28, 32, 74, 252–53, 296; and religion, 93–94, 123, 172, 185, 187, 221, 288; and revelation, 179, 281; scripture, in harmony with, 180; and the state, 20; and theology, 164; translation of, 16–18; and *ʿulama* (scientists), 164
Science (magazine), 114
Science among the Arabs (Sayhi), 280
Science and Civilization in Christianity in Islam (Abduh), 191
"Science and the Good of the Nation" (article), 83
Science and the New Humanism (Sarton), 305
La Sciénce Sociale (journal), 89
Scientific American (magazine), 28
Scientific Revolution, 256, 259
scientific socialism, 243
Scopes trial, 287
Scott, John Gordon, 333n102
Scottish Common Sense, 15
Scramble for Africa, 10, 87, 155
"Second Great Awakening," 73
Second International, 239, 250
Secret History (Blunt), 244
Secret Society, 62, 342n184
self-determination, 234
Self-Help (Smiles), 59, 82, 338n150
"Sensibility and Its Diverse Forms" (article), 114
Seyhülislam, 135
Sfayr, Jirjis, 354n109
Shadi, Ahmad Zaki Abu, 227, 294
Shadudi, Asʿad Ibrahim al-, 57
Shahin, ʿAbd al-Sabur, 307, 309, 387n1, 387n2; blasphemy trial of, 308
Shaltut, Shaykh Mahmud, 217, 304, 315
Shams al-Birr (Sun of Virtue), 62, 70
shariʿa (religious laws), 138–39, 147, 150, 156, 171, 192, 197, 276; and man's intellect v. animal intellect, 151–52; purpose of, as communal *maslaha* (interest), 139; and schools, 138
Shaw, George Bernard, 241–42, 244, 246, 257
Shawqi, Ahmad, 293–94
Shaykh, Sayyid, 123
Shidyaq, Ahmad Faris, 228
Shihabi, Mustafa al-, 292
Shirbini, ʿAbd al-Rahman al-, 214
Shumayyil, Amin, 33
Shumayyil, Shibli, 18, 31, 33, 69, 75, 78–79, 85, 128, 132, 140, 143–44, 224, 226, 228, 232–33, 238–39, 243–44, 247, 253, 273, 276, 278, 284–85, 310; Afghani, reaction to, 123; Arab autonomy, within the empire, movement for, 127; background of, 103–4; on Bastian, 105–7; Büchner, translation of by, 105, 107, 110–11, 116–17, 125, 138, 169–70, 264–65, 267; concept of unity (*tawhid*), 99, 110; as controversial, 116; criticism of, 117, 119–20, 125; Darwin, praise of, 108; and evolutionary socialism, 235; and fetishism, 111; Hurani, response to criticism of, 118; materialism (*al-maddiya*), defense of, 99–100, 109, 129; natural religion, call for by, 126; as outlier, 114; personal faith of, 113; progress, and pedagogy, 126; on religion, 111–13; spontaneous generation, defense of, 103, 107–8; state, role of, ambivalence toward, 236; supporters of, 125–26; Syrian diasporas, seeking assistance from, 127; on "true socialism," 235; university education, abolition of, call for, 126; West, imitating of, accusations of, 122–23
Shuqri, Naguib Bey, 83
Siam, 93
Sidgwick, Henry, 28
Sidon (Lebanon), 31
Silver Birch (Al-Ruh al-Aʿzam), 314
Simmel, Georg, 197
Sinott, Edmund, 263
Sirr taqaddum al-Inkiliz al-Saksuniyin (The secret of the progress of the Anglo-Saxons) (Zaghlul), 89
Sirr tatawwur al-umam (Secret of evolution of nations), 291. See also *Lois psychologiques de l'évolution des peuples* (Le Bon)
siyasa (politics), 133
Smiles, Samuel, 82
Smith, Adam, 320n14

Smith, Eli, 46–47, 62, 332n89, 333n102
Smith, Elliot, 255
Smyrna (Turkey), 47
social animals, 227
social evolution, 223; exchange of aid (*tabadul al-musa'ida*), as founded on, 231; and socialism, 224; and survival of the fittest, 231–32
Social Evolution (Kidd), 232
Social Gospel movement, 51
socialism (*al-ishtirakiya*), 219, 221, 224, 238, 242, 256, 258; anticolonial nationalism, associations with, 257; criticism of, 230–31, 298–99; and Darwin, 224; in Egypt, 225, 227, 229, 245, 247, 250–51, 254; and gradual change, 245; individualism, as opposite of, 234; industrialization, associations with, 229; Islam, as compatible with, 239; journal writing on, 226–27; liberty, as enemy of, 231; and Marx, 224; natural science, as based on, 235–36, 244; political implications of, 257; religion, compatibility with, 227–28, 239–40, 244; and social evolution, 224; as term, 228–29
Socialism (Haddad), 234
social organism, 86
"The Social Organism" (article), 81–82, 85
Social Question (*al-mas'alla al-ijtima'iya*), 219
social selection, 233. *See also* group selection
society: as social organism, 82, 224
Society of the Eastern Bond, 292
Society of the Egyptians for the Egyptians, 255–56
Socrates, 121, 266
soul: and spiritualist research, 282
Spectator (magazine), 28
speculative or dialectical theology (*'ilm al-kalam*), 181
Spencer, Herbert, 5, 8, 15, 28, 73–74, 82, 88, 90, 95, 97, 108, 118, 126, 140, 166–67, 214, 233, 236, 253, 279, 284, 297–98, 320n13, 320n14, 367n127; and 'Abdul, 161–64, 197, 281; appeal of, as global, 81; death of, 83; on education, 195–96; on family, 196; liberal utilitarianism, and moral collectivism, 227; materialism, charges of, 85; and moral development, 196; and moral education, 19, 196; popularizing ideas of, 84; and religion, 197–98; society, as social organism, 82; and "survival of the fittest," 85–86; synthetic philosophy of, 84, 175; universal progression, 84; Unknown, concept of, 85
Spencer and Spencerism (MacPherson), 84
Spinoza, Baruch, 84, 104, 297
spiritualism, 233, 283, 314
spontaneous generation, 101–3, 105–8, 116, 125, 276, 278, 280, 284
Sudan, 88–89, 300; Mahdiya state in, defeat of, 87, 155
Sudan Times (newspaper), 80, 88
Suez Canal, 95, 258–59
Sufism, 163–64, 211, 253, 314; and *dhikr* (incantations), 212
sunan, 173–75
sunan al-tabi'a (natural laws), 173
sunna, 137, 173, 298
Sunni Islam, 211
Surur, Taha 'Abd al-Bakhi, 203–4
susialism, 228
Swadeshi movement, 255
Syria, 11, 25, 30, 56, 58, 78, 96, 105, 113, 127, 133, 136, 145, 157, 212, 238, 304; literacy rate of, 329n23; missionaries in, 45–46, 48, 51, 135
Syrian Protestant College (Al-Madrassa al-Kulliya al-Injiliya), 4, 18, 25, 44, 60–61, 63, 65, 100, 102, 116–17, 141, 143, 147, 194, 216, 319n4; Arabic books, publishing of, 57–58; Board of Managers, evangelical principles of, 56; curriculum at, 57; "Declaration of Principles" manifesto, 69; Hurani, subsidizing of articles by, 118–19; and Lewis affair, 136; opening of, 52–53; resignations at, 69–70; science education in, 53, 55–56; student protests at, 70. *See also* American University of Beirut
Syrian Scientific Society, 49. *See also* Syrian Society of Arts and Sciences (Al-Jam'iya al-Surriya Ii-Iktisab al-'Ulum wa al-Funun),
Syrian Society of Arts and Sciences (Al-Jam'iya al-Surriya Ii-Iktisab al-'Ulum

wa al-Funun), 48, 62. *See also* Syrian Scientific Society
Syro-Egyptian Society of London, 333n102

tafsir (Qur'anic exegesis), 13, 161, 171, 177, 179–80, 218, 308, 312, 314–15
Tafsir al-Manar (Abduh), 171, 177, 218, 312
Tafsir al-Qur'an (Abduh), 180
tafsir ʿilmi tradition (scientific exegesis), 181, 295, 310, 312, 388n2
Taftazani, al-, 134
taghrib (Westernization), 170
Tahafut al-falasifa (A refutation of the philosophers) (al-Ghazali), 148, 186
Tahafut al-tahafut (The incoherence of the incoherence), (Ibn Rushd), 186
Tahdhib al-akhlaq (The refinement of character) (Ibn Miskawayh), 196, 269
Tahrir al-mar'a (The liberation of women) (Amin), 196
Tahta (Egypt), 31
Tahtawi, Rifaʿa Rafiʿ al-, 275, 323n38
tajdid (renewal), 165
Tajdid al-ʿArabiya (Revival of Arabic) (Mazhar), 302
Ta'ifa (group), 19
Takrir fi islah al-mahakim al-shariʿiya (Shariʿa court reforms regulations) (Abduh), 171
takwin (to bring into being), 280
takwin alkawn, 311
talfiq, 212, 213
Tanta (Egypt), 31, 194
Tanwir al-adhhan (Enlightenment of Minds) (Zilzal), 57
Tanzimat reforms, 212
Taqla, Bishara, 78–79, 95
Taqla, Salim, 78–79, 95
taqlid (imitation), 170, 205, 212, 265–66, 289, 290; and *ʿalim*, public role of, 206
tarbiya (general education and child-rearing), 19, 97, 133
Taʿrib li-sharh Bukhnir ʿala madhhab Darwin (A translation of Büchner's commentaries on Darwin) (Shumayyil), 110
Tarikh al-lugha al-ʿArabiya ka'in hay (History of the Arabic language as a living entity) (Zaydan), 273
Tarikh al-madhahab al-ishtirakiya (History of socialisms) (Al-Mansuri), 238, 239
tasalsala, 41–42, 153, 297
tatawwur (evolution), 17, 42
Tawfiq, Khedive, 32, 169, 199, 319n4
Tawfiq, Muhammad, 210, 213
tawhid (unity), 99, 110, 149, 211, 315
ta'wil (analogic reasoning), 13, 150
Taylor, Isaac, 155–56
Temple, Frederick, 40, 142
Thabit, Khalil, 231–32
Thamrat al-Funun (The fruits of knowledge), 134
theology: and science, 164
theory of evolution, 1, 40, 42, 117, 140, 176; as absurd, 65; and materialism, 310; and religion, 276, 278; social and cultural implications of, 253–54. *See also* evolution
Third International, 251
Third Way, 317
Thompson, William, 281, 383n44. *See also* Kelvin, Lord
Thomson, William, 52
Tolstoy, Leo, 220
Torah, 138
traditional schools (*madhahab*), 167
transcendentalism, 15, 279–80, 313
transcendental materialism, 279
transcendental positivism, 279, 281
translation, 4–6, 11–12, 16–18, 28, 32, 47, 59, 65, 68, 76, 82, 85, 89, 90, 92, 100, 105, 107, 110, 111, 114, 116, 119–22, 124, 134, 138, 154, 155, 169, 170, 184, 186, 189, 228, 229, 253, 261–76, 282, 286, 288, 291, 294, 299, 301, 305, 312, 316, 320n14, 321n20, 325n54, 325n57, 331n35, 331n40, 332n89, 338n145, 345n46, 350n21, 353n107, 376n40, 378n80, 380n1, 383n48
Transvaal war, 161
Treatises on Meteorology (Loomis), 61
Tripoli (Libya), 22, 31, 133–34, 157
True Socialism (*al-ishtirakiya al-haqiqaya*) (Shumayyil), 221, 235
The Truth about the Naturalists and Their School (Haqiqat-i mazhab-i nichiri va bayan hal nichariyan) (Afghani), 119–20. See also *A Refutation of the Materialists* (Afghani)
Tunis (Tunisia), 182

Turkey, 88, 127, 251, 254, 289–90
Tuwayrani, Hasan Husni al-, 203
Tylor, E. B., 28, 111
Tyndall, John, 28, 102, 105–7, 116, 118

ʿulama (scientists), 16, 92, 137, 139, 157–58, 166, 179, 181, 189–90, 197–98, 201, 217–18, 220, 240, 281, 283, 290, 301, 311, 314; blind adherence (*taqlid*) of, 167, 195; criticism of, 192–95; *iʿjaz al-Qurʾan*, popularity of, 295; and *kalam*, 182; Muslim pedagogy, 135–36; and natural selection, 176; new subjects, refusal to teach by, 192; Raziq affair, and role of in public life, 287; reason, as guardians of, 156; and science, 164–65, 170; universities, and independent thought in, 289
umma (community or state; nation or people), 19, 133, 139, 165, 217, 222, 233
Umma Party, 231, 241, 243
Université Saint-Joseph, 63
United Arab Republic, 304
United States, 31, 41, 45, 52, 81, 84, 141, 229, 240; Social Gospel movement in, 51
Unveiling the Luminous Secrets of the Qurʾan (Iskandarani), 179, 364n63
ʿUrabi, Husni al-, 250
ʿUrabi revolt, 133, 168–69, 172, 200, 206, 236
USSR, 258. *See also* Russia
Usul al-kimiya (Principles of Chemistry) (Van Dyck), 57–58
utopia, 221, 228
Utopia (More), 228

Van Dyck, Cornelius, 47, 57–61, 69, 103, 332n89
Van Dyck, William, 69–70
Victoria, Queen, 183
Vie de Jésus (Renan), 186
Voltaire, 12, 104, 297

Wafd movement, 249, 251–52, 257
Wafd Party, 231, 299–300
Wahhab, ʿAbd al-, 205, 211
Wahhabism, 157, 211, 213
Wajdi, Muhammad Farid, 93, 180, 281, 297, 311–12; critique of Western materialism, and theory of evolution, 283; and missing links, criticism of, 283
Wajdiyat (journal), 282
wajib al-wujud (necessary being), 152
Wallace, Alfred Russel, 28, 42, 68, 85, 118, 223–24, 239, 285
Wallas, Graham, 242
"The War of European Civilization Compared with the Arab Conquest of Muslim Civilization" (Rida), 233
Wasil, Muhammad, 120–21
watan (country or homeland), 19, 133
Wathbat al-sharq (Advance of the East) (Mazhar): three schools of thought in, 290
Webb, Beatrice, 242
Webb, Sidney, 242, 257
Weekly News (Al-Nashra al-Usbuʿiya) (newspaper), 117
Weismann, August, 247, 283
Wells, H. G., 246, 252–53, 257, 276
Westermarck, Edvard, 286
What Is Darwinism? (Hodge), 34
What Is the Nahda? (Musa), 256
White, Andrew Dickson, 1, 2, 5, 8, 263, 287–88, 311, 319n4, 319n6, 319n8, 320n8; and Darwinian hypothesis, 4
Wilberforce, Bishop, 190
Wilson, Woodrow, 234, 249
wisdom (*hikma*), 66
women: education of, as imperative, 196–97; liberation of, call for, 87, 90–91; and sexual equality, 247
World Union of Socialist Parties, 251
World War I, 128, 234, 244, 257; collectivism, turn to, 232; materialist philosophy, as end to, 233; as war of "materialist civilization" (*al-madaniya al-maddiya*), 233
Wortabet, John, 103

Yahya, Harun, 388n10. *See also* Oktar, Adnan
Yemen, 157
Yeni ilm-i kalam (The new theology) (Hakki), 156
YMCA, 62

Yokohama (Japan), 88
Young Ottoman movement, 104
Young Turk Revolution, 126, 213, 241, 244, 265; and pan-Turkism of, 127
Yusuf, Azhari shaykh Ali, 95

Zaghlul, Ahmad Fathi, 89–90, 231, 291
Zaghlul, Sa'ad, 89, 216–17, 231, 257, 298–99
zann (guesswork), 153
Zawahri, Shaykh Muhammad al-, 214
Zayd, Nasr Hamid Abu, 309
Zaydan, Jurji, 11, 31–32, 44, 70–71, 77–79, 88, 91, 104, 216, 273, 323n35; and reverse selection (*al-intikhab al-'aksi*), 233–34
Zaydi imam, 157
Ziadeh, Mayy, 104, 293, 321n25
Zilzal, Bishara, 18, 34, 38–39, 57, 125, 142, 326n59
Zionism, 252
Zoroastrianism, 248